NONDESTRUCTIVE TESTING HANDBOOK

Third Edition

Volume 9

Visual Testing

Technical Editors
Michael W. Allgaier
Robert E. Cameron

Editor
Patrick O. Moore

American Society for Nondestructive Testing

Copyright © 2010
American Society for Nondestructive Testing, Incorporated
All rights reserved.

ASNT is not responsible for the authenticity or accuracy of information herein. Published opinions and statements do not necessarily reflect the opinion of ASNT. Products or services that are advertised or mentioned do not carry the endorsement or recommendation of ASNT.

IRRSP, *Level III Study Guide, Materials Evaluation, NDT Handbook, Nondestructive Testing Handbook, The NDT Technician* and www.asnt.org are trademarks of the American Society for Nondestructive Testing. ACCP, ASNT, *Research in Nondestructive Evaluation* and *RNDE* are registered trademarks of The American Society for Nondestructive Testing, Inc.

ASNT exists to create a safer world by promoting the profession and technologies of nondestructive testing.

American Society for Nondestructive Testing, Incorporated
1711 Arlingate Lane
PO Box 28518
Columbus, OH 43228-0518
(614) 274-6003; fax (614) 274-6899
www.asnt.org

Library of Congress Cataloging-in-Publication Data

Visual testing / editor, Patrick O. Moore -- 3rd ed.
p. cm. -- (Nondestructive testing handbook ; v. 9)
Rev. ed. of: Visual and optical testing, 1993
Includes bibliographical references and index.
ISBN 978-1-57117-186-3 (alk. paper)
1. Nondestructive Testing. 2. Engineering inspection. 3. Optical measurements. I. Moore, Patrick O., II. American Society for Nondestructive Testing. III. Visual testing.

TA417.2.V57 2010
620.1'127--dc22

2010018505

first printing 05/10
second printing 11/15 with new impositions for pages ii, 28, 29, 53, 104, 105, 151, 183, 230, 233, 264, 319, 330.
ebook 12/15

ISBN-13: 978-1-57117-186-3 (print)
ISBN-13: 978-1-57117-207-5 (CD)
ISBN-13: 978-1-57117-266-2 (ebook)

Errata if available for this printing may be obtained from ASNT's Web site, <www.asnt.org>.

Published by the American Society for Nondestructive Testing
PRINTED IN THE UNITED STATES OF AMERICA

President's Foreword

ASNT exists to create a safer world by promoting the profession and technologies of nondestructive testing.

The dedicated efforts of the Technical and Education Council continue to advance NDT technology through their tireless efforts in creating new NDT education and resource materials. Their important achievements are a testimonial to the efforts of these dedicated volunteers.

One of the best ways to promote NDT technology is to update and maintain our handbooks as science and technology advances. The *NDT Handbook* series is one of ASNT's premier products. It is recognized both nationally and internationally as a valuable study and reference resource for NDT.

Visual Testing, Volume 9 of the third edition, is the result of the dedicated efforts of volunteers and ASNT staff to update the handbook and align with today's technological advancements.

Vision is an integral part of everyday life. It is not surprising that visual testing is usually the initial examination performed on components, parts and structures.

As the demand for inspectors continues to increase, there will be a significant demand to keep materials current and develop new NDT technology handbooks. As technology continues to advance, ASNT will continue to keep its library of resources current and useful as an essential resource to the NDT community.

The opportunities for the NDT professional are endless. Involvement on the Technical and Education Committee is an excellent way to give back to this proud profession. I encourage each ASNT member to become involved and give back to the profession of NDT. I guarantee that you will get more than you give.

Joel W. Whitaker
ASNT President, 2009-2010

Foreword

Aims of a Handbook

The volume you are holding in your hand is the ninth in the third edition of the *Nondestructive Testing Handbook*. In the beginning of each volume, ASNT has stated the purposes and nature of the *NDT Handbook* series.

Handbooks exist in many disciplines of science and technology, and certain features set them apart from other reference works. A handbook should ideally provide the basic knowledge necessary for an understanding of the technology, including both scientific principles and means of application. The third edition of the *NDT Handbook* provides this knowledge through method specific volumes.

The typical reader may be assumed to have completed a few years of college toward a degree in engineering or science and has the background of an elementary physics or mechanics course. Additionally, this volume allows for computer based media that enhance all levels of education and training.

Standards, specifications, recommended practices and inspection procedures are discussed for instructional purposes, but at a level of generalization that is illustrative rather than comprehensive. Standards writing bodies take great pains to ensure that their documents are definitive in wording and technical accuracy. People writing contracts or procedures should consult the actual standards when appropriate.

Those who design qualifying examinations or study material for them draw on ASNT handbooks as a quick and convenient way of approximating the body of knowledge. Committees and individuals who write or anticipate questions are selective in what they draw from any source. The parts of a handbook that give scientific background, for instance, may have little bearing on a practical examination except to provide the physical foundation to assist handling of more challenging tasks. Other parts of a handbook are specific to a certain industry. This handbook provides a collection of perspectives on its subject to broaden its value and convenience to the nondestructive testing community.

The present volume is a worthy addition to the third edition. The editors, technical editors, ASNT staff, many contributors and reviewers worked together to bring the project to completion. For their scholarship and dedication, I thank them all.

Richard H. Bossi
Handbook Development Director

Preface

The first visual testing report is found written in the book of Genesis, "He saw that it was good."

Visual testing is the test that precedes every other test. For years, a certification in magnetic particle testing or liquid penetrant testing would suffice to be the equivalent of a visual testing qualification.

The inspector had to "look" at the object, part, component or system before performing any other nondestructive testing (NDT) to "see" if the surface was suitable for further testing.

Its primary role as first test makes it the most important of all the methods of nondestructive testing. For years, *how* to look at something defined visual testing. *What* the inspector is looking at entails a broad spectrum of applications. This is probably why visual testing was formalized so late in industry — codified by the nuclear industry, in the 1980s, and appearing last in the sequence of *NDT Handbook* volumes, in 1993.

Its main limitation is that the test surface must be accessible. Direct visual testing has always addressed direct line of sight from the eyeball to the test surface. With the help of a candle and a mirror, otherwise inaccessible surfaces became accessible. As the light source progressed from a candle to a light bulb, to a fiber optic cable, to an illumination bundle, the limiting factor was the lens optic system and eventually the fiber optic system.

The main content difference of this edition of the visual volume of the *NDT Handbook* is the significant addition of the topic of indirect (or remote) visual testing. Coupling the recent advances in remote visual test techniques with modern image recording capabilities makes the recording and transferring of visual images a major advance in recording, transferring and retaining visual data of a test object. This technology is a major advantage over other NDT methods.

Visual testing allows direct interpretation of test results without encoding, decoding, extrapolating and evaluating data from other NDT methods. To assess the condition of the test object, what the inspectors see is what they get. Visual is the most directly useful test method to assess the condition of an object.

Michael W. Allgaier
Robert E. Cameron
Technical Editors

Editor's Preface

Early in 1986, Robert McMaster sat up in his hospital bed and handed me a piece of paper from a technical committee member. On the paper was scratched an outline for the book you are now reading.

This book on visual testing (VT) began with Robert McMaster. McMaster was ASNT's president from 1952 to 1953. He compiled and in 1959 published the first edition of the *NDT Handbook*. That edition was a milestone in the history of nondestructive testing (NDT).

McMaster is revered in ASNT because of two major visions that he imparted. First, he believed that NDT had a mission, an important role among applied sciences such as engineering: NDT's purpose was to improve the quality of products and services, to preserve not just the quality of life but to preserve life itself through public safety. He often compared nondestructive inspectors to physicians, saving lives. Without NDT, airplanes crash and buildings fall and boilers explode.

Second, McMaster wanted to ground NDT solidly as a material science. He had studied under Enrico Fermi and Robert Millikan at CalTech. McMaster believed in the nobility of science, that it improved our lives through understanding natural laws and applying that understanding.

His first edition of the *NDT Handboo*k was monumental, 54 sections in two volumes. There were fifteen sections for radiographic testing and two for visual testing. That the visual method was represented at all is remarkable, and reflects McMaster's scientific bent and the conviction that NDT should be represented in every band in the electromagnetic spectrum, even the visible radiation we call *light*. But on that winter afternoon in 1986, an exasperated McMaster pointed to the brief outline: "It's just a list of different kinds of borescopes! Just borescopes!"

The challenge for the writer of that outline, as for McMaster in 1959 and for others since, is precisely how the method is to be defined. For some, it was defined by its instruments, mainly the industrial endoscopes called *borescopes*. Others believed, wrongly, that the term *visual* denoted viewing unmediated by lenses and that another word, *optical,* was needed to include instruments such as borescopes. For McMaster, however, as for every volume of the third edition of the *NDT Handbook,* the word *visual* carved out a niche in the electromagnetic spectrum somewhere between infrared and X-rays (both of which, by the way, are also mediated through optics). Still, as late as the 1980s, some people assumed that the term *visual testing* meant only "vision acuity examination."

One of the intriguing things about VT is that very few publications have been dedicated to it as nondestructive testing, distinct from fields such as astronomy or medical endoscopy. By 1990, there were two books on VT, one on borescopes and one mainly on direct viewing.

The next step was taken by Michael Allgaier and ASNT's VT Committee. Allgaier collected available material, and in 1993 ASNT published it as Volume 8 in the second edition of the *NDT Handbook.* That volume defined the method. Henceforth, VT was to include both direct and indirect techniques. It would be scientifically grounded in the physics of light. Its study would include basic optometry, since the eye is the primary sensor. VT's representation in standards for industries such as energy and petroleum would be duly noted.

Before that book, the method would usually go unnoticed: inspectors would not even realize that their visual inspection was actually nondestructive testing. After that book, the foundation was laid for an ASNT method — with trainers, qualifying examinations and a literature for study.

The present volume builds on the success of that 1993 volume. Information has been added on digital capabilities that inspectors use routinely. The coverage of indirect techniques (sometimes called *remote inspection*) has been updated to reflect current technology for cameras and measurement. The discussions of optometry and physics are updated. The chapter on metals is completely revised with an eye for practicality. The material on direct techniques is presented in one chapter. References are updated throughout. The entire book has been revised to be clearly organized and functionally complete.

McMaster's stay in the hospital in the winter of 1986 was one of several that would end with his death in July. I like to think that, if he had lived to see it, he would have celebrated this book and VT's place as an NDT method.

Dozens of contributors and reviewers freely shared their expertise; in particular Technical Editors Michael Allgaier and Robert Cameron provided leadership and encouragement. On ASNT staff, Senior Manager of Publications Timothy Jones provided essential administrative support. My colleague, Technical Publications Supervisor Hollis Humphries, proofed the entire book and supervised all its graphics. A hearty thanks to them all.

Patrick Moore
NDT Handbook Editor

Acknowledgments

All contributors are also reviewers but are listed once, as contributors.

Handbook Development Committee

Richard H. Bossi, Boeing Aerospace
Michael W. Allgaier, Mistras
David R. Bajula, Acuren Inspection
Albert S. Birks, Naval Surface Warfare Center
Lisa Brasche, Iowa State University
James E. Cox, Zetec, Incorporated
David L. Culbertson, El Paso Corporation
James L. Doyle, Jr., NorthWest Research Associates
Nat Y. Faransso, KBR
Gerard K. Hacker, Teledyne Brown Engineering
Harb S. Hayre, Ceie Specs
Eric v.K. Hill, Embry-Riddle Aeronautical University
James W. Houf, American Society for Nondestructive Testing
Frank A. Iddings
Morteza K. Jafari, Fugro South
Timothy E. Jones, American Society for Nondestructive Testing
John K. Keve, DynCorp Tri-Cities Services
Doron Kishoni, Business Solutions USA, Canada
Xavier P.V. Maldague, University Laval
George A. Matzkanin, Texas Research Institute
Ronnie K. Miller, Mistras
Scott D. Miller
Mani Mina, Technology Resource Group
David G. Moore, Sandia National Laboratories
Patrick O. Moore, American Society for Nondestructive Testing
Stanislav I. Rokhlin, Ohio State University
Frank J. Sattler
Fred Seppi, Williams International
Kermit A. Skeie
Roderic K. Stanley, NDE Information Consultants
Stuart A. Tison, Millipore Corporation
Noel A. Tracy, Universal Technology Corporation
Satish S. Udpa, Michigan State University
Mark F.A. Warchol, Alcoa
Glenn A. Washer, University of Missouri — Columbia
George C. Wheeler
Gary L. Workman, University of Alabama, Huntsville

Contributors

Michael W. Allgaier, Mistras
David R. Atkins, Packer Engineering
David R. Bajula, Acuren Inspection
Bruce L. Bates
Thomas D. Britton, General Electric Sensing and Inspection Technologies
Brian P. Buske, General Electric Sensing and Inspection Technologies
Donald R. Christina, Boeing Company
John C. Duke, Jr., Virginia Polytechnic Institute and State University
Mohamed El-Gomati, University of York, United Kingdom
Nat Y. Faransso, KBR
Gregory W. Good, Ohio State University, College of Optometry
Doron Kishoni, Business Solutions USA, Canada
Douglas G. Krauss, Huddleston Technical Services, Redstone Arsenal
William J. Lang, Lenox Instrument Company
Trevor Liddell, General Electric Sensing and Inspection Technologies
Zheng Liu, Research Officer, National Research Council Canada
Joseph L. Mackin, Team Industrial Services
Stephen L. Meiley, Champion International
Richard T. Nademus, Exelon Corporation
Yoshihiro Ohno, National Institute of Standards and Technology
Donald Parrish, Southern Company Services
David A. Pasquazzi, David Pasquazzi and Associates
Stanislav I. Rokhlin, Ohio State University
Donald J. Roth, National Aeronautics and Space Administration, Glenn Research Center
Gregory C. Sayler, MD Helicopters
Roderic K. Stanley, NDE Information Consultants
Marvin W. Trimm, Savannah River National Laboratory
Hiroyuki Ukida, University of Tokushima, Japan
Michael A. Urzendowski, Valero Energy
Robert W. Warke, LeTourneau University

Reviewers

Steven E. Anderson, Canam Steel
Jerry D. Beasley, Omaha Public Power
Kenneth Becker, Sigma Transducers
James J. Bogner, GPR Testing and Inspection

Richard H. Bossi, Boeing Research and Technology
Lisa Brasche, Iowa State University
Robert H. Bushnell
James R. Cahill, General Electric Sensing and Inspection Technologies
Robert E. Cameron
Eugene J. Chemma, Arcelor Mittal Steel
David Clark, LightDancer Interactive Technologies
Christopher I. Collins, Olympus Industrial Systems Europa
Jackson R. Crissey, Jr., Plant Performance Services
Claude D. Davis, Unified Testing Services
Edward R. Generazio, NASA Langley Research Center
Lawrence O. Goldberg, Seatest
Jack K. Harper, Babcock and Wilcox, Oak Ridge
James W. Houf, American Society for Nondestructive Testing
Charles P. Longo, American Society for Nondestructive Testing
Walter R. Matulewicz, Tinker Air Force Base
Charles H. Mazel, BlueLine NDT
Eugene A. Mechtly, University of Illinois at Urbana-Champaign
John W. Miller
Scott D. Miller
Van B. Nakagawara, Federal Aviation Authority, Civil Aerospace Medical Institute
David K. Park, Olympus Industrial America
Bruce A. Pellegrino, General Electric Sensing and Inspection Technologies, Everest RVI
William C. Plumstead, Sr., PQT Services
Frank J. Sattler
David Sentelle, American Society for Nondestructive Testing
Robert E. Stevens, United Airlines
Mark F.A. Warchol, Alcoa
Stanley L. Weatherly, Boeing Company

CONTENTS

1

CHAPTER

Introduction to Visual Testing

Mohamed El-Gomati, University of York, Heslington, North Yorkshire, United Kingdom (Part 3)

William J. Lang, Lenox Instrument Company, Trevose, Pennsylvania (Part 3)

Marvin W. Trimm, Savannah River National Laboratory, Aiken, South Carolina (Part 2)

Part 1. Nondestructive Testing

Scope of Nondestructive Testing

Nondestructive testing is a materials science concerned with many aspects of quality and serviceability of materials and structures. The science of nondestructive testing incorporates all the technology for process monitoring and for detection and measurement of significant properties, including discontinuities, in items ranging from research test objects to finished hardware and products in service. Nondestructive testing examines materials and structures without impairment of serviceability and reveals hidden properties and discontinuities.

Nondestructive testing is becoming increasingly vital in the effective conduct of research, development, design and manufacturing programs. Only with appropriate nondestructive testing can the benefits of advanced materials science be fully realized. The information required for appreciating the broad scope of nondestructive testing is available in many publications and reports.

Definition

Nondestructive testing (NDT) has been defined as those methods used to test a part or material or system without impairing its future usefulness.[1] The term is generally applied to nonmedical investigations of material integrity.

Nondestructive testing is used to investigate specifically the material integrity or properties of a test object. A number of other technologies — for instance, radio astronomy, voltage and current measurement and rheometry (flow measurement) — are nondestructive but are not used specifically to evaluate material properties. Radar and sonar are classified as nondestructive testing when used to inspect dams, for instance, but not when used to chart a river bottom.

Nondestructive testing asks "Is there something wrong with this material?" In contrast, performance and proof tests ask "Does this component work?" It is not considered nondestructive testing when an inspector checks a circuit by running electric current through it. Hydrostatic pressure testing is a form of proof testing that sometimes destroys the test object.

A gray area in the definition of nondestructive testing is the phrase future usefulness. Some material investigations involve taking a sample of the test object for a test that is inherently destructive. A noncritical part of a pressure vessel may be scraped or shaved to get a sample for electron microscopy, for example. Although future usefulness of the vessel is not impaired by the loss of material, the procedure is inherently destructive and the shaving itself — in one sense the true test object — has been removed from service permanently.

The idea of future usefulness is relevant to the quality control practice of sampling. Sampling (that is, less than 100 percent testing to draw inferences about the unsampled lots) is nondestructive testing if the tested sample is returned to service. If steel bolts are tested to verify their alloy and are then returned to service, then the test is nondestructive. In contrast, even if spectroscopy in the chemical testing of many fluids is inherently nondestructive, the testing is destructive if the samples are poured down the drain after testing.

Nondestructive testing is not confined to crack detection. Other anomalies include porosity, wall thinning from corrosion and many sorts of disbonds. Nondestructive material characterization is a field concerned with properties including material identification and microstructural characteristics — such as resin curing, case hardening and stress — that directly influence the service life of the test object.

Methods and Techniques

Nondestructive testing has also been defined by listing or classifying the various techniques.[1-3] This approach to nondestructive testing is practical in that it typically highlights methods in use by industry.

In the *Nondestructive Testing Handbook,* the word *method* is used for a group of test techniques that share a form of probing energy. The ultrasonic test method, for example, uses acoustic waves at a frequency higher than audible sound. Infrared and thermal testing and radiographic testing are two test methods that use electromagnetic radiation, each in a defined wavelength range. The word *technique,* in contrast, denotes a way of adapting the method to the application. Through-transmission immersion testing is a technique of the ultrasonic method, for example.

Purposes of Nondestructive Testing

Since the 1920s, the art of testing without destroying the test object has developed from a laboratory curiosity to an indispensable tool of fabrication, construction, manufacturing and maintenance processes. No longer is visual testing of materials, parts and complete products the principal nondestructive test for quality. Nondestructive tests in great variety are in worldwide use to detect variations in structure, minute changes in surface finish, the presence of cracks or other physical discontinuities, to measure the thickness of materials and coatings and to determine other characteristics of industrial products. Scientists and engineers of many countries have contributed greatly to nondestructive test development and applications.

How is nondestructive testing useful? Why do thousands of industrial concerns buy the test equipment, pay the subsequent operating costs of the testing and even reshape manufacturing processes to fit the needs and findings of nondestructive testing? Modern nondestructive tests are used by manufacturers (1) to ensure product integrity and in turn reliability, (2) to avoid failures, prevent accidents and save human life (Figs. 1 and 2), (3) to make a profit for the user, (4) to ensure customer satisfaction and maintain the manufacturer's reputation, (5) to aid in better product design, (6) to control manufacturing processes, (7) to lower manufacturing costs, (8) to maintain uniform quality levels and (9) to ensure operational readiness.

These reasons for widespread and profitable nondestructive testing are sufficient in themselves but parallel developments have contributed to the technology's growth and acceptance.

Increased Demand on Machines

In the interest of greater performance and reduced cost for materials, the design engineer is often under pressure to reduce weight. Weight can be saved sometimes by substituting aluminum alloys, magnesium alloys or composite materials for steel or iron but such light parts may not be the same size or design as those they replace. The tendency is also to reduce the size. These pressures on the designer have subjected parts of all sorts to increased stress levels. Even such commonplace objects as sewing machines, sauce pans and luggage are also lighter and more heavily loaded than ever before. The stress to be supported is known as dynamic stress or dynamic loading, as opposed to static stress. It often fluctuates and reverses at low or high frequencies. Frequency of stress reversals increases with the speeds of modern machines, so components tend to fatigue and fail more rapidly.

Another cause of increased stress on modern products is a reduction in the safety factor. An engineer designs with certain known loads in mind. On the supposition that materials and workmanship are never perfect, a safety

FIGURE 1. Fatigue cracks caused damage to aircraft fuselage, causing death of flight attendant and injury to passengers (April 1988).

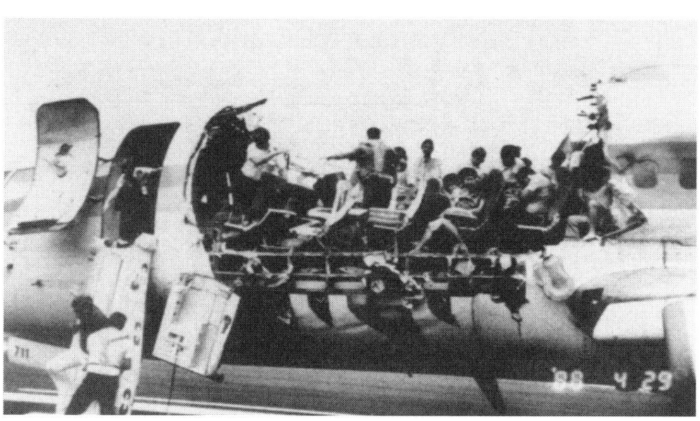

FIGURE 2. Boilers operate with high internal steam pressure. Material discontinuities can lead to sudden, violent failure with possible injury to people and damage to property.

factor of 2, 3, 5 or 10 is applied. However, a lower factor is often used that depends on considerations such as cost or weight.

New demands on machinery have also stimulated the development and use of new materials whose operating characteristics and performances are not completely known. These new materials could create greater and potentially dangerous problems. For example, an aircraft part was built from an alloy whose work hardening, notch resistance and fatigue life were not well known. After relatively short periods of service, some of the aircraft using these parts suffered disastrous failures. Sufficient and proper nondestructive tests could have saved many lives.

As technology improves and as service requirements increase, machines are subjected to greater variations and extremes of all kinds of stress, creating an increasing demand for stronger or more damage tolerant materials.

Engineering Demands for Sounder Materials

Another justification for nondestructive tests is the designer's demand for sounder materials. As size and weight decrease and the factor of safety is lowered, more emphasis is placed on better raw material control and higher quality of materials, manufacturing processes and workmanship.

An interesting fact is that a producer of raw material or of a finished product sometimes does not improve quality or performance until that improvement is demanded by the customer. The pressure of the customer is transferred to implementation of improved design or manufacturing. Nondestructive testing is frequently called on to confirm delivery of this new quality level.

Public Demands for Greater Safety

The demands and expectations of the public for greater safety are widespread. Review the record of the courts in granting high awards to injured persons. Consider the outcry for greater automobile safety as evidenced by the required automotive safety belts and the demand for air bags, blowout proof tires and antilock braking systems. The publicly supported activities of the National Safety Council, Underwriters Laboratories, the Occupational Safety and Health Administration, the Federal Aviation Administration and other agencies around the world are only a few of the ways in which this demand for safety is expressed. It has been expressed directly by passengers who cancel reservations following a serious aircraft accident. This demand for personal safety has been another strong force in the development of nondestructive tests.

Rising Costs of Failure

Aside from awards to the injured or to estates of the deceased and aside from costs to the public (because of evacuations occasioned by chemical leaks, for example), there are other factors in the rising costs of mechanical failure.

These costs are increasing for many reasons. Some important ones are (1) greater costs of materials and labor, (2) greater costs of complex parts, (3) greater costs because of the complexity of assemblies, (4) a greater probability that failure of one part will cause failure of others because of overloads, (5) the probability that the failure of one part will damage other parts of high value and (6) part failure in an integrated automatic production machine, shutting down an entire high speed production line. In the past, when production was carried out on many separate machines, the broken one could be bypassed until repaired. Today, one machine is often tied into the production cycles of several others. Loss of such production is one of the greatest losses resulting from part failure.

Classification of Methods

The National Materials Advisory Board (NMAB) Ad Hoc Committee on Nondestructive Evaluation classified techniques into six major method categories: visual, penetrating radiation, magnetic-electrical, mechanical vibration, thermal and chemical/electrochemical.[3] A modified version of their system is presented in Table 1.[1]

Each method can be completely characterized in terms of five principal factors: (1) energy source or medium used to probe the object (such as X-rays, ultrasonic waves or thermal radiation), (2) nature of the signals, image or signature resulting from interaction with the object (attenuation of X-rays or reflection of ultrasound, for example), (3) means of detecting or sensing resultant signals (photoemulsion, piezoelectric crystal or inductance coil), (4) means of indicating or recording signals (meter deflection, oscilloscope trace or radiograph) and (5) basis for interpreting the results (direct or indirect indication, qualitative or quantitative and pertinent dependencies).

The objective of each method is to provide information about one or more of the following material parameters: (1) discontinuities and separations (such as cracks, voids, inclusions and delaminations), (2) structure or malstructure (such as crystalline structure, grain size, segregation and misalignment), (3) dimensions and metrology (such as thickness, diameter, gap size and discontinuity size), (4) physical and mechanical properties (such as reflectivity, conductivity, elastic modulus and sonic velocity), (5) composition and chemical analysis (such as alloy identification, impurities and elemental distributions), (6) stress and dynamic response (such as residual stress, crack growth, wear and vibration), (7) signature analysis (such as image content, frequency spectrum and field configuration) and (8) heat sources.

Material characteristics in Table 1 are further defined in Table 2 with respect to specific objectives and specific attributes to be measured, detected and defined.

Methods that use electromagnetic radiation (Table 3) can be divided according to the segment of the spectrum each uses as interrogating energy: radar, thermography, visual testing and X-radiography (Fig. 3). Methods using vibration and ultrasound are in a different spectrum: the acoustic.

The limitations of a method include conditions (such as access, physical contact and surface preparation) and requirements to adapt the probe to the test object. Other factors limit the detection or characterization of discontinuities or attributes and limit interpretation of signals or images.

Classification by Test Object

Nondestructive test techniques may be classified according to how they detect indications relative to the surface of a test object. Surface methods include liquid penetrant testing, visual testing and moiré testing. Surface/near-surface methods include tap, holographic, shearographic, magnetic particle and electromagnetic testing. When surface or near-surface methods are applied during intermediate manufacturing, they provide preliminary assurance that volumetric methods performed on the completed object or component will reveal few rejectable discontinuities. Volumetric methods include radiography, ultrasonic testing and acoustic emission testing. Through-boundary techniques include leak testing, some infrared thermographic techniques, airborne ultrasonic testing and certain techniques of acoustic emission testing. Other less easily classified methods are material identification, vibration analysis and strain gaging.

No one nondestructive test method is all revealing. In some cases, one method or technique may be adequate for testing a specific object or component. However, in most cases, it takes a series of test methods to do a complete nondestructive test of an object or component. For example, if surface cracks must be detected and eliminated and if the object or component is made of ferromagnetic

TABLE 1. Nondestructive test method categories.

Categories	Test Objectives
Basic Categories	
Mechanical and optical	color; cracks; dimensions; film thickness; gaging; reflectivity; strain distribution and magnitude; surface finish; surface flaws; through-cracks
Penetrating radiation	cracks; density and chemistry variations; elemental distribution; foreign objects; inclusions; microporosity; misalignment; missing parts; segregation; service degradation; shrinkage; thickness; voids
Electromagnetic and electronic	alloy content; anisotropy; cavities; cold work; local strain, hardness; composition; contamination; corrosion; cracks; crack depth; crystal structure; electrical conductivities; flakes; heat treatment; hot tears; inclusions; ion concentrations; laps; lattice strain; layer thickness; moisture content; polarization; seams; segregation; shrinkage; state of cure; tensile strength; thickness; disbonds; voids
Sonic and ultrasonic	crack initiation and propagation; cracks, voids; damping factor; degree of cure; degree of impregnation; degree of sintering; delaminations; density; dimensions; elastic moduli; grain size; inclusions; mechanical degradation; misalignment; porosity; radiation degradation; structure of composites; surface stress; tensile, shear and compressive strength; disbonds; wear
Infrared and thermal	anisotropy; bonding; composition; emissivity; heat contours; plating thickness; porosity; reflectivity; stress; thermal conductivity; thickness; voids; cracks; delaminations; heat treatment; state of cure; moisture; corrosion
Chemical and analytical	alloy identification; composition; cracks; elemental analysis and distribution; grain size; inclusions; macrostructure; porosity; segregation; surface anomalies
Auxiliary Categories	
Image generation	dimensional variations; dynamic performance; anomaly characterization and definition; anomaly distribution; anomaly propagation; magnetic field configurations
Signal image analysis	data selection, processing and display; anomaly mapping, correlation and identification; image enhancement; separation of multiple variables; signature analysis

material, then magnetic particle testing would be the appropriate choice. If the material is aluminum or titanium, then the choice would be liquid penetrant or electromagnetic testing. However, if internal discontinuities are to be detected, then ultrasonic testing or radiography would be chosen. The exact technique in each case depends on the thickness and nature of the material and the types of discontinuities that must be detected.

TABLE 2. Objectives of nondestructive test methods.

Objectives	Attributes Measured or Detected
Discontinuities and Separations	
Surface anomalies	roughness, scratches, gouges, crazing, pitting, imbedded foreign material
Surface connected anomalies	cracks, porosity, pinholes, laps, seams, folds, inclusions
Internal anomalies	cracks, separations, hot tears, cold shuts, shrinkage, voids, lack of fusion, pores, cavities, delaminations, disbonds, poor bonds, inclusions, segregations
Structure	
Microstructure	molecular structure; crystalline structure and/or strain; lattice structure; strain; dislocation; vacancy; deformation
Matrix structure	grain structure, size, orientation and phase; sinter and porosity; impregnation; filler and/or reinforcement distribution; anisotropy; heterogeneity; segregation
Small structural anomalies	leaks (lack of seal or through-holes), poor fit, poor contact, loose parts, loose particles, foreign objects
Gross structural anomalies	assembly errors; misalignment; poor spacing or ordering; deformation; malformation; missing parts
Dimensions and Measures	
Displacement, position	linear measurement; separation; gap size; discontinuity size, depth, location and orientation
Dimensional variations	unevenness; nonuniformity; eccentricity; shape and contour; size and mass variations
Thickness, density	film, coating, layer, plating, wall and sheet thickness; density or thickness variations
Physical and Mechanical Properties	
Electrical properties	resistivity; conductivity; dielectric constant and dissipation factor
Magnetic properties	polarization; permeability; ferromagnetism; cohesive force, susceptibility
Thermal properties	conductivity; thermal time constant and thermoelectric potential; diffusivity; effusivity; specific heat
Mechanical properties	compressive, shear and tensile strength (and moduli); Poisson's ratio; sonic speed; hardness; temper and embrittlement
Surface properties	color, reflectivity, refraction index, emissivity
Chemical Composition and Analysis	
Elemental analysis	detection, identification, distribution and/or profile
Impurity concentrations	contamination, depletion, doping and diffusants
Metallurgical content	variation; alloy identification, verification and sorting
Physiochemical state	moisture content; degree of cure; ion concentrations and corrosion; reaction products
Stress and Dynamic Response	
Stress, strain, fatigue	heat treatment, annealing and cold work effects; stress and strain; fatigue damage and residual life
Mechanical damage	wear, spalling, erosion, friction effects
Chemical damage	corrosion, stress corrosion, phase transformation
Other damage	radiation damage and high frequency voltage breakdown
Dynamic performance	crack initiation, crack propagation, plastic deformation, creep, excessive motion, vibration, damping, timing of events, any anomalous behavior
Signature Analysis	
Electromagnetic field	potential; intensity; field distribution and pattern
Thermal field	isotherms, heat contours, temperatures, heat flow, temperature distribution, heat leaks, hot spots, contrast
Acoustic signature	noise, vibration characteristics, frequency amplitude, harmonic spectrum, harmonic analysis, sonic emissions, ultrasonic emissions
Radioactive signature	distribution and diffusion of isotopes and tracers
Signal or image analysis	image enhancement and quantization; pattern recognition; densitometry; signal classification, separation and correlation; discontinuity identification, definition (size and shape) and distribution analysis; discontinuity mapping and display

Nondestructive Testing's Value

In manufacturing, nondestructive testing may be accepted reluctantly because its contribution to profits may not be obvious to management. Nondestructive testing is sometimes thought of only as a cost item and can be curtailed by industry downsizing. When a company cuts costs, two vulnerable areas are quality and safety. When bidding contract work, companies add profit margin to all cost items, including nondestructive testing, so a profit should be made on the nondestructive testing. The attitude toward nondestructive testing is positive when management understands its value.

Nondestructive testing should be used as a control mechanism to ensure that manufacturing processes are within design performance requirements. When used properly, nondestructive testing saves money for the manufacturer. Rather than costing the manufacturer money, nondestructive testing should add profits to the manufacturing process.

Nondestructive Test Methods

To optimize nondestructive testing, it is necessary first to understand the principles and applications of all the methods. The following section briefly describes major methods and the applications associated with them.

Visual Testing

Visual testing is the subject of the present volume and of a volume in the previous edition.[4]

Principles. Visual testing (Fig. 4) is the observation of a test object, either directly with the eyes or indirectly using optical instruments, by an inspector to evaluate the presence of surface anomalies and the object's conformance to specification. Visual testing should be the first nondestructive test method applied to an item. The test procedure is to clear obstructions from the surface, provide adequate illumination and observe. A prerequisite necessary for competent visual testing of an object is knowledge of the manufacturing processes by which it was made, of its service history and of its potential failure modes, as well as related industry experience.

Applications. Visual testing is widely used on a variety of objects to detect surface discontinuities associated with various structural failure mechanisms. Even when other nondestructive tests are performed, visual tests often provide a useful supplement. When the eddy current testing of process tubing is performed, for example, visual testing is often performed to verify and more closely examine the

Table 3. Nondestructive test methods and corresponding parts of electromagnetic spectrum.

Interrogating Energy	Test Method	Approximate Wavelengths (m)	Approximate Frequencies (Hz)
X-rays or gamma rays	radiography (RT)	10^{-16} to 10^{-8}	10^{24} to 10^{17}
Ultraviolet radiation	various minor methods[a]	10^{-8} to 10^{-7}	10^{17} to 10^{15}
Light (visible radiation)	visual testing (VT)	4×10^{-7} to 7×10^{-7}	10^{15}
Heat or thermal radiation	infrared and thermal testing (IR)	10^{-6} to 10^{-3}	10^{15} to 10^{11}
Radio waves	radar and microwave methods	10^{-3} to 10^{1}	10^{11} to 10^{7}

a. Ultraviolet radiation is used in various methods: (1) viewing of fluorescent indications in liquid penetrant testing and magnetic particle testing; (2) lasers and optical sensors operating at ultraviolet wavelengths.

Figure 3. Electromagnetic spectrum.

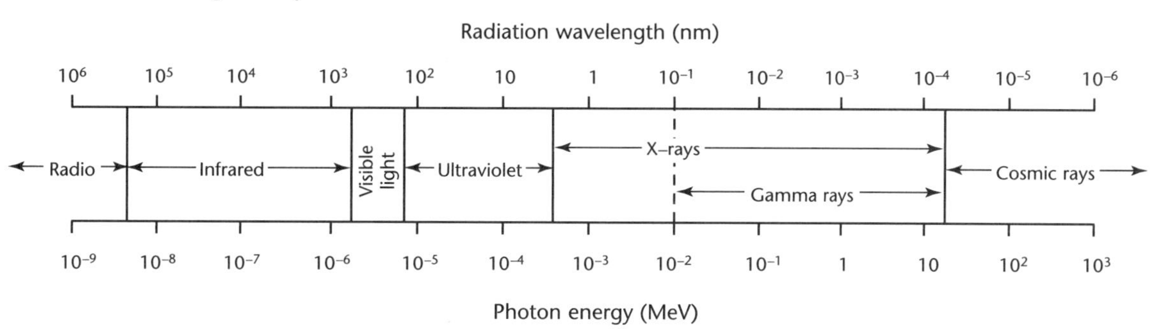

surface condition. The following discontinuities may be detected by a simple visual test: surface discontinuities, cracks, misalignment, warping, corrosion, wear and physical damage.

Magnetic Particle Testing

Principles. Magnetic particle testing (Fig. 5) is a method of locating surface and near-surface discontinuities in ferromagnetic materials. It depends on the fact that when the test object is magnetized, discontinuities that lie in a direction generally transverse to the direction of the magnetic field will cause a magnetic flux leakage field to be formed at and above the surface of the test object. The presence of this leakage field and therefore the presence of the discontinuity is detected with fine ferromagnetic particles applied over the surface, with some of the particles being gathered and held to form an outline of the discontinuity. This generally indicates its location, size, shape and extent. Magnetic particles are applied over a surface as dry particles or as wet particles in a liquid carrier such as water or oil.

Applications. The principal industrial uses of magnetic particle testing include final, receiving and in-process testing; testing for quality control; testing for maintenance and overhaul in the transportation industries; testing for plant and machinery maintenance; and testing of large components. Some discontinuities typically detected are surface discontinuities, seams, cracks and laps.

FIGURE 4. Visual test using borescope to view interior of cylinder.

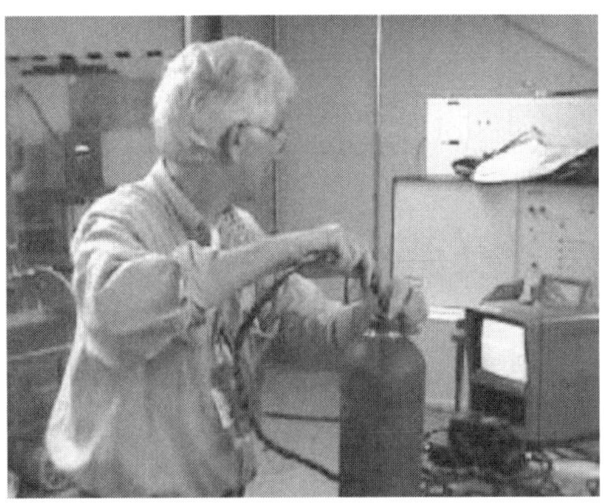

FIGURE 5. Test object demonstrating magnetic particle method.

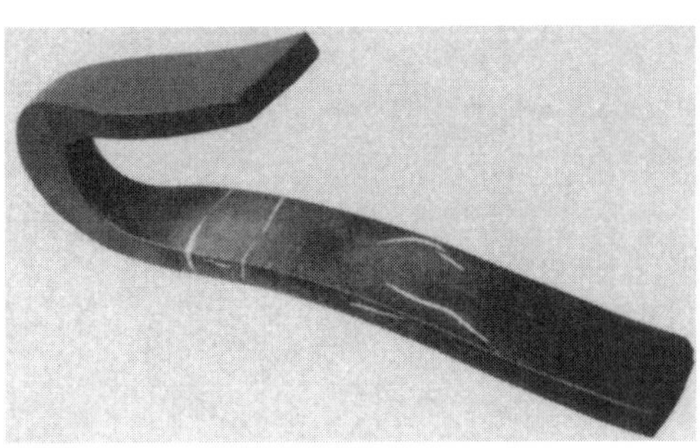

Liquid Penetrant Testing

Principles. Liquid penetrant testing (Fig. 6) reveals discontinuities open to the surfaces of solid and nonporous materials. Indications of a wide variety of discontinuity sizes can be found regardless of the configuration of the test object and regardless of discontinuity orientations. Liquid penetrants seep into various types of minute surface openings by capillary action. The cavities of interest can be very small, often invisible to the unaided eye. The ability of a given liquid to flow over a surface and enter surface cavities depends on the following: cleanliness of the surface, surface tension of the liquid, configuration of the cavity, contact angle of the liquid, ability of the liquid to wet the surface, cleanliness of the cavity and size of the surface opening of the cavity.

Applications. The principal industrial uses of liquid penetrant testing include postfabrication testing, receiving testing, in-process testing and quality control, testing for maintenance and overhaul in the transportation industries, in-plant and machinery maintenance testing and testing of large components. The following are some of the typically detected discontinuities: surface discontinuities, seams, cracks, laps, porosity and leak paths.

FIGURE 6. Liquid penetrant indication of cracking.

Eddy Current Testing

Principles. Based on electromagnetic induction, eddy current testing is perhaps the best known of the techniques in the electromagnetic test method. Eddy current testing is used to identify or differentiate among a wide variety of physical, structural and metallurgical conditions in electrically conductive ferromagnetic and nonferromagnetic metals and metal test objects. The method is based on indirect measurement and on correlation between the instrument reading and the structural characteristics and serviceability of the test objects.

With a basic system, the test object is placed within or next to an electric coil in which high frequency alternating current is flowing. This excitation current establishes an electromagnetic field around the coil. This primary field causes eddy currents to flow in the test object because of electromagnetic induction (Fig. 7). Inversely, the eddy currents affected by characteristics (conductivity, permeability, thickness, discontinuities and geometry) of the test object create a secondary magnetic field that opposes the primary field. This interaction affects the coil impedance and can be displayed in various ways.

Eddy currents flow in closed loops in the test object. Their two most important characteristics, amplitude and phase, are influenced by the arrangement and characteristics of the instrumentation and test object. For example, during the test of a tube, the eddy currents flow symmetrically in the tube when discontinuities are not present. However, when a crack is present, then the eddy current flow is impeded and changed in direction, causing significant changes in the associated electromagnetic field.

Applications. An important industrial use of eddy current testing is on heat exchanger tubing. For example, eddy current testing is often specified for thin wall tubing in pressurized water reactors, steam generators, turbine condensers and air conditioning heat exchangers. Eddy current testing is also used in aircraft maintenance. The following are some of the typical material characteristics that may affect conductivity and be evaluated by eddy current testing: cracks, inclusions, dents and holes; grain size; heat treatment; coating and material thickness; composition, conductivity or permeability; and alloy composition.

Radiographic Testing

Principles. Radiographic testing (Fig. 8) is based on the test object's attenuation of penetrating radiation — either electromagnetic radiation of very short wavelength or particulate radiation (X-rays, gamma rays and neutrons). Different portions of an object absorb different amounts of penetrating radiation because of differences in density and variations in thickness of the test object or differences in absorption characteristics caused by variation in composition. These variations in the attenuation of the penetrating radiation can be monitored by detecting the unattenuated radiation that passes through the object.

This monitoring may be in different forms. The traditional form is through radiation sensitive film. Radioscopic sensors provide digital images. X-ray computed tomography is a three-dimensional, volumetric radiographic technique.

Applications. The principal industrial uses of radiographic testing involve testing of castings and weldments, particularly

FIGURE 7. Electromagnetic testing: (a) representative setup for eddy current test; (b) inservice detection of discontinuities.

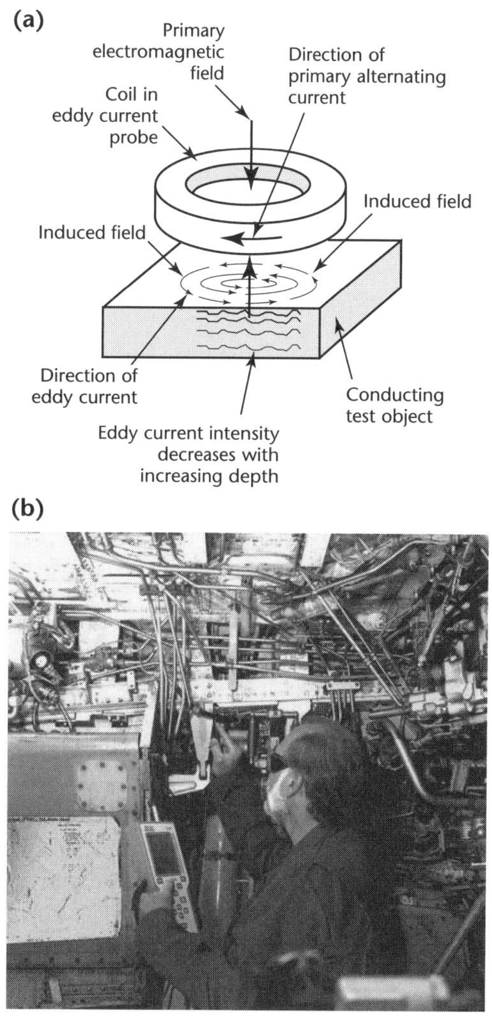

where there is a critical need to ensure freedom from internal discontinuities. Radiographic testing is often specified for thick wall castings and for weldments in steam power equipment (boiler and turbine components and assemblies). The method can also be used on forgings and mechanical assemblies, although with mechanical assemblies radiographic testing is usually limited to testing for conditions and proper placement of components. Radiographic testing is used to detect inclusions, lack of fusion, cracks, corrosion, porosity, leak paths, missing or incomplete components and debris.

Acoustic Emission Testing

Principles. Acoustic emissions are stress waves produced by sudden movement in stressed materials. The classic sources of acoustic emission are crack growth and plastic deformation. Sudden movement at the source produces a stress wave that radiates out into the test object and excites a sensitive piezoelectric sensor. As the stress in the material is raised, emissions are generated. The signals from one or more sensors are amplified and measured to produce data for display and interpretation.

The source of acoustic emission energy is the elastic stress field in the material. Without stress, there is no emission. Therefore, an acoustic emission test (Fig. 9) is usually carried out during a controlled loading of the test object. This can be a proof load before service; a controlled variation of load while the structure is in service; a fatigue, pressure or creep test; or a complex loading program. Often, a structure is going to be loaded hydrostatically anyway during service and acoustic emission testing is used because it gives valuable additional information about the expected performance of the structure under load. Other times, acoustic emission testing is selected for reasons of economy or safety and loading is applied specifically for the acoustic emission test.

Applications. Acoustic emission is a natural phenomenon occurring in the widest range of materials, structures and processes. The largest scale events observed with acoustic emission testing are seismic; the smallest are microscopic dislocations in stressed metals.

The equipment used is highly sensitive to any kind of movement in its operating frequency (typically 20 to 1200 kHz). The equipment can detect not only crack growth and material deformation but also such processes as solidification, friction, impact, flow and phase transformations. Therefore, acoustic emission testing is also used for in-process weld monitoring, for detecting tool touch and tool wear during automatic machining, for detecting wear and loss of lubrication in rotating equipment, for detecting loose parts and loose particles, for preservice proof testing and for detecting and monitoring leaks, cavitation and flow.

Ultrasonic Testing

Principles. In ultrasonic testing (Fig. 10), beams of acoustic waves at a frequency too high to hear are introduced into a material for the detection of surface and subsurface discontinuities. These acoustic waves travel through the material with some energy loss (attenuation) and are reflected and refracted at interfaces. The echoes are then analyzed to define and locate discontinuities.

FIGURE 8. Representative setup for radiographic testing.

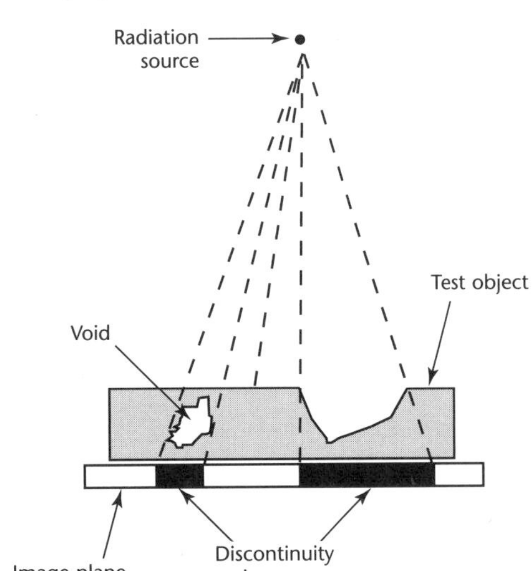

FIGURE 9. Acoustic emission monitoring of floor beam on suspension bridge.

Applications. Ultrasonic testing is widely used in metals, principally for thickness measurement and discontinuity detection. This method can be used to detect internal discontinuities in most engineering metals and alloys. Bonds produced by welding, brazing, soldering and adhesives can also be ultrasonically tested. In-line techniques have been developed for monitoring and classifying materials as acceptable, salvageable or scrap and for process control. Also tested are piping and pressure vessels, nuclear systems, motor vehicles, machinery, railroad stock and bridges.

Leak Testing

Principles. Leak testing is concerned with the flow of liquids or gases from pressurized components or into evacuated components. The principles of leak testing involve the physics of liquids or gases flowing through a barrier where a pressure differential or capillary action exists.

Leak testing encompasses procedures that fall into these basic functions: leak location, leakage measurement and leakage monitoring. There are several subsidiary methods of leak testing, entailing tracer gas detection (Fig. 11), pressure change measurement, observation of bubble formation, acoustic emission leak testing and other principles.

Applications. Like other forms of nondestructive testing, leak testing affects the safety and performance of a product. Reliable leak testing decreases costs by reducing the number of reworked products, warranty repairs and liability claims. The most common reasons for performing a leak test are to prevent the loss of costly materials or energy, to prevent contamination of the environment, to ensure component or system reliability and to prevent an explosion or fire.

Infrared and Thermal Testing

Principles. Conduction, convection and radiation are the primary mechanisms of heat transfer in an object or system. Electromagnetic radiation is emitted from all bodies to a degree that depends on their energy state.

Thermal testing involves the measurement or mapping of surface temperatures when heat flows from, to or through a test object. Temperature

FIGURE 10. Classic setups for ultrasonic testing: (a) longitudinal wave technique; (b) transverse wave technique.

(a)

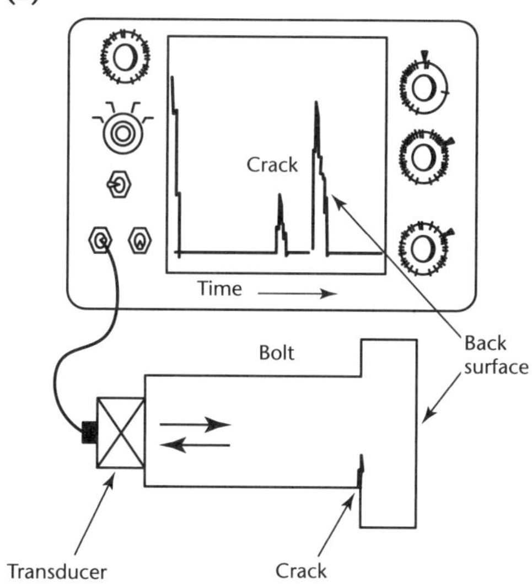

(b)

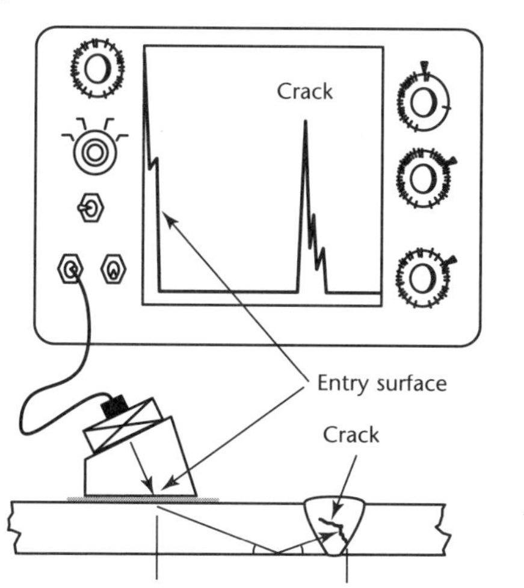

FIGURE 11. Leakage measurement dynamic leak testing using vacuum pumping: (a) pressurized system mode for leak testing of smaller components; (b) pressurized envelope mode for leak testing of larger volume systems.

(a)

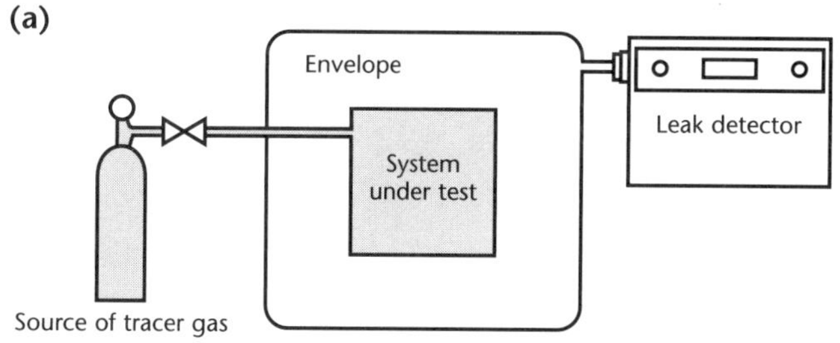

(b)

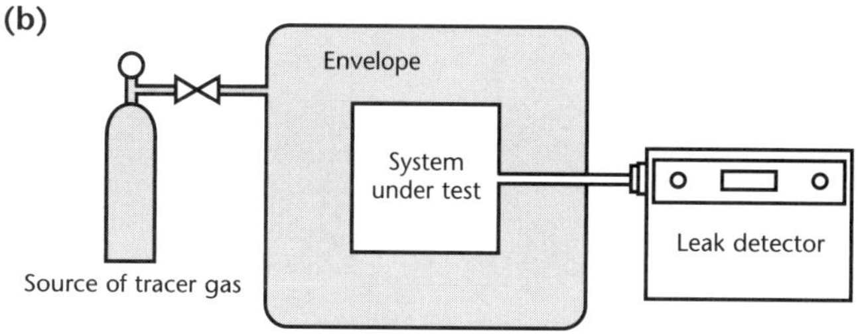

differentials on a surface, or changes in surface temperature with time, are related to heat flow patterns and can be used to detect discontinuities or to determine the heat transfer characteristics of an object. For example, during the operation of an electrical breaker, a hot spot detected at an electrical termination may be caused by a loose or corroded connection (Fig. 12). The resistance to electrical flow through the connection produces an increase in surface temperature of the connection.

Applications. There are two basic categories of infrared and thermal test applications: electrical and mechanical. The specific applications within these two categories are numerous.

Electrical applications include transmission and distribution lines, transformers, disconnects, switches, fuses, relays, breakers, motor windings, capacitor banks, cable trays, bus taps and other components and subsystems.

Mechanical applications include insulation (in boilers, furnaces, kilns, piping, ducts, vessels, refrigerated trucks and systems, tank cars and elsewhere), friction in rotating equipment (bearings, couplings, gears, gearboxes, conveyor belts, pumps, compressors and other components) and fluid flow (steam lines; heat exchangers; tank fluid levels; exothermic reactions; composite structures; heating, ventilation and air conditioning systems; leaks above and below ground; cooling and heating; tube blockages; environmental assessment of thermal discharge; boiler or furnace air leakage; condenser or turbine system leakage; pumps; compressors; and other system applications).

Other Methods

There are many other methods of nondestructive testing, including optical methods such as holography, shearography and moiré imaging; material identification methods such as chemical spot testing, spark testing and spectroscopy; strain gaging; and acoustic methods such as vibration analysis and tapping.

Figure 12. Infrared thermography of automatic transfer switches for emergency diesel generator. Hot spots appear bright in thermogram (inset).

PART 2. Management of Visual Testing

Selection of Visual Testing

Visual testing is an important method in the broad field of nondestructive testing. Visual testing is used to locate surface anomalies in most materials and subsurface discontinuities in translucent materials. Visual testing is performed either by a direct technique or by a remote (that is, indirect) technique. One definition of the direct technique is to place the eye within 600 mm (24 in.) and not less than 30 degrees from the test surface. Mirrors may be used to improve the angle of vision, and aids such as magnifying lenses may be used to assist examinations. The remote, or indirect, technique may include accessories such as mirrors, borescopes, video probes or cameras to correct for the distance or angles of view. With a remote (indirect) technique, resolution must be equivalent to that of the direct technique.

Visual test equipment is designed to detect structural characteristics of a part. These characteristics range from simple surface discontinuities on flat surfaces to various fabrication or inservice discontinuities in complex geometries.

As a result, specific applications have been developed using visual testing: detecting discontinuities in fabricated structures such as airframes, piping and pressure vessels, ships, bridges, motor vehicles and machinery and predicting the impending failure in highly stressed components exposed to the various modes of fatigue.

Advantages

The visual method is a sensitive means of locating surface anomalies in various materials. There is little or no limitation on the size or shape of the part being inspected. Indications provide a graphic representation of the actual discontinuity. Precleaning may be necessary if the surface cleanliness impairs an adequate view of the test surface, but discontinuities filled with foreign material may be detected. The need for precleaning will largely depend on the size and type of discontinuities specified by acceptance criteria. The following are the primary advantages typically associated with visual testing: (1) economy, (2) speed, (3) sensitivity, (4) versatility, (5) applicability to irregular shapes, (6) field mobility, (7) minimal training requirements and (8) minimal equipment requirements.

Limitations

Visual testing requires a line of sight to the test surface and lighting adequate to detect and interpret anomalies of interest. Visual testing may be limited by component geometry: size, contour, surface roughness, complexity and discontinuity orientation. Remote visual equipment may be required to access interior surfaces and remote equipment providing adequate viewing angles, sensitivity, resolution and illumination may be costly. For proper interpretation of indications, the inspector needs skill with the technique used, experience using the visual equipment and knowledge of the test object.

Management of Visual Testing Programs

Management of a visual testing program requires consideration of many items before it can produce the desired results. Some basic questions must be answered before a program can be implemented effectively.

1. Is the program needed?
2. Are qualified personnel available?
3. Are qualified and approved procedures in place? Are regulatory requirements in place that mandate program characteristics?
4. What is the magnitude of the program that will provide desired results?
5. What provisions must be made for personnel safety and for compliance with environmental regulations?
6. What is the performance date for a program to be fully implemented?
7. Is there a cost benefit of visual testing?
8. What are the available resources in material, personnel and money?

Once these questions are answered, then a recommendation can be made to select the type of inspection agency. Three primary types of agencies responsible for inspection are (1) service companies, (2) consultants and (3) in-house programs.

Although these are the main agency types, some programs may, routinely or as needed, require support personnel from a combination of two or more of these sources. Before a final decision is made, advantages and disadvantages of each agency type must be considered.

Service Companies

Once a service company is selected, responsibilities need to be defined.

1. Who will identify the components within the facility to be examined?
2. Will the contract be for time and materials or have a specific scope of work?
3. If a time and materials contract is awarded, who will monitor the time and materials charged?
4. If a scope of work is required, who is technically qualified to develop and approve it?
5. What products or documents (test reports, trending, recommendations, root cause analysis and others) will be provided once the tests are completed?
6. Who will evaluate and accept the product (test reports, trending, recommendations, root cause analysis and others) within the service company?
7. Do the service company workers possess qualifications and certifications required by contract and by applicable regulations?
8. Do the service company workers require site specific training (confined space entry, electrical safety, hazardous materials and others) or clearance to enter and work in the facility?
9. Does the service company retain any liability for test results?

Consultants

1. Will the contract be for time and materials or have a specific scope of work?
2. If a scope of work is required, who is technically qualified to develop and approve it?
3. Who will identify the required qualifications of the consultant?
4. Is the purpose of the consultant to develop or update a program or is it to oversee and evaluate the performance of an existing program?
5. Will the consultant have oversight responsibility for tests performed?
6. What products or documents (trending, recommendations, root cause analysis and others) are provided once the tests are completed?
7. Who will evaluate the consultant's performance (test reports, trending, recommendations, root cause analysis and other functions) within the sponsoring company?
8. Does the consultant possess qualifications and certifications required by contract and by applicable regulations?
9. Does the consultant require site specific training (confined space entry, electrical safety, hazardous materials and others) or clearance to enter and work in the facility?
10. Does the consultant retain any liability for test results?

In-House Programs

1. Who will determine the scope of the program, such as which techniques will be used?
2. What are the regulatory requirements (codes and standards) associated with program development and implementation?
3. Who will develop a cost benefit analysis for the program?
4. How much time and what resources are available to establish the program?
5. What are the qualification requirements (education, training, experience and others) for personnel?
6. Do program personnel require additional training (safety, confined space entry or others) or qualifications?
7. Are subject matter experts required to provide technical guidance during personnel development?
8. Are procedures required to perform work in the facility?
9. If procedures are required, who will develop, review and approve them?
10. Who will determine the technical specifications for test equipment?

Visual Test Procedures

The conduct of test operations (in-house or contracted) should be performed in accordance with specific instructions from an expert. Specific instructions are typically written as a technical procedure. In many cases, codes and specifications will require that a technical procedure be developed for each individual test. In other cases, the same procedure is used repeatedly.

The procedure can take many forms. A procedure may comprise general instructions that address only major aspects of test techniques. Or a procedure may be written as a step-by-step process requiring a supervisor's or a qualified/certified worker's signature after each step. The following is a typical format for an industrial procedure.

1. The *purpose* identifies the intent of the procedure.
2. The *scope* establishes the items, tests and techniques covered and not covered by the procedure.
3. *References* are specific documents from which criteria are extracted or are documents satisfied by implementation of the procedure.
4. *Definitions* are needed for terms and abbreviations that are not common knowledge to people who will read the procedure.
5. Statements about *personnel requirements* address specific requirements to perform tasks in accordance with the procedure — issues such as personnel qualification, certification and access clearance.
6. Calibration requirements and model numbers of qualified *equipment* must be specified.
7. The test *procedure* provides a sequential process to be used to conduct test activities.
8. A *system performance* check is needed before a test. The check might be daily or detailed.
9. *Acceptance criteria* establish component characteristics that will identify the items suitable for service (initial use or continued service).
10. *Reports* (records) document specific test techniques, equipment used, personnel, activity, date performed and test results.
11. *Attachments* may include (if required) items such as report forms, instrument calibration forms, qualified equipment matrix, schedules and others.

Once the procedure is written, an expert in the subject evaluates it. If the procedure meets requirements, the expert will approve it for use. Some codes and standards also require the procedure to be qualified — that is, demonstrated to the satisfaction of a representative of a regulatory body or jurisdictional authority.

Visual Test Specifications[4]

A visual test specification must anticipate issues that arise during testing. A specification is specific to a component or product and may be tailored to comply with one or more standards. A specification can require more stringent limits than the standard(s) it was written to satisfy. In practice, a specification provides a list of testing parameters that describes the techniques for locating and categorizing discontinuities in a specific test object. A typical specification includes acceptance criteria and is required by the designer, buyer or manufacturer of the article it covers.

Specifications are written to eliminate variables of human operators and system designs, to produce an accurate result regardless of who performs the visual test. Specifications must be written with a full knowledge of (1) visual test techniques, (2) a technique's individual sensitivities, (3) the test object design, (4) its material characteristics and (5) the discontinuities critical to the test object's service life. In most mature manufacturing applications, nondestructive tests are considered during design and such specifications are specified on the test object's original drawing.

Visual specifications are produced to standardize test results, not to eliminate the initiative of the technician. There is no substitute for an experienced inspector who assumes personal responsibility for the quality and accuracy of the test.

Testing specifications are working documents that tell how to locate discontinuities in a specific test object. Even well established and successful specifications need periodic review and revision. It is very important that relevant knowledge of field proven techniques and advances in inspection technologies be incorporated as quickly as possible into industry specifications.

Interpretation

Interpretation may be complex, especially before a procedure has been established. The interpreter must have a knowledge of the following: (1) the underlying physical process, (2) techniques and equipment, (3) details about the test object (configuration, material properties, fabrication process, potential discontinuities and anticipated service conditions) and (4) possible sources of false indications that might be mistaken for meaningful visual indications.

After interpretation, acceptance criteria and rejection criteria are applied in a phase called *evaluation*.

Reliability of Test Results

When a test is performed, there are four possible outcomes: (1) a rejectable discontinuity can be found when one is present, (2) a rejectable discontinuity can be missed even when one is present, (3) a rejectable discontinuity can be indicated when none is present and (4) no rejectable discontinuity is found when none is present. A reliable testing process and a qualified inspector should find all discontinuities of concern with no discontinuities missed (no errors as in case 2 above) and no false calls (case 3 above).

To approach this goal, the probability of finding a rejectable discontinuity must be high and the inspector must be both proficient in the testing process and motivated to perform with maximum efficiency. An ineffective inspector may accept test objects that contain discontinuities, with the result of possible inservice part failure. The same inspector may reject parts that do not contain rejectable discontinuities, with the result of unnecessary scrap and repair. Neither scenario is desirable.

Visual Test Standards

Traditionally, the purpose of specifications and standards has been to define the requirements that goods or services must meet. As such, they are intended to be incorporated into contracts so that both the buyer and provider have a well defined description of what one will receive and the other will deliver.

Standards have undergone a process of peer review in industry and can be invoked with the force of law by contract or by government regulation. In contrast, a specification represents an employer's instructions to employees and is specific to a contract or workplace. Many a specification originates as a detailed description either as part of a purchaser's requirements or as part of a vendor's offer. Specifications may be incorporated into standards through the normal review process. Standards and specifications exist in three basic areas: equipment, processes and personnel.

1. Standards for visual equipment include criteria that address surface accessibility, sensitivity, degree of magnification, field of view, depth of field, minimum lighting requirements and other matters.
2. ASTM International and other organizations publish standards for test techniques. Some other standards are for quality assurance procedures and are not specific to a test method or even to testing in general. Table 4 lists standards used in visual testing. The United States Department of Defense has replaced most military specifications and standards with industry consensus specifications and standards. A source for nondestructive test standards is the *Annual Book of ASTM Standards*.[5]
3. Qualification and certification of testing personnel are discussed below with specific reference to recommendations of ASNT Recommended Practice No. SNT-TC-1A.[6]

Personnel Qualification and Certification

One of the most critical aspects of the test process is the qualification of testing personnel. Nondestructive testing is sometimes referred to as a *special process,* special in that it is difficult to determine the adequacy of a test by merely observing the process or the documentation it generates. The quality of the test largely depends on the skills and knowledge of the inspector.

The American Society for Nondestructive Testing (ASNT) has been a world leader in the qualification and certification of nondestructive testing personnel since the 1960s. (*Qualification* demonstrates that an individual has the required training, experience, knowledge and abilities; *certification* provides written testimony that an individual is qualified.) By the twenty-first century, the American Society for Nondestructive Testing had instituted three avenues and four major documents for the qualification and certification of nondestructive testing personnel.

1. Recommended Practice No. SNT-TC-1A, *Personnel Qualification and Certification in Nondestructive Testing,* provides guidelines to employers for personnel qualification and certification in nondestructive testing. This recommended practice identifies the attributes that should be considered when qualifying nondestructive testing personnel. It requires the employer to develop and implement a written practice, a procedure that details the specific process and any limitation in the qualification and certification of nondestructive testing personnel.[6]
2. ANSI/ASNT CP-189, *Standard for Qualification and Certification of Nondestructive Testing Personnel,* resembles SNT-TC-1A but establishes specific requirements for the qualification and certification of Level I and II nondestructive testing personnel. For Level III, CP-189 references an examination administered by the American Society for Nondestructive Testing. CP-189 is a consensus standard as defined by the American National Standards Institute (ANSI). It is recognized as the American standard for nondestructive testing. It is not considered a recommended practice; it is a national standard.[7]

Table 4. Some standards specifying visual testing.

American Concrete Institute

ACI 201.1R, *Guide for Conducting a Visual Inspection of Concrete in Service* (2008).

American National Standards Institute

ANSI B3.2, *Rolling Element Bearings — Aircraft Engine, Engine Gearbox, and Accessory Applications — Surface Visual Inspection* (1999).

ANSI/EIA 699, *Test Method for the Visual Inspection of Quartz Crystal Resonator Blanks* (1997).

American Petroleum Institute

API 5D, *Specification for Drill Pipe* (2001).

API 5L, *Specification for Line Pipe* (2008).

API 570, *Piping Inspection Code: Inspection, Repair, Alteration, and Rerating of In-Service Piping Systems* (2006).

API 620, *Design and Construction of Large, Welded, Low-Pressure Storage Tanks* (2008).

API 650, *Welded Tanks for Oil Storage* (2007).

API RP-5A5 [ISO 15463-2003], *Recommended Practice for Field Inspection of New Casing, Tubing and Plain End Drill Pipe* (2005).

API RP-5L8, *Recommended Practice for Field Inspection of New Line Pipe* (1996).

API RP-7G, *Recommended Practice for Drill Stem Design and Operating Limits* (2003).

API SPEC 5CT [ISO 11960], *Specification for Casing and Tubing* (2006).

API SPEC 7, *Specification for Rotary Drill Stem Elements* (2008).

API STD 1104, *Welding of Pipelines and Related Facilities* (2005).

API STD 5T1, *Imperfection Terminology* (2003).

API STD 653, *Tank Inspection, Repair, Alteration, and Reconstruction* (2008).

ASME International

ASME Boiler and Pressure Vessel Code: Section I, *Rules for Construction of Power Boilers* (2007).

ASME Boiler and Pressure Vessel Code: Section III, *Rules for Construction of Nuclear Power Plant Components* (2007).

ASME Boiler and Pressure Vessel Code: Section IV, *Rules for Construction of Heating Boilers* (2007).

ASME Boiler and Pressure Vessel Code: Section V, *Nondestructive Examination.* Article 9, *Visual Examination* (2009).

ASME Boiler and Pressure Vessel Code: Section VI, *Recommended Rules for the Care and Operation of Heating Boilers* (2007).

ASME Boiler and Pressure Vessel Code: Section VII, *Recommended Guidelines for the Care of Power Boilers* (2007).

ASME Boiler and Pressure Vessel Code: Section VIII, *Rules for Construction of Pressure Vessels* (Divisions 1, 2 and 3) (2007).

ASME Boiler and Pressure Vessel Code: Section X, *Fiber Reinforced Plastic Pressure Vessels* (2007).

ASME Boiler and Pressure Vessel Code: Section XI, *Rules for Inservice Inspection of Nuclear Power Plant Components* (2007).

ASME Boiler and Pressure Vessel Code: Section XII, *Rules for Construction and Continued Service of Transport Tanks* (2007).

ASME B 31.1, *Power Piping* (2007).

ASME B 31.3, *Process Piping* (2008).

ASME B 31.4, *Pipeline Transportation Systems for Liquid Hydrocarbons and Other Liquids* (2006).

ASME B 31.5, *Refrigeration Piping and Heat Transfer Components* (2006).

ASME B 31.8, *Gas Transmission and Distribution Piping Systems* (2007).

ASTM International

ASTM A 802M, *Standard Practice for Steel Castings, Surface Acceptance Standards, Visual Examination* (2006).

ASTM D 2562, *Standard Practice for Classifying Visual Defects in Parts Molded from Reinforced Thermosetting Plastics* (2008).

ASTM D 2563, *Standard Practice for Classifying Visual Defects in Glass-Reinforced Plastic Laminate Parts* (2008).

ASTM D 4385, *Standard Practice for Classifying Visual Defects in Thermosetting Reinforced Plastic Pultruded Products* (2008).

ASTM E 1799, *Standard Practice for Visual Inspections of Photovoltaic Modules* (1999).

ASTM F 1236, *Standard Guide for Visual Inspection of Electrical Protective Rubber Products* (2007).

ASTM F 584, *Standard Practice for Visual Inspection of Semiconductor Lead-Bonding Wire* (2006).

American Welding Society

AWS B1.11, *Guide for the Visual Examination of Welds* (2000).

AWS D1.1M, *Structural Welding Code — Steel* (2008).

AWS D8.1M, *Specification for Automotive Weld Quality — Resistance Spot Welding of Steel* (2007).

AWS D18.2, *Guide to Weld Discoloration Levels on Inside of Austenitic Stainless Steel Tube* (1999).

AWS G1.6, *Specification for the Qualification of Plastics Welding Inspectors for Hot Gas, Hot Gas Extrusion, and Heated Tool Butt Thermoplastic Welds* (2006).

AWS QC1, *Standard for AWS Certification of Welding Inspectors* (2007).

Association Connecting Electronics Industries

IPC-OI-645, *Standard for Visual Optical Inspection Aids* (1993).

Compressed Gas Association

CGA C-13, *Guidelines for Periodic Visual Inspection and Requalification of Acetylene Cylinders* (2006).

CGA C-6, *Standards for Visual Inspection of Steel Compressed Gas Cylinders* (2007).

CGA C-6.1, *Standards for Visual Inspection of High Pressure Aluminum Compressed Gas Cylinders* (2006).

CGA C-6.2, *Guidelines for Visual Inspection and Requalification of Fiber Reinforced High Pressure Cylinders* (2005).

CGA C-6.3, *Guidelines for Visual Inspection and Requalification of Low Pressure Aluminum Compressed Gas Cylinders* (1999).

CGA C-6.4, *Methods for External Visual Inspection of Natural Gas Vehicle (NGV) and Hydrogen Vehicle (HV) Fuel Containers and Their Installations* (2007).

European Committee for Standardization

CEN EN 13508 [DIN 13508] P2, *Conditions of Drain and Sewer Systems Outside Buildings* — Part 2: *Visual Inspection Coding* System (2007).

CEN EN 13018 [BS 13018], *Non-Destructive Testing — Visual Testing — General Principles* (2007).

CEN EN 13100-1 [BS 13100-1], *Non-Destructive Testing of Welded Joints of Thermoplastics Semi-Finished Products* — Part 1: *Visual Examination* (2000).

CEN EN 3841-201 [BS 3841-201], *Circuit Breakers — Test Methods* — Part 201, *Visual Inspection* (2005).

Federal Aviation Administration

FAA AC 43-204, *Visual Inspection for Aircraft* (1997).

International Electrotechnical Commission

IEC 60748-23-2, *Semiconductor Devices — Integrated Circuits —* PART 23-2: *Hybrid Integrated Circuits and Film Structures — Manufacturing Line Certification – Internal Visual Inspection and Special Tests* (2002).

International Organization for Standardization

ISO 11960 [API SPEC 5CT], *Petroleum and Natural Gas Industries — Steel Pipes for Use as Casing or Tubing for Wells* (2006).

ISO 17637, *Non-Destructive Testing of Welds — Visual Testing of Fusion-Welded Joints* (2003).

ISO 3058, *Non-Destructive Testing — Aids to Visual Inspection — Selection of Low-Power Magnifiers* (1998).

Japanese Institute of Standards

JIS H 0613, *Non-Ferrous Metals and Metallurgy — Visual Inspection for Sliced and Lapped Silicon Wafers* (1978).

JIS H 0614, *Non-Ferrous Metals and Metallurgy — Visual Inspection for Silicon Wafers with Specular Surfaces* (1996).

JIS Z 3090, *Visual Testing Method of Fusion-Welded Joints* (2005).

Manufacturers Standardization Society

MSS SP-55, *Quality Standard for Steel Castings for Valves, Flanges and Fittings and Other Piping Components — Visual Method for Evaluation of Surface Irregularities* (2006).

South African Bureau of Standards

SAA AS 3978, *Non-Destructive Testing — Visual Inspection of Metal Products and Components* (2003).

SAA AS/NZS 3894.8, *Surface Treatment and Coating — Site Testing of Protective Coatings — Visual Determination of Gloss* (2006).

3. ANSI/ASNT CP-105, *ASNT Standard Topical Outlines for Qualification of Nondestructive Testing Personnel,* is a standard that establishes the minimum topical outline requirements for the qualification of nondestructive testing (NDT) personnel. The outlines in this single standard are referenced by both SNT-TC-1A and CP-189. CP-105 is a consensus standard of the American National Standards Institute (ANSI) and is recognized as an American standard for nondestructive testing. It is not considered a recommended practice; it is a national standard.[8]
4. The ASNT Central Certification Program (ACCP), unlike SNT-TC-1A and CP-189, is a third party certification process that identifies qualification and certification attributes for Level II and Level III nondestructive testing personnel. The American Society for Nondestructive Testing certifies that the individual has the skills and knowledge for many nondestructive test method applications. It does not remove the responsibility for the final determination of personnel qualification from the employer. The employer evaluates an individual's skills and knowledge for application of company procedures using designated techniques and equipment identified for specific tests. ACCP is not a standard or recommended practice; it is a service administered by the American Society for Nondestructive Testing.[9]

Excerpts from Recommended Practice No. SNT-TC-1A

To give an idea of the contents of these documents, the following items are excerpted from Recommended Practice No. SNT-TC-1A.[6] The original text is arranged in outline format and includes recommendations that are not specific to visual testing.

> **Scope** … This Recommended Practice has been prepared to establish guidelines for the qualification and certification of NDT personnel whose specific jobs require appropriate knowledge of the technical principles underlying the nondestructive tests they perform, witness, monitor, or evaluate. … This document provides guidelines for the establishment of a qualification and certification program. …
>
> **Written Practice** … The employer shall establish a written practice for the control and administration of NDT personnel training, examination, and certification. … The employer's written practice should describe the responsibility of each level of certification for determining the acceptability of materials or components in accordance with the applicable codes, standards, specifications and procedures. …
>
> **Education, Training, and Experience Requirements for Initial Qualification** … Candidates for certification in NDT should have sufficient education, training, and experience to ensure qualification in those NDT methods in which they are being considered for certification. … Table 6.3.1A [see Table 5 in this *Nondestructive Testing Handbook* chapter, for visual testing] lists recommended training and experience factors to be considered by the employer in establishing written practices for initial qualification of Level I and Level II individuals. …
>
> **Training Programs** … Personnel being considered for initial certification should complete sufficient organized training to become thoroughly familiar with the principles and practices of the specified NDT method related to the level of certification desired and applicable to the processes to be used and the products to be tested. …
>
> **Examinations** … For Level I and II personnel, a composite grade should be determined by simple averaging of the results of the general, specific and practical examinations … Examinations administered for qualification should result in a passing composite grade of at least 80 percent, with no individual examination having a passing grade less than 70 percent. …
>
> **Practical [Examination] (for NDT Level I and II)** … The candidate should demonstrate … ability to operate the necessary NDT equipment, record, and analyze the resultant information to the degree required. … At least one flawed specimen should be tested and the results of the NDT analyzed by the candidate. …
>
> **Certification** … Certification of all levels of NDT personnel is the responsibility of the employer. … Certification of NDT personnel shall be based on demonstration of satisfactory qualification in accordance with [sections on education, training, experience and examinations] as described in the employer's written practice. … Personnel certification records shall be maintained on file by the employer. …
>
> **Recertification** … All levels of NDT personnel shall be recertified periodically in accordance with one of the [following:] continuing satisfactory technical performance [or reexamination] in those portions of the examinations … deemed necessary by the employer's NDT Level III. … Recommended maximum recertification intervals are 5 years for all certification levels.

These recommendations from the 2006 edition of Recommended Practice No. SNT-TC-1A are cited only to provide an idea of items that must be considered in the development of an in-house nondestructive testing program. Because the text above is excerpted, those developing a personnel qualification program should consult the complete text of SNT-TC-1A and other applicable procedures and practices. If an outside agency is contracted for visual test services, then the contractor must have a qualification and certification program to satisfy the codes and standards in force.

The minimum number of questions that should be administered in the written examination for visual test personnel is as follows: 40 questions in the general examination and 20 questions in the specific examination. The number of questions is the same for Level I and Level II personnel. Table 5 shows required hours of training for Level I and Level II.

Central Certification

Another standard that may be a source for compliance is published by the International Organization for Standardization (ISO). The work of preparing international standards is normally carried out through technical committees of this worldwide federation of national standards bodies. Each ISO member body interested in a subject for which a technical committee has been established has the right to be represented on that committee. International organizations, governmental and nongovernmental, in liaison with the International Organization for Standardization, also take part in the work.

Technical Committee ISO/TC 135, Non-Destructive Testing Subcommittee SC 7, Personnel Qualification, prepared international standard ISO 9712, *Non-Destructive Testing — Qualification and Certification of Personnel.*[10] In its statement of scope, ISO 9712 states that it "specifies the qualification and certification of personnel involved in non-destructive testing ... in one or more of the following methods: acoustic emission testing; eddy current testing; infrared thermographic testing; leak testing (hydraulic pressure tests excluded); magnetic particle testing; penetrant testing; radiographic testing; strain testing; ultrasonic testing; visual testing (direct unaided visual tests and visual tests carried out during the application of another NDT method are excluded)."

The International Organization for Standardization also publishes a standard for something called *limited certification.*[11] Inspectors whose actions are limited sometimes have limited training requirements. Limited certification would not be applicable to visual testing inspectors in the field, for example, but may be desired for assembly line operators of remote visual testing equipment to detect debris inside fabricated products.

Table 5. Recommended training and experience (in hours) for visual testing personnel according to Recommended Practice No. SNT-TC-1A.[6]

	Level I	Level II
High school graduate[a]	8 h	16 h
Two years of college[b]	4 h	8 h
Work experience in method[c]	70 h	140 h
Total hours in nondestructive testing	130 h	270 h

a. Or equivalent.

b. Completion with a passing grade of at least two years of engineering or science study in a university, college or technical school.

c. Minimum work experience in method, per level. Note: For Level II certification, the experience shall consist of time as Level I or equivalent. If a person is being qualified directly to Level II with no time at Level I, the required experience shall consist of the sum of the times required for Level I and Level II and the required training shall consist of the sum of the hours required for Level I and Level II.

Safety in Visual Testing[12]

This information is presented solely for educational purposes and should not be consulted in place of current safety regulations. Note that units of measure have been converted to this book's format and are not those commonly used in all industries. Human vision can be disrupted or destroyed by improper use of any radiation, including light. Consult the most recent safety documents and the manufacturer's literature before working near any radiation source.

Need for Safety

Developments in optical testing technology have created a need for better understanding of the potential health hazards caused by high intensity light sources or by artificial light sources of any intensity in the work area. The human eye operates optimally in an environment illuminated directly or indirectly by sunlight, with characteristic spectral distribution and range of intensities that are very different from those of most artificial sources. The eye can handle only a limited range of night vision tasks.

Evidence has accumulated that photochemical changes occur in eyes under the influence of normal daylight illumination — short term and long term visual impairment and exacerbation of retinal disease have been observed and it is important to understand why this occurs. Periodic fluctuations of visible and ultraviolet radiation occur with the regular diurnal light-and-dark cycles and with the lengthening and shortening of the cycle as a result of seasonal changes. These fluctuations are known to affect all biological systems critically. The majority of such light/dark effects is based on circadian cycles and controlled by the pineal system, which can be affected directly by the transmission of light to the pineal gland or indirectly by effects on the optic nerve pathway.

Also of concern are the results of work that has been done demonstrating that light affects immunological reactions *in vitro* and *in vivo* by influencing the antigenicity of molecules, antibody function and the reactivity of lymphocytes.

Given the variety of visual tasks and illumination that confronts the visual inspector, it is important to consider whether failures in performance might be a result of excessive exposure to light or other radiation or even a result of insufficient light sources. A myth exists that 20/20 foveal vision, in the absence of color blindness, is all that is necessary for optimal vision. In fact, there may be visual field loss in and beyond the fovea centralis for many reasons; the inspector may have poor stereoscopic vision; visual ability may be impaired by glare or reflection; or actual vision may be affected by medical or psychological conditions.

Visual Safety Recommendations

The American Conference of Governmental Industrial Hygienists (ACGIH) has proposed two threshold limit values (TLVs) for noncoherent visible light, one covering damage to the retina by a thermal mechanism and one covering retinal damage by a photochemical mechanism. Threshold limit values for visible light, established by the American Conference of Governmental Industrial Hygienists, are intended only to prevent excessive occupational exposure and are limited to exposure durations of 8 h or less. They are not intended to cover photosensitive individuals.[13,14]

Laser Hazards

Loss of vision resulting from retinal burns following observation of the sun has been described throughout history. Common technological equivalents to this problem are coherent light sources: lasers. In addition to the development of lasers, further improvement in other high radiance light sources (a result of smaller, more efficient reflectors and more compact, brighter sources) has presented the potential for chorioretinal injury. It is thought that chorioretinal burns from artificial sources in industrial situations have been very much less frequent than similar burns from the sun.

Because of the publicity of the health hazard caused by exposure to laser radiation, awareness of such hazards is probably much greater than the general awareness of the hazard from high intensity noncoherent visible sources which may be as great or greater. Generally, lasers are used in specialized environments by technicians familiar with the hazards and trained to avoid exposure by the use of protective eyewear and clothing. Laser standards of manufacture and use have been well developed and probably have contributed more than anything else to a heightened awareness of safe laser operation.

Laser hazard controls are common sense procedures designed to (1) restrict personnel from entering the beam path and (2) limit the primary and reflected beams from occupied areas. Should an individual be exposed to excessive laser light, the probability of damage to the retina is high because of the high energy pulse capabilities of some lasers. However, the probability of visual impairment is relatively low because of the small area of damage on the retina. Once the initial flash blindness and pain have subsided, the resulting scotomas (damaged unresponsive areas) can sometimes be ignored by the accident victim.

The tissue surrounding the absorption site can much more readily conduct away heat for small image sizes than it can for large image sizes. In fact, retinal injury thresholds for less than 0.1 to 10 s exposure show a high dependence on the image size, 0.01 to 0.1 $W \cdot mm^{-2}$ for a 1000 µm wide image up to about 0.01 $kW \cdot mm^{-2}$ for a 20 µm image. In contrast, the sun produces merely a 160 µm diameter image on the retina.

Consensus standards provide guidance for the safe use of lasers.[15,16]

High Luminance Light Sources

The normal reaction to a high luminance light source is to blink and look away from the source. The probability of overexposure to noncoherent light sources is higher than the probability of exposure to lasers, yet extended (high luminance) sources are used in a more casual and possibly more hazardous way. In the nondestructive testing industry, extended sources are used as general illumination and in many specialized applications. Unfortunately, there are comparatively few guidelines for the safe use of extended sources of visible light.

Infrared Hazards

Infrared radiation comprises that invisible radiation beyond the red end of the visible spectrum up to about 1 mm wavelength. Infrared is absorbed by many substances and its principal biological effect is known as hyperthermia, heating that can be lethal to cells. Usually, the response to intense infrared radiation is pain and the natural reaction is to move away from the source so that burns do not develop.

Ultraviolet Hazards

Before development of the laser, the principal hazard in the use of intense light sources was the potential eye and skin injury from ultraviolet radiation. Ultraviolet radiation is invisible radiation beyond the violet end of the visible spectrum with wavelengths down to about 185 nm. It is strongly absorbed by the cornea and the lens of the eye. Ultraviolet radiation at wavelengths shorter than 185 nm is absorbed by air, is often called vacuum ultraviolet and is rarely of concern to the visual inspector. Many useful high intensity arc sources and some lasers may emit associated, potentially hazardous, levels of ultraviolet radiation. With appropriate precautions, such sources can serve very useful visual testing functions.

Studies have clarified the spectral radiant exposure doses and relative spectral effectiveness of ultraviolet radiation required to elicit an adverse biological response. These responses include keratoconjunctivitis (known as welder's flash), possible generation of cataracts and erythema or reddening of the skin. Longer wavelength ultraviolet radiation can lead to fluorescence of the eye's lens and ocular media, eyestrain and headache. These conditions lead, in turn, to low task performance resulting from the fatigue associated with increased effort. Chronic exposure to ultraviolet radiation accelerates skin aging and possibly increases the risk of developing certain forms of skin cancer.

It should also be mentioned that some individuals are hypersensitive to ultraviolet radiation and may develop a reaction following, what would be for the average healthy human, suberythemal exposures. However, it is unusual for these symptoms of exceptional photosensitivity to be elicited solely by the limited emission spectrum of an industrial light source. An inspector is typically aware of such sensitivity because of earlier exposures to sunlight.

In industry, the visual inspector may encounter many sources of visible and invisible radiation: incandescent lamps, compact arc sources (solar simulators), quartz halogen lamps, metal vapor (sodium and mercury) and metal halide discharge lamps, fluorescent lamps and flash lamps among others. Because of the high ultraviolet attenuation afforded by many visually transparent materials, an empirical approach is sometimes taken for the problem of light sources associated with ultraviolet: the source is enclosed and provided with ultraviolet absorbing glass or plastic lenses. If injurious effects continue to develop, the thickness of the protective lens is increased.

The photochemical effects of ultraviolet radiation on the skin and eye are still not completely understood. Records of ultraviolet radiation's relative spectral effectiveness for eliciting a particular biological effect (referred to by photobiologists as action spectra) are generally available. Ultraviolet irradiance may be measured at a point of interest with a portable radiometer and compared with the ultraviolet radiation hazard criteria.

For purposes of determining exposure levels, it is important to note that most inexpensive, portable radiometers are not equally responsive at all wavelengths throughout the ultraviolet spectrum and are usually only calibrated at one wavelength with no guarantees at any other wavelength. Such radiometers have been designed for a particular application using a particular lamp.

A common example in the nondestructive testing industry is the ultraviolet radiometer used in fluorescent liquid penetrant and magnetic particle applications. These meters are usually calibrated at 365 nm, the predominant ultraviolet output of the filtered 100 W medium pressure mercury vapor lamp commonly used in the industry. Use of the meter at any other wavelength in the ultraviolet spectrum may lead to significant errors. To minimize problems in assessing the hazard presented by industrial lighting, it is important to use a radiometer that has been calibrated with an ultraviolet spectral distribution as close as possible to the lamp of interest.

If the inspector is concerned about the safety of a given situation, ultraviolet absorbing eye protection and facewear is readily available from several sources. An additional benefit of such protection is that it prevents the annoyance of lens fluorescence and provides the wearer considerable protection from all ultraviolet radiation. In certain applications, tinted lenses can also provide enhanced visibility of the test object.

Damage to Retina

Although ultraviolet radiation from most of the high intensity visible light sources may be the principal concern, the potential for chorioretinal injury from visible radiation should not be overlooked.

It is possible to multiply the spectral absorption data of the human retina by the spectral transmission data of the eye's optical media at all wavelengths to arrive at an estimate of the relative absorbed spectral dose in the retina and the underlying choroid for a given spectral radiant exposure of the cornea. In practice, the evaluation of potential chorioretinal burn hazards depends on the maximum luminance and spectral distribution of the source; possible retinal image sizes; the image quality; pupil size; spectral scattering and absorption by the cornea, aqueous humor, the lens and the vitreous humor; and absorption and scattering in the various retinal layers.

Calculation of the permissible luminance from a permissible retinal illuminance for a source breaks down for very small retinal image sizes or for very small hot spots in an extended image caused by diffraction of light at the pupil, aberrations introduced by the cornea and lens and scattering from the cornea and the rest of the ocular media. Because the effects of aberration increase with increasing pupil size, greater blur and reduced peak retinal illuminance are noticed for larger pupil sizes and for a given corneal illumination.

Thermal Factor

Visible and near infrared radiation up to about 1400 nm (associated with most optical sources) is transmitted through the eye's ocular media and absorbed in significant doses principally in the retina. These radiations pass through the neural layers of the retina. A small amount is absorbed by the visual pigments in the rods and cones, to initiate the visual response, and the remaining energy is absorbed in the retinal pigment epithelium and choroid. The retinal pigment epithelium is optically the most dense absorbent layer (because of high concentrations of melanin granules) and the greatest temperature changes arise in this layer.

For short (0.1 to 100 s) accidental exposures to the sun or artificial radiation sources, the mechanism of injury is generally thought to be hyperthermia resulting in protein denaturation and enzyme inactivation. Because the large, complex organic molecules absorbing the radiant energy have broad spectral absorption bands, the hazard potential for chorioretinal injury is not expected to depend on the coherence or monochromaticity of the source. Injury from a laser or a nonlaser radiation source should not differ if image size, exposure time and wavelength are the same.

Because different regions of the retina play different roles in vision, the functional loss of all or part of one of these regions varies in significance. The greatest vision acuity exists only for central (foveal) vision, so that the loss of this retinal area dramatically reduces visual capabilities. In comparison, the loss of an area of similar size located in the peripheral retina could be subjectively unnoticed.

The human retina is normally subjected to irradiances below 1 $\mu W \cdot mm^{-2}$, except for occasional momentary exposures to the sun, arc lamps, quartz halogen lamps, normal incandescent lamps, flash lamps and similar radiant sources. The natural aversion or pain response to bright lights normally limits exposure to no more than 0.15 to 0.2 s. In some instances, individuals can suppress this response with little difficulty and stare at bright sources, as commonly occurs during solar eclipses.

Fortunately, few arc sources are sufficiently large and sufficiently bright to be a retinal burn hazard under normal viewing conditions. Only when an arc or hot filament is greatly magnified (in an optical projection system, for example) can hazardous irradiance be imaged on a sufficiently large area of the retina to cause a burn. Visual inspectors do not normally step into a projected beam at close range or view a welding arc with binoculars or a telescope.

Nearly all conceivable accident situations require a hazardous exposure to be delivered within the period of a blink reflex. If an arc is struck while an inspector is located at a very close viewing range, it is possible that a retinal burn could occur. At lower exposures, an inspector experiences a short term depression in photopic (daylight) sensitivity and a marked, longer term loss of scotopic (dark adapted) vision. That is why it is so important for visual inspectors in critical fluorescent penetrant and magnetic particle test environments to undergo dark adaptation before actually attempting to find discontinuities. Not only does the pupil have to adapt to the reduced visible level in a booth but the actual retinal receptors must attain maximum sensitivity. This effect may take half an hour or more, depending on the preceding state of the eye's adaptation.

Blue Hazard

The so-called blue hazard function has been used with the thermal factor to calculate exposure durations, to avoid damaging the retina.

The blue hazard is based on the demonstration that the retina can be damaged by blue light at intensities that do not elevate retinal temperatures sufficiently to cause a thermal hazard. It has been found that blue light can produce 10 to 100 times more retinal damage (permanent decrease in spectral sensitivity in this spectral range) than longer visible wavelengths. Note that there are some common situations in which both thermal and blue hazards may be present.

Photosensitizers

Over the past few decades, a large number of commonly used drugs, food additives, soaps and cosmetics have been identified as phototoxic or photoallergenic agents even at the longer wavelengths of the visible spectrum.[17] Colored drugs and food additives are possible photosensitizers for organs below the skin because longer wavelength visible radiations penetrate deeply into the body.

Eye Protection Filters

Because continuous visible light sources elicit a normal aversion or pain response that can protect the eye and skin from injury, visual comfort has often been used as an approximate hazard index and eye protection and other hazard controls have been provided on this basis.

Eye protection filters for various workers were developed empirically but now are standardized as shades and specified for particular applications.

Other protective techniques include use of high ambient light levels and specialized filters to further attenuate intense spectral lines. Laser eye protection is designed to have an adequate optical density at the laser wavelengths along with the greatest visual transmission at all other wavelengths.

Always bear in mind that hazard criteria must not be considered to represent fine lines between safe and hazardous exposure conditions. To be properly applied, interpretation of hazard criteria must be based on practical knowledge of potential exposure conditions and the user, whether a professional inspector or a general consumer. Accuracy of hazard criteria is limited by biological uncertainties including diet, genetic photosensitivity and the large safety factors required to be built into the recommendations.

Part 3. History of Visual Testing

Optics

Early physicists offered explanations of vision and light that have informed later understanding and made possible the development of optical devices: sextants, corrective eyewear, periscopes, telescopes, microscopes, cameras and borescopes. These scientists offered mathematical proofs of optical principles, including perspective, reflection and refraction.

1. In perspective, a near object appears larger than a distant object of the same size.
2. In reflection, light bounces off a surface. If the surface is shiny, the viewer sees a reversed, or *mirror,* image and the shiny surface is called *specular,* from the Latin *speculum,* "mirror."
3. Refraction bends the path of light as it moves from one medium into another, for example, from air into water. Refraction makes it possible for a convex lens to magnify an image.

With these concepts about the nature of light were others — for example, that light travels in a straight line and that it does not emanate from the viewer's eye.

The optical principles were not merely explained but were proven mathematically. For this reason, the pages of early optical treatises have diagrams like those in modern geometry books.

Greeks

The word *optics* comes from the Greek word ὀπτική, *optike,* "sight." For the Greeks, optics was part of the study of geometry. In Greek, the word *geometry* literally means "earth measurement." Geometry was a practical science, used to calculate distances and estimate the height of objects.

Writing around BCE 300, Euclid, a Greek, wrote a mathematical treatise that has dominated geometry for more than 2000 years. He also wrote *Optics,* a treatise that described behaviors of light, including perspective.[18]

Ptolemy, who lived in Alexandria in the second century, also touched on optical principles in his exhaustive astronomical treatise, called the *Almagest,* "great work."[19]

After the Roman Empire, Europe entered a period often called the Dark

Figure 13. Ibn Sahl's tenth century description of diffraction: (a) manuscript; (b) simplified enlargement of upper left corner.[25]

(a)

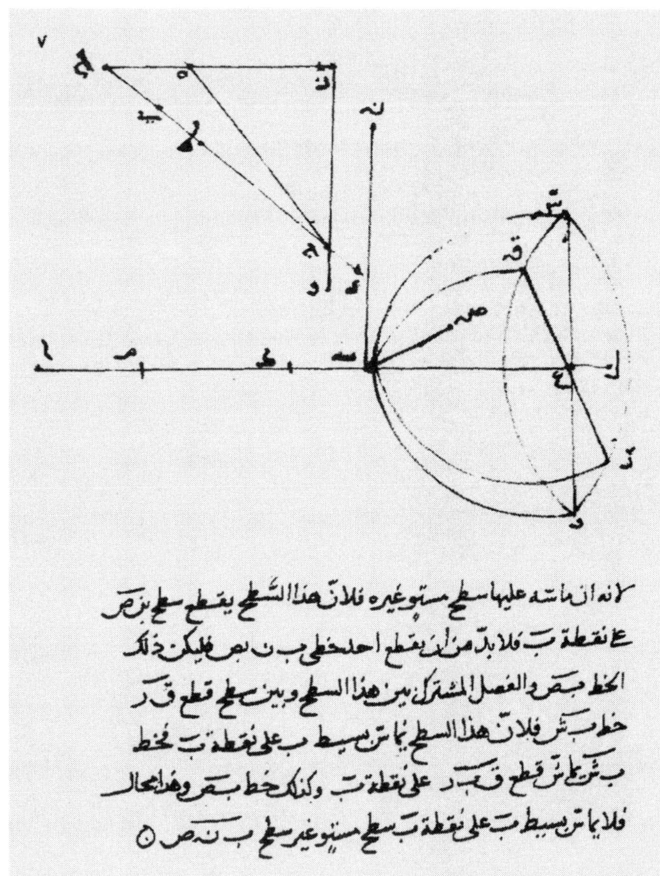

(b)

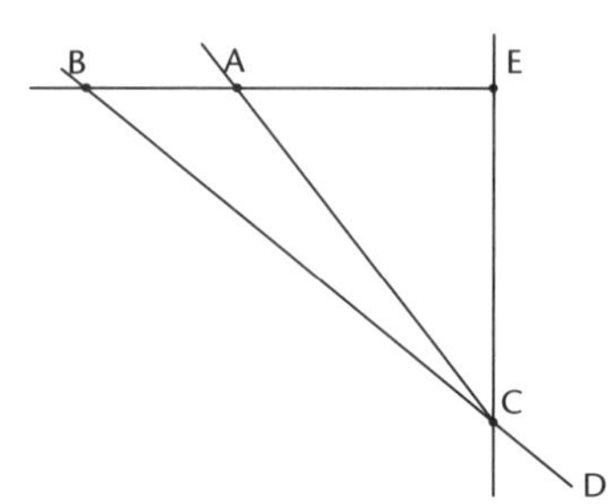

Legend

A. Light source.
B. Point where extension of line CD meets extension of line AE.
C. Point on illuminated surface.
D. Point in line of refracted ray of light.
E. Point on surface CE such that AEC forms right angle.

Ages, when much ancient learning was lost. Some Greek philosophy survived because it had been translated into Arabic. Much later, the works of Ptolemy and Aristotle were translated from Arabic into Latin and so came to European scientists such as Roger Bacon and Johannes Kepler.[20]

Medieval Arab Optics

The Greek era of science was followed by the Arab scientific Golden Age, from the eighth to the sixteenth century. Nearly all of the writing was in Arabic, the scientific language before the twelfth century. This period began with an intensive period of translation of Greek books brought to Baghdad, the imperial and scientific center.

Although early Arab scientists contributed much to other disciplines such as chemistry, biology, medicine and engineering, their enduring legacy was in mathematics, astronomy and optics. They were intrigued by the mechanism of vision and the function of the eye and brain in processing this information.[21,22]

The early Arab scientists were fascinated by what they read in the Greek books and wanted to understand such phenomena, but the respect these Arabs had for the Greek authorities did not stop them questioning their theories in a new way, the scientific method known today. The observation and measurement of data were followed by the formulation and testing of hypotheses to explain the data.

Ibn Sahl

Ibn Sahl (CE circa 940-1000) was an Arab mathematician and physicist. His predecessors and contemporaries researched designs of military mirrors for burning targets at a distance. Ibn Sahl departed from his predecessors in studying reflection and refraction of the Sun's rays. The interest in refraction led him to the study of lenses and their shapes in great detail. In these studies, Ibn Sahl discovered the relationship between the incident and refracted rays of light, the relationship rediscovered by Willebrord Snellius some 650 years later and now referred to as *Snell's law.*[23-26] In Fig. 13, light from point A enters a new medium at point C and refracts along the line CD. If the line CD is extended to point B, the ratio of length AC to length BC is the index of refraction.

Lens and mirror shapes Ibn Sahl considered were the elliptical, parabolic, hyperbolic and biconvex. Ibn Sahl went further and designed machines for the precise drawing of mathematical shapes. Ibn Sahl informed the work of another optical physicist, Ibn al-Haytham.

Ibn al-Haytham

Ibn al-Haytham (CE 965-1039), also known as *Alhacen* or *Alhazen,* was born in Basra, Iraq, and studied in Baghdad (Fig. 14). In pursuit of knowledge, he traveled to Iran and Syria and settled in Egypt. He wrote more than 90 books and treatises on optics, astronomy, mathematics, philosophy, medicine and logic.[26-29]

His most important work was a critique of Ptolemy's *Almagest.* Ibn al-Haytham prefaced this critique by stating that his methods will criticize premises and exercise caution in drawing conclusions, not to follow authorities blindly. On the mechanism of vision, he was able to reject the two competing Greek theories favored by Euclid and Ptolemy. To test these theories in experiments, Ibn al-Haytham invented the *camera obscura* (literally the

FIGURE 14. Ibn al-Haytham's portrait on Iraqi currency, with optics diagram next to him.

"dark chamber"), or pinhole camera, the basis of photography. Ibn Haytham wrote a detailed account of all his experimental setups and the data he measured. This book served as the textbook on optics for centuries throughout Europe (Fig. 15).[30] He dissected the eye and named its parts (lens, cornea, retina). He explained for the first time the imperfection of the eye's lenses, introducing the concept of spherical aberration (Fig. 15c).

The early Arab interest in the physiology of the eye together with the mechanism of vision led a later scientist, Hunayn ibn Ishaq, to write that "it is a prerequisite for whoever wants to understand the function of the eye to be cognizant of the function of the brain, since the process of vision begins and ends therein" (a translation of the Arabic text in Fig. 16).[31] Ibn al-Haytham's understanding of the relationship between the eye and the brain enabled him to recognize an optical illusion, where the Moon appears larger on the horizon than at its zenith. Some have tried to explain the Moon's apparent size as diffraction of sunlight through the atmosphere; some try to explain with other models. Ibn al-Haytham simply identified it as one of many phenomena where light plays tricks on the brain.

Ibn al-Haytham's analysis of his data led him to put forward or question models. He was a scientist, using mathematics to formulate physical theories and to conduct careful experiments. His writings were transmitted to western Europe in Latin and founded the technology of optics.

Boiler Inspection, 1870-1920[32,33]

The first nondestructive test method was visual testing, and the term *visual testing* here refers, not to a caveman's inspection of his spearhead (although that is indeed nondestructive testing) but rather to documented inspection of a product according to a particular procedure or specification designed to recognize material defects. Most specifications for visual testing ask various quality questions.

1. Are the contracted steps in processing or fabrication performed completely and in the correct sequence?

FIGURE 15. Sixteenth century edition of Ibn al-Haytham's treatise, in Latin: (a) cover page; (b) caption and engraving on "three parts of vision, direct, reflected and refracted"; (c) engraved diagram of eye with parts labeled.[30]

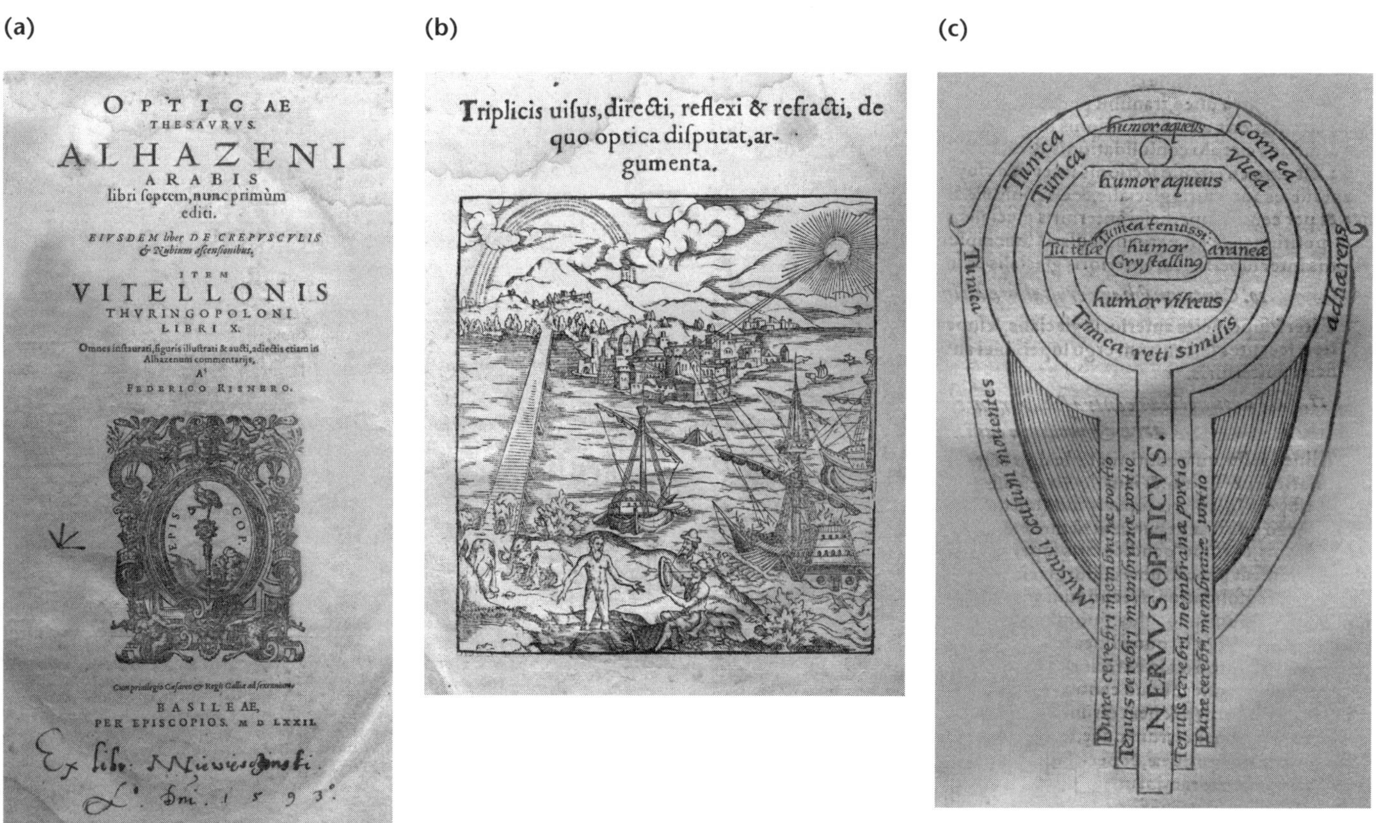

2. Are the right materials and components used throughout? Are bolts the right size, for instance?
3. Are fasteners and supports spaced and installed according to specification?
4. Are protective lubricants, weather strips and coatings applied according to specification?
5. Are there signs of damage, such as wear, corrosion, dents, strain, buckling or visible cracking?

These visual checks are, however, not necessarily *nondestructive tests*: the questions except for the last address fabrication and maintenance quality rather than material discontinuities.

The introduction of steam power in the nineteenth century led to a rash of boiler explosions and to the need for inspection (Fig. 17). The 1860s saw the introduction of boiler inspection combined with boiler insurance in the United States and the United Kingdom.[32,33]

FIGURE 16. Thirteenth century manuscript page from Hunayn ibn Ishaq, *Book of Ten Treatises on the Eye.*[22,31]

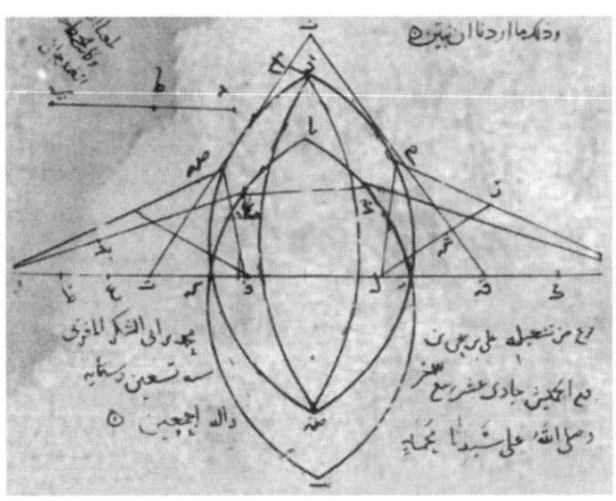

Boiler inspection was an early application of visual testing. Insurance inspectors would, of course, look for corrosion in the inservice boilers they insured. Early editions of the *ASME Boiler Code* asked the inspector to *inspect* components, that is, to *look* at them.[34] A half century would pass before other methods of nondestructive testing would provide the context needed to make it clear that this aspect of the boiler inspector's job was the visual test method of nondestructive testing.

The earliest standards of the American Society of Mechanical Engineers (ASME), although they emphasized proof tests and destructive tests, say that the boilers must be free of gross surface blemishes and other signs of poor workmanship. In 1915, the first edition of the *Boiler Code* expected the inspector to look at malleable castings to determine that they were "true to pattern, free from blemishes, scale or shrinkage cracks. A variation of 1/16 in. per foot [1.6 mm per 0.3 m] shall be permissible." The finish of flat bars had "to be smoothly rolled and free from slivers, depressions, seams, crop ends," and burns. The inspector examined all parts to be sure that "the finished material shall be free from injurious defects and shall have a workmanlike finish."[35] Twenty-first century versions of the *Boiler Code,* although briefly, explicitly treat visual testing as nondestructive testing.[36]

FIGURE 17. Drawing of steam boiler explosion in nineteenth century.

Medical Endoscopy[37,38]

The development of self illuminated telescopic devices can be traced back to early interest in exploring the interior human anatomy without invasive procedures.[38] The first borescopes were medical endoscopes turned to industrial applications, for an endoscope does not care what aperture it is interrogating. Medical endoscopes and industrial borescopes share several features: (1) a source of illumination, (2) a means of delivering an image to the viewer's eye and (3) adjustability to view a surface of interest. Early endoscopes for looking down the esophagus were called *gastroscopes*; endoscopes for looking at the bladder were called *cystoscopes*.

Devices for viewing the interior of objects are called *endoscopes*, from the Greek words for "inside view." Today the term *endoscope* in the United States denotes a medical instrument. Nearly all endoscopes have an integral light source; some incorporate surgical tweezers or other devices. Industrial endoscopes are called *borescopes* because they were originally used in machined apertures and holes such as gun bores. There are both flexible and rigid, fiber optic and direct light borescopes.

In 1806, Philipp Bozzini of Frankfurt announced the invention of his *Lichtleiter* (German for "light guide"). Having served as a surgeon in the Napoleonic wars, Bozzini envisioned using his device for medical research. It is considered the first endoscope.[39,40]

In 1876, Max Nitze, a urologist, developed a practical cystoscope to view the human bladder. A platinum loop in its tip furnished a bright light when heated with galvanic current. Two years later, Thomas Edison introduced an incandescent light in the United States. Within a short time, scientists in Austria made and used a minute electric bulb in Nitze's cystoscope, even before the electric light was in use in America.

The early cystoscopes contained simple lenses; these were soon replaced by achromatic combinations. In 1900, Reinhold Wappler revolutionized the optical system of the cystoscope and produced the first American models. The forward oblique viewing system was later introduced and has proved very useful in both medical and industrial applications. Direct vision and retrospective systems were also first developed for cystoscopy.

Borescopes and related instruments for nondestructive testing have followed the same basic design used in cystoscopic devices. The range of borescope sizes has increased, sectionalized instruments have been introduced and other special devices have been developed for industrial applications.

An early inventor and manufacturer was a German, Georg Wolf, who cofounded an optical equipment company in 1906.[42] He filed patents for medical endoscopes in the United States in 1922.[43,44] A few months later, a Robert Wolf filed a patent for a cystoscope.[45] When Georg Wolf died in 1938, his son Richard Wolf continued the family business, which has continued with his name into the twenty-first century.

Georg Wolf in 1932 produced a flexible gastroscope, developed by Rudolph Schindler for observing the interior of the stomach wall.[46] The instrument consisted of a rigid section and a flexible section. Many lenses of small focal distance were used to allow bending of the instrument to an angle of 34 degrees in several planes. The tip of the device contained the objective and the prism, causing the necessary axial deviation of the bundle of rays coming from the illuminated gastric wall. The size of the image depended on the distance of the objective from the observed surface. The sharp image could be magnified or reduced. Later in the century, flexible gastroscopes had rubber tubes over the flexible portion, in diameters of about 14 mm (0.55 in.) and 8 mm (0.3 in.).[42]

Early Patents to Automate Visual Testing

A patent was filed in June 1920 for an inspection table that would lift rolled plates so that the underside could be visually tested. The invention was addressing the steel industry's need for visual testing.[47] In 1922, another inspection table was patented that made test objects turn over as they were conveyed past the inspector. This table was designed especially for fruits and vegetables but could be used, the inventor said, for anything that should be rotated for inspection on all sides.[48]

In July 1925, Floyd Firestone of the University of Michigan, Ann Arbor, filed a patent for automated scanning and flaw detection. (This is the same Firestone who later invented the Supersonic Reflectoscope®, an ultrasonic instrument widely used in the United States in the 1940s.) The optical scanning invention was envisioned for bearing rollers or "other articles with surfaces of revolution, and even to plane surfaces, so long as the surface of the article, or as much thereof as needs inspection, may be moved within the field of view."[49] How could optical inspection have been automated in the years before computers facilitated decision making? Small areas would successively be brought into view to a microscope, and a light sensitive cell would detect brightness variations below a selected threshold and trigger a sorting armature. It is not known if this scheme was ever implemented by industry. A later design was advanced in 1938 for sheet metal.[50] In the 1980s, microprocessing made automated vision easier to implement.[51]

Industrial Endoscopy: Borescopy[37]

Patents for endoscopes specifically for industrial applications appeared in the 1920s and 1930s. A patent was filed in 1922 for the inspection of rivets inside tubing in, for example, a boiler or airplane. The device resembled a periscope like those seen in old movies about submarines, with several differences: (1) it was portable and small enough to fit inside tubing; (2) it included light bulbs for illumination; (3) it provided for rotation of the objective end while the eyepiece remained stationary.[52]

A patent was filed in 1927 literally for a *bore scope* — to look inside gun bores (Fig. 18).[53] Another patent to look inside gun bores was filed on behalf of the Carl Zeiss company, Jena, Germany, in 1932 in Germany and in 1933 in the United States.[54]

The visual technology for tubing was represented by a patent filed in 1938; the invention, which could generically be called a *tube scope,* became important for the inspection of petroleum drill pipe in the United States.[55] A service using the instrument rather than the instrument itself was provided to the petroleum industry. Figure 19 shows the design and application. The patent also provided for a separate attachment to scour the tube's inside surface before visual testing.

FIGURE 18. Drawing from patent for borescope for gun barrels.[53]

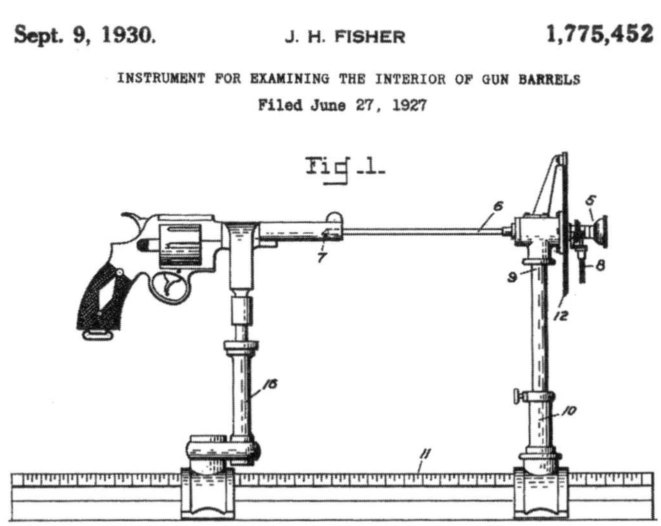

FIGURE 19. Borescopy of tubing: (a) drawing from 1941 patent[55]; (b) photograph of application.

(a)

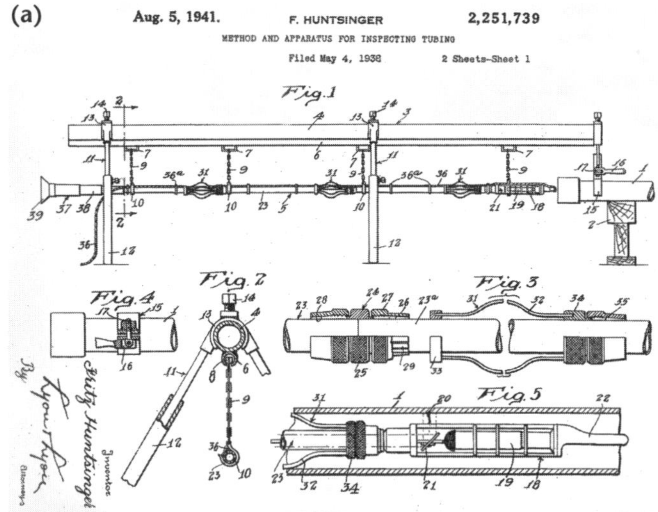

(b)

Legend

1. End of stand of pipe.
2. Pipe rack.
3. Frame.
4. Strong back.
5. Viewing apparatus.
6. Track.
7. Trolleys.
8. Rollers.
9. Chain.
10. Hook.
11. A frames.
12. Legs.
13. Saddles.
14. Set screw.
15. Clamp.
16. Lever with cam head.
17. Tightening bolt.
18. Lighting guard.
19. Electric lamp.
20. Crack.
21. Reflector.
22. Guide horn.
23. Barrel.
24. Joint clamp.
25. Body of joint sleeve.
26. Sleeve split to fit over barrel.
27. Coupling nut.
28. Tapered bores.
29. Tapered or conical outer face of sleeve.
31. Guide sleeves.
32. Resilient bow or bar of sleeve.
33. Loose collar.
34. Collar clamped to barrel.
35. Tightening nut threaded onto sleeve.
36. Electric lamp cord.
37. Telescoping means.
38. Telescoping sections.
39. Eye piece.

Flexible borescopes for industrial use are more rugged than gastroscopes, having flexible steel tubes instead of rubber for the outer tube of the flexible portion. A typical flexible borescope is 13 mm (0.5 in.) in diameter and has a 1 m (40 in.) working length, with flexibility in about 500 mm (20 in.) of the length. Extension sections are available in 1, 2 or 3 m (40, 80 or 120 in.) lengths, permitting assembly of borescopes up to 10 m (30 ft) in length. In such flexible instruments, the image remains round and sharp when the tube is bent to an angle of about 34 degrees. Beyond that limit, the image becomes elliptical but remains clear until obliterated at about 45 degrees of total bending.

Crampton

After the early medical developments, certain segments of American industry needed visual testing equipment for special inspection applications. One of the first individuals to help fill this need was George Sumner Crampton. George Crampton (Fig. 20) was born in Rock Island, Illinois, in 1874. He was said to have set up a small machine shop by the age of 10 and his first ambition was to become an electrical engineer. He chose instead to study medicine and received his M.D. from the University of Pennsylvania in 1898. While he was interning at Pennsylvania Hospital, Crampton's mechanical and engineering abilities were recognized and he was advised to become an oculist. He returned to the university, took a degree in ophthalmology and later practiced in Philadelphia, Pennsylvania and Princeton, New Jersey.[56]

In 1921, the Westinghouse Company asked Crampton to make a device that could be used to check for discontinuities inside the rotor of a steam turbine (Fig. 21). Crampton developed the instrument in his Philadelphia shop and delivered the prototype within a week — it was the first borescope produced by his company. Crampton continued to supply custom borescopes for testing inaccessible and often dark areas on power turbines, oil refinery piping, gas mains, soft drink tanks and other components. Crampton soon was recognized for his ability to design and manufacture borescopes, periscopes and other optical equipment for specific testing applications.

After retiring as emeritus professor of ophthalmology at the university, Crampton continued private practice in downtown Philadelphia. At the same time, he worked on borescopes and other instruments in a small shop he had established in a remodeled nineteenth century coach house.

After World War II began, Crampton devoted much of his energy to the war effort, filling defense orders for borescopes (Fig. 22). Crampton practiced medicine until noon, then went to the nearby

FIGURE 20. George Crampton, developer of borescope.

FIGURE 21. Tests of forgings for steam turbine generator shaft in 1920s.

workshop where he visually tested the bores of 37 mm antiaircraft guns and other weapons.[56] During the war, borescopes were widely used for testing warship steam turbines (particularly their rotating shafts). The United States Army also used borescopes for inspecting the barrels of tank and antiaircraft weapons produced in Philadelphia. An even more challenging assignment lay ahead.

The scientists working to develop a successful nuclear chain reaction in the top secret Manhattan Project asked Crampton to provide a borescope for inspecting tubes near the radioactive pile at its guarded location beneath the stadium seats at the University of Chicago's Stagg Field. Crampton devised an aluminum borescope tube 35 mm (1.4 in.) in diameter and 10 m (33 ft) long. The device consisted of 2 m (6 ft) sections of dual tubing joined by bronze couplings which also carried an 8 V lighting circuit. The inspector standing directly in front of the bore was subject to radioactive emissions from the pile, so Crampton mounted the borescope outside of a heavy concrete barrier. The operator stood at a right angle to the borescope, looking through an eyepiece and revolving the instrument manually. The borescope contained a prism viewing head and had to be rotated constantly. It was supported in a steel V trough resting on supports whose height could be varied. Crampton also mounted a special photographic camera on the eyepiece.

The original Manhattan Project borescope was later improved with nondarkening optics and a swivel-joint eyepiece that permitted the operator to work from any angle (this newer instrument did not require the V trough). It also was capable of considerable bending to snake through the tubes in the reactor. Three borescopes were supplied for this epochal project and they are believed to be the first optical instruments to use glass resistant to radioactivity.

Aircraft inspection soon became one of the most important uses of borescope technology. In 1946, an ultraviolet light borescope was developed for fluorescent testing of the interior of hollow steel propeller blades. The 100 W viewing instrument revealed interior surface discontinuities as glowing green lines.[41]

Later, in 1958, the entire United States' B-47 bomber fleet was grounded because of metal fatigue cracks resulting from low level simulated bombing missions. Visual testing with borescopes proved to be the first step toward resolving the problem. The program became known as Project Milk Bottle, a reference to the bottle shaped pin that was a primary connection between the fuselage and wing (Fig. 23).

In the late 1950s, a system was developed for automatic testing of helicopter blades. The borescope, supported by a long bench, could test the blades while the operator viewed results on a television screen (Fig. 24). The system was used extensively during the

FIGURE 22. Using a borescope, inspector at a converted automobile plant during World War II checks interiors of gun tubes for 90 mm antiaircraft guns.

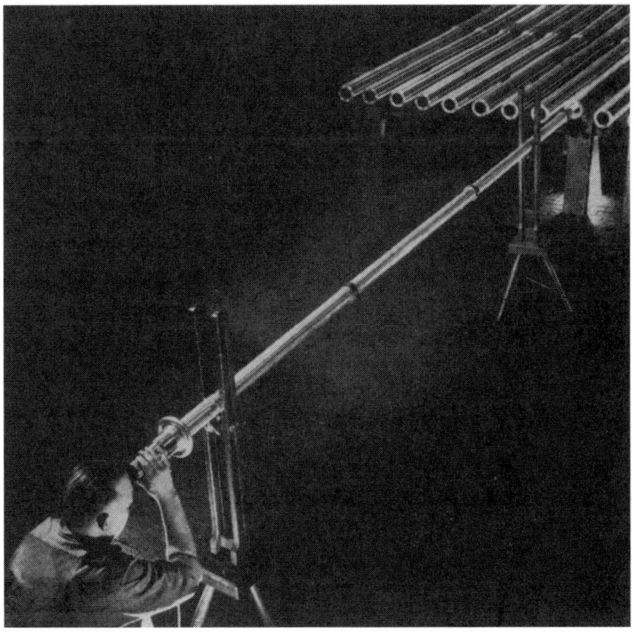

FIGURE 23. Inspector uses borescope to check for metal fatigue cracks in B-47 bomber during grounding of bomber fleet in 1958.

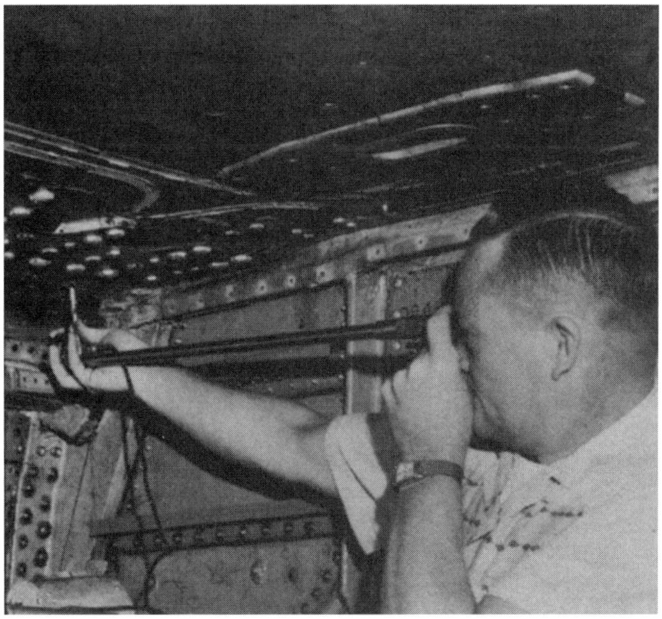

Vietnam conflict and helicopter manufacturers continue to use borescopes for such critical tests.

In 1962, Crampton sold his borescope business to John Lang of Cheltenham, Pennsylvania.[56,57] Lang had developed the radiation resistant optics used in the Manhattan Project borescope, as well as a system for keeping it functional in high temperature environments. Lang also helped pioneer the use of closed circuit television with borescopes for testing the inner surfaces of jet engines and wings, hollow helicopter blades and nuclear reactors. In 1965, the company received a patent on a borescope whose mirror could be precisely controlled. This borescope could zoom to high magnification and could intensely illuminate the walls of a chamber by means of a quartz incandescent lamp containing iodine vapor.

The basic design of the borescope has been in use for many decades and it continues to develop, accommodating advances in video, illumination, robotic and computer technologies.

Certification of Visual Inspectors

The recognition of the visual testing technique and the development of formal procedures for educating and qualifying visual inspectors were important milestones in the history of visual inspection. Because visual testing can be performed without any intervening apparatus, it was certainly one of the first forms of nondestructive testing. In its early industrial applications, visual tests were used simply to verify compliance to a drawing or specification. This was basically a dimensional check. The soundness of the object was determined by liquid penetrant, magnetic particle, radiography or ultrasonic testing.

Following World War II, few inspection standards included visual testing. By the early 1960s, visual tests were an accepted addition to the American Welding Society's code books. In NAVSHIPS 250-1500-1, the US Navy included visual tests with its specifications for other nondestructive testing techniques for welds.[58]

By 1965, there were standards for testing, and criteria for certifying the inspector had been established in five test methods: liquid penetrant, magnetic particle, eddy current, radiographic and ultrasonic testing. These five were cited in ASNT Recommended Practice No. SNT-TC-1A, introduced in the late 1960s. The broad use of visual testing hindered its addition to this group as a specific method — there were too many different applications on too many test objects to permit the use of specific acceptance criteria. It also was reasoned that visual testing would occur as a natural result of applying any other nondestructive test method.

Expanded Need for Visual Certification

In the early 1970s, the need for certified visual inspectors began to increase. Nuclear power construction was at a peak, visual certification was becoming mandatory and nondestructive testing was being required. In 1976, the American Society for Nondestructive Testing began considering the need for certified visual inspectors. ASNT had become a leading force in nondestructive testing and American industry had accepted its Recommended Practice No. SNT-TC-1A as a guide for certifying other NDT inspectors.[6]

In the spring of 1976, ASNT began surveying industry about their inspection needs and their position on visual testing. Because of the many and varied responses to the survey, a society task force was established to analyze the survey data. In 1977, the task force recommended that visual inspectors be certified and that visual testing be made a supplement to Recommended Practice No. SNT-TC-1A (1975). At this time, the American

Figure 24. Visual testing of frame of 10 m (32 ft) long helicopter blade using 10 m (32 ft) borescope. Inspector could view magnified results on television at bottom left.

Welding Society implemented a program that, following the United States Navy, was the first to certify inspectors whose sole function was visual weld testing.

During 1978, ASNT subcommittees were formed for the eastern and western halves of the United States. These groups verified the need for both visual standards and trained, qualified and certified inspectors. In 1980, a Visual Methods Committee was formed in ASNT's Technical Council and the early meetings defined the scope and purpose of visual testing (dimensional testing was excluded). In 1984, the Visual Personnel Qualification Committee was formed in ASNT's Education and Qualification Council. In 1986, a training outline and a recommended reference list was finalized and the Board of Directors approved incorporation of visual testing into ASNT's Recommended Practice No. SNT-TC-1A.

PART 4. Measurement Units for Visual Testing

Origin of International System

In 1960, the General Conference on Weights and Measures established the International System of Units. *Le Systéme International d'Unités* (SI) was designed so that a single set of measurement units could be used by all branches of science, engineering and the general public. Without SI, the *Nondestructive Testing Handbook* series would contain a confusing mix of obsolete centimeter gram second (CGS) units, inch pound units and the units preferred by certain localities or scientific specialties.

SI is the modern version of the metric system and ends the division between metric units used by scientists and metric units used by engineers and the public. Scientists have given up their units based on centimeter and gram and engineers have abandoned the kilogram-force in favor of the newton. Electrical engineers have retained the ampere, volt and ohm but changed all units related to magnetism.

Table 6 lists the seven SI base units. Table 7 lists derived units with special names. In SI, the unit of time is the second (s) and hour (h) is recognized for use with SI.

For more information, the reader is referred to the information available through national standards organizations and specialized information compiled by technical organizations.[59-61]

Multipliers

In science and engineering, very large or very small numbers with units are expressed by using the SI multipliers, prefixes of 10^3 intervals (Table 8). For example, a millimeter (mm) is 0.001 meter (m). The volume unit cubic centimeter (cm^3) is $(0.01\ m)^3$ or $10^{-6}\ m^3$. Unit submultiples such as the centimeter, decimeter, dekameter and hectometer are avoided in scientific and technical uses of SI because of their variance from the convenient 10^3 or 10^{-3} intervals that make equations easy to manipulate.

In SI, the distinction between upper and lower case letters is meaningful and should be observed. For example, the meanings of the prefix *m* (milli) and the prefix *M* (mega) differ by nine orders of magnitude.

TABLE 6. SI base units.

Quantity	Unit	Symbol
Length	meter	m
Mass	kilogram	kg
Time	second	s
Electric current	ampere	A
Temperature	kelvin	K
Amount of substance	mole	mol
Luminous intensity	candela	cd

TABLE 7. SI derived units with special names.[a]

Quantity	Units	Symbol	Relation to Other SI Units[b]
Capacitance	farad	F	$C{\cdot}V^{-1}$
Catalytic activity	katal	kat	$s^{-1}{\cdot}mol$
Conductance	siemens	S	$A{\cdot}V^{-1}$
Energy	joule	J	N·m
Frequency (periodic)	hertz	Hz	$1{\cdot}s^{-1}$
Force	newton	N	$kg{\cdot}m{\cdot}s^{-2}$
Inductance	henry	H	$Wb{\cdot}A^{-1}$
Illuminance	lux	lx	$lm{\cdot}m^{-2}$
Luminous flux	lumen	lm	cd·sr
Electric charge	coulomb	C	A·s
Electric potential[c]	volt	V	$W{\cdot}A^{-1}$
Electric resistance	ohm	Ω	$V{\cdot}A^{-1}$
Magnetic flux	weber	Wb	V·s
Magnetic flux density	tesla	T	$Wb{\cdot}m^{-2}$
Plane angle	radian	rad	1
Power	watt	W	$J{\cdot}s^{-1}$
Pressure (stress)	pascal	Pa	$N{\cdot}m^{-2}$
Radiation absorbed dose	gray	Gy	$J{\cdot}kg^{-1}$
Radiation dose equivalent	sievert	Sv	$J{\cdot}kg^{-1}$
Radioactivity	becquerel	Bq	$1{\cdot}s^{-1}$
Solid angle	steradian	sr	1
Temperature	degree celsius	°C	K
Time[a]	hour	h	3600 s
Volume[a]	liter	L	dm^3

a. Hour and liter are not SI units but are accepted for use with SI.
b. Number one (1) expresses a dimensionless relationship.
c. Electromotive force.

Units for Visual Testing

Terms for some quantities have been replaced. Brightness is now luminance; illumination is illuminance; transmission factor is transmittance. Names of some units have changed: the meter candle is now lux; the nit is now candela per square meter ($cd \cdot m^{-2}$).

TABLE 8. SI prefixes and multipliers.

Prefix	Symbol	Multiplier
yotta	Y	10^{24}
zetta	Z	10^{21}
exa	E	10^{18}
peta	P	10^{15}
tera	T	10^{12}
giga	G	10^{9}
mega	M	10^{6}
kilo	k	10^{3}
hecto[a]	h	10^{2}
deka[a]	da	10
deci[a]	d	10^{-1}
centi[a]	c	10^{-2}
milli	m	10^{-3}
micro	μ	10^{-6}
nano	n	10^{-9}
pico	p	10^{-12}
femto	f	10^{-15}
atto	a	10^{-18}
zepto	z	10^{-21}
yocto	y	10^{-24}

a. Avoid these prefixes (except in dm^3 and cm^3) for science and engineering.

Old units should not be used in science and engineering; Table 9 gives some conversions to SI units. Footcandle (ftc) and phot convert to lux (lx). Stilb (sb), footlambert and lambert convert to candela per square meter ($cd \cdot m^{-2}$).

In visual testing, units express measurements of visible light as part of the electromagnetic spectrum. Nanometer (nm) is used rather than angstrom (Å) for wavelength. The velocity c of light is expressed as a ratio of distance in meters (m) to time in seconds (s): in a vacuum, $2.997\,924\,58 \times 10^8\ m \cdot s^{-1}$.

Illumination

The intensity of visible radiation — that is, of light — was formerly measured in footcandles (ftc) and is now expressed in lux (lx): 1 ftc = 10 lx. A typical indoor office has illumination of about 400 lx. Daylight ranges from 1 to 25 klx; direct sunlight, several times more.

Vision requires a source of illumination. The light source is measured in candela (cd), defined as the luminous intensity in a given direction of a source that emits monochromatic radiation of 540×10^{12} hertz (Hz) at a radiant intensity of 1.46×10^{-3} watt per steradian ($W \cdot sr^{-1}$).

The luminous flux in a steradian (sr) is measured in lumens (lm). The measurement in lumens is the product of candela and steradian (1 lm = 1 cd·sr). A light flux of one lumen (1 lm) striking one square meter (1 m^2) on the surface of the sphere around the source illuminates it with one lux (1 lx), the unit of illuminance. If the source itself is scaled to

TABLE 9. Examples of conversions to SI units for visual testing.

Quantity	Measurement in Non-SI Unit	Multiply by	To Get Measurement in SI Unit
Angle	minute (min)	$2.908\,882 \times 10^{-4}$	radian (rad)
	degree (deg)	$1.745\,329 \times 10^{-2}$	radian (rad)
Area	square inch (in.^{2})	645	square millimeter (mm^2)
Distance	angstrom (Å)	0.1	nanometer (nm)
	inch (in.)	25.4	millimeter (mm)
Power	British thermal unit per hour ($BTU \cdot h^{-1}$)	0.293	watt (W)
Illuminance	footcandle (ftc)	10.76	lux (lx)
	phot (ph)	10 000	lux (lx)
Luminance	candela per square foot ($cd \cdot ft^{-2}$)	10.76	candela per square meter ($cd \cdot m^{-2}$)
	candela per square inch ($cd \cdot in.^{-2}$)	1 550	candela per square meter ($cd \cdot m^{-2}$)
	footlambert (ftl)	3.426	candela per square meter ($cd \cdot m^{-2}$)
	lambert	3 183 (= 10 000 ÷ π)	candela per square meter ($cd \cdot m^{-2}$)
	nit (nt)	1	candela per square meter ($cd \cdot m^{-2}$)
	stilb (sb)	10 000	candela per square meter ($cd \cdot m^{-2}$)
Temperature (increment)	degree fahrenheit (°F)	0.556	kelvin (K) or degree celsius (°C)
Temperature (scale)	degree fahrenheit (°F)	(°F – 32) ÷ 1.8	degree celsius (°C)
Temperature (scale)	degree fahrenheit (°F)	(°F – 32) ÷ 1.8 + 273.15	kelvin (K)

one square meter (1 m^2) and emits one candela (1 cd), the luminance of the source is 1 $cd \cdot m^{-2}$.

Quantities and units for photometric measurement of light are discussed in the chapter on light.

Optometric Units

The diopter is a variable used to express the refracting power of curved mirrors, lenses and the eye. The diopter is the inverse of the distance (in meters) from the lens (or mirror) to an image of a distant object; that is, the diopter is the inverse of the focal distance of the lens or mirror (where that distance is measured in meters).

To express retinal illuminance, the troland (Td) is most often used. It is not a true unit of illumination but is the product of the target luminance (in candela per square meter) and the pupil area (in square millimeters).

Ultraviolet Radiation

Ultraviolet radiation is of concern because some visual inspectors also document the vision acuity and color discrimination of personnel who use ultraviolet lamps to perform liquid penetrant and magnetic particle testing.

The term *light* is widely used for electromagnetic radiation in the visible part of the spectrum. The term *black light,* however, should not be used for ultraviolet radiation, because (1) the term has become ambiguous, denoting sometimes the ultraviolet lamp and sometimes its radiation, (2) the term *black* here means merely invisible and not a color and (3) ultraviolet radiation is not light, any more than X-rays are.

Although both light and ultraviolet radiation are measured in watts per square meter, their wavelengths have distinct ranges. Because ultraviolet radiation is invisible, photometric measurement units such as the lumen and lux should never be applied to ultraviolet radiation.

Ultraviolet radiation is divided into three ranges: UV-A (320 to 400 nm), UV-B (280 to 320 nm) and UV-C (100 to 280 nm). This is analogous to the segmentation of visible light into the wavelengths that produce the colors. Blue light, for example, generally has wavelengths between 455 and 492 nm. Yellow light is between 577 and 597 nm. The analogy to visible radiation might help those first learning to measure ultraviolet radiation. A certain intensity of yellow light will produce on a surface a certain illuminance measured in lux. In the same way, a certain amount of ultraviolet radiation will produce an irradiance on a test surface.

Ultraviolet irradiance is a time dependent measure of the amount of energy falling on a prescribed surface area and is expressed in watts per square meter ($W \cdot m^{-2}$) or (to avoid exponents) microwatts per square centimeter ($\mu W \cdot cm^{-2}$). One unit of irradiance (1 $\mu W \cdot cm^{-2}$) is the power (microwatt) falling on one square centimeter (cm^{-2}) of surface area. At higher irradiance, the milliwatt per square centimeter ($mW \cdot cm^{-2}$) is sometimes used: 1000 $\mu W \cdot cm^{-2}$ = 1 $mW \cdot cm^{-2}$, and 1 $\mu W \cdot cm^{-2}$ = 10^{-2} $W \cdot m^{-2}$.

More information on the physics and safe use of ultraviolet radiation can be found in literature about liquid penetrant and magnetic particle testing.

References

1. *Nondestructive Testing Handbook,* second edition: Vol. 10, *Nondestructive Testing Overview.* Columbus, OH: American Society for Nondestructive Testing (1996).
2. Wenk, S.A. and R.C. McMaster. *Choosing NDT: Applications, Costs and Benefits of Nondestructive Testing in Your Quality Assurance Program.* Columbus, OH: American Society for Nondestructive Testing (1987).
3. TO33B-1-1 (NAVAIR 01-1A-16) TM43-0103, *Nondestructive Testing Methods.* Washington, DC: Department of Defense (June 1984).
4. *Nondestructive Testing Handbook,* second edition: Vol. 8, *Visual and Optical Testing.* Columbus, OH: American Society for Nondestructive Testing (1993).
5. *Annual Book of ASTM Standards:* Section 3, *Metals Test Methods and Analytical Procedures.* Vol. 03.03, *Nondestructive Testing.* West Conshohocken, PA: ASTM International (2009).
6. Recommended Practice No. SNT-TC-1A, *Personnel Qualification and Certification in Nondestructive Testing.* Columbus, OH: American Society for Nondestructive Testing (2006).
7. ANSI/ASNT CP-189, *Standard for Qualification and Certification of Nondestructive Testing Personnel.* Columbus, OH: American Society for Nondestructive Testing (2006).
8. ASNT Standard CP-105, *Topical Outlines for Qualification of Nondestructive Testing Personnel.* Columbus, OH: American Society for Nondestructive Testing (2006).
9. *ASNT Central Certification Program* (ACCP), Revision 4 (March 2005). Columbus, OH: American Society for Nondestructive Testing (2005).
10. ANSI/ASNT CP-106 (ISO 9712, modified), *Non-Destructive Testing — Qualification and Certification of Personnel,* third edition. Geneva, Switzerland: International Organization for Standardization (2008).
11. ISO 20807, *Non-Destructive Testing — Qualification of Personnel for Limited Application of Non-Destructive Testing.* Geneva, Switzerland: International Organization for Standardization (2007).
12. Ness, S. "Safety for Visual and Optical Tests." *Nondestructive Testing Handbook,* second edition: Vol. 8, *Visual and Optical Testing.* Columbus, OH: American Society for Nondestructive Testing (1993): p 22-26.
13. *Documentation of the Threshold Limit Values for Physical Agents,* seventh edition. Cincinnati, OH: American Conference of Governmental Industrial Hygienists (2008).
14. *Guide to Occupational Exposure Values.* Cincinnati, OH: American Conference of Governmental Industrial Hygienists (2009).
15. ANSI Z 136, *Safe Use of Lasers.* New York, NY: American National Standards Institute (2007).
16. IEC 60825, *Safety of Laser Products.* Geneva, Switzerland: International Electrotechnical Commission (2008)
17. *Threshold Limit Values [for Chemical Substances and Physical Agents] and Biological Exposure Indices.* Cincinnati, OH: American Conference of Governmental Industrial Hygienists (2001, 2009).
18. Euclid. *Great Books of the Western World,* second edition: Vol. 10, *Elements.* Chicago, IL: Encyclopaedia Britannica (1990).
19. Ptolemy. *Great Books of the Western World,* second edition: Vol. 15, *Almagest.* Chicago, IL: Encyclopaedia Britannica (1990).
20. *L'Ottica de Claudio Tolomeo* [*The Optics of Ptolemy: Rendered in Latin from an Arabic Translation of an Incomplete Greek Text*]. Turin, Italy: Stamperia Reale della Ditta G.B. Paravia E.C. de I. Violarde (1885).
21. Salih, H., M. Al-Amri and M. El Gomati. "The Miracle of Light." *A World of Science: Natural Sciences Quarterly Newsletter.* Vol. 3, No. 4. Paris, France: United Nations Educational, Scientific and Cultural Organization (October-December 2005): p 2-7.
22. Turner, H.R. *Science in Medieval Islam: An Illustrated Introduction.* Austin, TX: University of Texas (1997).

23. Rashed, R. "A Pioneer in Anaclastics: Ibn Sahl on Burning Mirrors and Lenses." *Isis.* Vol. 81. Chicago, IL: University of Chicago Press (1990): p 464-491.
24. Rashed, R. "Ibn Sahl et al-Quhî: Les Projections. Addenda & Corrigenda." *Arabic Sciences and Philosophy.* Vol. 10, No. 1. Cambridge, United Kingdom: Cambridge University Press (2000): p 79-100.
25. Kwan, A., J. Dudley and E. Lantz. "Who Really Discovered Snell's Law?" *Physics World.* London, United Kingdom: Institute of Physics (April 2002): p 64.
26. Zghal, M., H.-E. Bouali, Z. ben Lakhdar and H. Hamam. "The First Steps for Learning Optics: Ibn Sahl's, Al-Haytham's and Young's Works on Refraction as Typical Examples." *Education and Training in Optics and Photonics Conference (ETOP) 2007 Proceedings* (Ottawa, Ontario, Canada, June 2007). Bellingham WA: SPIE (2007).
27. Marshall, O.S. "Alhazen and the Telescope." *Astronomical Society of the Pacific Leaflets.* Vol. 6, No. 251. San Francisco, CA: Astronomical Society of the Pacific (February 1950): p 4-11.
28. Rashed, R. *Ibn al-Haytham: Astronomie, Géométrie Sphérique et Trigonométri les Mathématiques Infinitésimales du IXe au XIe Siècle.* Vol. 5. London, United Kingdom: al-Furqæn Islamic Heritage Foundation (2006).
29. Omar, Saleh Beshara. *Ibn-Al-Haytham's Optics: A Study of the Origins of Experimental Science.* Minneapolis, MN: Bibliotheca Islamica (1977).
30. *Opticae thesavrvs. Alhazeni Arabis libri septem, nuncprimum editi. Eivsdem liber de crepvscvlis & nubium ascensionibus. Item Vitellonis Thvringopoloni libri x. Omnes instaurati, figuris illustrati & aucti, adiectis etiam in Alhazenum commentarijs, a Federico Risnero.* Posner Collection, manuscript 535 A17O 1572. Pittsburgh, PA: Carnegie Mellon University.
31. Figure 9.4 in Turner [q.v.], reproducing page from Hunayn ibn Ishaq, *Book of Ten Treatises on the Eye.* MS 100 Tibb Taimur (thirteenth century manuscript). Cairo, Egypt: Egyptian National Library. This manuscript is credited by Turner (p 250) and is translated here by El Gomati.
32. Moore, P.O. "The Development of Boiler Inspection, 1866-1947." *Materials Evaluation.* Vol. 43, No. 7. Columbus, OH: American Society for Nondestructive Testing (June 1985): p 800-805.
33. Moore, P.O. "The Heritage of Nondestructive Testing." *Proceedings of National Board of Boiler and Pressure Vessel Inspectors* [May 1986]. Columbus, OH: National Board of Boiler and Pressure Vessel Inspectors (May 1986): p 25-43.
34. Sinclair, B. *A Centennial History of the American Society of Mechanical Engineers: 1880-1980.* New York, NY: ASME (1980).
35. *Rules for the Construction of Stationary Boilers and for Allowable Working Pressures* (Report of the Committee to Formulate Standard Specifications for the Construction of Steam Boilers and Other Pressure Vessels and for Their Care in Service, known as the Boiler Code Committee, ed. of 1914 with index). New York, NY: ASME International (1915).
36. Article 9, "Visual Examination." *2007 ASME Boiler and Pressure Vessel Code:* Section V, *Nondestructive Examination.* New York, NY: American Society of Mechanical Engineers (2007): p 154-156.
37. Lang, W.[J.], P. Sigmund and H. Swain. "History of the Borescope." *Nondestructive Testing Handbook,* second edition: Vol. 8, *Visual and Optical Testing.* Columbus, OH: American Society for Nondestructive Testing (1993): p 4-8.
38. Modlin, I.M. *A Brief History of Endoscopy.* New Haven, CT: Yale University School of Medicine (2000).
39. Rathert, P., W. Lutzeyer and W.E. Goddwin, "Philipp Bozzini (1773-1809) and the Lichtleiter." *Urology.* Volume 3, No. 1. New York, NY: Cahners Publishing (January 1974): p 113-118.
40. Bush, Ronnie Beth, Hanna Leonhardt, Irving Bush and Ralph Landes. "Dr. Bozzini's Lichtleiter: A Translation of His Original Article (1806)." *Urology.* Volume 3, No. 1. New York, NY: Cahners Publishing (January 1974): p 119-123.
41. Lang, W.J. "George S. Crampton and the Origins of Industrial Endoscopy." *Materials Evaluation.* Vol. 46, No. 13. Columbus, OH: American Society for Nondestructive Testing (December 1988): p 1639-1642.
42. "1906-2006 Wolf Familiengeschichte" ["Wolf Family History, 1906-2006"]. Web site. Knittlingen, Germany: Richard Wolf GmbH (2009).
43. Wolf, G. United States Patent 1 431 902, *Instrument for Examining Cavities of the Body* (1922).
44. Wolf, G. United States Patent 1 606 861, *Gastroscope* (1926).
45. Wolf, R. United States Patent 1 595 050, *Cystoscope and Similar Instrument* (1926).

46. Schindler, R. *Gastroscopy.* Chicago, IL: University of Chicago (1937).
47. Huston, C.L. United States Patent 1 454 677, *Inspection Lift* (1923).
48. Contant, W.A. United States Patent 1 468 803, *Inspection Table* (1923).
49. Firestone, F. United States Patent 1 640 567, *Machine for Detecting Flaws* (1927).
50. Gulliksen, F.H. United States Patent 2 229 451, *Photosensitive Apparatus for Inspecting Materials* (1941).
51. Pryor, T.R.P. United States Patent 4 576 482, *Electro-Optical Inspection* (1986).
52. Lyon, S.P. United States Patent 1 602 233, *Inside Rivet Inspection Device* (1926).
53. Fisher, J.H. United States Patent 1 775 452, *Instrument for Examining the Interior of Gun Barrels* (1930).
54. Helfer, D. United States Patent 2 010 301, *Device for Examining the Diameter of Bores* (1935).
55. Huntsinger, F. United States Patent 2 251 739, *Method and Apparatus for Inspecting Tubing* (1941).
56. "Crampton to Get Original of His Famous 'Scope." *Philadelphia Inquirer.* Philadelphia, PA (1946).
57. Spencer, S.M. "Amiable Oculist Is Expert on Device for U.S. Cannon." *Philadelphia Bulletin.* Philadelphia, PA (July 1941).
58. NAVSHIPS 250-1500-1, *Welding Standard.* Washington, DC: United States Department of Defense, for Naval Sea Systems Command.
59. IEEE/ASTM SI 10, *Standard for Use of the International System of Units (SI): The Modern Metric System.* New York, NY: IEEE (2002).
60. Thompson, A. and B.N. Taylor. NIST SP 811, *Guide for the Use of the International System of Units (SI).* Washington, DC: United States Government Printing Office (2008).
61. Taylor, B.N. and A. Thompson. NIST SP 330, *The International System of Units (SI).* Washington, DC: United States Government Printing Office (2008).

2

Chapter

Light

Yoshihiro Ohno, Gaithersburg, Maryland (Part 3)

John C. Duke, Jr., Virginia Polytechnic Institute and State University, Blacksburg, Virginia

Doron Kishoni, Centennial, Colorado

Gregory C. Sayler, MD Helicopters, Mesa, Arizona

Portions of this chapter are reprinted with permission from the *IES Lighting Handbook: Reference Volume,*[2] © Copyright 1981, Illuminating Engineering Society of North America (IESNA). Reference numbers superscripted in headings indicate sections adapted. ASNT has revised the text in 1993 and 2010, and deficiencies are not the responsibility of the IESNA.

Portions of Part 3 are reprinted from NIST SP 250-37, *Photometric Calibrations*[8] — a work of the United States government and not subject to copyright.

Part 1. Physics of Light

Overview of Light[1-3]

Most nondestructive tests are performed by a probing energy such as X-rays, ultrasound or magnetism to a test object. In visual and optical testing, the probing medium is light — that portion of the electromagnetic spectrum that has a wavelength of 360 to 830 nm and that can excite the human retina. The upper and lower limits of visible radiation vary from one individual to another.

Radiant energy at the proper wavelength makes visible anything from which it is emitted or reflected in sufficient quantity to activate the receptors in the eye. The quantity of such radiant energy may be evaluated in many ways, including: radiant flux (measured in joules per second or in watts) and luminous flux (measured in lumens).

Radiant Energy Theories[1,2]

Several theories describing radiant energy have been proposed.[4] The text below briefly discusses the primary theories.

Particle Theory

The particle theory, or corpuscular theory, was advanced by Isaac Newton and is based on the following premises.

1. Luminous bodies emit radiant energy in particles.
2. These particles are intermittently ejected in straight lines.
3. The particles act on the retina of the eye, stimulating the optic nerves to produce the sensation of light.

Wave Theory

The wave theory of radiant energy was championed by Christian Huygens and is based on these premises.

1. Light results from the molecular vibration in luminous material.
2. The vibrations are transmitted through the ether in wavelike movements (comparable to ripples in water).
3. The vibrations act on the retina of the eye, stimulating the optic nerves to produce visual sensation. The velocity of a wave is the product of its wavelength and its frequency.

Electromagnetic Theory

The electromagnetic theory was advanced by James Clerk Maxwell and is based on these premises.[5]

1. Luminous bodies emit light in the form of radiant energy.
2. This radiant energy is propagated in the form of electromagnetic waves.
3. The electromagnetic waves act on the retina of the eye, stimulating the optic nerves to produce the sensation of sight.

Quantum Theory

The quantum theory is an updated version of the corpuscular theory. It was advanced by Max Planck and is based on these premises.

1. Energy is emitted and absorbed in discrete quanta (photons).
2. The energy E in each quantum is the product of Planck's constant h and frequency ν:

$$E = h\nu \tag{1}$$

where $h = 6.626 \times 10^{-34}$ J·s and ν (Greek letter nu) is in hertz.

Unified Theory

The unified theory of radiant energy was proposed by De Broglie and Heisenberg and is based on the premise that every moving element of mass is associated with a wave of length λ:

$$\lambda = \frac{h}{mv} \tag{2}$$

where m is the mass of the particle (kilograms), v (Roman letter vee) is the velocity of the particle (meters per second), λ is the wavelength of the wave motion (meters) and h is Planck's constant or 6.626×10^{-34} J·s.

It is impossible to determine all of the properties distinctive of a wave or a particle simultaneously, for the energy to do so changes one of the properties being determined.

The explanations of radiant energy provided by the quantum theory and the electromagnetic wave theory are appropriate for the purposes of nondestructive testing. Whether it behaves like a wave or like a particle, light is radiation produced by atomic or molecular processes. That is, in an incandescent body, a gas discharge or a solid state device, light is produced when excited electrons have just reverted to more stable positions in their respective atoms, thereby releasing energy.

The general limits of the radiant energy spectrum extend over a range of wavelengths varying from 10^{-16} to over 10^5 m. Radiant energy if visible has wavelengths between 380 and 770 nm.

Generation of Light[3]

Light is created at the atomic level. An atom consists of a nucleus with a positive charge orbited by electrons with a negative charge. In the atom's normal state, the orbits of the electrons are stable and no electromagnetic radiation is emitted. The orbits, or shells, allowed around the nucleus are a characteristic of each element.

When an electron is excited, it moves out to a higher, unstable orbit or it is removed from the atom's orbit. Eventually, the excited electron moves back into the original orbit or into another lower energy, more stable orbit. The energy the electron loses during this move is given off as a discrete particle of radiant energy. Planck's equation gives the wavelength of the emitted radiation:

$$E_1 - E_2 = h\nu \tag{3}$$

where E_1 is energy of the excited orbit, E_2 is energy of the normal orbit, h is Planck's constant (6.626×10^{-34} J·s) and ν (Greek letter nu) is frequency in hertz (Hz).

All light is produced by this change in the orbit of electrons. Natural and artificial light sources are referred to as luminous bodies. Natural light sources include sunlight, the aurora borealis, and bioluminescence. Most other apparent light sources, such as the moon, only reflect the light of a luminous body.

Principles of Light Measurement[1-3]

Light measurement involves a consideration of the inverse square law and the cosine law.[6,7] These laws apply to all light, but their simplest expressions are for point sources.

Inverse Square Law[1-3]

The inverse square law (Fig. 1a) states that the illumination E (lux) at a point on a surface varies directly with the luminous intensity I (candela) of the source and inversely as the square of the distance d (meter) between the source and the point. If the surface at the point is normal to the direction of the incident light, the law may be expressed as:

$$E = \frac{I}{d^2} \tag{4}$$

This equation is accurate within 0.5 percent when d is at least five times the longest line transecting the illuminating source's area.

The inverse square law means that an increase in distance from the radiation source diminishes geometrically the irradiation, or illumination, at the test surface. The inverse square law is important also for calculating exposures to X-rays. The common sensical application of this law is as follows: to make a surface brighter, move a lamp closer.

Cosine Laws[1-3]

Lambert's cosine law (Fig. 1b) states that the illuminance E_2 at a point on a surface varies with the cosine of the angle θ of incidence:

$$E_2 = E_1 \cos\theta \tag{5}$$

where E_1 is illuminance (lux) at the point where the angle of incidence is normal and where θ is the angle (degree) of incidence. The angle of incidence is the angle between the normal to the surface and the direction of the incident light.

The inverse square law and the cosine law can be combined to yield the following relationship (in lux):

$$E = \frac{I}{d^2}\cos\theta \tag{6}$$

where I is the source luminous intensity (candela).

An extension of the cosine law is the cosine cubed law, a convenient alternative in certain calculations. Substituting $a \cdot (\cos\theta)^{-1}$ for d (Fig. 1c) lets Eq. 6 be rewritten:

$$E = \frac{I \cos^3\theta}{a^2} \tag{7}$$

where a is the distance normal to the surface. The common sensical application of the cosine laws is that the light will spread out on a surface if the lamp is held off to one side. The lighted area will be larger but less bright than if the lamp were directly above the surface.

FIGURE 1. Principles of photometry: (a) inverse square law, illustrating how same quantity of light flux is distributed over greater area as distance from source to surface is increased; (b) cosine law, showing that light flux striking surface at angles other than normal is distributed over greater area; (c) cosine cubed law, explaining transformation of inverse square formula.

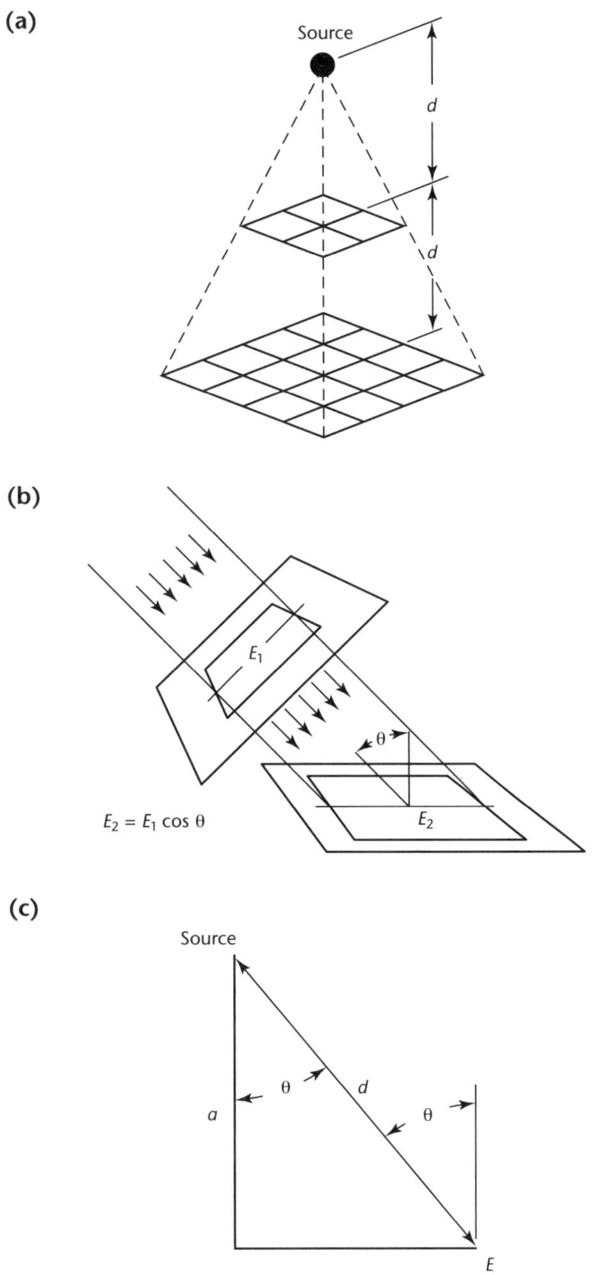

Wavelength and Frequency[3]

The important variables in light propagation include wavelength, frequency, reflection and refraction. The concern of the nondestructive test engineer revolves around the effects of light based on the intensity or the wavelength and frequency of the light. Optical radiation covers the spectrum from 10^{-9} to 10^{-3} m. Visible radiation is generally considered to be that portion of the spectrum from 360×10^{-9} to 830×10^{-9} m. The nanometer (10^{-9} m) is the unit for measuring the wavelength of light. Figure 2 shows the place of light in the electromagnetic spectrum.

All forms of electromagnetic radiation travel through a vacuum at the same speed, 299 793 km·s^{-1} (186 282 mi·s^{-1}). When light travels through any other medium, the velocity is altered. The frequency remains fixed and is independent of the medium. The difference in the time it takes light to travel through different media is responsible for the operating principles of optical instruments.

The following formula gives the relationship between the velocity, frequency and wavelength:

$$v = \frac{\lambda \nu}{n} \tag{8}$$

where n is the medium's index of refraction, v (Roman letter vee) is velocity (m·s^{-1}) of light waves in the medium, λ is the wavelength of the light wave in a vacuum and ν (Greek letter nu) is the frequency (in hertz). In the International System, the nanometer (1 nm = 10^{-9} m) and the micrometer (1 µm = 10^{-6} m) are units of wavelength in the visible region.

FIGURE 2. Electromagnetic wavelength spectrum.[3]

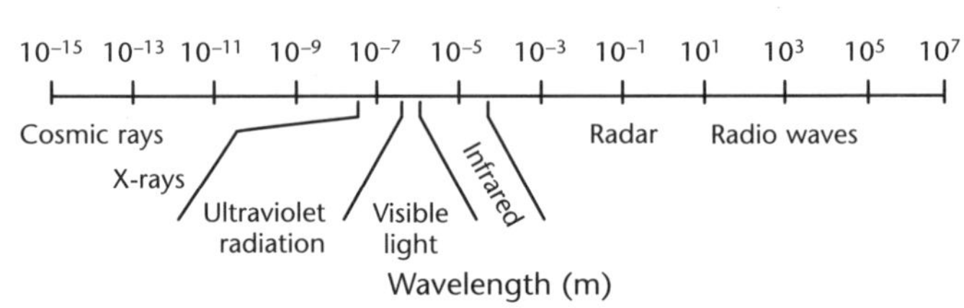

PART 2. Refraction and Color

Reflection and Refraction[3]

Light is reflected when it strikes a surface. Regular, or specular, reflection is caused when light strikes a smooth surface. The direction of the reflected beam can be determined by constructing a line perpendicular to the reflecting surface. The angle of reflection is the same as the angle of incidence compared to the perpendicular, or normal line, as illustrated in Fig. 3.

Diffuse reflection is caused when light strikes a rough surface. The rough surface has many different plane surfaces. Each incident light beam strikes a different reflecting plane and is reflected at an angle that corresponds to the relative angle of the plane surface. The difference from a regular reflection forms the basis of some optical techniques.

Refraction is the bending or pivoting of light from its original direction at the interface of two different media. If the optical density of the new medium is greater than the density of the original, the path of light is bent, or refracted, toward a line normal to the object's surface. If the optical density of the new medium is less than that of the original, the light is refracted away from the normal.

The amount of refraction depends on the angle of the incident light and the index of refraction. The index of refraction is given by the ratio of the velocity of light in a vacuum to the velocity of light in the medium:

$$(9) \quad n = \frac{V_v}{V_m}$$

where n is index of refraction, V_m is the velocity (meters per second) in a medium and V_v is the velocity (meters per second) in a vacuum.

Based on Eq. 9, the actual angle of refraction is determined by a mathematical relation, discovered by Ibn Sahl in the tenth century and later called *Snell's law.* The index of refraction is equal to the ratio of the sine of the angle of incidence to the sine of the angle of refraction. Equation 10 applies the law of refraction:

$$(10) \quad n_0 \sin\theta_0 = n_1 \sin\theta_1$$

Figure 4 demonstrates the angles in Eq. 10.

The science of optics provides an explanation for the operation of many visual and optical tools from simple magnifying glasses to metallographs. Light is commonly focused in the mirrors and lenses of optical devices using the principles of reflection and refraction built into the device. Classical optics explains the manipulation of light to produce an image for human viewing by

FIGURE 3. Angles of incidence and reflection: $\phi = \theta$.[3]

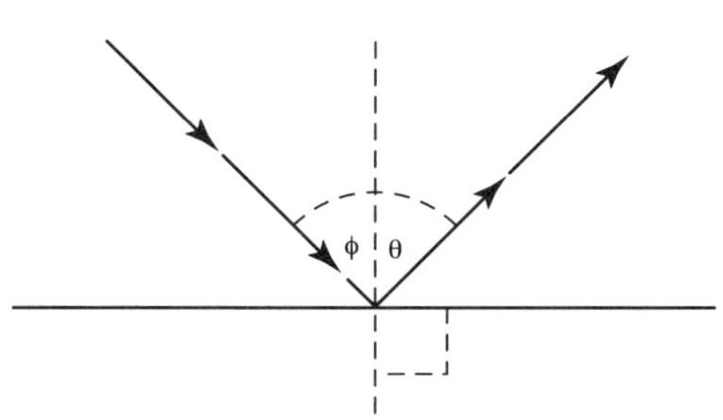

Legend
ϕ = Angle of incidence
θ = Angle of reflection

FIGURE 4. Refraction of light.[3]

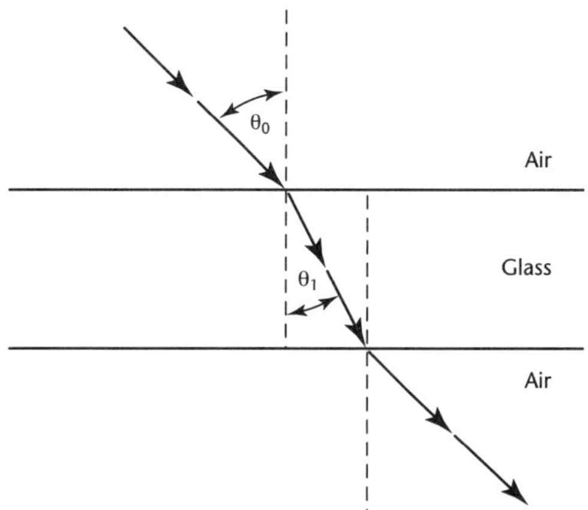

Legend
θ_0 = angle of incidence
θ_1 = angle of refraction

mechanical devices: lenses, mirrors and prisms.

Lenses

A lens is a device that converges or disperses light by refraction. Converging lenses focus light on a single point while divergent lenses disperse light. When describing lenses, the conventional standard is to describe the shape of the lens surface from left to right using the following terminology. Plano describes a flat surface. Convex lenses are converging lenses; they are thicker in the center than at the edge. Concave lenses are divergent; they are thinner in the center than at the edge. Figure 5 shows examples of these lenses.

Thin lenses are those where the thickness of the lens is small compared to the focal length. The properties of thin lenses are described using the lens law. This law relates the image distance, object distance and focal length of a lens as follows:

$$\text{(11)} \quad \frac{1}{f} = \frac{1}{d} + \frac{1}{u}$$

where d is image distance, f is focal length and u is object distance. The meter is the unit of length.

The focal length or focal distance of a lens is the distance from the principal plane to the focal plane. The principal focus or focal plane is where parallel incident light rays converge after being refracted by the lens. Single converging lens magnifiers, as shown in Fig. 6, allow the eye to be positioned closer to the test object when the retinal focus plane is at the focal plane of the lens. Because a simple lens uses the same principles of refraction as a prism, there is chromatic aberration.

Divergent lenses, as shown in Fig. 7, diffract light outward, causing minification. Magnification increases as the object gets closer to the focal plane.

FIGURE 5. Shapes of convex and concave lenses: (a) biconvex; (b) plano-convex; (c) convexo-concave; (d) biconcave; (e) plano-concave; (f) concavo-convex.[3]

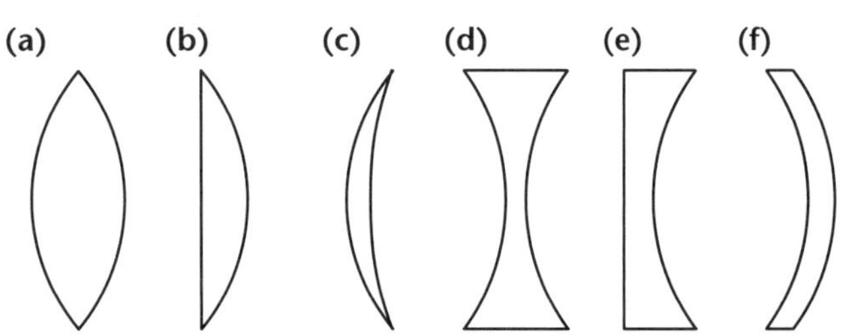

The magnification produced by a diverging lens may be calculated by manipulating the lens law. Magnification is the size of the image divided by the size of the object:

$$\text{(12)} \quad M = \frac{D_i}{D_o} = \frac{S_i}{S_o}$$

where D_i is image distance from principal plane, D_o is object distance from principal plane, M is magnification, S_i is image size and S_o is object size.

Thick lenses may consist of a single thick lens, combinations of thin lenses, or compound arrangements of thin lenses. Compound lenses (Fig. 8) are used in most optical systems from doublet and triplet magnifiers to sophisticated zoom camera lenses. Compound lenses provide for high magnification and close control of the focal plane. When properly

FIGURE 6. Converging lens geometry.[3]

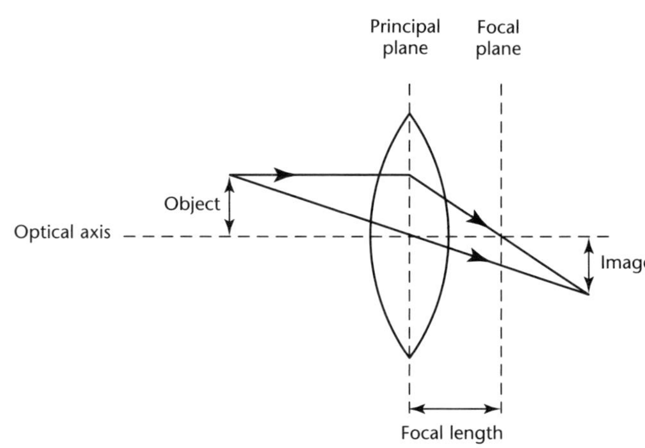

FIGURE 7. Diverging lens geometry.[3]

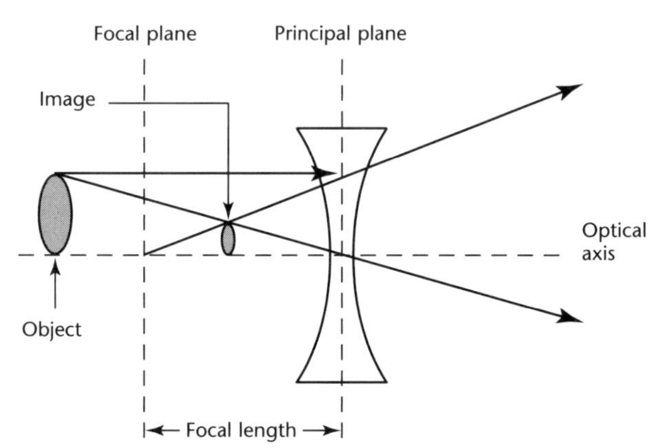

designed, compound lenses also can correct the chromatic and spherical aberrations inherent to a single lens.

Mirrors

Mirrors change the direction of light by reflection. Mirrors can be flat, convex, concave, or parabolic. Flat or plane mirrors are arranged singularly or in series to transmit an image or light. Convex mirrors provide an enlarged field of view of the reflected image.

A concave or spherical mirror has a reflecting focal point. If light is projected onto a spherical mirror normal to the curve of the surface, the light will be focused slightly in front of the mirror. If a point light source is placed at its focal point, the light will be reflected from the mirror so that it is parallel to the normal of the curve.

A concave mirror can also be used for image enlargement. An image that is small compared to the width of the mirror will be reflected back in a diverging and optically reversed image.

Prisms

On most surfaces, incident light is partially reflected and partially refracted. The greater the angle of incidence and the difference in the refractive indices of the material, the more light will be reflected instead of refracted. The angle above which all light is reflected is known as the *critical angle*. Prisms use the critical angle to change the direction or the orientation of the image produced by light rays. Two common types of prisms are the right angle prism and the porro prism (Fig. 9). The right angle prism deflects the light rays 90 degrees. The porro prism produces a 180 degree reflection. Both prisms are common in optical instruments.

Prisms are also used to separate the frequencies of a chromatic light source by diffraction. Because the two refracting surfaces of the prism are not parallel, the distance the light paths travel varies from the top to the bottom of the prism. Because the index of refraction changes with the frequency of the light, the higher frequency portions of the spectrum emerge from the base of the prism.

FIGURE 8. Compound lenses: (a) combination lens; (b) doublet; (c) triplet.[3]

(a) (b) (c)

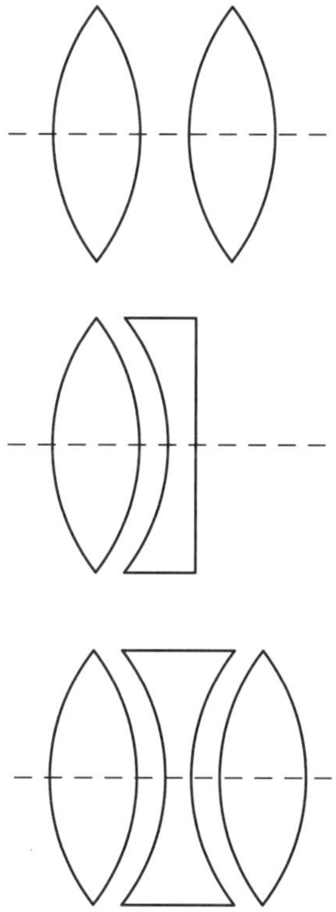

FIGURE 9. Common prisms: (a) right angle prism; (b) porro prism.[3]

(a) (b)

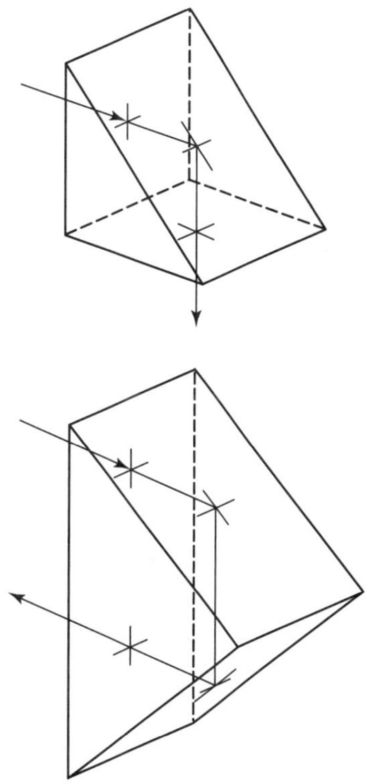

Spectral Emissivity[1,2]

The wave theory permits a convenient representation of radiant energy in an arrangement based on the light's wavelength or frequency. This arrangement is called a *spectrum* and is useful for indicating the relationship between various radiant energy wavelength regions. Such a representation should not be taken to mean that each region of the spectrum is physically divided from the others — actually there is a small but discrete transition from one region to the next.

All forms of radiant energy are transmitted at the same speed in a vacuum: 299 793 km·s^{-1} (299 792.458 mi·s^{-1}). Table 1 gives the speed of light in different media for a frequency corresponding to a wavelength of 589 nm in air.

Each form of energy differs in wavelength and therefore in frequency. The wavelength and velocity may be altered by the medium through which the radiation passes, but the frequency is fixed independently of the medium. Equation 8 above shows the relationship between radiation speed, frequency, wavelength and the medium's index of refraction.

Blackbody[1-3]

Light sources are frequently compared to a theoretical light source known as a *blackbody*. A blackbody absorbs all of the radiant energy that falls on it. A blackbody is also a perfect radiator, radiating more total power at any wavelength than any other source for an equivalent area.

Most real approximations of theoretical blackbodies emit in the infrared range and are often used to calibrate instruments that measure infrared radiation. Nevertheless, the concept of the blackbody is of value for inspectors in visual testing for two reasons.

1. The concept of the blackbody is used to define and measure emissivity, a physical characteristic of light sources. Emissivity is a material's ability to radiate energy and is measured as a ratio of source emissivity to blackbody emissivity, the blackbody having an emissivity of one for a given wavelength.
2. The theoretical blackbody is used to define colors, and working blackbodies are used to generate colors.

Some inspection devices such as lamps and gloss meters include emissivity in their technical data.

A perfect blackbody is a theoretical ideal; no perfect blackbody instrument exists in the real world. The theoretically perfect performance of the blackbody can be approximated in the laboratory by constructing a hollow cylinder or sphere, with a small hole in its wall (Fig. 10). Radiation at a controlled wavelength is introduced into the hole and reflects until completely absorbed by the wall. The absorption of the light's energy produces uniform heating of the wall. Radiation that exits through the small hole will have the characteristics of a perfect radiator for that specific temperature.

TABLE 1. Speed of light for wavelength of 589 nm (fraunhofer D lines for sodium).

Medium	Speed (10^6 m·s^{-1})
Vacuum	299.792 458
Air (100 kPa at 0 °C)	299.724
Crown glass	197.368 421
Water	225.563 010

Graybody[1-3]

No known radiator has the emissive power of a blackbody. The spectral emissivity $\varepsilon(\lambda)$ of a light source is the ratio of the light source output to the output of the theoretical blackbody. The blackbody's ideal emissivity equals one.

When the spectral emissivity is uniform for all wavelengths, the radiator is known as a *graybody*. No known radiator has a uniform spectral emissivity for all visible, infrared and ultraviolet wavelengths. In the visible region, a carbon filament exhibits very nearly

FIGURE 10. Small aperture in an enclosure exhibits blackbody characteristics.[1,2]

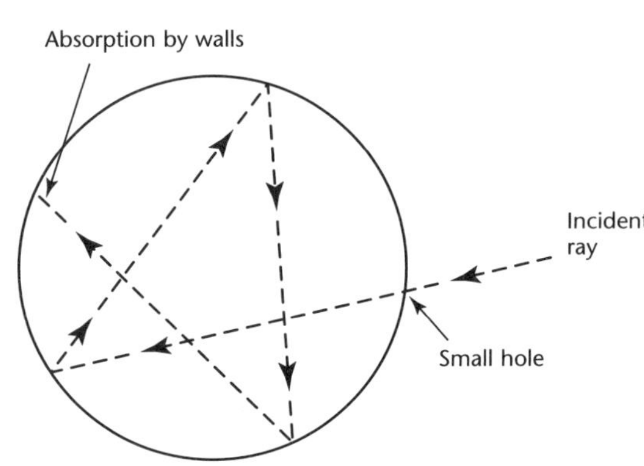

uniform emissivity and is nearly a graybody.

The emissivity of all known materials varies with wavelength. In Fig. 11, the radiation curves for a blackbody, a graybody and a selective radiator (tungsten), all operating at 3000 K, are plotted on the same logarithmic scale to show differences in output.

Physical Laws for Blackbody Radiation[1,2]

Planck's Radiation Law. Data describing blackbody radiation curves have been obtained by using a specially constructed and uniformly heated tube as a blackbody source. Planck, introducing the concept of discrete quanta of energy, developed an equation depicting these curves. It gives the spectral radiance of a blackbody as a function of wavelength and temperature.

Wien's Displacement Law. Wien's displacement law gives the relationship between the wavelength at which a blackbody at temperature T (kelvin) emits maximum power per unit wavelength at that temperature. In fact, the product of absolute temperature T and the peak wavelength is a constant. It gives the relationship between blackbody distributions at various temperatures only with this important limitation.

FIGURE 11. Radiation curves for blackbody, graybody and selective radiators operating at 3000 K.[1,2]

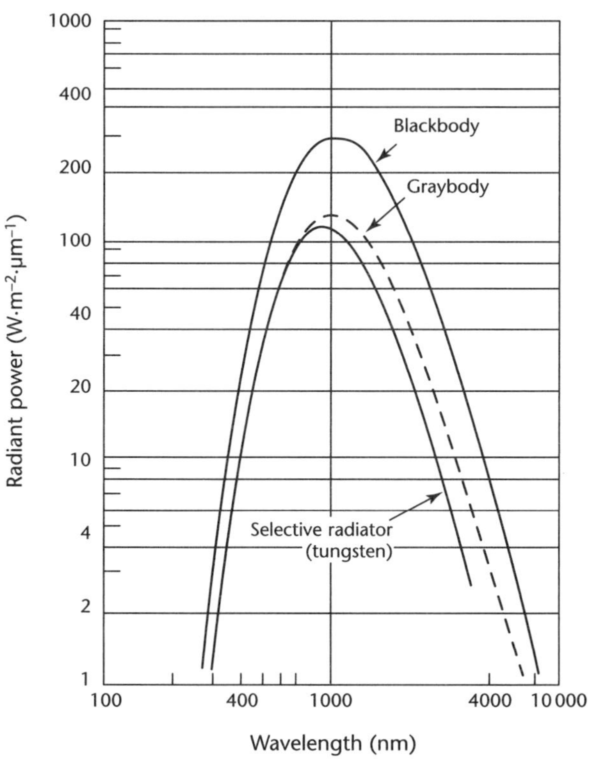

Stefan-Boltzmann Law. The Stefan-Boltzmann law is obtained by integrating Planck's expression for the spectral radiant exitance from zero to infinite wavelength. The law states that the total radiant power per unit area of a blackbody varies as the fourth power of the absolute temperature. The Stefan-Boltzmann law is explained in introductory physics texts. This law applies to the total power in the entire spectrum, not only in the visible portion.

Color[3]

The color temperature of a given light source is determined by the temperature at which a blackbody must be operated to produce the spectral distribution that most closely matches the spectral distribution of the light source. The color temperature may be used to measure the color from incandescent light sources. For example, as a blackbody source heats up, the color temperature of emitted light increases as the visible color changes from dull red to bluish white. In many photographic and electronic imaging applications, the color of the light source must be known to get accurate color response in the imager. Figure 12 illustrates the color temperature of some natural and artificial light sources.

FIGURE 12. Color temperatures of light sources.[3]

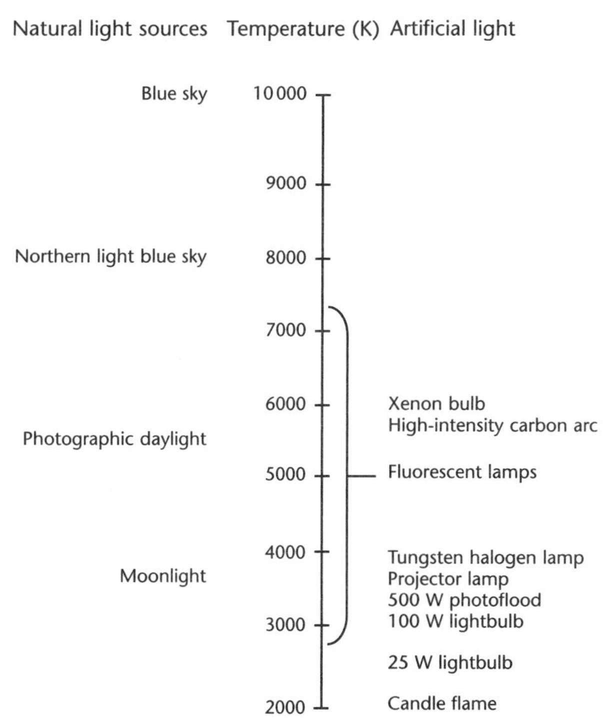

Rigorously speaking, color temperature is used only for light sources that have planckian-like spectral distributions. The term *correlated color temperature* is used for sources (such as fluorescent lamps) that have different spectral distributions.

The color of light is determined by its wavelength components. (There is no characteristic wavelength for white light.) Figure 13 illustrates the wavelengths of the colors of the visible portion of the spectrum. A spectral color is monochromatic, consisting entirely of light of one wavelength. Most light sources are not monochromatic, meaning they are comprised of a variety of wavelengths. Sunlight and white light are mixtures of the entire visible light spectrum.

All of the visible colors can be created by mixing the proper amounts of the primary colors. Emitted light of any color can be created by mixing the additive primaries (red, green and blue components) in color displays or projectors. The color of objects created by reflected light is created using subtractive primaries. Each subtractive primary absorbs one of the additive primaries and reflects the other two. The subtractive primaries are magenta, yellow, and cyan. The rest of the spectrum is created by mixing pigments of these three subtractive primary colors, as shown in Fig. 14.

FIGURE 13. Wavelength of light colors: (a) refraction of colors at different wavelengths; (b) color wavelengths.[3]

(a)

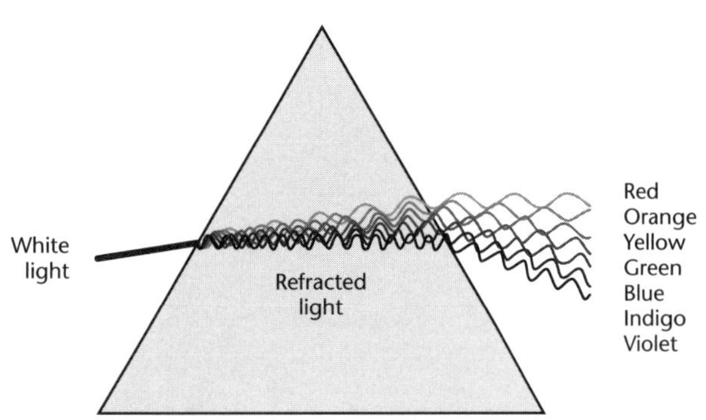

(b)

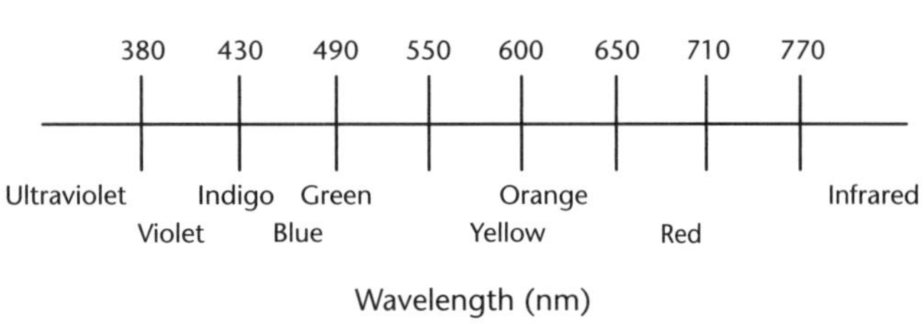

FIGURE 14. Primary colors and formation of secondary colors: (a) additive primary colors; (b) subtractive primary colors.[3]

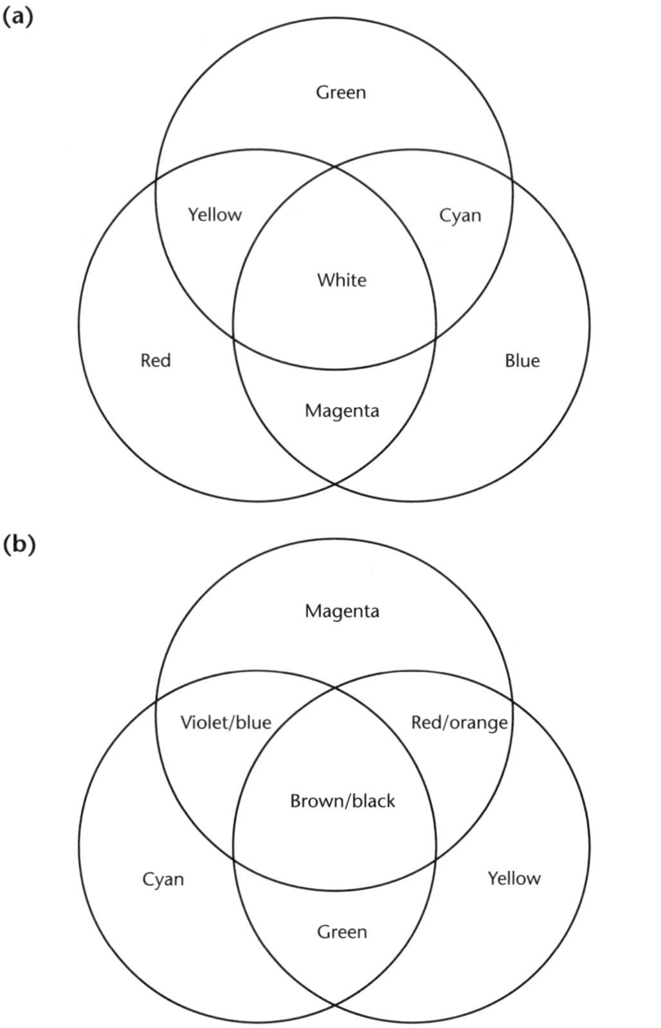

PART 3. Photometry[1-3,8]

Luminous Efficacy[1,2]

The luminous efficacy of a light source is described by the ratio of the total luminous flux (lumens) to the total power input (watts).

Many apparent differences in intensity between radiant energy of various wavelengths are in fact differences in the ability of various sensing devices to detect them.[9] The reception characteristics of the human eye have been subject to extensive investigations and the results may be summarized as follows.

1. The spectral response characteristic of the human eye varies between individuals, with time, with viewing conditions and with the age and health of an individual.
2. From the available data, a luminous efficacy curve can be selected to represent an observer.

The standard spectral luminous efficiency curve for photopic (light adapted) vision represents a typical characteristic, adopted to help solve photometric problems. Most observers yield only a slight response to ultraviolet radiation at the nearly visible wavelengths because the lens of the eye absorbs nearly all of it.

Photopic, Scotopic and Mesopic Vision[1,2,8]

The human eye contains two basic types of retinal receptors known as *rods* and *cones*. They differ not only in relative spectral response and other properties but by orders of magnitude in responsivity. The rods are more sensitive than cones and respond more to the blue and less to the red end of the spectrum. However, they do not actually give the sensation of color as the cones do.[1,2]

Luminance is measured in candelas per square meter ($cd{\cdot}m^{-2}$). When the eye has been subjected to a field luminance of more than 3.0 $cd{\cdot}m^{-2}$ (~0.3 $cd{\cdot}ft^{-2}$) for more than a few minutes, the eye is said to be in a light adapted state in which only the cones are responsible for vision; the state is known as *photopic* vision.[1,2] The spectral luminous efficiency function $v(l)$ for photopic vision assumes additivity of sensation and a 2 degree field of view at luminance levels above ~3 $cd{\cdot}m^{-2}$.

The spectral responsivity of human vision deviates significantly at low luminance, less than 10^{-3} $cd{\cdot}m^{-2}$. Vision at low luminance is called *scotopic* vision.

At light levels five or more orders of magnitude below this, at or below ~10^{-3} $cd{\cdot}m^{-2}$ (10^{-4} $cd{\cdot}ft^{-2}$), the cones no longer function and the responsivity is that of the rods. This is known as *scotopic* vision. After being light adapted, the eye usually requires time to become dark adapted when the light level is lowered. The time needed to adapt completely depends on the initial luminance of the starting condition but is usually achieved in 30 to 45 minutes.[1,2]

Between the levels at which the eye exhibits photopic and scotopic responses, the spectral and other responses of the eye vary continuously. In this state, known as *mesopic*, properties of both cone and rod receptors contribute.[1,2] Mesopic light levels cover a range of luminance from ~10^{-3} to ~3 $cd{\cdot}m^{-2}$. Many visual tests are made under photopic conditions, but most measurements of fluorescent and phosphorescent materials are made under scotopic and mesopic conditions. Because of the changes in the eye's spectral response at these levels, it is necessary to take the adaption luminance into account when evaluating the results of such measurements.[12]

Mesopic conditions may be considered twilight, such as shaded or interior surfaces not well illuminated. Ordinary photopic photometers are often used for the measurement of light at mesopic levels but can lead to misleading evaluations because the luminous efficiency of the eye shifts toward shorter wavelengths at mesopic levels. Techniques of mesopic photometry should give a better assessment of light at mesopic levels than does photopic photometry alone.[13] In current practice, almost all photometric quantities are given in terms of photopic vision, even at low light levels.[3]

Functions of Photometers[1-3]

A photometer is a device for measuring radiant energy in the visible spectrum (Fig. 15). Photometers are of two types.

1. Laboratory photometers are fixed in position and yield very accurate results.
2. Portable photometers are of lower accuracy for making measurements in the field.

Both types of meters may be grouped according to the quantity measured: luminous intensity (candelas), luminous flux, illuminance, luminance and light distribution. Devices measuring reflectance and transmittance, color, spectral distribution and visibility are not considered photometers.

In the twenty-first century, most photometers contain solid state devices such as photovoltaic cells. A photovoltaic cell converts radiant energy directly into electrical energy. It provides a small current proportional to the incident illumination and also produces a small electromotive force that forces this current through a low resistance circuit.

With modern metering techniques, electronic alterations can be accomplished to keep the output of a receiver and amplifier combination in range of linearity and readability.

Photometric Measurement[8]

Photometry is the measurement of light and its properties, and the measuring instrument is a photometer. The primary aim of photometry is to measure visible optical radiation, light, in such a way that the results correlate with visual sensation to a normal human observing that radiation. Before 1940, visual comparison dominated photometry: an observer was required to match the brightness of two visual fields viewed simultaneously or sequentially. The earliest photometers depended on visual appraisal by the operator as the means of measurement, and such meters are rarely used now.

This method of photometry is called *visual photometry.* Such meters are seldom used in the twenty-first century; they have been replaced by photodetectors, quantitative meters sensitive to light's physical properties. The technology of measuring radiant energy incident on a receiver and measuring electrical quantities is referred to as *physical photometry.* Physical photometers are more accurate and simpler to operate than their earlier counterparts.

To achieve the aim of photometry, the characteristics of human vision must be taken into account. The relative spectral responsivity of the human eye is called *the spectral luminous efficiency function for photopic vision*, or the $V(\lambda)$ function, and is defined by the International Commission on Illumination (Commission Internationale de l'éclairage, International Commission on Illumination, CIE). The $V(\lambda)$ function is in the domain 360 to 830 nm and is normalized at its peak, 555 nm (Fig. 16). This model is used by the International Committee for Weights and Measures (Comité International des Poids et Mesures, CIPM) to help define the candela. The values of the function at 1 nm increments are tabulated and referenced.[8,13] A photodetector having a spectral responsivity matched to the $V(\lambda)$ function replaces human eyes in photometry.

Radiometry concerns physical measurement of optical radiation as a function of its wavelength. As specified in the definition of the candela by CIPM, a photometric quantity X_v is defined in relation to the corresponding radiometric quantity $X_{e,\lambda}$ by the equation:

$$(13) \quad X_v = K_m \int_{360\,\text{nm}}^{830\,\text{nm}} X_{e,\lambda} V(\lambda)\, d\lambda$$

FIGURE 15. Geometry for detector based candela realization.[8]

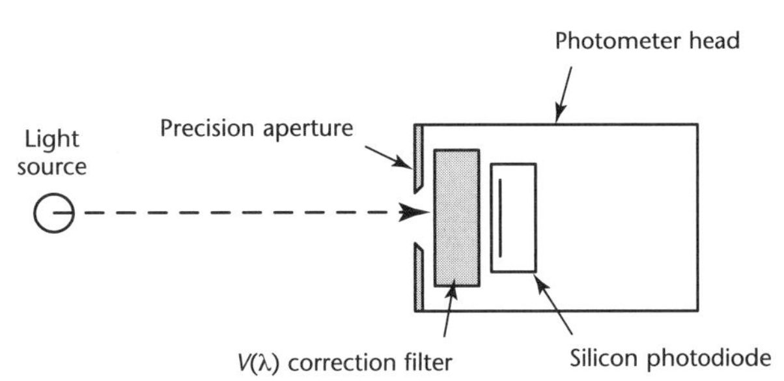

FIGURE 16. Luminous efficiency function.[8]

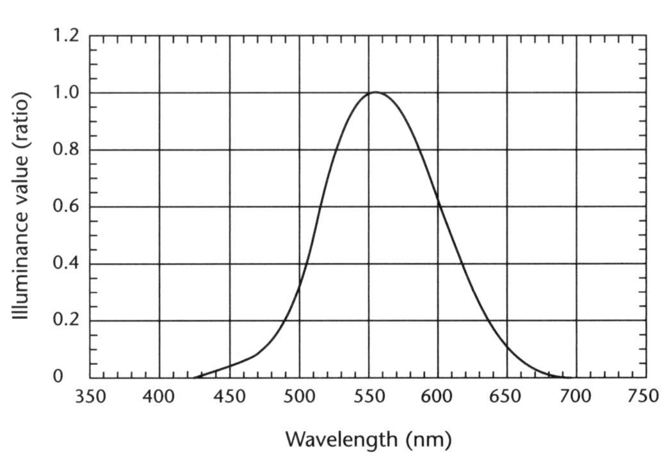

The constant K_m relates the photometric quantities and radiometric quantities and is the maximum spectral luminous efficacy (of radiation) for photopic vision. K_m is rounded to 683 $lm \cdot W^{-1}$ with negligible errors.[8]

Photometric Quantities

Radiometry is the measurement of radiant energy, both visible and invisible. *Photometry* in Greek means simply *light measurement.* Radiometry and photometry have the same principles but different units of measure (Table 2). As indicated in Table 3, many characteristics of light, light sources, lighting materials and lighting installations may be measured, including (1) illuminance, (2) luminance, (3) luminous intensity, (4) luminous flux, (5) contrast, (6) color appearance and color rendering, (7) spectral distribution, (8) electrical characteristics and (9) radiant energy. Photometric quantities are defined below.

The basis of all photometric quantities is luminous intensity, measured in candelas. The candela (cd) is based on the radiation from platinum at solidification temperature, emitting monochromatic

TABLE 2. Photometric quantities and corresponding radiometric quantities.

Photometric Quantity	Photometric Unit	Radiometric Quantity	Radiometric Unit
Color temperature	kelvin (K)	radiance temperature	kelvin (K)
Illuminance	lux·(lx)[b]	irradiance	watt per square meter ($W \cdot m^{-2}$)
Luminance	candela per square meter ($cd \cdot m^{-2}$)[c]	radiance	watt per steradian per square meter ($W \cdot sr^{-1} \cdot m^{-2}$)
Luminous energy	lumen second (lm·s)	radiant energy	joule (J)
Luminous exitance	lumen per square meter ($lm \cdot m^{-2}$)	radiant exitance	watt per square meter ($W \cdot m^{-2}$)
Luminous exposure	lux second (lx·s)	radiant exposure	watt per square meter second ($W \cdot m^{-2} \cdot s$)
Luminous flux	lm (lumen)	radiant flux	watt (W)
Luminous intensity	candela (cd)[a]	radiant intensity	watt per steradian ($W \cdot sr^{-1}$)

a. 1 cd = 1 $lm \cdot sr^{-1}$.
b. 1 lx = 1 $lm \cdot m^{-2}$.
c. 1 $cd \cdot m^{-2}$ = 1 $lm \cdot sr^{-1} \cdot m^{-2}$.

TABLE 3. Measurable characteristics of light, light sources and lighting materials.

Characteristic	Unit	Instrument
Light		
Color[a]	none	spectrophotometer and colorimeter
Illuminance (flux density)[b]	lux	photometer
Polarization degree[a]	percent (dimensionless ratio)	polarization photometer
Polarization orientation[a]	degree (angle)	analyzing prism
Wavelength[a]	meter	spectrometer
Light Sources		
Angle, direction or zonal distribution[a]	lumen or candela per direction	goniophotometer
Color temperature[b]	kelvin	colorimeter or filtered photometer
Energy radiated[a]	joule per square meter	calibrated radiometer
Luminance[b]	candela per square meter	photometer or luminance meter
Luminous flux (light output)[a]	lumen	integrating sphere photometer
Luminous intensity[b]	candela	photometer
Power consumption[b]	watt	watt meter, or volt meter and ammeter[c]
Spectral power distribution[a]	watts per nanometer	spectroradiometer
Lighting Materials		
Optical density	dimensionless number	densitometer
Reflectance[b]	percent (dimensionless ratios)	reflectometer
Spectral reflectance and transmittance[a]	percent (at specific wavelengths)	spectrophotometer
Transmittance[b]	percent (dimensionless ratios)	photometer

a. Can be measured in laboratory.
b. Can be measured in field or laboratory.
c. For direct current and for unity power factor alternating current.

radiation of frequency 540 THz with a radiant intensity in that direction of 1/683 $W \cdot sr^{-1}$.

Luminous Flux. Although the candela is a base unit, luminous flux is a more fundamental photometric quantity in that the four other photometric quantities are defined in terms of lumen with appropriate geometric factors. Luminous flux Φ_v is measured in lumens and is the time rate of flow of light as weighted by $V(\lambda)$:

$$(14) \quad \Phi_v = K_m \int_\lambda \Phi_{e,\lambda} V(\lambda) d\lambda$$

where $\Phi_{e,\lambda}$ is the spectral concentration of radiant flux ($W \cdot nm^{-1}$) as a function of wavelength λ in nanometers.

Luminous Intensity. Luminous intensity I_v is the luminous flux from a point source and emitted per unit solid angle in a given direction:

$$(15) \quad I_v = \frac{d\Phi_v}{d\Omega}$$

where $d\Phi_v$ is the luminous flux leaving the source and propagating in an element of solid angle $d\Omega$ containing the given direction.

Illuminance. Illuminance E_v is the density (in lux) of luminous flux incident on a given area of a planar surface at a given instant:

$$(16) \quad E_v = \frac{d\Phi_v}{dA}$$

where $d\Phi_v$ is the luminous flux incident on an element dA of the surface.

Luminance. Luminance L_v (in candelas per square meter) is the luminous flux at a given surface where the angle of incidence or refraction is considered. The luminous flux may be exiting, passing through, or arriving at the surface. Historically, luminance has been referred to as *photometric brightness*:

$$(17) \quad L_v = \frac{d^2\Phi_v}{d\Omega \, dA \cos\theta}$$

where d^2F_v given in Eq. 17 is the luminous flux emitted (reflected or transmitted) by an elementary beam passing through the given point and propagating in the solid angle $d\Omega$ containing the given direction; dA is the area of a section of that beam containing the given point; and θ is the angle between the normal to that section and the direction of the beam.

Luminous Exitance. Luminous exitance M_v is the density of luminous flux leaving a surface at a point. The equation is the same as Eq. 16, with $d\Phi_v$ meaning the luminous flux leaving a surface. This quantity is rarely used in photometry.

Luminous Exposure. Luminous exposure is the time integral of illuminance over a given duration and is not discussed here.

Color Temperature. Color temperature T_c is the temperature of a planckian radiator with radiation of the same chromaticity as that of the light source. Because the chromaticity coordinates of most lamps do not fall on the planckian locus, actual lamp calibrations use either distribution temperature or correlated color temperature.

Distribution Temperature. Distribution temperature T_d is the temperature of a blackbody with a spectral power distribution closest to that of the light source in question and it is useful for quasiplanckian sources.

Correlated Color Temperature. This quantity is used for sources (for example, discharge lamps) whose spectral power distribution differs significantly from that of planckian radiation. The correlated color temperature T_{cp} is the temperature of the planckian radiator whose perceived color most closely resembles that of the light source in question. Informally, the phrase *color temperature* can denote correlated color temperature.

General information (definitions, symbols, and expressions) on many other physical quantities and units, including photometric and radiometric quantities, are given elsewhere.[8,14,15]

Calibration[8]

Luminous Flux Calibration

The NIST luminous flux unit is based on the candela and is derived by using an integrating sphere (Fig. 17) and an external source. An integrating sphere is an optical device, a hollow sphere to control light for testing and calibration of optical instruments such as luminous flux measuring devices or standard lamp emitters. The total flux of a lamp is measured inside the sphere and compared to a known amount of flux introduced into the sphere from a source outside the sphere. The measurement device is at an aperture cut into the sphere. The light may be introduced through another aperture or by a source positioned in the center of the sphere. The interior of the sphere is painted white with a high reflectance paint such as barium sulfate, so that light of all colors is diffusely reflected, not absorbed by the interior walls. The cavity is spherical so that light is diffusely reflected and distributed uniformly in the cavity.

Precautions for Photometers and Illuminance Meters

An understanding of how a photometer is contructed can help in understanding its operation.

A standard photometer head should have either a limiting aperture (whose area is much smaller than the photodiode area) or a flat diffuser such as an opal glass in front of the $V(\lambda)$ correction filter so that the reference plane of the photometer is accurately and clearly defined. Some commercial photometer heads only have a $V(\lambda)$ correction filter attached in front of the silicon photodiode. If a photometer head does not have an aperture or a diffuser, the photodiode surface might be used as the reference plane of the photometer head. In this case, because of the refraction index of the $V(\lambda)$ correction filter, which is usually several millimeters in thickness, the effective reference plane can be several millimeters from the photodiode surface.

Sometimes, the front surface of the filter is simply defined as the reference plane of such photometer heads, in which case the true reference plane can be more than 10 mm from the filter surface. When the reference plane is not correctly defined, the departure from the inverse square law causes the responsivity of the photometer to vary with distance to the source, and serious errors may occur when the photometer is close to the source. The same problem occurs with a large lamp at close distances. To avoid these difficulties, standard photometers having a limiting aperture or a flat diffuser on the front are recommended.

Figure 17. Geometry (top view) of integrating sphere for luminous flux measurement.

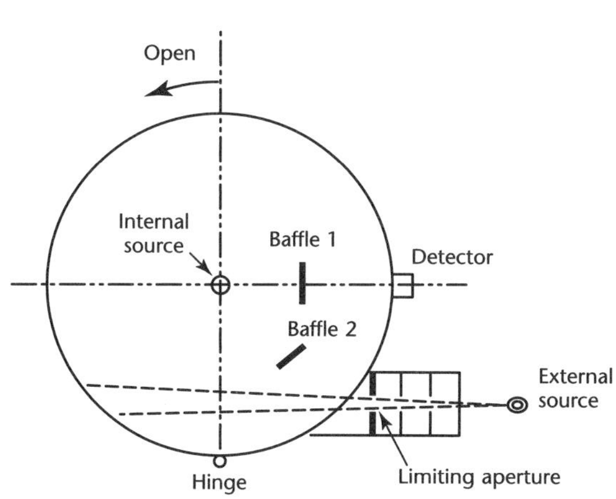

Some illuminance meter heads have a dome shaped or mesa shaped diffuser in the front. In this case, it is usually difficult to define the correct reference plane. Such illuminance meters should not be used as standard photometers unless they are always used at the same distance from the source. Also, illuminance meters with poor spectral match or with only a three-digit display are not adequate as standard photometers. Illuminance meters may have various structures of the light receiving surface for cosine correction. The reference plane of an illuminance meter head can be used during calibration. The reference plane of photometer heads (without an aperture or a diffuser) and illuminance meter heads can be experimentally determined.

Standard Photometers

At the National Institute of Standards and Technology, the candela is realized and maintained on a group of standard photometers (referred to as the *NIST standard photometers*) calibrated for illuminance responsivity in amperes per lux ($A \cdot lx^{-1}$). A standard photometer consists basically of a silicon photodiode, a $V(\lambda)$ correction filter, and a precision aperture (Fig. 15).

These standard photometers also embody the NIST illuminance unit and allow luminous intensity to be determined from measured illuminance and distance. The realization and maintenance of the photometric units at NIST are shown in Fig. 18. The NIST cryogenic radiometer serves as the starting point at the top of the chain.

The illuminance responsivity ($A \cdot lx^{-1}$) of each photometer is calculated from the spectral responsivity, the aperture area and other correction factors. The standard photometers are recalibrated annually by using the detector spectral responsivity scale. The details of the candela realization are described elsewhere.[8] Table 4 provides a list of the photometric calibration services currently available at NIST. The complete description of these services is reported in a user's guide.[16] The guide covers details of optical radiation, of calibration procedures and of measurement ranges and uncertainties. NIST issues calibrated artifacts to customers and calibrates artifacts submitted by customers.

Related Instruments[1,2]

Reflectometers and Gloss Meters

Reflectometers are instruments used to measure reflectance of materials or surfaces in specialized ways. The reflectometer measures diffuse, specular and total reflectance. Those instruments designed to determine specular reflectance are known as *gloss meters*.

One traditional reflectometer used a collimated beam. The beam source and receiving cell were mounted in a fixed relationship in the same housing. The housing had an aperture through which the beam travels. This head (aperture) was set on a standard reflectance reference with the aperture against the standard. The head was placed on the test surface and the reading is recorded.

Two cautions are recommended with reflectometers. The reference standard should be in the range of the value expected for the surface to be measured. Also, if the area to be considered is large, several measurements should be taken and averaged to obtain a representative value.

Integrating Sphere Photometer

The total luminous flux from a source can be measured by an integrating sphere.[1-2,6,17-20] Other geometric shapes have also been used. The theory of the integrating sphere assumes an empty sphere whose inner surface is perfectly diffusing and of uniform nonselective reflectance. Every point on the inner surface reflects to every other point and the illuminance at any point is made up of two components: the flux coming directly from the source and that reflected from other parts of the sphere wall. With these assumptions, it follows that, for any part of the wall, the illuminance and the luminance from reflected light is proportional only to the total flux from the source, regardless of its distribution.

The luminance of a small area of the wall or the luminance of the outer surface of a diffusely transmitting window in the wall, carefully screened from direct light from the source but receiving light from other parts of the sphere, is therefore a relative measurement of the flux output of the source. The integrating sphere is not perfect: its interior contains a light source, its supports, electric wires, an aperture and a baffle between the light

FIGURE 18. Flow chart of photometer calibration.[8]

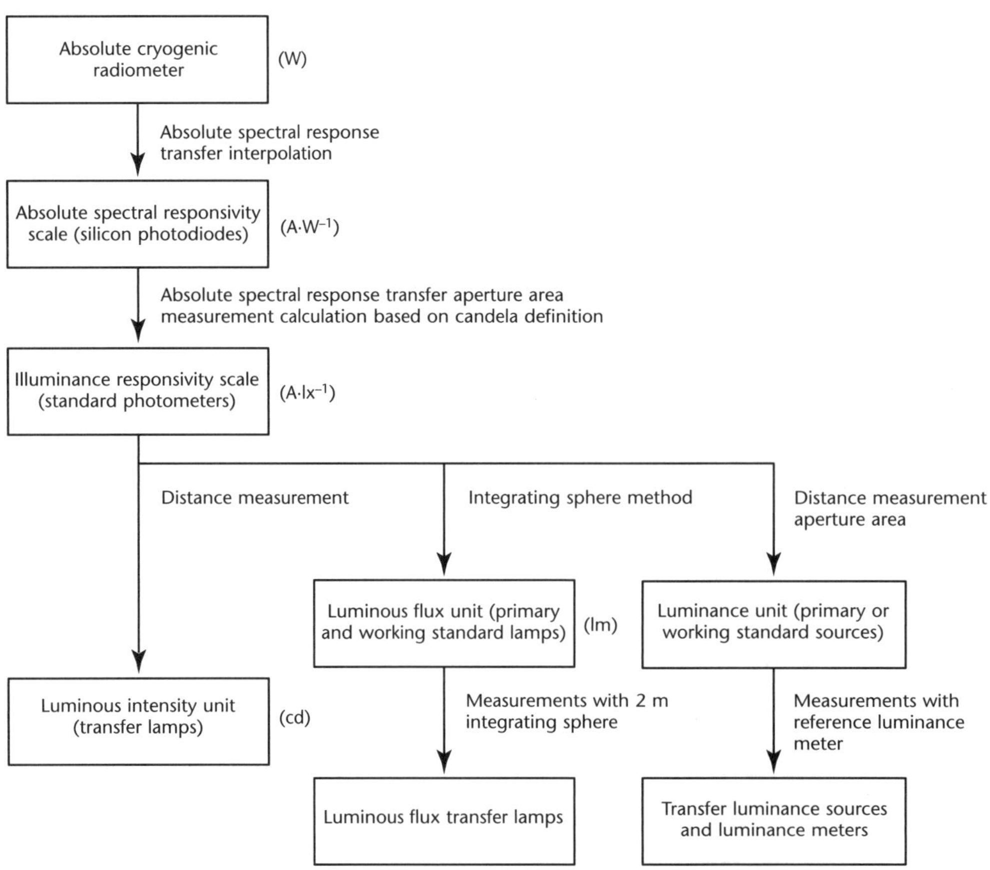

sources and the aperture. The various elements entering into the considerations of a sphere, as an integrator, make it difficult to use a sphere for absolute measurement of flux without correction factors.

Equivalent Sphere Illumination Photometers

Equivalent sphere illumination may be used to evaluate lighting systems. The equivalent sphere illumination of a visual test object at a specific location in a room illuminated with a specific lighting system is defined as that level of perfectly diffuse (spherical) illuminance that makes the test object as visible in the equivalent sphere as it is in the real lighting environment.

TABLE 4. Light measuring instruments calibrated by National Institute of Standards and Technology (NIST). This list is subject to agency revision. Restrictions may apply.

Devices Issued by NIST

- 1000 W tungsten quartz halogen lamp in two-post base
- 30 W deuterium arc lamp in two-post base
- Color temperature standard lamps
- Luminous intensity standard lamps
- Opal glass luminance coefficient standards
- Spectral radiance standard, integrating sphere source
- Spectral radiance standard, tungsten strip lamp
- Standard reference photometer

Devices accepted for test or calibration

- Color measuring instruments for displays
- Flashing light photometers
- Illuminance meters
- Incandescent and florescent lamps, for total luminous flux, luminous Intensity or color temperature
- Lamps, for color temperature
- Light emitting diodes, for luminous intensity and luminous flux
- Luminance meters
- Luminance sources and transmitting diffusers
- Material, for spectral reflectance
- Material, for spectral transmittance
- Material, for specular gloss
- Material, for surface color (illuminated at normal and measured at 45 degrees to surface)
- Photodetectors, for spatial uniformity of responsivity
- Photometers
- Spectral transmittance filters (carbon yellow glass)
- Spectral transmittance filters (cobalt blue glass)
- Spectral transmittance filters (copper green glass)
- Spectral transmittance filters (selenium orange glass)

Radiometers

Radiometers are used to measure radiant power over a wide range of wavelengths, including the ultraviolet, visible or infrared spectral regions.

The overall response of such detectors can be modified by using appropriate filters to approximate some desired function.

Spectrophotometers[1-3]

Spectrophotometers[21] and spectroradiometers include two main components.

1. A monochromator separates or disperses the wavelengths of the spectrum using a prism (diffraction grating) to disperse the luminous flux into a spectrum for analysis. Any given wavelength of light is isolated by an exit slit. Monochromators often use optical glass for the visible spectrum. Measurement in the ultraviolet or infrared spectrum requires specialized quartz optical components.
2. A receptor measures the power contained within a certain wavelength range of the dispersed light.

When a spectroradiometer measures the spectral power distribution of a radiant energy source, the radiation enters the entrance window (entrance slit) and is diffracted by the diffraction grating. The exit and focusing lens are positioned to isolate the desired wavelength. This wavelength passes through the exit slit and is measured by a photosensitive device.

Spectrophotometers measure the light reflected from the test surface. The concept and operation is very similar to that of a spectroradiometer. The spectrophotometer compares the reflected or transmitted light to the incident light. Many spectrophotometers have a built-in light source. The values measured by any photometer depend on the quality of the incident light.

Goniophotometer

A goniometer is an instrument designed to measure an angle precisely — for example, the angle of a cutting blade, the surface tension on a drop of liquid or the range of motion of an artificial limb. Goniometers are used to triangulate the sources of radar or radio signals. If the goniometer is also a photometric instrument, it is called a *goniophotometer* and measures precisely the direction of light — that is, its angles of propagation and incidence.[22]

References

1. Section 2, "The Physics of Light." *Nondestructive Testing Handbook,* second edition: Vol. 8, *Visual and Optical Testing.* Columbus, OH: American Society for Nondestructive Testing (1993): p 29-50.
2. *IES Lighting Handbook: Reference Volume.* New York, NY: Illuminating Engineering Society of North America (1981). Superseded by *IESNA Lighting Handbook: Reference and Application,* ninth edition. New York, NY: Illuminating Engineering Society of North America (2000).
3. Sayler, G.[C.] *ASNT Level III Study Guide: Visual and Optical Testing Method.* Columbus, OH: American Society for Nondestructive Testing (1998, revised 2006).
4. Born, M. and E. Wolf. *Principles of Optics: Electromagnetic Theory of Propagation, Interference and Diffraction of Light,* seventh edition. Cambridge, United Kingdom: Cambridge University (1999).
5. Maxwell, J.C. *A Treatise on Electricity and Magnetism,* third edition (1891). New York, NY: Dover (1960).
6. Walsh, J.W.T. *Photometry,* third edition. London, England: Constable (1958).
7. "General Guide to Photometry." *Illuminating Engineering.* Vol. 50. New York, NY: Illuminating Engineering Society of North America (March 1955): p 147.
8. Ohno, Y. NIST SP 250-37, *Photometric Calibrations.* Gaithersburg, MD: National Institute of Standards and Technology (1997).
9. Forsythe, W.E. *Measurement of Radiant Energy.* New York, NY: McGraw-Hill (1937).
10. CIE 81, *Mesopic Photometry: History, Special Problems, and Practical Solutions* (1989). Vienna, Austria: Commission Internationale de l'éclairage [International Commission on Illumination] (CIE) (1989).
11. IESNA TM-12, *Spectral Effects of Lighting on Visual Performance at Mesopic Light Levels.* New York, NY: Illuminating Engineering Society of North America (2006).
12. Wright, W.D. *Photometry and the Eye.* London, United Kingdom: Hatton Press (1949): p 31, 80, 123, 124.
13. CIE 18.2, *The Basis of Physical Photometry,* second edition. Vienna, Austria: Commission Internationale de l'éclairage [International Commission on Illumination] (CIE) (1983).
14. ISO 31, *Quantities and Units.* Geneva, Switzerland: International Organization for Standardization (2005).
15. Appendix. *IES Lighting Handbook,* eighth edition. Washington, DC: Illuminating Engineering Society of North America (1993): p 946-949.
16. NIST Special Publication 250, *NIST Calibration Program: Calibration Services Users Guide.* Gaithersburg, MD: National Institute of Standards and Technology (2009).
17. Rosa, E.B. and A.H. Taylor. Paper 447, "Theory Construction and Use of the Photometric Integrating Sphere." *Bulletin of the NBS.* Washington, DC: National Bureau of Standards [National Institute of Standards and Technology] (September 1921).
18. Buckley, H. "The Effect of Non-Uniform Reflectance of the Interior Surface of Spherical Photometric Integrators." *Transactions of the IES.* Vol. 41. London, United Kingdom: Illuminating Engineering Society (July 1946): p 167.
19. Hardy, A.C. and O.W. Pineo. "The Errors Due to the Finite Size of Holes and Sample in Integrating Spheres." *Journal of the Optical Society of America.* Vol. 21. Washington, DC: Optical Society of America (August 1931): p 502.
20. IESNA LM-78, *Approved Method for Total Luminous Flux Measurement of Lamps Using an Integrating Sphere Photometer.* New York, NY: Illuminating Engineering Society of North America (2007).
21. Larason, T.C. and J.M. Houston. NIST SP 250-41, *Spectroradiometric Detector Measurements: Ultraviolet, Visible, and Near-Infrared Detectors for Spectral Power.* Gaithersburg, MD: National Institute of Standards and Technology (2008).
22. IESNA LM-75, *Goniophotometer Types and Photometric Coordinates.* New York, NY: Illuminating Engineering Society of North America (2001).

Bibliography

Bureau of Naval Personnel. *Basic Optics and Optical Instruments*. New York, NY: Dover (1969).

CIE 17.4, *International Lighting Vocabulary*. Vienna, Austria: Commission Internationale de l'éclairage [International Commission on Illumination] (CIE) (1987).

CIE 81, *Mesopic Photometry: History, Special Problems and Practical Solutions*. Vienna, Austria: Commission Internationale de l'Éclairage [International Commission on Illumination] (CIE) (1989).

Fowles, G.R. *Introduction to Modern Optics,* second edition. New York, NY: Dover (1989).

Hecht, E. *Optics,* second edition. Reading, MA: Addison Wesley (2001).

Hecht, E. *Schaum's Outline of Theory and Problems of Optics*. New York, NY: McGraw-Hill (2007).

IESNA Lighting Handbook: Reference and Application, ninth edition. New York, NY: Illuminating Engineering Society of North America (2000).

Johnson, B.K. *Optics and Optical Instruments: An Introduction,* third edition. New York, NY: Dover (1960).

Smith, W.J. *Modern Optical Engineering,* fourth edition. New York, NY: McGraw-Hill (2007).

Calibration

IESNA LM-45, *Electrical and Photometric Measurements of General Service Incandescent Filament Lamps*. New York, NY: Illuminating Engineering Society of North America (2000).

IESNA LM-46, *Photometric Testing of Indoor Luminaires Using High Intensity Discharge or Incandescent Filament Lamps*. New York, NY: Illuminating Engineering Society of North America (2004).

IESNA LM-51, *Electrical and Photometric Measurements of HID Lamps*. New York, NY: Illuminating Engineering Society of North America (2000).

IESNA LM-59, *Electrical and Photometric Measurement of Low Pressure Sodium Lamps*. New York, NY: Illuminating Engineering Society of North America (2000).

IESNA LM-115, *Guide for Reporting General Lighting Equipment Engineering Data for Indoor Luminaires*. New York, NY: Illuminating Engineering Society of North America (2003).

Ohno, Y. and Yuqin Zong. "Detector-Based Integrating Sphere Photometry." *Proceedings, 24th Session of the CIE* [Warsaw, Poland, 1999]. Vol. 1-1. Wien, Österreich [Vienna, Austria]: Commission Internationale de l'Éclairage [International Commission on Illumination] (CIE) (1999); p 155-160.

Color

CIE Report 15-2004, *Colorimetry,* third edition. Wien, Österreich [Vienna, Austria]: Commission Internationale de l'Éclairage [International Commission on Illumination] (CIE) (2004).

CIE S 014-2; ISO 1164-2, *CIE Standard Illuminants for Colorimetry*. Wien, Österreich [Vienna, Austria]: Commission Internationale de l'Éclairage [International Commission on Illumination (CIE)] (2006).

Schanda, J. *Colorimetry: Understanding the CIE System*. New York, NY: Wiley (2007).

Measurement Units

IEEE/ASTM SI 10, *Standard for Use of the International System of Units (SI): The Modern Metric System*. New York, NY: IEEE (2002).

Taylor, B.N. and A. Thompson. NIST SP 330, *The International System of Units (SI)*. Washington, DC: United States Government Printing Office (2008).

Taylor, B.N. and C.E. Kuyatt. NIST TN 1297, *Guidelines for Evaluating and Expressing the Uncertainty of NIST Measurement Results*. Gaithersburg, MD: National Institute of Standards and Technology (1994).

Thompson, A. and B.N. Taylor. NIST SP 811, *Guide for the Use of the International System of Units (SI)*. Washington, DC: United States Government Printing Office (2008).

3

CHAPTER

Vision Acuity for Nondestructive Testing

Gregory W. Good, College of Optometry, Ohio State University, Columbus, Ohio

PART 1. Vision

Visual Abilities

The identification of objects across the field of vision requires several different elements of discrimination. In visual testing tasks, vision discriminates primarily three elements: space, contrast and color.

1. Spatial discrimination enables the visual system to resolve fine details in the visual scene. It is typically expressed in terms of vision acuity. This requires both a sharply focused retinal image and a sufficient retinal grain to resolve the resulting optical image.
2. Luminous contrast sensitivity refers to being able to identify differences in the intensity of light for adjacent scene elements. Optical aberrations can greatly affect this visual attribute for high spatial frequency targets (images with fine critical detail). For large targets with low spatial frequency, neural processing across the retina limits resolution.
3. For color recognition, normal discrimination requires that normal photopigments are present within the retinal photoreceptors (cones and rods) and that the visual processing neurons transmit an image of high fidelity to the higher visual centers within the brain.

Visual Data Collection[1]

Human visual processing occurs in two steps. First the entire field of vision is processed in an automatic function of the brain, sometimes called *preattentive processing*. Secondly, focus is localized to a specific object in the processed field. Segregating specific items from the general field is the foundation of identification. In this process, various light patterns reaching the eyes are simplified and encoded, as lines, spots, edges, shadows, colors, orientations and locations in the field of view.

The first step in identification is the comparison of visual data with the long term memory of previously collected data. Some researchers have suggested that this comparison procedure is a physiological cause of deja vu, the uncanny feeling of having seen something before.[2]

The accumulated data are then processed through a series of specific systems. Certain of our light sensors receive and respond only to certain stimuli and transmit their data to particular areas of the brain for translation. One kind of sensor accepts data on lines and edges; other sensors process only directions of movement or color. Processing of these data discriminates complex views by analyzing their components.[3]

Experiments show that these areas of sensitivity have a kind of persistence. This persistence can be illustrated by staring at a lit candle, then diverting the eyes to a blank wall. For a second or two, the image of the candle is retained. The same persistence occurs with motion detection and can be illustrated by staring at a moving object, such as a waterfall, then at a stationary object like the river bank. The bank will seem to flow because the visual memory of motion is still present.

Differentiation in Field of View[1]

Boundary and edge detection can be illustrated by the pattern changes in Fig. 1. When scanning the figure from left to right, the block of reversed *L*s is difficult to separate from the upright *T*s in the center but the boundary between the normal *T*s and the tilted *T*s is apparent. The difficulty in differentiation occurs because horizontal and vertical lines comprise the *L* and upright *T* groups, creating a similarity that the brain momentarily retains as the eye moves

FIGURE 1. Pattern changes facilitate boundary detection.[1]

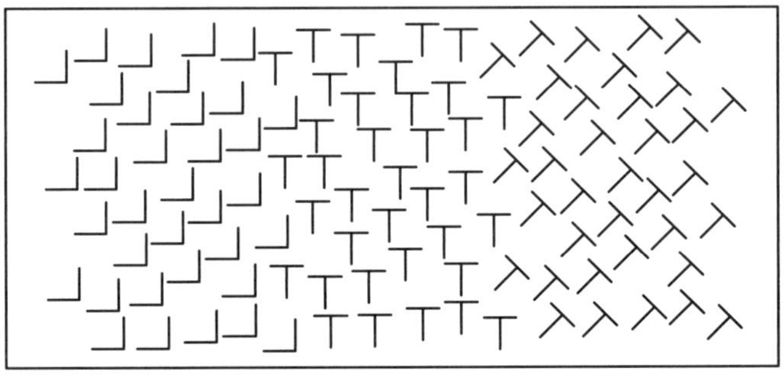

from one group to the other. On the other hand, the tilted *T*s share no edge orientations with the upright *T*s, making them stand out in the figure.

Differentiation of colors is more difficult when the different colors are in similarly shaped objects in a pattern. The recognition of geometric similarities tends to overpower the difference in colors, even when colors are of interest. Also, in a group of different shapes of unlike colors, where no one form is dominant, a particular form may hide within the varied field of view. However, if the particular form contains a major color variance, it is very apparent. Experiments have shown that such an object may be detected with as much ease from a field of thirty as it is from a field of three.[4]

Searching in Field of View[1]

The obstacles to differentiation discussed above indicate that similar objects are difficult to identify individually. During preattentive processing, particular objects that share common properties such as length, width, thickness or orientation are not different enough to stand out. If the differences between a target object and the general field is dramatic, then a visual inspector requires little knowledge of what is to be identified. When the target object is similar to the general field, the inspector needs more specific detail about the target. In addition, the time required to detect a target increases linearly with the number of similar objects in its general field.

When an unspecified target is being sought, the entire field must be scrutinized. If the target is known, it has been shown statistically that only about half of the field must be searched.

The differences between a search for simple features and a search for conjunctions or combinations of features have implications in nondestructive testing. For example, visual inspectors may be required to take more time to check a manufactured component when the possible errors in manufacturing are characterized by combinations of undesired properties. Less time could be taken for a visual test if manufacturing errors always changed a single property.[5]

Another aspect of searching the field of view addresses the absence of features. The presence of a feature is easier to locate than its absence. For example, if a single letter *O* is introduced to a field of many *Q*s, it is more difficult to detect than a single *Q* in a field of *O*s. The same difficulty is apparent when searching for an open *O* in a field of closed *O*s. In this case statistics show that the apparent similarity in the target objects is greater and even more search time is necessary.

Experimentation in the area of visual search tasks encompasses several tests of many individuals. Such experiments start with studies of features that should stand out readily, displaying basic elements of early vision recognition. The experiments cover several categories, including quantitative properties such as length or number. Also included are search tasks concentrating on single lines, orientation, curves, simple forms and ratios of sizes. All these tests verify that visual systems respond more favorably to targets that have something added (*Q* versus *O*) rather than something missing.

In addition, it has been determined that the ability to distinguish differences in intensity becomes more acute with a decreasing field intensity. This is the basis of Weber's law. The features it addresses are those involved in the early visual processes: color, size, contrast, orientation, curvature, lines, borders, movement and stereoscopic depth.

Fluorescent Materials[1]

Fluorescence is the absorption of light at one wavelength and reemission of this light at another wavelength. Fluorescence is a complex phenomenon that occurs in gases, liquids and solids. For the purpose of visual nondestructive testing, fluorescence is used with long wave ultraviolet radiation as an excitation source.

Studies show that the intensity of fluorescence in most situations is directly proportional to the intensity of the ultraviolet radiation that excites it. Care must be taken when using short wave or wide bandwidth ultraviolet sources. A safety habit is to hold the lamp so the light is directed away from you. Long wave ultraviolet is generally safe, but individuals should use adequate protection if they are photosensitive or subjected to long exposure times.

Fluorescence and ultraviolet radiation are discussed in this volume's chapter on physics and in the discussion of safety in the book's introductory chapter.

Components of the Human Eye

The components of the human eye (Fig. 2) are often compared to those of a camera. The front window (the cornea) and the interior crystalline lens are the primary focusing elements of the eye. These two elements together work like the lens of a camera.

The cornea is the highly curved front surface that provides most of the focusing power for the eye. Optical, refractive power is measured in diopters. A diopter is the reciprocal of the focal length in meters: for example, a +4 diopter lens has a focal distance of 0.25 m, and +2 diopter lens has a focal distance of 0.50 m. The cornea has a refractive power of about 43 diopters.

The crystalline lens is positioned just behind the iris and adds additional refractive power. When relaxed, it has a refractive power of approximately 19 diopters. When added together, taking into account their relative positions in the eye, the total refracting power of an average eye is +58.64 diopters.[6] As an aside, the crystalline lens has an elastic capsule that allows it to change shape, increasing its refracting power. This elasticity allows the eye to change its point of focus for better retinal imaging, depending on the distance from the eye to the viewed object. The point at the center of the lens is the nodal point, the vertex of the visual angle.

According to the normal eye's design, when the internal ocular musculature is at rest, distant targets are sharply focused onto the retina. The eye can increase its refracting power, however, by constriction of the ciliary muscle inside the eye.[7] Constriction of the ciliary muscle allows the flexible crystalline lens to fatten and become more convex, increasing the focusing power of the eye. With age, however, the crystalline lens naturally loses flexibility, and hence focusing power. There is a predictable decrease in focus ability such that most people require reading glasses by the age of 45 years. The overall focusing power of the eye at different ages is shown in Table 1.[8] The inverse of the maximum focusing power of the eye indicates the shortest distance at which the eyes can focus clearly. The loss of sufficient focusing power such that reading glasses are required is termed *presbyopia.* By age 50, most eyes cannot focus comfortably on their own to an object closer than 500 mm (20 in.).

FIGURE 2. Cross section of the eye (horizontal section of right eye from above).

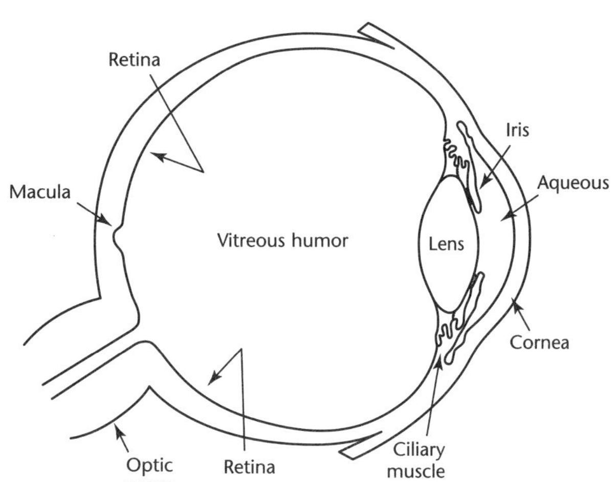

The iris is the colored ring around the pupil, in the front of the eye. It contains a sphincter muscle to constrict and a dilator muscle to dilate the pupil. This constriction and dilation corresponds to the action of the aperture of a camera. The iris is heavily pigmented across its interior surface to block light transmission to the retina except for that which passes through the clear pupil. In bright sunshine, the iris sphincter constricts to create a very small pupil. In low light conditions, the sphincter relaxes whereas the dilator muscle increases the light reaching the retina.

The retina of the eye is a very thin neural structure that lies against the inner surface of the eye. It is analogous to the film of a camera, receiving and documenting an image of the world. Unlike the film of a camera, however, the retina does not have a uniform grain across its surface. Whereas a photograph is uniformly clear across the image, the retina has its best resolving power only in the small, central area of the macula called the *fovea centralis.*[9] Thus, six extraocular muscles are required to rotate the eyes to align the fovea centralis of each eye onto the proper section of the retinal image.[10]

The rods and cones are the light receptor elements of the retina. These cells absorb the light energy from the retinal image to begin the visual process. The cones are contained in the macula, are responsible for central vision, function best in bright light and enable color vision. The rods are spread across the peripheral retina and are responsible for peripheral and night vision.

TABLE 1. Focusing power *D* (in diopters) of eye versus age.[8]

Age (years)	Minimum[a] (diopters)	Average[b] (diopters)	Maximum[c] (diopters)
10	12.5	15.5	21.0
20	10.0	12.5	17.0
30	7.5	9.5	13.0
40	5.0	6.5	9.0
50	2.5	3.5	5.0
60	0.0	0.5	1.0
70	0.0	0.0	0.0

a. Minimum power $D = 15.0 - 0.25 \times \text{age}$ and $D \geq 0$
b. Average power $D = 18.5 - 0.30 \times \text{age}$ and $D \geq 0$
c. Maximum power $D = 25.0 - 0.4 \times \text{age}$ and $D \geq 0$

Signals from the rods and cones are transmitted to bipolar cells before transmission to the retinal ganglion cells. Axons of the retinal ganglion cells comprise the optic nerve and transmit visual signals to the brain. It is interesting to note that light first passes through retinal cells and elements and around retinal blood vessels before forming the retinal image at the layer of the rods and cones. Vision is improved centrally, however, when the inner retinal layers are pulled aside at the fovea centralis to allow straight passage directly to the cones in the central retina.

To focus an image sharply on the retina, the refracting power of the eye must be precisely tuned to the overall eye length.[11] If the eye power is too strong, or if the eye is too long, the optical focal point falls short of the retina. Thus, the image is poorly focused on the retina when viewing a distant object. This eye is called *myopic;* a person with this condition is called *nearsighted* and experiences blurred far vision. If the refracting power is too weak, or if the eye is too short, the best optical focus is behind the retina. An eye with this condition is called *hyperopic;* a person with this condition is called *farsighted* and experiences blurred near vision. In these cases, lenses can be prescribed either in spectacles or contact lenses to modify the overall refracting power to get a sharply focused image of the world onto the retina. Additionally, if the shape of the cornea or crystalline lens surface is toric (flat on one side) and not spherical, a clear image is not found at any location. In this case, astigmatism is present. To correct for astigmatism, the correcting lenses must be toric in an opposite orientation.

Additional information about the eye is available elsewhere.[1,9-13]

Part 2. Vision Acuity

Object Discrimination

Vision acuity represents a measure of the ability of the eye to resolve fine detail. It is usually measured by determining the size of the smallest letter that can be correctly identified. The letter is to be of high contrast, a black letter on a white background. With proper correction, a normal eye should be able to identify a letter that subtends an angle to the eye of five minutes of arc or less.

Figure 3 illustrates the relationship of visual angle to object size. An object when moved closer to the viewer subtends a larger visual angle, occupying a greater part of the field of view (Fig. 3a). The corollary is that another object subtending the same visual angle must, if closer, also be smaller (Fig. 3b). (To give an example of the size of visual angle, both the sun and the moon subtend an angle of approximately 0.5 degrees to observers on earth.)

Figure 4 illustrates the designs of letters typically used for vision acuity testing. In the United States vision acuity letters are typically designed using a 5 × 5 grid.[13] That is, the letters are 5 units high by 5 units wide (Fig. 4a). The letter strokes are then one unit wide. Critical detail (such as the gap in the ring of the letter C or the separation between the legs on the letter E) is also represented by one unit. Using this design (5 unit × 5 unit with 1 unit critical detail), a letter subtending five minutes of arc would have one minute of arc critical detail. An eye that can resolve a small letter with only 1 minute of arc critical detail is said to have *20/20* vision acuity.

The snellen fraction (such as 20/20 or 20/40) is routinely used to express vision acuity. The numerator expresses the testing distance at which an individual test subject can distinguish two adjacent objects; often 6 m or 20 ft is used as this testing distance. The denominator is the actual distance at which the smallest letter would subtend an angle of 5 minutes of arc — a distance discernable by a standard subject (that is, someone with 6/6 [20/20] vision). When someone has 6/12 (20/40) vision, he or she must be at 6 m (20 ft) or closer to see the same detail that a normally sighted person sees at 12 m (40 ft).[14]

The familiar snellen chart uses snellen letters (Fig. 5).

Figure 3. Vision acuity letter and test distance: (a) letter of given size moved to one third of original test distance subtends angle three times larger and therefore represents vision acuity three times worse (for example, 6/6 at 6 m but 2/6 = 6/18 at 2 m [20/20 at 20 ft but 6.67/20 = 20/60 at 6.67 ft]); (b) small, near object and large, distant object subtend same visual angle, so nearer letter must be smaller to subtend same angle to the eye as larger, more distant letter.

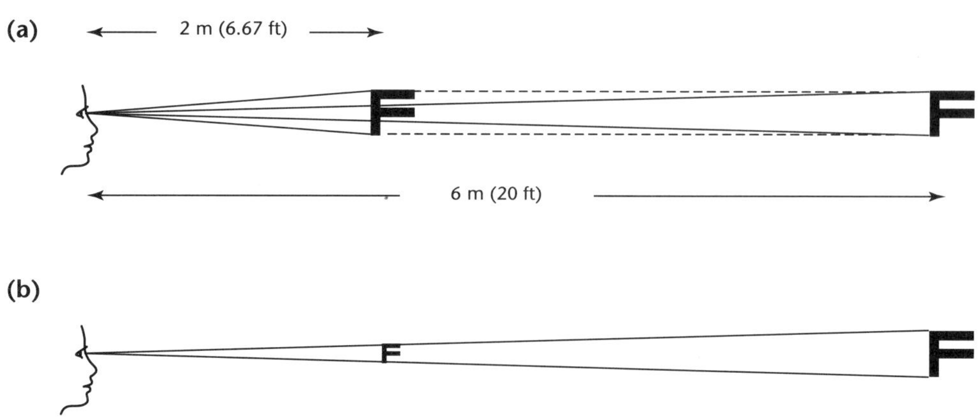

Angles Subtended by Snellen Letters

For example, at 6 m (20 ft) distance, a letter 25 mm (1 in.) tall would subtend an angle of 14.4 minutes of arc to a test subject, where 60 min = 1 deg. The calculations are shown below:

$$\text{(1)} \quad \text{Letter size} = 25 \text{ mm} = 0.0254 \text{ m}$$

$$= 1 \text{ in.} = 0.0833 \text{ ft}$$

$$\text{(2)} \quad \text{Test distance} = 6096 \text{ mm} = 6 \text{ m}$$

$$= 240 \text{ in.} = 20 \text{ ft}$$

$$\text{(3)} \quad \tan \alpha = \frac{0.0254 \text{ m}}{6 \text{ m}} = \frac{0.0833 \text{ ft}}{20 \text{ ft}}$$

$$\text{(4)} \quad \alpha = 0.24 \text{ deg} = 14.4 \text{ min}$$

$$\text{(5)} \quad \frac{14.4 \text{ min}}{5 \text{ min}} = 2.88$$

Therefore, the letter read is 2.88 times larger than a 6/6 (20/20) letter, which subtends an angle of five minutes of arc. If this is the smallest letter which could be read at this distance, vision acuity is worse by 2.88 times. The measured vision acuity would be as follows in the International System:

FIGURE 4. Vision acuity letter: 5 × 5 matrix, like that illustrated by Bailey.[4]

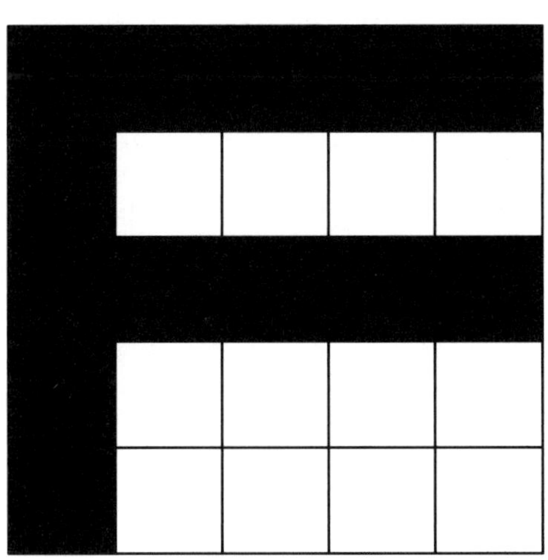

$$\text{(6)} \quad \frac{6}{6 \times 2.88} = \frac{6}{17.28} \cong \frac{6}{18}$$

or as follows in English units:

$$\text{(7)} \quad \frac{20}{20 \times 2.88} = \frac{20}{57.6} \cong \frac{20}{60}$$

These same calculations hold regardless of the viewing distance. That is, whether tested at 6 m (20 ft), 3 m (10 ft) or 400 mm (16 in.), a 6/6 (20/20) letter subtends 5 minutes of arc to the eye. Of course, at 400 mm (16 in.), a 6/6 (20/20) must be very, very small. In fact, at

FIGURE 5. Example of snellen chart. Vision test images in this book are provided for educational purposes and must not be used for vision tests.

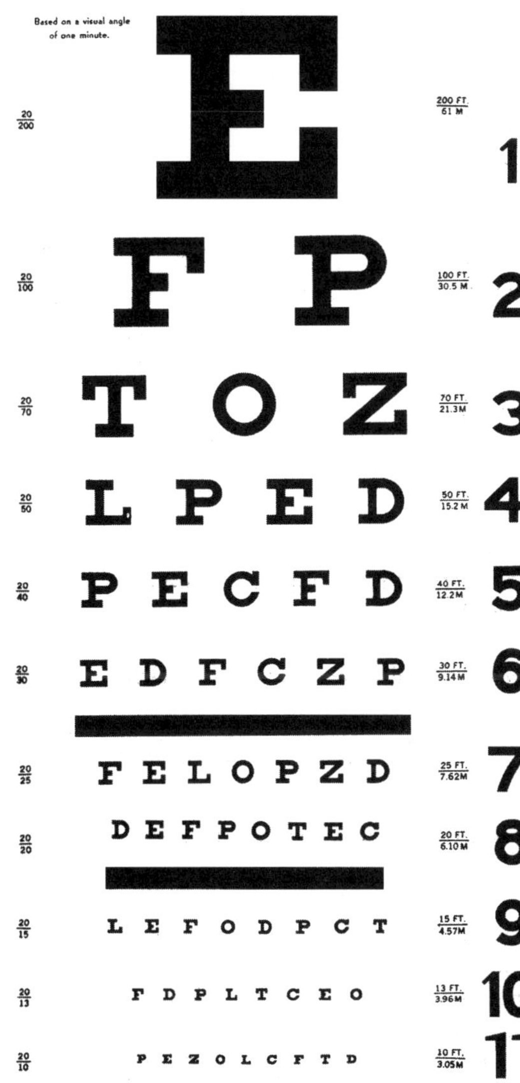

400 mm (16 in.), a 6/6 (20/20) letter is only 0.6 mm (0.023 in.) high (Fig. 4).

Corrective Lenses

An individual may have reduced vision acuity for many different reasons. The most frequent reason, however, is uncorrected refractive error. A nearsighted person needs to wear proper glasses, or distance vision acuity will be reduced. The retinal image will be blurred such that small letters cannot be identified from afar. For a farsighted individual, near vision acuity may be reduced without proper reading glasses. To determine if an inspector has adequate vision for a particular task, it is important that testing be accomplished (1) with the same glasses typically worn for that task and (2) at the distance appropriate for the task in question. When proper correcting glasses are worn, far and near vision acuity measures should agree. A vision acuity line difference of only 0.5 is observed[15] for subjects aged 25 to 77 years when far vision acuity was compared to near vision acuity. A difference in far vision acuity versus near vision acuity can be an important indicator that a refractive correction can make a significant difference in visual performance.

Additional reasons for reduced vision acuity include cloudy ocular media (such as a scarred cornea or a cataractous crystalline lens), glaucoma, or a scarred central retina from an intraocular infection or hemorrhage (from, for example, diabetic retinopathy or macular degeneration).

Contrast Sensitivity

Vision is conventionally graded using measures of vision acuity. Visual objects of high contrast (that is, black letters on a white background) are presented in decreasing sizes until visual detail is just barely resolved. There are various formulas and resulting scales to express contrast. One kind of contrast, called *weber contrast*, is often applied to single letters on a uniform background and is defined[16] as the difference in luminance (brightness) between the background luminance L_B and the target luminance L_D divided by the background luminance (L_B):

$$\text{(8)} \quad \text{Contrast} = \frac{L_B - L_D}{L_B}$$

For a typical vision acuity chart with black letters, contrast is greater than 85 percent. The measurement of vision acuity is, therefore, an angular measure of the smallest detail that can be resolved when presented with high contrast. Vision acuity measurements may not be good predictors, however, of functional vision performance on the job, for most objects in daily life are larger in size and often of lower contrast.[17] As an example, when viewing an X-ray image, a separation of metallic layers in a surface may be indicated only by a subtle difference in brightness of adjacent elements. The visual attributes required for identification of the separation would depend more on contrast sensitivity than on vision acuity. Also, sensitivity to contrast depends on the size of the object used for testing. Often grating targets (patterns of light and dark bars) of various sizes are used for testing contrast sensitivity. For these repetitive targets, michelson contrast[16] is often used for calculations. At each spatial frequency (defined as number of complete grating cycles per degree of visual angle) the contrast of the grating is modified to the threshold of visibility. By plotting the threshold contrast values at different spatial frequencies, a contrast sensitivity function is determined.

The contrast sensitivity of the normal visual system is highest with medium spatial frequencies of between 2 and 6 cycles per degree of visual angle.[16] The sensitivity to contrast then drops off for higher and lower spatial frequency targets. The high frequency drop appears to be a result of diffraction by the optical aberrations of the eye and by the neural gain of the retina. Sensitivity also drops at spatial frequencies below four cycles per degree. This drop is thought to be a function of the lateral inhibition mechanism of the visual system.[16]

There are several methods for measuring someone's contrast sensitivity.

1. The Pelli-Robson™ letter sensitivity chart presents letters of constant size but with decreasing contrast from 90 percent to less than one percent at a 1 m test distance.[18] The letter size and test distance were chosen to indicate the subject's overall sensitivity to contrast. Although this test will not provide a detailed assessment of contrast sensitivity to all target sizes, it is argued[18] that, if measuring high contrast vision acuity, sufficient additional information is provided by this test to determine if a subject's vision falls within the normal range.

2. The Mars™ letter contrast sensitivity test (Fig. 6) is similar in design to the Pelli-Robson™ letter sensitivity chart but calls for a normal reading distance. It also uses letters of varying contrast. Both the Pelli-Robson™ and Mars™ letter charts are easily understood by patients and require little testing time, and their repeated results appear to have excellent reliability.[19,20] The Mars™ chart, however, is hand held and is easily stored and carried for use at short notice.
3. The Vistech™ chart presents grating targets of five different spatial frequencies.[21] Subjects report the orientation of the gratings (tilted left, right, vertical, or target is blank) as contrast is slowly decreased. The results provide a measure of contrast sensitivity to various spatial frequency targets. It is reported,[22] however, that the Vistech™ chart has lower reliability (that is, repeatability from test to test) than other contrast sensitivity tests.

Color Discrimination

Human color vision is possible because of the three chromatic components of the color processing system in a normal visual system.[23] In bright light, the retinal cone cells are the predominant photoreceptors for vision. The perception of color is based on differences in wavelength responses among three types of retinal cone cells: short wavelength (blue/yellow), middle wavelength (green) and long wavelength (red).

Dichromatism

About 92 percent of males and over 99 percent of females have normal cone systems.[24] Of the 8 percent of males and 0.45 percent of females who have inherited abnormal components, color discrimination can vary from nearly normal to severely inferior. Deficient color discrimination is commonly and erroneously called *color blindness*. About two percent of males and 0.05 percent of women are dichromats, limited by a dichromatic (two-color) system. These individuals must judge all colors using only the system sensitive to short wavelengths and one of the two systems sensitive to longer wavelengths. Although these individuals are not totally color blind, their color discrimination is substantially reduced. Colors that are easily perceived as different by color vision normals can look identical and be confused by dichromats, so color naming errors are commonplace. Common color confusions for severe red/green defectives are brown with dark green, purple with navy blue, orange with yellow and blue/green with white (or gray).

These dichromats are different from red/green defectives: those with protanopia, whose long wavelength (red) sensitive cones are missing or defective, and those with deuteranopia, whose middle wavelength (green) sensitive cones are missing or defective.

Red/green color vision defects are inherited with sex linked recessive transmission. The controlling genes for the red and green photopigments are located on the X chromosome, so males have red/green color deficiencies much more often than females. Inherited blue/yellow defects involve cones sensitive to short wavelengths and are rare, affecting 0.003 percent of people. They are as likely to affect women as men (Table 2).

Anomalous Trichromatism

Six percent of males and 0.35 percent of women have all three component systems but with an abnormal sensitivity in one of the two longer wavelength sensitive components. The abnormal red/green photopigments have sensitivities shifted

FIGURE 6. Mars™ letter contrast sensitivity chart. Vision test images in this book are provided for educational purposes and must not be used for vision examinations. Contrasts have been altered in printing and are not accurate for vision examinations.

relative to normal so that there is less difference between the two photopigment ranges. These persons are called *anomalous trichromats* and can have widely varying color discrimination. Mild anomalous trichromats can perform most color discrimination tasks as well as those with normal color vision, whereas severe anomalous trichromats have very reduced color discrimination and often cannot be differentiated from dichromats.

Color Deficiencies: Inherited versus Acquired

Individuals with inherited color abnormalities have color discrimination deficiencies that are stable, predictable and easy to diagnose. Color discrimination is diminished similarly in both eyes, so tests often screen color vision with both eyes viewing. Except for color discrimination, individuals with inherited color vision deficiencies have normal vision functioning.

Individuals with ocular disease often show acquired color deficits.[24] Acquired deficits are less predictable in their effects on vision and typically progress in severity over time (Table 3). The two eyes can show very different effects, so each eye is tested separately. Very often, acquired defects initially show discrimination loss more typical of blue/yellow defects. Additionally, because these defects result from damage in the visual system, changes in vision acuity or other visual measures may accompany loss of color discrimination. Although damage to the vision may be confined to a single color channel (for example, the blue/yellow system) at first, later vision losses may be difficult to categorize.

Age Effect[12]

The lens does not transmit light of the shortest wavelengths and is largely responsible for the termination of response at the low end of the spectrum. As age increases, the lens yellows, increasing the absorption in the blue region and tending to increase the shortest wavelength that can be seen. This spectral variation is one factor in color differences between observers of different age, especially for tasks involving shorter wavelength perceptions.

Table 2. Comparative prevalence of color vision deficiencies for individuals of European descent. Noncaucasians show about half of percentages. Red/green deficiencies are inherited as sex linked recessive traits. Blue/yellow (tritanomalous) deficiencies are inherited as autosomal dominant traits.

Color Vision	Cones	Males (percent)	Females (percent)
Normal	normal	92	99.55
Anomalous deficiencies (abnormal cone)			
Protanomalous	long wavelength sensitive	1	0.02
Deuteranomalous	middle wavelength sensitive	5	0.40
Tritanomalous	short wavelength sensitive	unknown	unknown
Dichromatic deficiencies (missing cone)			
Protanopic	long wavelength sensitive	1	0.02
Deuteranopic	middle wavelength sensitive	1	0.01
Tritanopic	short wavelength sensitive	0.003	0.003

Table 3. Acquired color vision deficiencies from diseases affecting eyes.

Structure Affected	Pathology	Primary Axis of Confusion
Crystalline lens	aging cataracts	blue/yellow
Retina	age related macular degeneration	blue/yellow
	retinitis pigmentosa	blue/yellow
Optic nerve	glaucoma	blue/yellow
	Leber's hereditary optic atrophy	red/green
	optic neuritis	red/green
Systemic diseases	diabetic retinopathy	blue/yellow
	hypertensive retinopathy	blue/yellow

PART 3. Vision Testing

Vision Examination Frequency

Periodic vision examinations are required to ensure that workers and inspectors continue to have the visual skills required to perform their duties safely and efficiently. Simple changes in the focusing elements of the eyes can require a change in glasses. And eye disease can decrease visual functions so much that optimal vision is no longer possible even with an updated glasses prescription. Professional eyecare organizations recommend that adults of working age benefit from eye examinations every two years.[25] Optimal performance of visual inspectors may require vision examinations of greater frequency. Vision testing using a reexamination cycle of six or twelve months may be required, depending on the criticality of the inspection. Additionally, it is important that proper examination methodology document the vision of a worker. Improper examination techniques can let a worker with inadequate vision pass the screening.

Vision Acuity Standardization

In a typical vision screening examination, vision acuity is measured for a far (3 m [10 ft] or greater) and a near (0.5 m [20 in.] or closer) observation distance. In the twentieth century, standardized testing changed only slightly; however, scoring methodology changed significantly after 1994.

Chart Design

In 1994 the Committee on Vision of the National Research Council[26] recommended that vision acuity charts be designed such that there is general uniformity from one row of letters to the next (Fig. 7). That is, the only difference between rows should be the size of the optotypes (chart symbols) on the row. The spacing between optotypes at each level and the spacing between successive rows should be proportional to the size of the optotypes on the row in question. Also, all optotypes should have about the same difficulty of recognition and each row of optotypes should have the same number of characters, a design called *logMAR* (logarithm of minimum angle of resolution).[27,28] These charts have five letters in each row. The change in letter size from one row to the next is logarithmic so that each successive row on the chart is about 20 percent smaller — that is, about 80 percent of the preceding row in size. This design allows for a single chart to be used at any distance required by the geometry of the testing location.

Scoring Method

In their 1980 recommendations, the Committee on Vision of the National Research Council[29] recommended that vision acuity be defined as the smallest size at which seven of ten optotypes could be correctly identified. Because many

FIGURE 7. Bailey-Lovie logMAR vision acuity chart, with optotypes using 5 × 5 grid of Fig. 4. Vision test images in this book are provided for educational purposes and must not be used for vision examinations.

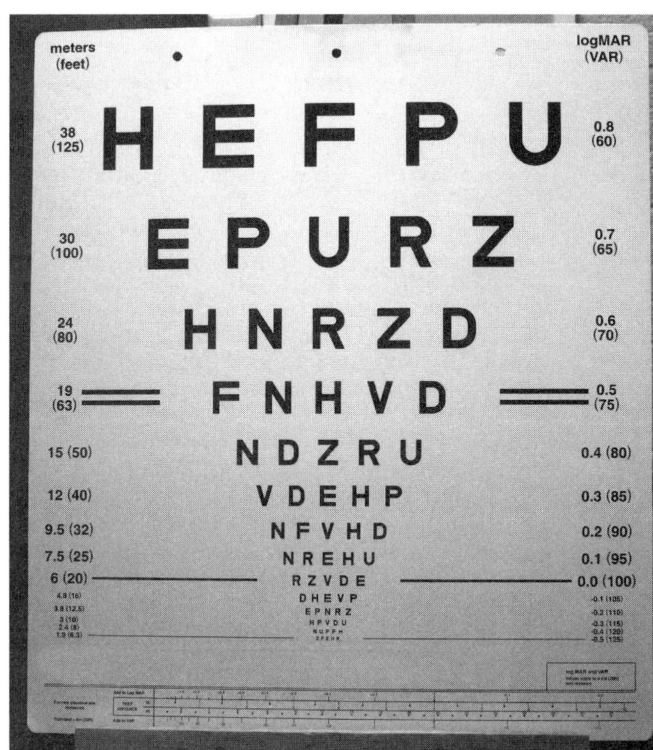

charts did not have ten optotypes in each row, an equivalent percentage of correct responses was often used (three of four letters or four of five letters) — at least 70 percent.

In 1994, the recommendation[26] for scoring method underwent a fundamental change. Research had shown that, for logarithmically designed charts with uniform letters per row, letter-by-letter scoring yielded greater accuracy in measurement and less variation in retests.[30] Thus, instead of giving credit for a given acuity score if 70 percent of the letters are read correctly, the new recommendation required that the equivalent of 100 percent of the letters in a given row must be correct. This means that if only one letter is missed in the 20/25 row but only one letter is read correctly in the next, smaller, 20/20 row, then the acuity is 20/25 (notated as $20/25^{-1+1}$, with test results in superscript).

The exact recommendations for testing methodology are as follows.

1. Read every letter in each row beginning with the largest letters.
2. Count every letter read correctly.
3. Move on to the next, smaller row if two or more letters (40 percent or more) are read correctly in a given row.
4. Encourage the subject to guess for each letter.
5. Stop when only one or zero letters are read correctly in a given row.
6. Letters correctly identified in a smaller row can *offset* letters misread in a larger row.

Chart Lighting

Vision acuity charts should be presented in high contrast at moderate light levels. A recommendation from the Committee on Vision is that chart luminance should ideally be set 160 $cd \cdot m^{-2}$, but in no case should the luminance be less than 80 $cd \cdot m^{-2}$. In this brightness range, a young adult worker would show on average an increase in acuity of one letter with every twofold increase in chart luminance. Older workers may show even greater acuity differences with light level. For a clean white chart to have a background luminance of 160 $cd \cdot m^{-2}$, the illumination onto the chart should be 600 lx. Fluorescent lighting of the charts is allowed; however, the extra yellow and red light from incandescent lamps often makes the reading more comfortable for the examinee.

When projected images are used, the parameters for the size of the characters, the background luminance and the contrast ratio are the same as those specified for charts. In no case should the contrast or illumination of the projected image be changed. A projection lamp of appropriate wattage should be used. When a projected chart image is used, room lighting is subdued. This is to eliminate possible changes in chart luminance or in the contrast ratio of the chart background and characters.

Many of the lighting conditions for vision acuity examinations can be met by using tabletop, professional examination units for vision screening. With one such piece of equipment, the examinee views slides under controlled, ideal light conditions.

General Methodology

The administration of a vision acuity examination does not necessarily require medical personnel, provided the administrator has been trained and qualified to standard and approved methods. In some instances, specifications may require medically approved personnel. In these cases, the administrator of the examination may be trained by medically approved personnel for this application. In no instance, however, should any of these administrators try to evaluate the examination results.

Vision acuity is generally measured in each eye individually before measuring with both eyes viewing. As vision acuity is generally superior with both eyes, workers are tested initially with each eye singly so that the smallest letters on the chart are less likely to be memorized with repeated use. When accurate measurement of acuity in each eye is required, testing with different charts using different letter sequences can help reduce the chance of memorization.

Near versus Far Vision Testing

Normal vision examinations typically include vision acuity testing at both far and near distances. Young workers who are well corrected for far vision should have equivalent vision acuity measures for the two test distances. If 6/6 (20/20) vision acuity is present at 3 m (10 ft) or greater, the same acuity should be present for near working distances. Workers under 40 years of age should have natural focusing ability sufficient to get a clear retinal image, and hence, good vision acuity at normal near working distances. If unequal vision acuity measures are found for far and near test distances, this is an indication that clear imagery is not present for at least one of the test distances and that some sort of diffractive anomaly or pathology is present.

The most common natural cause of poor near vision is presbyopia, the natural loss of focusing ability with age. Workers over the age of 45 years need reading glasses or bifocals to ensure clear vision for work inside 500 mm (20 in.). Additionally, it is important to note that reading glasses or bifocals do not provide clear vision at all near distances: depending upon the power of the reading glasses, vision may not be clear at the specific distance required for near inspection tasks. For example, a bifocal power of +2.50 diopters (a diopter is the reciprocal of the focusing distance in meters) provides clear vision at 400 mm (about 16 in.). If visual inspection is required at an intermediate distance (for example at arm's length, about 660 mm), the material may be blurred through both the top and lower segment of the bifocal glasses. Care must be taken to wear corrective eyewear based upon the working distances of a worker's inspection tasks to ensure vision is good at all required distances. As an example, the Federal Aviation Administration (FAA)[31] requires commercial pilots over the age of 50 to demonstrate good vision acuity at far distance, near distance (400 mm [16 in.]) and an intermediate distance of 800 mm (32 in.). These requirements ensure that pilots can read paperwork, charts, gages and controls of aircraft at near and intermediate distances.

Near point vision acuity testing is most often accomplished at a distance of 400 mm (16 in.): the distance from the plane of the eyeglasses (or the bridge of the nose) to the center of the reading material. Because vision acuity is a measure of the visual angle of the smallest letters read, it is imperative that the test distance be accurate for proper documentation.

Near vision testing for nondestructive testing personnel can be challenging because of nonstandardized wording of vision requirements set by different agencies overseeing nondestructive testing personnel training and qualification. A sample requirement is as follows.

> The NDT inspector shall have natural or corrected near distance acuity in at least one eye capable of reading the Jaeger Number 1 Test Chart or equivalent at a distance of not less than 30 cm (12 in.).[32]

The jaeger near point chart (Fig. 8) is widely used in the United States for performance examinations of near vision acuity. English language text is arranged into groups of gradually increasing size. Letter size is designated as J1, J2 and so forth, according to the font size of the text. The vision acuity equivalent of each classic jaeger designation depends upon the distance at which the chart is held. J1 corresponds to 6/6 (20/20) when held at 356 mm (14 in.) but corresponds to 6/7 (20/23) when held at 300 mm (12 in.) and 6/5 (20/17) when held at 400 mm (16 in.). The snellen equivalent acuity, for example 6/6 (20/20), may be specified on the jaeger test card; however, this specification applies only when the test chart is held at the distance for which the chart was designed. It has been recommended[25] that jaeger notation should not be used for standardized testing because jaeger letters on charts from one manufacturer may be twice as large as those from another.[33]

The height of letters on reading cards varies according to the reading distance specified on the individual card. There is usually a set of numbers down the center of a card, one for each jaeger size and one for each distance at which that size text is to be read. At that distance, the letters

FIGURE 8. Jaeger reading chart for performance examination of near vision acuity. Vision test images in this book are provided for educational purposes and must not be used for vision examinations.

READING CARD

15 J1

I walked up the street, gazing about, until near the market house I met a boy with bread. I had made many a meal on bread, and asked him where he got it. I then went to the baker's and asked for biscuit such as we had in Boston. I asked for a three peny loaf and was told that they had none such. Not knowing the difference of money and the greater cheapness, I bade him give me three penny worth of any sort. He gave me three great puffy rolls. I was Surprised at the quantity but took it, and walked off with a roll under each arm.

20 J2

Thus I went up Market Street as far as Fourth Street, passing by the house of Mr. Reed, my future wife's father. She, standing at the door, saw me and thought I made a most awkward appearance, as I certainly did. Then I turned and went down Chestnut Street and a part of Walnut Street, and found myself again at the wharf. Being filled with one of my rolls, I gave the other two to a woman and her child.

25 J3

By this time the street had many clean and well dressed people in it, all walking the same way. I joined them and was led into the great meeting house of the Quakers. I sat down among them and after looking around a while and hearing nothing said, I fell fast asleep. This was the first house I was in, or slept in, in Philadelphia.

30 J4

Looking in the faces of people, I met a young man whose countenance I liked, and asked if he would tell me where a stranger could get lodging. "Here", said he, "is one place that entertains strangers, but it is not a reputable house. If thee wilt walk with me, I will show thee a better." He brought me then to a place in Water Street, where I engaged a room and got dinner.

40 J7

While I was eating it several sly questions were asked me, as it seemed to be suspected from my youth and appearance that I might be some runaway. After dinner, my sleepiness returned, and being shown to a bed, I lay down without undressing and slept soundly till six in the evening.

50 J8

Our city, though laid out with beautiful regularity, the streets crossing each other at right angles, had the disgrace of allowing those streets to remain long unpaved. The wheels of heavy carriages plowed them into a quagmire.

60 J10

I saw the people wading in the mud while purchasing their provisions. A strip of ground down the middle of the market was at length paved with brick, so that they had firm footing.

The above letters subtend the visual angle of 5′ at the designated distance in inches.

will subtend the standard visual angle of 5 minutes of arc.

The visual angle is the angle formed by two lines drawn from the center of the eye lens to the top and the bottom of one of the tall letters as shown in Fig. 3. As you can see in Fig. 3b, both letters "F" subtend the same visual angle, but because one is twice as far away from the eye as the other, the letter height is different. This shows that if the distance from the eye is greater, the letter will be taller, even though both letters meet the 5 min visual angle requirement.

Some specifications state that near distance visual acuity shall be J1 or J2 at a distance of not less than 300 mm (12 in.). Some people believe that any reading card can be used at 300 mm (12 in.) for a J1 or J2 eye examination. This is not so. The subject must read the text at the distance specified on the card for the text to match the jaeger number shown for that text on that card.

The American Society for Nondestructive Testing has addressed near vision acuity testing distance in its recommended practice:

> The examination should ensure natural or corrected near-distance acuity in at least one eye such that the applicant is capable of reading a minimum of Jaeger Number 2 or equivalent type and size letter at the distance designated on the chart but not less than 12 inches (30.5 cm) on a standard Jaeger test chart.[34]

and in its central certification program document:

> The candidate shall [demonstrate near vision acuity] by reading a minimum of Jaeger J-1 or equivalent with one or both eyes, either corrected or uncorrected. The distance for this examination shall be dictated by the reading card being used but may not be less than 12 inches (30 cm).[35]

Here, the letters equivalent to jaeger are Times Roman near point and snellen letters, cited elsewhere.[14,35] It should be remembered that the jaeger test is not strictly a vision acuity test but rather a vision performance test.

Reduced snellen near point charts are also frequently used. These charts are designed like the distance vision acuity charts but are reduced in size so that the 20/*xx* designations are appropriate for a near examination distance. Typically, reduced snellen charts are calibrated for a 400 mm (16 in.) testing distance. The Federal Aviation Administration near point card (Fig. 9) provides two calibrations, one for a 400 mm (16 in.) testing distance and one for an 800 mm (32 in.) testing distance, as required by United States vision standards for commercial pilots. Additionally, two sets of letters are provided to help reduce unintentional memorization of the letters when testing the eyes individually.

Color Vision Testing

Clinical tests of color vision are designed to evaluate different aspects of color discrimination. For example, pseudoisochromatic plates are typically designed to determine if an individual has "normal" color vision. That is, does the individual possess the three, normal retinal receptors? Other tests (such as arrangement tests) are designed to determine color discrimination ability regardless of the normalcy status. It is a basic tenet of color vision testing that, when trying to predict the ability to perform a vocational color discrimination task, it is best to use a test that simulates the vocational task as closely as possible. Therefore, when determining a subject's overall color vision aptitude, a battery of tests is typically administered.

The testing of color discrimination requires standardized test conditions.

FIGURE 9. Near vision acuity card used by United States Federal Aviation Administration. Vision test images in this book are provided for educational purposes and must not be used for vision examinations.

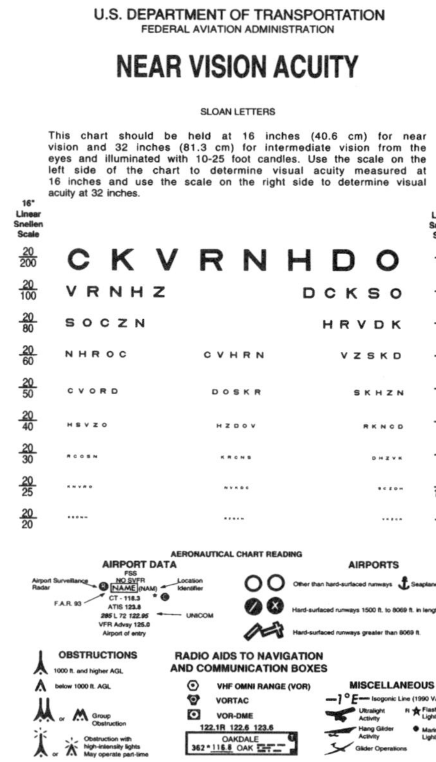

Color vision tests are designed with precise color and brightness characteristics. If test conditions are not controlled, test symbols may differ from their backgrounds in brightness as well as color. Thus, an inspector with defective color vision may erroneously pass the test not by having normal color discrimination but by having normal brightness discrimination. All testing should be accomplished using standard illuminant C, the International Commission on Illumination's standard light source for color vision testing.[36,37] This source simulates daylight: sunlight plus skylight. Incandescent lamps will have too much red light for testing and typical fluorescent bulbs may have too much blue light for testing. Fluorescent daylight bulbs are available that are acceptable for color vision testing.

Standardized testing procedures are also required especially when tests have strict pass/fail criteria. For example, pseudoisochromatic plates should be viewed for only 3 to 5 s.[37] An inspector even with a severe color deficiency may be able to identify test symbols if a longer viewing time is allowed. Also, shorter viewing times (1 to 2 s) are not encouraged even if a correct response is provided. This may give an unfair cue to whether a correct or an incorrect response was provided. Testers should display the plates unhurriedly and in a consistent manner that provides no feedback to the examinee.

Several standardized color vision tests are described below along with their vocational uses.

Pseudoisochromatic Plates

These tests are the most used tests in clinical practice. Subjects must report which figure made from dots or spots is seen against a similar background. The tests are typically designed to identify subjects with congenital red/green deficiency.[37] For this purpose, they usually do an exceptional job. Nearly all subjects with defective color vision fail the tests, while nearly all subjects with normal color vision pass: few false positives or negatives are found. Also, in general, whereas mild defectives make relatively few errors and severe defectives make many, the number of errors made on these tests is not highly correlated with functional performance. Pseudoisochromatic plate tests are easy to administer, however, and are frequently used as color vision screening instruments. Individuals who fail the pseudoisochromatic plate screening test are referred for further testing.

The Ishihara® color plates have high screening efficiency; however, they are designed only to identify inherited red/green discrimination deficiencies. As many acquired color deficits begin by showing decreased blue/yellow discrimination, the Ishihara® plates may need to be supplemented if all types of color discrimination are important.

Several instructions help use of isochromatic plates.

1. Use standard illuminant C (or its equivalent) to illuminate the test plates with a minimum of 200 lx illumination.
2. Allow only 3 to 5 s per plate; more time leads to correct identification.
3. Subjects who fail the plates can be referred for further testing to grade the type and severity of the apparent color deficiency.
4. Do not allow subjects to touch the plates as this will greatly decrease plate life, invalidating the relationship of color and brightness between the test figures and the background.
5. Close the test booklet and store it in a closed drawer. Exposure to ultraviolet radiation and visible light will, over time, bleach out the plate colors.

Farnsworth-Munsell 100 Hue Test

The farnsworth-munsell 100 hue test was developed to measure the fine color discrimination in persons with normal color vision and to evaluate losses in those with defective color vision.[37] The test consists of 85 caps of color from around the color circle. The colors of the caps were chosen so that approximately equal perceptual steps of color are represented from one cap to the next. Subjects work with one-fourth of the caps at a time and must replace the caps into a tray in order of color. The test can be very difficult for a person with defective color vision because the difference in color between adjacent caps is very small. The test is rarely used for vocational evaluations, because it requires 15 to 20 min for administration. This test, however, does present colors around the entire color circle so that any color discrimination deficiency can be identified (red/green, blue/yellow or overall discrimination loss). A total error score is calculated for the test which provides an indication of the subject's ability to see fine differences in color. There is a moderately wide variation in total error score even for subjects showing normal color vision by passing a pseudoisochromatic plate test. It has been reported that 95 percent of adults with normal color vision will have an error score of 100 or less; however, 78 percent scored 60 or less and 41 percent scored 40 or less.[38] The overall mean error score for their 126 color vision normals was 37.4 percent. Using these results for the farnsworth-munsell 100 hue test, it is logical to require an error score of 40 or less for occupations that require excellent discrimination of fine color.

Farnsworth D 15

The farnsworth D 15 test is an arrangement test using 16 colored disks, or color caps (Fig. 10). One cap is affixed to the test tray.[37] The other 15 caps must be replaced into the tray according to color (similar to the farnsworth-munsell 100 hue test). The difference in color between adjacent caps is much larger than for the farnsworth-munsell 100 hue test. The test was designed for vocational evaluation to fail only those with moderate to severe color vision deficiency.

The order of cap replacement is graphed on a color circle[12] for ease in identifying specific deficiencies of color recognition. Individuals fail the test when two or more major errors of replacement demonstrate gross color confusions across the color circle. When caps from across the color circle are replaced immediately next to each other, the orientation of the line joining these caps illustrates the color confusions and diagnoses the deficiency (Fig. 11a) The line in Fig. 11b shows the zig zag sequence of a red blind subject, based on brightness rather than hue.

The test is simple to administer and typically takes less than five minutes for administration and grading. Approximately 50 percent of congenital color deficient subjects fail the test.[39] Subjects with normal color vision will pass the test, showing that basic colors will not be confused. As with the farnsworth-munsell 100 hue test, the farnsworth D 15 test can identify both red/green and blue/yellow color deficiencies.

Figure 10. Color caps for farnsworth D 15 test. Vision test images in this book are provided for educational purposes and must not be used for vision tests. For more examples, see Bailey.[4]

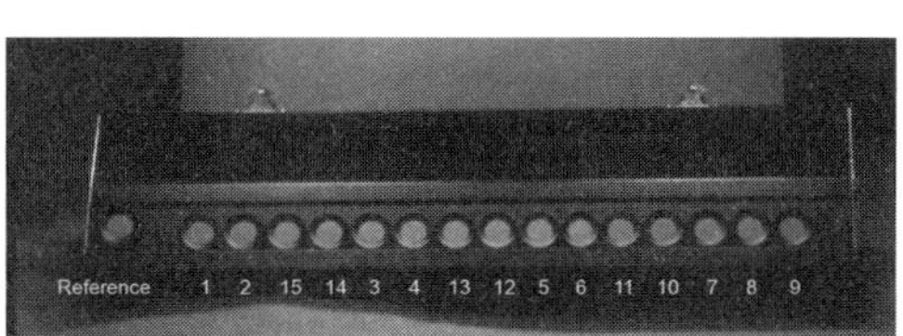

Lanthony Desaturated D 15 Test

The lanthony desaturated D 15 (dD 15) is similar in design and administration to the farnsworth D 15. Colors from around the entire color circle are represented. The principle differences between the two tests are the boldness of the colored caps and the degree of color difference between adjacent caps. The dD 15 has colors lighter and less intense than the farnsworth D 15 colors. It is much more difficult to complete and was designed

Figure 11. Circular arrangement for evaluation of farnsworth D 15 test: (a) axes of confusion for protan (long wavelength sensitive cone defect), deutan (middle wavelength sensitive cone defect) and tritan (short wavelength sensitive cone defect), with sample cap replacement for presumed medium deutan defective; (b) sequence indicating red blindness. Vision test images in this book are provided for educational purposes and must not be used for vision tests.

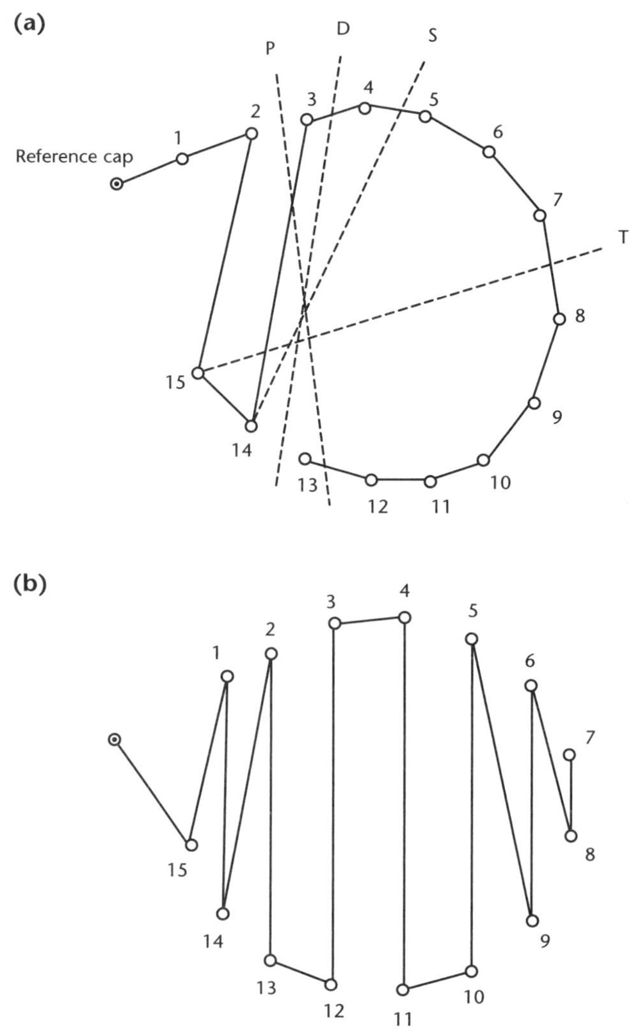

Legend
D = deutanomalous
P = protanomalous
S = sample, presumed median deutanomalous
T = tritanomalous

specifically to evaluate subtle color discrimination changes as a result of acquired color vision deficiencies. The test has been also used as an indicator of fine color discrimination. It is quickly administered and the results are relatively easy to evaluate. Approximately 75 percent of congenital color deficient subjects fail this test.[40]

Occupational Vision Requirements

Many professions require good visual functioning for safe and efficient accomplishments of their essential tasks. This is especially true when visual error can result in injury or death. For example within the aerospace industry, if a crack or area of corrosion is missed during a visual inspection of an airframe, a catastrophic event could occur.

To ensure good vision in workers, vision requirements must be met during the application for certain occupations. Periodic vision evaluations may then be required to ensure good vision in workers during their entire tenure.

Within industry, it has been reported[41] that vision testing for inspectors goes back about 40 years. In spite of the varied visual tasks inspectors in different occupations face, the vision requirements they list from various standards are surprisingly similar. The report states that, in spite of new technology and changes in specifications of inspections, vision testing changed little in the twentieth century. Standards have been shared and not based upon essential tasks of individual occupations, nor have standards been empirically determined.

What constitutes the minimum acceptable vision for an inspector in the aviation industry is difficult to determine. In terms of vision acuity, the standard should be based upon the angular size of the smallest detail for which detection is required.

Rummel[42] generated probability of detection (POD) curves to standardize nondestructive testing by the National Aeronautics and Space Administration for the space shuttle. This and other research led to the use of an anomaly size of 1.3 mm (0.05 in) as the 90/95 level — the size that operators performing special nondestructive testing procedures must detect 90 percent of the time with 95 percent confidence. In a benchmark study of probability of detection,[43] inspectors were to identify cracks visually in an out-of-service Boeing 737. In this study, the 90 percent detection point was found for cracks around 7.5 mm (0.3 in.). This value is much larger than the 90/95 value of 1.3 mm (0.05 in.) for nondestructive testing specialty procedures. For the visual inspection, the length of the crack, crack width, contrast and inspector accessibility all affected detection performance. These data suggest that calculation of a minimum acceptable vision acuity limit is not possible given the many variables at work. Discontinuity length, width, and contrast, light level and viewing distance are all factors contributing to the vision acuity demand of a given discontinuity. Also, in none of the studies mentioned did the researchers attempt to manipulate, restrict or even document viewing distances. With a greater viewing distance, a discontinuity of a given size subtends a smaller angle and hence will demand greater vision acuity.

One study[44] analyzed the visual task for inspections in terms of distinguishing a signal from background noise. The study concluded that the greater the strength of a signal (that is, the visibility of a crack) relative to the background noise, the more likely detection will occur on site. Relative signal strength can be increased by decreasing the viewing distance (crack subtends larger angle to the observer), by ensuring a focused retinal image (proper correcting lens for the specific working distance), or by improving the quality and quantity of light, that is, eliminate glare and increase illumination. Additionally, just as performance is enhanced by increasing target size and contrast above threshold levels, requiring vision better than predicted from calculation of minimum target detail is advisable. Better vision is important particularly because sensitivity decreases with prolonged searching, especially when discontinuities are rare. This phenomenon is known as *vigilance decrement.*[45]

Another study[46] investigated the vision requirement for the identification of a small crack in an airframe as a representative task for aviation assembly. It is an interesting example of an empirical determination of vision acuity required for crack detection in aviation maintenance. Using a computer model of a standard observer as the inspector, a probability of detection curve was generated for a representative crack. Results showed that if the probability of detection is 0.99 with normal vision, probability of detection drops to 0.90 with 20/25 vision acuity and to 0.60 with vision acuity of 20/30. For cracks more difficult to see (that is, those with a probability of detection of only 0.90 with 20/20 vision acuity), a drop in vision acuity has an even greater effect on inspection performance. In this case, probability of detection drops to 0.60 with 20/25 vision acuity and to 0.20 with

20/30 vision acuity. The authors summarize by saying that these data do not point to an absolute level of vision acuity for crack detection. Instead, the data provide information for responsible entities (the Federal Aviation Administration for aviation maintenance inspectors in this case) to set a vision acuity standard based on an acceptable degree of risk, the probability of missing a defect.

The study[46] generally supports a near vision requirement for nondestructive testing. It also suggests that all inspectors will benefit from maximum correction at the proper working distances required for all inspection processes.

Illumination Angle

Small variations in surface roughness and contour cast tiny shadows that can help visual testing. For this reason, discontinuity detectability is greatly affected by the angle of incidence of the illumination.

References

1. Bailey, W.[H.]. Part 3, "Vision and Light." Section 1, "Fundamentals of Visual and Optical Testing." *Nondestructive Testing Handbook,* second edition: Vol. 8, *Visual and Optical Testing.* Columbus, OH: American Society for Nondestructive Testing (1993): p 9-21.
2. Frisby, J.P. *Seeing: Illusion, Brain and Mind.* New York, NY: Oxford University Press (1980).
3. Masland, R.H. *The Functional Architecture of the Retina.* New York, NY: Scientific American Books (December 1986).
4. Bailey, W.H. "Why Color Vision Tests?" *Materials Evaluation.* Vol. 42, No. 13. Columbus, OH: American Society for Nondestructive Testing (December 1984): p 1546-1550.
5. Treisman, A. *Features and Objects in Visual Processing.* New York, NY: Scientific American Books (November 1986).
6. Emsley, H.H. *Visual Optics.* Vol. 1. London, United Kingdom: Haton Press (1963): p 343-348.
7. Glasser A. and P.L. Kaufman. "Accommodation and Presbyopia." *Adler's Physiology of the Eye: Clinical Application.* St. Louis, MO: C.V. Mosby (2003): p 197-233.
8. Hofstetter, H.W. "A Comparison of Duane's and Donders' Tables of the Amplitude of Accommodation." *American Journal of Optometry and Archives of the American Academy of Optometry.* Vol. 21. Minneapolis, MN: American Journal of Optometry Publishing Association (1944): p 345-363.
9. Sharma R.K. and B.E.J. Ehinger. Chapter 10, "Development and Structure of the Retina." *Adler's Physiology of the Eye: Clinical Application.* St. Louis, MO: C.V. Mosby (2003): p 319-47.
10. Porter J.D., F.H. Andrade and R.S. Baker. Chapter 34, "The Extraocular Muscles." *Adler's Physiology of the Eye: Clinical Application.* St. Louis, MO: C.V. Mosby (2003): p 787-817.
11. Rosenfield, M. "Refractive Status of the Eye." *Borish's Clinical Refraction,* second edition. St Louis, MO: Butterworth-Heinemann (2006): p 3-34.
12. Dunn, D., C.[N.] Sherlock, S. Ness and L.[O.] Goldberg. Section 3, "The Visual and Optical Testing Environment." *Nondestructive Testing Handbook,* second edition: Vol. 8, *Visual and Optical Testing.* Columbus, OH: American Society for Nondestructive Testing (1993): p 51-71.
13. Bailey, I.L. "Visual Acuity." *Borish's Clinical Refraction,* second edition. Saint Louis, MO: Butterworth-Heinemann (2006): p 217-246.
14. Houf, J.W. "Understanding Near Vision Eye Tests." *The NDT Technician.* Vol. 8, No. 4. Columbus, OH: American Society for Nondestructive Testing (October 2009): p 1-3, 5.
15. Lovie-Kitchin, J.E. and B. Brown. "Repeatability and Intercorrelations of Standard Vision Tests as a Function of Age." *Optometry and Vision Science.* Vol. 77, No. 8. Philadelphia, PA: Lippincott Williams and Wilkins, for American Academy of Optometry (August 2000): p 412-420.
16. Elliott, D.B. "Contrast Sensitivity and Glare Testing." *Borish's Clinical Refraction,* second edition. St. Louis, MO: Butterworth-Heinemann (2006): p 247-288.
17. Owsley C. "Contrast Sensitivity." *Ophthalmology Clinics of North America.* Vol. 16, No. 2. Philadelphia, PA: W.B. Saunders (2003): p 171-177.
18. Pelli, D.G., J.G. Robson and A.J. Wilkins. "The Design of a New Letter Chart for Measuring Contrast Sensitivity." *Clinical Vision Science.* Vol. 2, No. 3. London, UK: Pergamon (1988): p 187-199.
19. Dougherty, B.E., R.E. Flom and M.A. Bullimore. "An Evaluation of the Mars Letter Contrast Sensitivity Test." *Optometry and Vision Science.* Vol. 82, No. 11. Philadelphia, PA: Lippincott Williams and Wilkins, for American Academy of Optometry (November 2005): p 970-975.
20. Thayaparan, K., M.D. Crossland and G.S. Rubin. "Clinical Assessment of Two New Contrast Sensitivity Charts." *British Journal of Ophthalmology.* Vol. 91, No. 6. London, UK: BMJ Publishing Group (June 2007): p 749-52.

21. Ginsburg, A.P. "A New Contrast Sensitivity Vision Test Chart." *American Journal of Optometry and Physiological Optics*. Vol. 61, No. 6. Baltimore, MD: Lippincott Williams and Wilkins, for the American Academy of Optometry (June 1984): p 403-407.
22. Rubin, B.S. "Reliability and Sensitivity of Clinical Contrast Sensitivity Tests." *Clinical Vision Science*. Vol. 2, No. 3. London, UK: Pergamon Press (1988): p 169-177.
23. Sakmar, T.P. "Color Vision." *Adler's Physiology of the Eye: Clinical Application*. St. Louis, MO: C.V. Mosby (2003): p 578-585.
24. Pease, P. "Color Vision." *Borish's Clinical Refraction,* second edition. Saint Louis, MO: Butterworth-Heinemann (2006): p 289-346.
25. *Optometric Clinical Practice Guideline: Comprehensive Adult Eye and Vision Examination*, second edition. St. Louis, MO: American Optometric Association (2005).
26. Committee on Vision, Commission on Behavioral and Social Sciences and Education, National Research Council. *Measurement of Visual Field and Visual Acuity for Disability Determination.* Washington, DC: National Academy Press (1994).
27. Bailey, I.L. and J.E. Lovie. "New Design Principles for Visual Acuity Letter Charts." *American Journal of Optometry and Physiological Optics*. Vol. 53, No. 11. Baltimore, MD: Lippincott Williams and Wilkins, for the American Academy of Optometry (November 1976): p 740-745.
28. Ferris, F.L., A. Kassoff, G.H. Bresnick and I.L. Bailey. "New Visual Acuity Charts for Clinical Research." *American Journal of Ophthalmology.* Vol. 94, No. 1. New York, NY: Elsevier Science (July 1982): p 91-96.
29. National Academy of Sciences, National Research Council Committee on Vision. "Report of Working Group 39: Recommended Standard Procedures for the Clinical Measurement and Specification of Visual Acuity." *Advances in Ophthalmology*. Vol. 41. Basel, Switzerland: S. Karger (1980): p 103-148.
30. Bailey, I.L., M.A. Bullimore, T.W. Raasch and H.R. Taylor. "Clinical Grading and the Effect of Scaling." *Investigative Ophthalmology and Visual Science*. Vol. 32, No. 2. Rockville, MD: Association for Research in Vision and Ophthalmology (February 1991): p 422-432.
31. Nakagawara, V.B. and K.J. Wood. "Clinical Application of the New Civil Airman Vision Standards and Certification Procedures." *Optometry — Journal of the American Optometric Association*, Vol. 69, No. 3. New York, NY: Elsevier, for the American Optometric Association (March 1998): p 144-150.
32. Good, G.W., J.J. Nichols, M.V. Subbaram, V.B. Nakagawara, R.W. Montgomery and R. Collins. AAR-100, *Visual Functioning of Aviation Maintenance Inspectors*. FY04 Aviation Maintenance, General Aviation, and Vertical Flight. Washington, DC: Federal Aviation Administration, Office of the Chief Scientist for Human Factors (2004).
33. Jose, R.T. and R.M. Atcherson. "Type Size Variability for Near Point Acuity Tests." *American Journal of Optometry and Physiological Optics*. Vol. 54, No. 9. Baltimore, MD: Lippincott Williams and Wilkins, for the American Academy of Optometry (September 1977): p 634-638.
34. Recommended Practice No. SNT-TC-1A, *Personnel Qualification and Certification in Nondestructive Testing*. Columbus, OH: American Society for Nondestructive Testing (2006).
35. ACCP-CP-1, *ASNT Central Certification Program*, Revision 5. Columbus, OH: American Society for Nondestructive Testing (October 2008).
36. CIE Report 15-2004, *Colorimetry,* third edition. Wien, Österreich [Vienna, Austria]: Commission Internationale de l'Éclairage [International Commission on Illumination] (CIE) (2004).
37. Birch, J. *Diagnosis of Defective Colour Vision,* second edition. Boston, MA: Butterworth-Heinemann (2001).
38. Mahon, L.E. and A.J. Vingrys. "Scoring the Farnsworth-Munsell 100-Hue for Vocational Guidance." *Optometry and Vision Science*. Vol. 72, No. 8. Philadelphia, PA: Lippincott Williams and Wilkins, for American Academy of Optometry (August 1995): p 547-551.
39. Cole, B.L. and A.J. Vingrys. "Who Fails Lantern Tests?" *Documenta Ophthalmologica*. Vol. 55, No. 3. Amsterdam, Netherlands: Springer, for the International Society for Clinical Electrophysiology of Vision (May 1983): p 157-175.

40. Atchison, D.A., K.J. Bowman and A.J. Vingrys. "Quantitative Scoring Methods for D15 Panel Tests in the Diagnosis of Congenital Color Vision Deficiencies." *Optometry and Vision Science*. Vol. 68, No. 1. Philadelphia, PA: Lippincott Williams and Wilkins, for American Academy of Optometry (January 1991): p 41-48.
41. Kleven, S. and L. Hyvärinen. "Vision Testing Requirements for Industry." *Materials Evaluation*. Vol. 57, No. 8. Columbus, OH: American Society for Nondestructive Testing (August 1999): p 797-803.
42. Rummel, W.[L.] "Probability of Detection as a Quantitative Measure of Nondestructive Testing End-to-End Process Capabilities." *Materials Evaluation*. Columbus, OH: American Society for Nondestructive Testing (January 1998): p 29-35.
43. Spencer F.[W.], D. Schurman, C.G. Drury, R. Smith and G. Worrall. DOT/FA/AR-96/65, *Visual Inspection Research Project Report on Benchmark Inspections*. Springfield, VA: National Technical Information Service, for United States Department of Transportation, Federal Aviation Administration, Office of Aviation Research (October 1996).
44. Drury, C.G. "Good Practices in Visual Inspection." *Human Factors in Aviation Maintenance,* Phase Nine Progress Report. Washington, DC: United States Department of Transportation, Federal Aviation Administration, Office of Aerospace Medicine (2001).
45. Mackworth, N.H. "The Breakdown of Vigilance During Prolonged Visual Search." *Quarterly Journal of Experimental Psychology*. Vol. 1, No. 1. London, UK: Taylor and Francis, for Psychology Press (April 1948): p 6-21.
46. Beard, B.L., K.M. Jones, C. Chacon and A.J. Ahumada [Jr.]. DRFA-2045, *Empirical Basis for an Occupational Vision Standard*. Washington, DC: United States Department of Transportation, Federal Aviation Administration (2002).

Bibliography

Bailey, W.H. "Fluorescent Penetrant Testing and Color Vision Deficiencies." *Materials Evaluation*. Vol. 61, No. 6. Columbus, OH: American Society for Nondestructive Testing (June 2003): p 660-663.

Bailey, W.H. "Requirements for Visual Acuity in the Aerospace Industry." *Materials Evaluation*. Vol. 46, No. 13. Columbus, OH: American Society for Nondestructive Testing (December 1988): p 1625-1630.

Bailey, W.H. "Color Vision Testing — Who Needs It?" *Materials Evaluation*. Vol. 43, No. 10. Columbus, OH: American Society for Nondestructive Testing (September 1985): p 1194-1197.

Bailey, W.H. "The Case for Eye Test Standardization." *Materials Evaluation*. Vol. 40, No. 8. Columbus, OH: American Society for Nondestructive Testing (July 1982): p 826-829.

Bailey, W.H. "Charts for Visual Testing of Inspection Personnel." *Materials Evaluation*. Vol. 41, No. 7. Columbus, OH: American Society for Nondestructive Testing (June 1983): p 849-853.

Batey, J.E. "Jaeger vs. Computer." *Materials Evaluation*. Vol. 54, No. 9. Columbus, OH: American Society for Nondestructive Testing (September 1996): p 974–979. Erratum, Vol. 54, No. 11, November 1996, p 1284.

Beard, B.L., W.A. Hisle and A.J. Ahumada, Jr. *Occupational Vision Standards: A Review*. Moffett Field, CA: NASA Ames Research Center [2002].

Functional Vision: A Practitioner's Guide to Evaluation and Intervention, first edition. New York, NY: American Foundation for the Blind (2004).

Good, G.W. and V.B. Nakagawara. Final Report [FAA Grant 02-G-031], *Vision Standards and Testing Requirements for Nondestructive Inspection (NDI) and Testing (NDT) Personnel and Visual Inspectors*. Washington, DC: Federal Aviation Administration, (AAR-100) Aviation Maintenance and Human Factors (2005).

Riley, J.N., E.P. Papadakis and S.J. Gorton. "Availability of Training in Visual Inspection for the Air Transport Industry." *Materials Evaluation*. Vol. 54, No. 12. Columbus, OH: American Society for Nondestructive Testing (December 1996): p 1368-1372, 1374-1375.

Weatherly, S.L. "Testing Visual Acuity with the Jaeger Eye Chart." *Materials Evaluation*. Vol. 60, No. 8. Columbus, OH: American Society for Nondestructive Testing (August 2002): p 928-929.

Weatherly, S.[L.]. "The Equivalency of Vision Acuity Exams for NDT Personnel." *Materials Evaluation*. Vol. 65, No. 3. Columbus, OH: American Society for Nondestructive Testing (March 2007): p 289-290.

4

CHAPTER

Visual Test Imaging

Michael W. Allgaier, Mistras Group, Princeton Junction, New Jersey (Parts 1 and 3)

Thomas D. Britton, General Electric Sensing and Inspection Technologies, Skaneateles, New York (Parts 1 and 3)

Brian P. Buske, General Electric Sensing and Inspection Technologies, Skaneateles, New York (Part 3)

Trevor Liddell, General Electric Sensing and Inspection Technologies, Lewistown, Pennsylvania (Parts 1 and 3)

David A. Pasquazzi, David Pasquazzi and Associates, Cranston, Rhode Island (Part 1)

Donald J. Roth, National Aeronautics and Space Administration, Glenn Research Center, Cleveland, Ohio (Part 3)

Portions of Parts 1 and 3 are reprinted with permission from *Nondestructive Testing: Remote Visual Inspection,* training modules © [2005] General Electric Corporation, Lewistown, PA. Reference numbers superscripted in headings indicate sections adapted. ASNT has revised the text in 2010, and deficiencies are not the responsibility of General Electric Corporation.

PART 1. Photography in Visual Testing[1]

Supersession

Several important technologies — photography, videography and image processing — practiced by visual inspectors have since 1980 been revolutionized by advances in microelectronics, microprocessing and telecommunications. The pace of technology is likely to continue through the twenty-first century. One result of these innovations is *supersession*, here meaning the process by which one technology becomes obsolete and is superseded, or replaced, by another. The issue of supersession is not unique for engineering but affects many aspects of life. The process of supersession will repeat in other particulars, so the inspector must continually learn new things and adapt.

There are several areas where supersession affects the nondestructive inspector.

1. Published standards and regulations impose requirements. It is not unusual for a standard to undergo changes every year or two, and the inspector must be aware of changes that affect contracted tasks.
2. Test equipment and software may need to be upgraded. Such improvements do not necessarily cause expenses but in fact may save money. Computers for example are now smaller, faster and less expensive than ever and permit wireless connections. Photographic film and analog videotape take longer and cost more to make and consume more space and time for viewing, sorting and storage than do the digital images that have superseded them.
3. Inspectors need to be trained on how to use new instruments and new techniques, equipment and applications. Inspectors need to keep their own logs and document their educational activities, some of which are required for certification. And management needs to allot time for training of the inspection staff.
4. The inspection procedures that Level IIIs write for applications at their workplaces need to be reviewed periodically and revised to accommodate new instruments and test materials. These written procedures may be referenced by contracts.

Readers of the present discussion should be aware that any discussion of digital technology is incomplete and obsolete before it is published. The present text needs to be supplemented with up-to-date information about file formats, media platforms and other subjects — especially about digital technology.

Photographs as Permanent Record for Visual Testing[1]

The eyes are sensors that can scan and evaluate objects large or small. There is no substitute for a direct view of a test object, so visual testing — the oldest of the nondestructive testing methods — continues to be valued. Trying to recall details such as the exact size, location and orientation of indications long after a visual test is at best difficult, at worst reckless. A permanent visual record is often needed for engineering purposes, personnel training, failure analysis and crack growth monitoring (Fig. 1). Photography is often the most practical and least expensive way to preserve visual test results permanently.

Photographs are historical records with many scientific and technical applications. A permanent record of a test makes it possible to subsequently review, scrutinize and evaluate a discontinuity. Photographs can be used as reference documents (Fig. 2), just as other nondestructive test methods use radiographic film or plotted data. The main use of photography is to record and to archive visual tests.

Digital photography is a valuable inspection tool. It can be used to record the test object in any method but is especially valuable to preserve test results in qualitative surface methods — liquid penetrant, magnetic particle and visual testing. Test results can be saved as an archive of the inspection, as is done with radiographs and ultrasonic displays.

In the twentieth century, articles about photography of surface tests featured advice specific to photographic film.[2-4] Digital techniques have obviated much of this information.[5,6]

Also, in liquid penetrant and magnetic particle testing, surface indications have been documented by using transfers, strips of cellophane or transparent adhesive tape, to lift indications and attach them to reports.[2,7] These transfers had drawbacks: photocopy machines picked up glare from the tape, and lifting was impractical for wet techniques. Fluorescence itself would not be visible later and could not be photocopied. Film photography also kept the inspector from judging the adequacy of an image until the film was developed. Digital photography provides an immediate image, and many more photographs can be taken if an image is unsatisfactory, without having to return to the site or repeat a test.

Documentation[5,6]

Inspectors need to save images of test objects for various reasons — comparison, process control, training, documentation and quality assurance. Photographs demonstrate that the job was performed. The following measures can make photography a useful part of the inspection.

1. Few cameras let the operator name files, so keep a written log of photograph file names, who performed the tests, when, test parameters and any circumstances of interest.
2. Write information such as the test number, location and time — information like that used to identify radiographs — on a small card or piece of paper and place it by the test object within the camera's field of view. This step makes it easier to identify and sort test photographs later. A ruler or tape measure in the photograph will provide scale.
3. Photograph the reference standard at the time of inspection and include its image with the documentation.
4. Specify or describe photography in the written procedures.

The visible characteristics of photographs are affected by the photographer's choice of settings. Some options are hardware specific and can be selected at the moment a test is documented — zoom, for example. Camera models change every year: for desired settings, inspectors should consult the owner's manual of the camera used. A good rule of thumb is that the resulting image should provide visible information like that in the reference standards used for discontinuity evaluation.

Some conditions are controlled at the time of photography — illumination, for example. Some photographic settings can be modified later — a color image may be changed to gray scale for printing in a manual, for example.

Photographing of Indications[5,6]

Photography is art. Digital technology has made it faster, easier and less expensive than it was in the twentieth century, but doing it well requires training and practice. Basic principles still apply, with

FIGURE 1. Cracked engine truss mount showing critical areas and crack orientation.

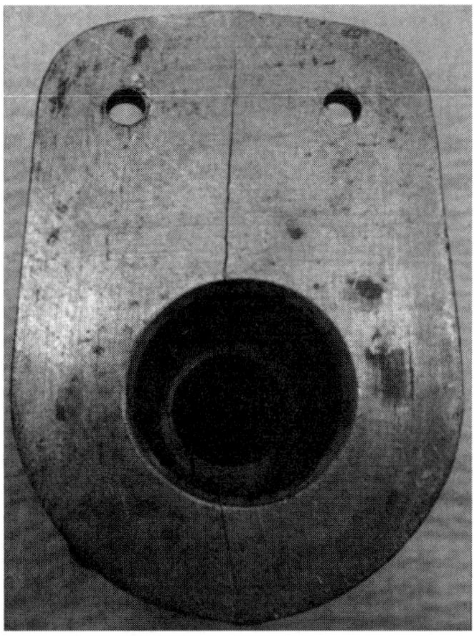

FIGURE 2. Archival photograph of exfoliation corrosion damage on aircraft wing spar.

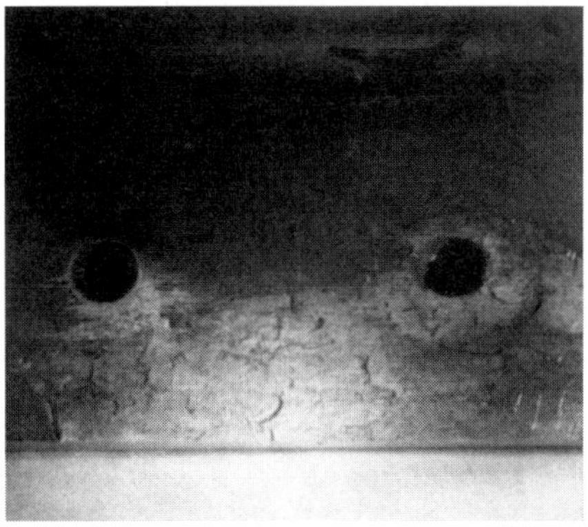

changes due to the digital medium. Books are available on the subject.[8,9]

Compact Camera versus Reflex Cameras

There are two general types of digital cameras in use today. They are the single-lens reflex and the compact camera. The single-lens reflex is more sophisticated and more versatile. Its special feature is that the scene is viewed through the lens by a system of mirrors and prisms. This gives a very clear view of 95 percent or more of the picture area. These cameras have a liquid crystal display on the back where images already in the camera and operational menus can be viewed. These cameras use interchangeable lenses allowing a large selection of focal lengths. Most professionals use this type of camera.

The other type of camera is the compact camera. Here, the scene is viewed on the liquid crystal display on the back of the camera. Some of these cameras also allow viewing with another liquid crystal display in an eyepiece or by an optical viewer above the lens. This optical eyepiece shows only about two thirds of the picture area and does not exactly align with the lens, so this feature should not be used for scene selection in indication photography. Compact cameras are the normal consumer's choice because they are smaller, simpler and less expensive than the digital single-lens reflex. They can often do an adequate job of indication photography.

Illumination

Direct lighting from incandescent lamps can cause artifacts — light reflecting from the test object, for example, or refracted light creating glare in the camera lens. Indirect lighting and overhead illumination from fluorescent lamps help minimize this problem.

Anybody taking photographs outdoors is aware of light and avoids dark shadows. A portable flood lamp or camera flash may remedy dim lighting for a test.

Digital cameras adjust automatically to dim illumination, and subsequent image processing may salvage a dark photograph. However, a test should be performed only if reference standard indications are visible on site and at the moment of inspection.

Digital cameras are so sensitive that flash is not often necessary. Flash illumination with digital cameras should be avoided at close range, less than 0.6 m (2 ft). Because most indication photographs will be made at this distance or closer, flash photographs will be overexposed. Furthermore, in most cases the flash is very close to the camera lens and cannot be removed. Flash from such equipment is sometimes reflected directly back into the camera, producing glare. Therefore flash photographs of indications will normally be made only if all else fails.

Exposure Duration and Sensitivity

Because a macro image is enlarged, camera motion during exposure will be magnified and proper procedures are needed to avoid blurring. Briefer exposures and image stabilization help but do not eliminate the problem. Camera motion can be eliminated by using a tripod or by bracing the camera against a solid object. If this is not possible, one must use the highest practical shutter speed. Even with image stabilization, shutter speed should not exceed 0.03 s. With a tripod, it is wise to use a cable release or self timer to avoid moving the camera while pressing the shutter release. The self timer function lets the camera release the shutter without motion from hand pressure.

Digital cameras let the sensitivity of the camera be adjusted over a very wide range. With film, sensitivity or exposure index could only be changed by changing the film, and even the most sensitive films were not as sensitive as the digital cameras. In more sensitive films, increased grain size causes spots in the picture. Unfortunately, digital cameras have the same problem. Photographs made with high exposure sensitivities also show graininess, called *noise.* This noise gets worse as sensitivity increases. True, one can make digital photographs in the dark, but the photographs may be so grainy as to be unacceptable. Thus, the photographer must compromise all the above factors to get acceptable indication photographs.

Effect of Exposure Duration on Contrast

The purpose of an indication is to make a discontinuity visible. This means making the indication contrast strongly with the background. Usually there is enough contrast, but for emphasis it is often desirable to increase this contrast. Most digital cameras can be set to give more contrast. However, it is often easier and more practical to adjust contrast with an image editing program in the computer.

Similarly, the inspector should try to expose the photograph properly. In indication photography, a bad exposure may be redone unless the inspector or the test object moves on before the problem is noticed. A great aid to getting proper exposure is the histogram display, usually available via the menu. The histogram appears as a mountain and valley profile in the display. The right side of the display indicates the amount of bright tones; the left side, dark tones. Ideally, the display should look like a mountain more or less in the middle. If the mountain is off to either side of the display, the photograph is either overexposed or underexposed. Should this be the case, reset the camera and retake the shot if possible.

Badly underexposed shots can often be brightened to a usable degree with image editing software, but excessive graininess may be present. Software can often make slightly overexposed shots usable with some loss of detail. Badly overexposed areas cannot be corrected and are lost. Therefore, if one must err in exposure with digital photography, underexposure is better than overexposure.

White Balance

Photographs can have unnatural tints. When the picture is taken, this effect can be mitigated through a control called *white balance* on digital cameras. Unrealistic tints can be removed so that objects that appear white on site are white in the image. If skin appears too pink or too green, for instance, then the problem may be corrected through white balance. The white balance control lets the photographer adjust the tint of an image by scrolling through a spectrum of tint options in a single dial and selecting the most realistic one.

Most digital cameras also have a way for the user to tell the camera under what type of illumination a photograph is being taken. White balance takes into account the colors of the light source and adjusts the relative amounts of red, green and blue primary colors so that neutral colors are correct before the image is recorded.

Color can be balanced again later, in image processing. The selection of the RAW file format, discussed below, makes this easier.

When photographing indications, color is often unimportant. Therefore, white balance is usually not critical as it is in most photography. In fact, gray scale (black and white) indication photographs may be preferable. Digital cameras normally record in color, but in some cases they can be set to produce gray scale. And it is easy for software to convert images to gray scale.

Photography is a process which has many variables. It is necessary to do some experimentation to get good results. As equipment becomes better, the process should become still easier and more accurate.

Fluorescent Indications

In some cases, the indications to be photographed will be fluorescent. A magnetic particle or liquid penetrant inspector does not want to photograph ultraviolet radiation as such but instead wants to photograph the nonultraviolet radiation known as *fluorescence*. Ultraviolet radiation is invisible to the human eye; to be seen, its energy needs to be converted into visible light. A fluorescent magnetic particle indication is not ultraviolet, or else no one could see it. Similarly, photographs called *ultraviolet* are usually visible translations of ultraviolet originals.

As in night photography, the ambient area will be partially dark and bright indications will stand out in the photograph. The following measures may help the inspector photograph fluorescent indications.

1. Try night camera settings if the image is taken in a dark area.
2. Notice if wavelengths or colors are being filtered or enhanced, either with hardware (by a screen at the lens, for instance) or digitally. A test object viewed with protective goggles may look different in a photograph.
3. Consult reference standards with known discontinuities to select camera settings and filters.
4. Keep notes about circumstances that might affect discontinuity evaluation.

To prevent haze on a dark background, the ultraviolet radiation must be filtered. The background and indications should be visible. Brightness in the image can sometimes be adjusted with image processing software.

Photographic Equipment

Basic Equipment[1]

Typical equipment for producing a photographic record of visual tests includes a camera, a stable tripod, an automatically regulated flash and a shutter release. Digital photography requires the above items plus a computer and associated equipment including a power source, connections, file storage media and software for image processing and file management.

A macro lens is useful for closeups (Fig. 3), particularly in analog photography. The macro lens with a 28 to 80 mm zoom lens can be made to focus on objects some 300 mm (12 in.) from the lens or to magnify objects difficult or dangerous to approach. A tripod and shutter release are used to stabilize the camera for long exposures in dim light.

Usually, getting a 28 to 80 mm (1 to 3 in.) zoom lens means that the camera body and lens are bought separately. A camera, no matter how costly, is no better than its lens. An inexpensive lens can drastically reduce an expensive camera's image reproduction quality.

Lens[10]

The optical component of the camera is the lens. At its simplest, a lens is an optical device made of transparent material: glass or plastic. A lens transmits light and makes it converge to form an image that looks like the scene in front of the lens. This image is focused on a sensor at the camera's focal point or plane. A convex lens or mirror causes light to come together to a focal point; a concave lens or mirror causes light to spread out. A lens system is a series of two or more lenses or mirrors working together to transmit one beam of light.

As light travels from one medium to another, it changes speed. Light travels more quickly through air than it does through glass, so a lens slows it down. When light waves enter a piece of glass at an angle, one part of the wave will reach the glass before another and will start slowing down first. Light's change of direction can be described by the following analogy. When a shopping cart is pushed from pavement into grass at an angle, if the right wheel hits the grass first, it slows down while the left wheel is still on the pavement. Because the left wheel is briefly moving more quickly than the right wheel, the shopping cart turns to the right as it moves onto the grass (Fig. 4). The analogy fails in that the cart experiences mechanical resistance; light does not.

Focus[10]

Focal Point

The focal point is where light rays from a lens or mirror come together. The focal point of a lens is the point of convergence of light or a point from which the light diverges. The focal distance is the distance from the lens system to the focal point. When using a lens system for indirect visual testing, it is very important that the focal point is appropriate for the inspection parameters (Fig. 5).

FIGURE 3. Macro lens view revealing cracks and corrosion pitting on casting.

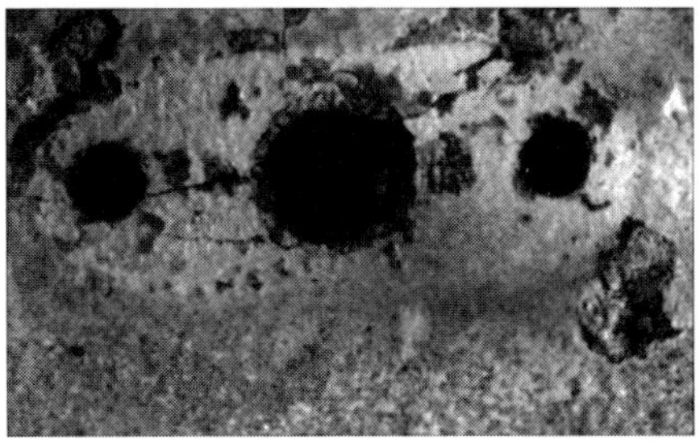

Depth of Field

Depth of field is the distance within which the object is sharply resolved (Fig. 5). Depth of field is the amount of

FIGURE 4. Refraction explained by analogy of shopping cart in grass.[10]

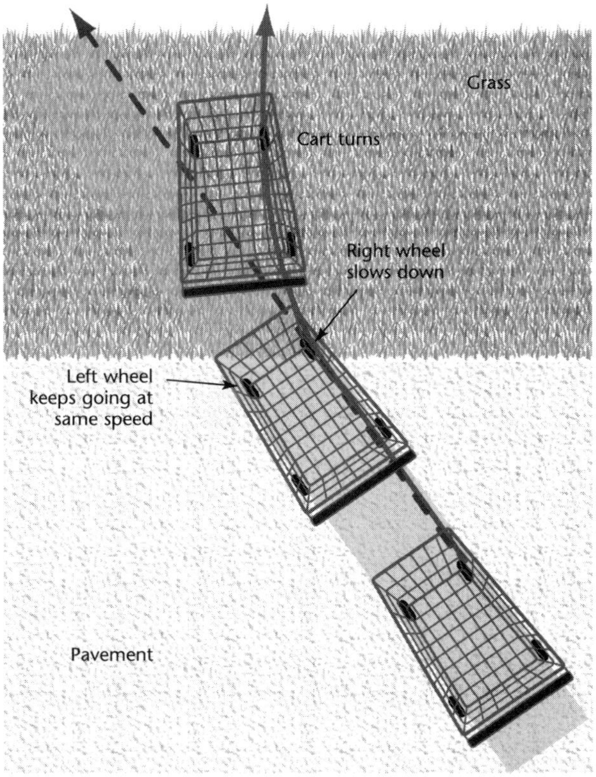

FIGURE 5. Depth of field.[10]

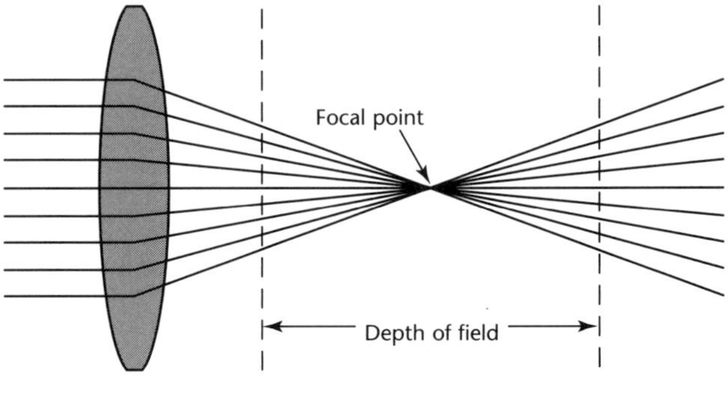

distance that the subject is still in focus and is expressed as a range from a certain minimum point close to the camera to a certain maximum point as distant from the camera. Everything in this range will be in focus and everything beyond, closer or farther, will be out of focus (Fig. 6).

Field of View

The field of view is the entire area that can be seen through an optical system as light is received from the conical angle subtended by the system's optics. An astronomical telescope's field of view is the area of the sky that can be seen with a given eyepiece. Theoretically, a field of view is three-dimensional, like a room, and not two-dimensional, like a wall. The area of interest in a field of view, however, is often a flat surface.

Different lenses can be attached to an instrument to achieve different fields of view. Figure 7 represents the field of view as represented by a rigid borescope. In this example, the field of view of the system is 60 degrees.

For different lenses, the grinding specific for each side results in desired features: closeup, close focus and high magnification; or short focus, wide angle views and high magnification. Table 1 and Fig. 8 show the interrelationship of depth of field and field of view.

FIGURE 6. Objects closer and farther than focal point are blurred.

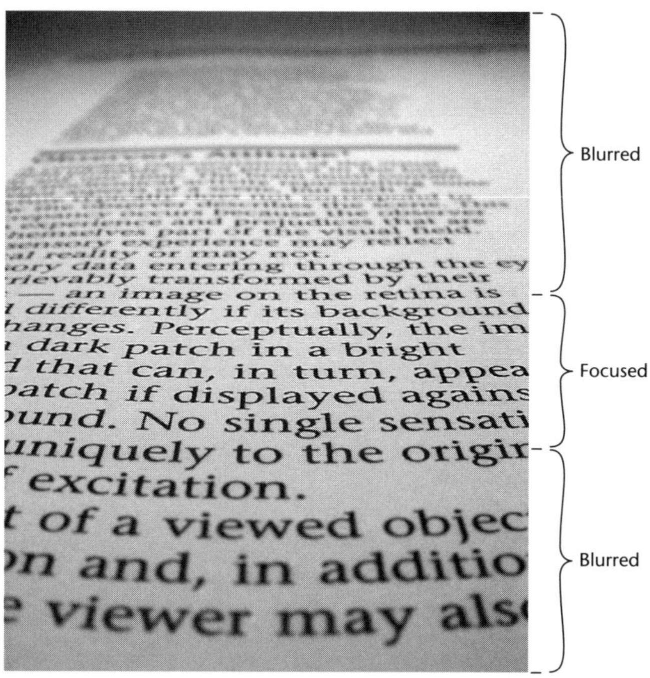

FIGURE 7. Field of view of one rigid borescope.[10]

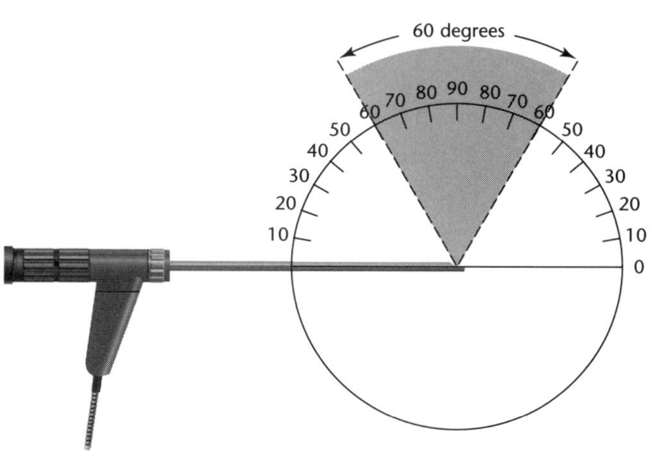

Capturing Details[5,6]

To capture maximum detail, the inspector wants the area of interest to be as large as possible within the camera's field of view. Many cameras offer optical zoom, letting

TABLE 1. Relationships among focal length, magnification, depth of field and field of view (see Fig. 8).

Focal Length	Magnification	Depth of Field	Field of View
increase	increase	decrease	decrease
decrease	decrease	increase	increase

FIGURE 8. Relationship of aperture to field of view: (a) narrow aperture; (b) wide aperture (sere Table 1).

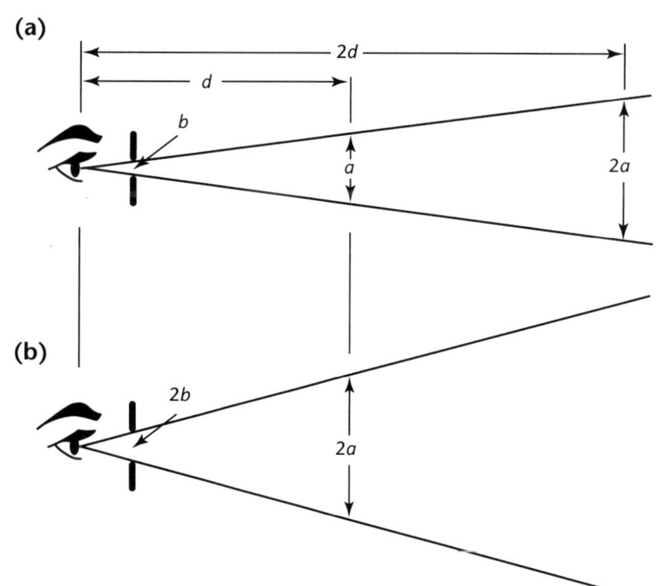

Legend
a = target in field of view
b = aperture diameter
d = distance to target

the user move the lens telescopically to "zoom in on" a subject and capture more detail in a closeup. The same advantage can be achieved simply by moving closer to the subject. Some digital cameras achieve zoom inexpensively by something called *digital zoom,* or *interpolation,* which crops the image and sacrifices details because not all the pixels are retained.

Camera focal length determines the angle of view and the depth of field. Short focal lengths give wide angles of view; long focal lengths give narrow angles of view. Short focal lengths give large depth of field; long focal lengths give small depth of field. The closer the subject is to the camera, the smaller the depth of field needs to be.

Indication photography is basically closeup or macro photography, *macro* meaning that the image is viewed in a scale larger than the original subject. For indication photography at close range, depth of field is a problem. Small depth of field may cause parts of the subject to be out of focus, making the photographic record unacceptable. Therefore, camera equipment and settings should be used that keep the indication in focus. Many cameras have a setting for macro photography that changes the lens to improve focus at close range. The macro setting does not alter the number of pixels.

In general, focal lengths should be short for indication photography. Short focal lengths maximize depth of field and permit closer working distances. On a digital lens single-reflex camera with typical midrange lens, the minimum focusing distance is around 500 mm (20 in.). Closer work needs a macro lens, closeup attachable lenses or extension tube. Most compact cameras will focus down to 75 mm (3 in.) so special attachments are not necessary.

Very closeup photography usually means very limited depth of field. If a test exposure shows some desired areas are out of focus, depth of field can be increased by using a smaller lens aperture (higher *f* stop). A smaller aperture may require a slower shutter speed and cause blurring from camera movement.

Digital cameras focus automatically, but the object of interest must be centered in the camera's field of view. Also, the range finder used for focusing in many cameras is acoustic and will not work through transparent obstacles such as window glass.

Zoom Lens[10]

Zooming is a lens adjustment that permits seeing detailed, close shots of a subject (scene target) or a broad, overall view. They allow a smooth, continuous change in the angular field of view so the angle of view can be made narrower or wider depending on their setting. As a result, a scene can be made to appear close or far, giving the impression of camera movement even though the camera remains in a fixed position. Zoom lenses that accomplish this have variable focal length.

Zoom lenses have variable focal lengths, an advantage over lenses of fixed focal length. The lens components in these assemblies are moved to change their relative physical positions, thereby varying the focal length and angle of view through a specified range of magnifications.

The zoom lens is a cleverly designed assembly of lens elements that can be moved to change the focal length from a wide angle to a narrow angle while the image on the sensor remains in focus. The telephoto lens for long distances is a zoom lens (Fig. 9). To achieve a variable focal length lens at least three groups of elements are needed. The front focusing objective group can be adjusted over limited distance with an external focus ring to finely focus the lens. Between the front and rear groups is a movable zoom group, which is moved appreciably (front to back) using a separate external zoom ring. The zoom group also contains corrective elements to optimize the image over the full zoom focal length range. Connected to this group are other lenses that are moved a small amount to automatically adjust and keep the image on the sensor in sharp focus, thereby minimizing the external adjustment of the front focusing group. At the camera end of the zoom lens is the rear stationary relay group, which determines the final image size when it comes to a focus on the camera sensor. Each group normally consists of several elements. When the

FIGURE 9. Telephoto field of view.[10]

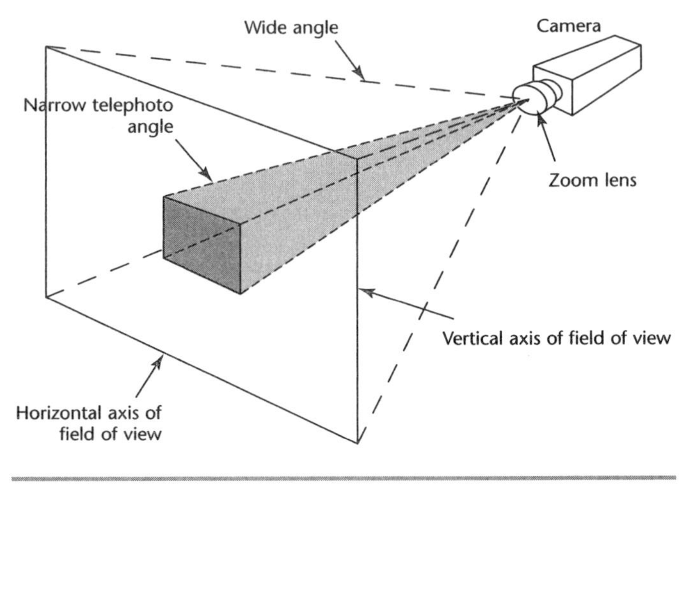

zoom group is positioned correctly, it sees the image produced by the objective group and creates a new image from it. The rear relay group picks up the image from the zoom group and relays it to the camera sensor. In a well designed zoom lens, a scene in focus at the wide angle (short focal length) setting remains in focus at the narrow angle (telephoto) setting and everywhere in between.

Figure 10 shows the continuously variable nature of the zoom lens. The field of view of the camera can be changed without replacing the lens. Surveillance elements in the lenses are physically moved to vary the focal length and thereby vary the angular field of view and magnification. By adjusting its zoom ring setting, the zoom lens permits viewing of narrow, medium and wide angle scenes. This allows a person to initially view a scene with a wide angle perspective and then use close telephoto viewing of one portion of the scene that is of particular interest.

FIGURE 10. Operation of zoom: (a) lens system; (b) wide view; (c) maintaining focus while zooming; (d) closeup. Notice that lens 1 moves to maintain focus.[10]

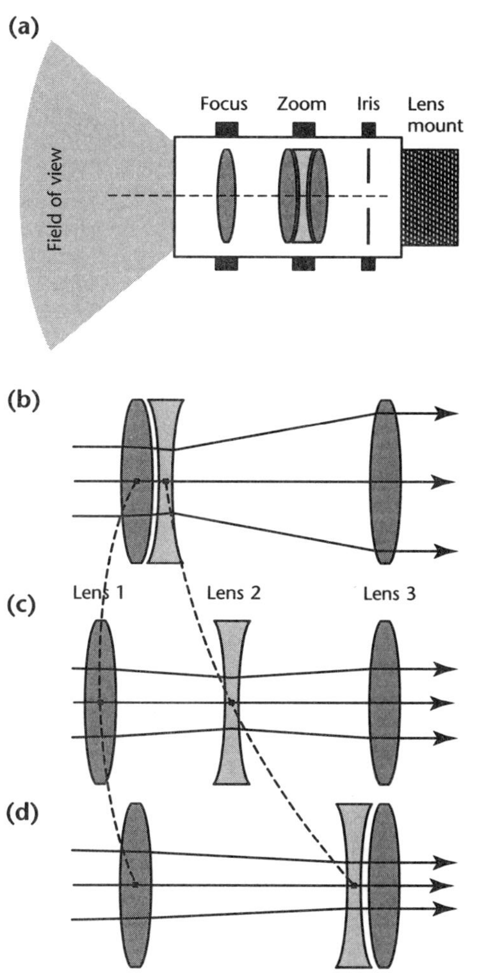

Effect of Magnification[11]

The benefits of magnification come with a price — not only does magnification increase the size of the image, it also increases the effect of camera motion and noise. A camera that has magnification is difficult to position accurately. To minimize the effect of motion, it is important to have as many adjustments as possible within the camera itself. For scanning where no specific fixturing is involved, a zoom lens is very useful: the camera can be positioned with the lens in the wide angle mode and then focused on the area of interest.

Some underwater cameras have integral panning. Such units let the lens scan over a 180 degree segment, looking forward and to each side of the camera — a useful feature in restricted locations.

If a camera with internal focusing is not available, it is especially important to provide a stable operating platform. The camera mount should let the camera freely rotate so that accessible areas can be viewed from all angles. In nuclear reactor applications, the camera is mounted on the refueling bridge and may be working at depths up to 30 m (100 ft). At these depths, water pressure increases at roughly 10 kPa·10 m^{-1} (about 0.5 $lb_f \cdot ft^{-1}$).

Because most discontinuities are three-dimensional, there is another factor to consider. Magnification is exact only at the principal plane of focus. Where measurements of overall discontinuity size are made directly off a photograph, the principal plane of focus must be at the widest part of the subject.

Aperture and Exposure[10]

Considerations affecting the brightness and contrast of the image are illumination, exposure time, lens aperture and image processing.

Illumination[11]

The photographer should exercise care to ensure that the illumination is sufficient for the inspection. In general, orientation of the light source is a first consideration. Lighting should originate from one direction on most three-dimensional objects to avoid ambiguity in relief. Where possible, lighting should originate from above the subject. If more light is required, it should be slightly weaker and more diffuse than the main light source.

If surfaces of interest are obscured in shadow, another light source may be added. A single light source may create patches of glare, a problem that may be solved by side lighting. Too much light, on the other hand, may reduce contrast and make it difficult to see indications.

When photographing certain test objects (pipe welds, for example), unwanted reflections from the flash unit are a common problem. These can usually be eliminated by moving the flash unit to direct the specularly reflected light away from the lens. Another effective method of eliminating subject reflection is to bounce the flash off a white surface.

Exposure Time and Numerical Aperture (Focal Number)[10,12]

To change the amount of light striking a sensor, one or two things may be changed: (1) the amount of time (shutter speed) the recording medium is exposed to the light or (2) the size of the aperture through which the light is passing. The aperture is the opening in the lens; in advanced cameras, the aperture can be adjusted to control the amount of light reaching the film or digital sensor.

To increase or decrease the amount of light passing through the lens, the size of the aperture in the lens can be changed (Fig. 11). This change can be achieved by the iris diaphragm, a ring of overlapping plates that contracts inward to shrink the aperture or expands outward to enlarge it. Essentially, this mechanism works the same way as the iris in your eye — it opens or closes in a circle, to shrink or expand the diameter of the lens. When the lens is smaller, it captures less light; when it is larger, it captures more light.

Reducing the aperture also increases the depth of field, the extent to which subject matter lying closer than or farther from the actual plane of focus appears to be in focus. In general, the smaller the aperture, the greater the distance from the plane of focus the subject matter may be while still appearing in focus. Figure 8 shows the interrelationship of aperture, focal distance and field of view.

FIGURE 11. Common *f* stops for popular focal ratios.[10]

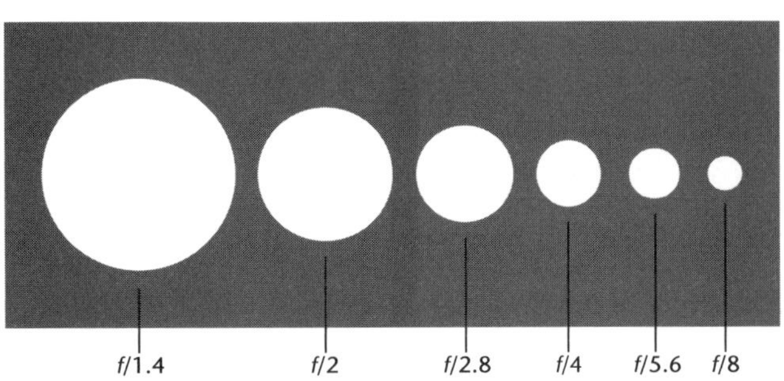

In combination with variation of the shutter speed, the aperture will regulate the photograph's exposure to light. Typically, a fast shutter speed will require a larger aperture to ensure a sufficient exposure to light, just as a slow shutter speed will typically require a smaller aperture to prevent excessive exposure to light.

Numerical aperture is defined and measured differently in different branches of optics. In photography, the ratio of the focal length of a lens or lens system to the effective diameter describes the effective focal length of the lens numerically and is called the *focal number* (or f *number*). It is used to rate the speed of the lens and is the ratio of the widest aperture allowed by the iris diaphragm relative to the focal length. For example, *f*/16 represents a diaphragm aperture diameter that is one 0.0625× the focal length — that is, the focal length is 16 times the aperture. The greater the *f* number, the less light per unit area reaches the focal plane of the camera film or sensor plane. Adjusting the *f* number affects the depth of field. Decreasing the focal length or increasing the aperture increases the device's efficiency: the range of angles over which it gathers light. This efficiency comes at the expense of reduced magnification and/or increased optical distortion.

Photogrammetry[1]

Photogrammetry is the science of obtaining quantitative measurements of physical objects through processes of recording, measuring and interpreting photographic images.[13-16] Photogrammetry is used to make topographic maps and surveys based on measurements and information obtained from aerial photographs. Much photogrammetry has been for nontopographic purposes such as architecture, civil and mechanical engineering and structural analysis. Close range, terrestrial applications are used for solving problems in remote measurement and permanent documentation of deformation, deflection or damage to a wide variety of large and small objects with surprising accuracy.

Close range photogrammetry generally involves camera-to-object distances less than 30 m (100 ft). Specially designed calibrated cameras are available for this work, but available amateur models can work too. Most of the close range applications use an analog approach that involves taking a pair of photographs with the camera oriented normal to the object at a known distance. By viewing this pair of stereo photographs through a stereoscope, a three-dimensional image of the object is reconstructed.

Instead of looking directly at the object, it can be photographed from two points analogous to the location of the two eyes. If these pictures are then viewed, left picture with left eye and right picture with right eye, the object appears again in three dimensions.

For various points on an object, relative differences in distance from the camera can be found by determining their parallax differences. With the aid of a parallax bar or stereo plotter, parallaxes can be measured with a precision of 10 µm (0.0004 in.). If the distance between camera stations (base), camera focal distance and object-to-camera distance are known, absolute measurements on the object can be made.

In petrochemical furnaces, close range photogrammetry can be applied to noncontact, on-stream monitoring of the condition of furnace tubes through measurement of bulging, bowing and creep as well as deterioration of components such as cracks in hangers and tubesheets or spalling of refractory and brickwork. Photogrammetry could prove useful as a quantitative tool applied periodically for charting the condition of furnace internals.

There are two kinds of photogrammetry: monocular and stereo.

FIGURE 12. Stereo effect: (a) schematic; (b) view from left eye; (c) view for right eye.

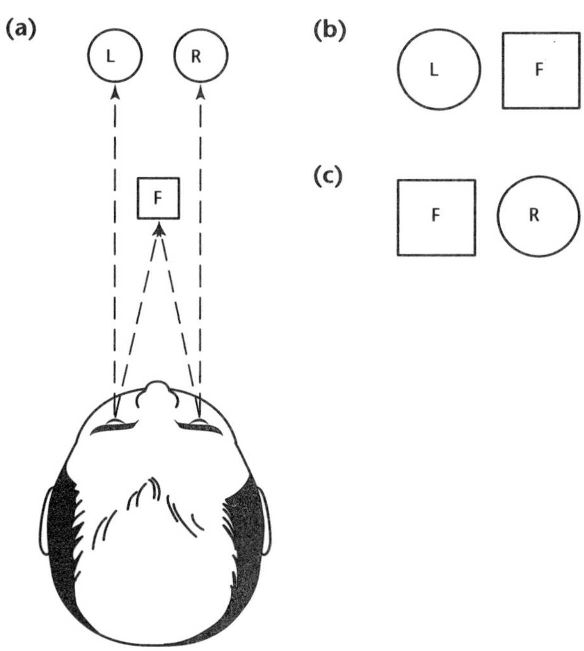

Legend
F = object in foreground
L = left object
R = right object

Monocular Photogrammetry

Monocular photogrammetry superimposes a transparent layer on a photographic image. The transparent material has a reticle, a calibrated scale for measurement. The distance between the reticle and the image surface is part of the measurement calculation. Unless the reticle is flush with the test surface, magnification is also a factor. This monocular photogrammetry resembles viewing with a loupe such as the measuring magnifier described in the chapter on direct visual testing.

The visual inspector using monocular photogrammetry needs to account for several factors.

1. When held touching the test surface, a contact reticle can measure surface features directly. If the reticle is positioned at a distance, the differences in focal distance (distance to image surface versus distance to reticle) must be known if features of the test surface are to be measured.
2. If the surface measured is a photograph, then its scale relative to the depicted surface must also be known before features on the test surface can be measured.
3. A further consideration for precise measurement is viewing angle. The viewing angle from feature to feature varies from one spot on the image to another. The visual inspector, however, needs to know if the camera has introduced distortion by using a fish eye lens, for instance, or by panning the scene.
4. The photograph is flat, but is the test surface flat or curved? If curved, like the inside of a pipe, did the camera rotate to pan the scene?

These factors that affect measurements must be weighed in relation to the demands of rejection criteria. Visual tests typically call for verification by other methods before rejection. In many cases, these factors will be trivial and can be ignored.

Stereo Photogrammetry

Stereo viewing is related to the optical phenomenon called *parallax,* in which the position of an observed object appears to vary with the viewer's line of sight (Fig. 12). In binocular vision, two overlapping fields of views enable depth perception and distance estimates. For photogrammetry, the purpose of measuring parallax difference between two object points is to develop the contour or shape of a surface so that information on deformation, relief (contour) or movement of the component can be determined.

Parallax difference between corresponding points on stereo photographs can be measured by means of a calibrated gage called a *parallax bar,* essentially an optical micrometer with one fixed and one movable measuring mark.

After appropriate positioning under a stereoscopic viewer, the parallax bar is placed on a stereo pair of photographs, aligning the two reference plates on the parallax bar to exactly coincide with one point on the two photographs. When this condition is achieved on the stereoscope, both marks on the parallax bar appear to fuse together into one point apparently floating into space above the model of the object. If two points of known elevation can be identified in a stereo model, the distance of other points on the model can be calculated. This technique is also used in radiography. Calculations and examples are illustrated elsewhere.[1]

Greater measuring accuracies are probably required for determining creep growth of high alloy castings where the total creep growth to failure is on the order of 5 percent. For evaluating photogrammetry as a means of determining the condition of furnaces, tube sheets, hangers or refractories, an assessment should be made of the minimum distortion, deformation, cracking or other significant precursor to failure. This size estimate can in turn be compared to the calculated accuracy limits of a particular photogrammetric setup used to decide whether results are acceptable.

High Temperature Limitations

Cameras for documenting conditions in high temperatures are subject to several constraints and limitations. The most serious problem in furnaces, for example, is the protection of the camera from high intensity radiant energy emitted through the viewing ports. This high temperature environment usually limits the placement period and exposure times to a few minutes. The relatively small size of the viewing ports cuts down the angle of view and limits the area that can be pictured within the furnace. Lighting conditions significantly affect the production of shadows and modeling of the surface of subjects. In this respect, a single lighting source is desirable. However, in a furnace the many burners, as lighting sources, tend to produce diffuse lighting effects that reduce overall shadow contrast and surface modeling. Flame luminosity, particularly from oil fired burners, is undesirable because the unburned incandescent carbon particles tend to hide details of the subject if it is located behind the flame. Uneven heat distribution also creates problems resulting from local variations in flue gas density causing heat waves that distort images of objects pictured in the firebox.

Because furnace firing conditions can be varied, some control can be exerted on the visual environment. More favorable periods, such as when gas rather than oil is being burned, offer lighting conditions more suitable for taking photographs. The stereo photographs shown in Fig. 13 were taken in a refinery furnace. Even though this furnace was oil fired at the time, good definition and image quality are apparent.

Still Photography and Videography

Movies, including digital videos, consist of successive frames of still photographic images. Still photography and moving videography share much of the same digital technology: hardware, processes and terminology. Digital cameras designed for still photography can also make movies, and vice versa. The charge coupled device, a sensor common in digital cameras, is discussed in the section on video, below.

FIGURE 13. Refinery pipestill furnace: (a) stereo photography setup; (b) stereo views.

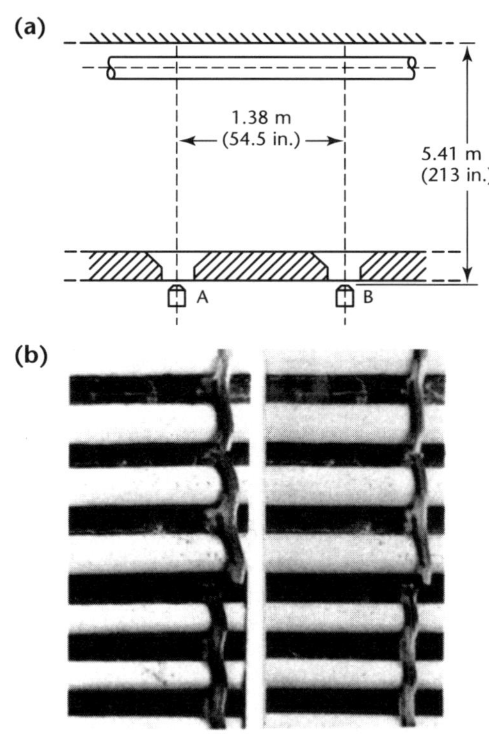

PART 2. Digital Processing and Archiving for Visual Testing[5,6]

Pixels

Digital photographs are designed for a computer display. A display consists of a grid of squares called *picture elements*, or *pixels* (Fig. 14).[10] Displays range from 320 × 240 pixels to over 3840 × 2400. A standard display of 800 × 600 pixels is called a *super video graphics array.* It is superseded in most computers but remains popular in mobile devices.

Each pixel is part of a larger image, and more pixels provide a more accurate representation of the original. Pixels are rectangular and on computer monitors are usually square. The square shape is an artifact of the display; the original electronic datum is dimensionless. Pixels may be translated into dots for printing. *Dot* and *pixel* are interchangeable terms: for example, dots per inch (DPI) and pixels per inch (PPI) are equivalent measures of resolution, discussed below.

There are two classes of graphic file formats. Line drawings use mathematical calculations called *vectors* to plot lines and are not considered here. Raster images, or bit mapped images, assign hue and intensity values to each pixel and are used for photography. Table 2 lists common file formats for digital images. Some cameras use proprietary formats similar to these common file formats.

A *bit map* defines an image and color for all pixels in the image. TIF and JPG files are bit mapped. A bit map does not need to contain color coded information for each pixel in every row. It needs to contain information indicating a new color as the display scans along the row.

FIGURE 14. Array of pixels in digital display.[10]

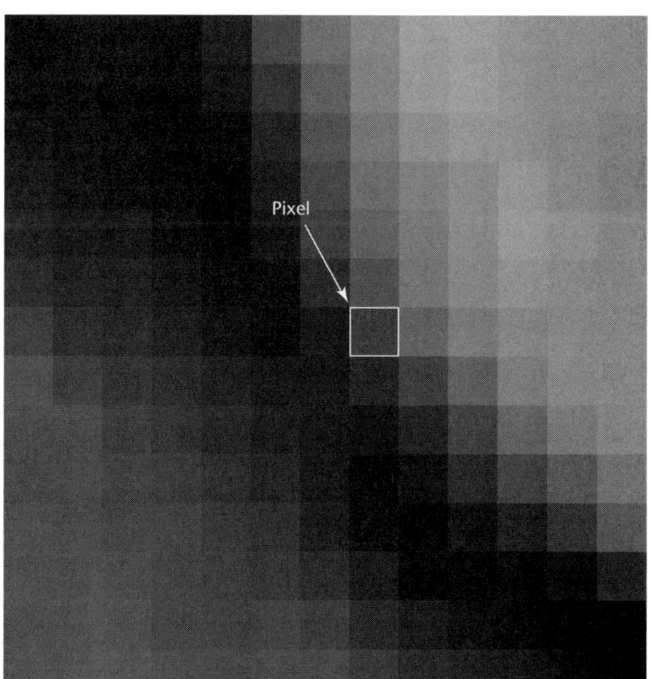

TABLE 2. Digital file formats of still images.

Designation	Format
Lossless Raster Formats	
BMP	bit mapped picture
DIB	device independent bitmap
PIC	three-letter file extension for PICT
PCT	three-letter file extension for PICT
PICT	picture metafile
RAW	extension for undeveloped image from digital camera (virtually lossless)
TIF	three-letter file extension for TIFF
TIFF	tagged image file format
Lossy Raster Formats	
GIF	graphics interchange format, native to CompuServe® internet service provider
IFF	Interchange file format
JFF	JPEG file format
JFIF	JPEG file interchange format
JIF	JPEG image format
JPEG	Joint Photographic Experts Group, a standards organization
JPG	three-letter file extension for JPEG
PBM	portable bit map
PGM	portable gray map
PNG	portable network graphics
RAW	extension for undeveloped image from digital camera (virtually lossless)
XBM	X bit map
Vector Formats (for line drawings)	
CGM	computer graphics metafile
EMF	enhanced metafile, version of WMF
EPS	encapsulated postscript, native to Adobe® software
PDF	portable document format, native to Adobe® software
SVG	scalable vector graphics
SWF	Shockwave Flash®
WMF	Windows® metafile, native to Microsoft® software

For this reason, an image with much solid color requires a small bit map.

The term *pixel* can also mean an element in a sensor array. An array is a group of adjacent receptors each of which records a single point of radiation and which together provide a composite picture. Such arrays are in cameras and other radiation detectors (X-ray fluoroscopes and infrared cameras) at various wavelengths.

File Formats

File formatting has been evolving rapidly with advances in digital technology. Several standards may be consulted or referenced for digital image archives.[8-11]

The file formats for digital photographs fall into two classes: lossy and lossless (Table 2). Lossy formats compress files to save hard disk space. A format is said to be *lossy* when digital information and hence image detail are lost. JPG files are lossy and are used in many digital cameras and web pages because their compression makes small files and because they can be viewed on personal computers. Lossless TIF and PICT and nearly lossless RAW formats provide similar files but use more disk memory.

Inspectors who archive digital photographs need to be aware that converting files from a lossless to a lossy format (for example, from a TIF to a JPG) can degrade images so that some details, such as crack tips, are harder to see. There are several occasions when this sort of image degradation can take place: (1) when a file is changed from a lossless to a lossy format, such as TIF to JPG; (2) when a lossy image is converted to a file with lower size or resolution; (3) when an image is imported into a document created by a word processing or layout program. This reduction may be inadvertent, caused by program settings. Of course, an image conversion may have no effect on the usefulness of the resulting image: though degraded, it may still show all features of interest.

Procedures need to be clear so that they can be followed. The procedures describing how photographs are archived should include definitions of *lossy* and *lossless* and of file formats so that the results can be interpreted later.

The purpose of the camera RAW format is to record all of the exact data from the camera's sensor and any other metadata, such as time and camera settings. Camera RAW files are much larger than JPG files. Some RAW formats do not use compression; others implement lossy data compression to reduce file size without affecting image quality. Camera RAW formats avoid the compression artifacts inherent in JPG files. RAW files have many advantages over JPG files, including the following:

1. Image quality is higher.
2. RAW conversion software gives the user fine control of more parameters such as lightness, color balance, hue and saturation.
3. Edits are performed nondestructively, without degrading image quality.
4. If the file is not compressed, the maximum amount of image detail is always kept within it.

Camera manufacturers use different versions of RAW format. As of 2009, there is no widely accepted standard for RAW format, so more specialized software may be required to open RAW files than to open standardized formats like JPG or TIF. Cameras that support RAW files typically come with proprietary software for conversion of their RAW format to TIF or JPG. To retrieve an image from a RAW file, the data must first be converted into a red, green and blue (RGB) image.

Because of proprietary formatting, some formats listed in Table 2 are in fact families of related formats. There are several forms of TIF and of PNG, for example, and a given version may not work with a given program. Before using new equipment, the inspector will want to check that test images can be transferred to and viewed with hardware at the archiving location.

Image Characteristics

Many computer programs are widely available for processing digital images. They let the user adjust settings that, in the twentieth century, used to rely on the skill of a photographer at the moment of image capture or film development.

Color

Hue. A color is described by its hue, saturation and value (HSV). Hue is what is normally thought of as color. A digital color in a computer display is a blend of three primary hues: red, green and blue. Saturation is the intensity or dominance of a hue: a hue can be increased or diminished in an image just as black versus white can be adjusted in an image through all shades of gray. Value is the lightness or darkness of a hue and is usually called *brightness* when the three hues are combined in white light. Image processing software permits the intensity and value of these hues to be adjusted independently within each image.

Conversion to Gray. If the image is converted to gray scale, as in a report printout, care must be taken that features of interest remain visible: a fluorescent nondestructive test indication in brilliant green may disappear entirely if it changes to a shade of gray like that in an adjacent area.

Color Balance. The camera setting of white balance, discussed above, can improve the realism of color photographs by preventing unnatural tints. Image processing programs also enable improvement of color balance. Program menus may refer to it by other terms, such as *color intensity* or *hue saturation.*

If a photograph is taken in a digital camera's RAW file format, image colors can be fixed without information loss. Most cameras take RAW photos in twelve-bit color (4096 shades per color) instead of eight-bit color (256 shades per color), enabling powerful balance adjustments without visible loss in quality.

Size

The more pixels in an image, the more information it carries. An image's quantity of pixels is its true size in the dimensionless electronic space of the computer.

To be viewed, the image must be translated to a medium such as a computer display or printed page, where details become visible to human eyes in order to be interpreted. In a process known as *scaling*, the pixels can be compressed or spread out to fit the desired viewing surface (Fig. 15). Scaling affects resolution, discussed below.

Image size and resolution are usually expressed in terms of image width, in the horizontal dimension. The size and resolution of the vertical scale can be manipulated independently but to prevent distortion usually undergo the same processes as the horizontal.

The number of pixels in an image can be decreased or increased in a step called *resampling* or *conversion*. Resampling cannot add more data to a photograph once it has been shot.

FIGURE 15. Example of scaling of digital images.

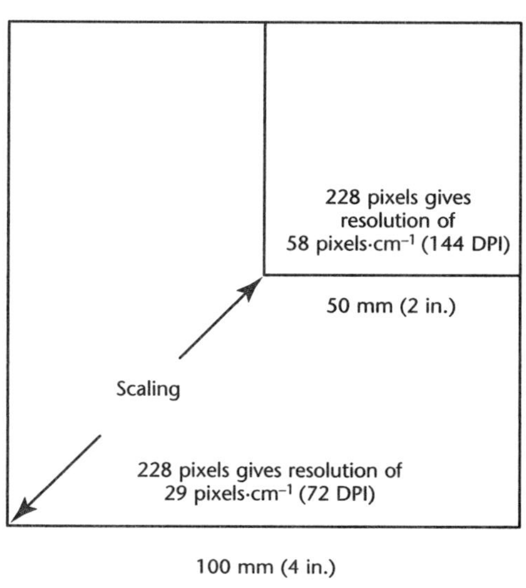

Resolution

The resolution of an image is its ability to distinguish small adjacent objects. Two thin adjacent lines appear as two thin lines, for instance, and not as one thicker line. Resolution depends on the number of pixels in the image and hence on its file size. In Fig. 16, two versions of the same image show the difference between high and low resolution. Details in an image cannot be restored simply by converting it to a higher resolution or from a lossy to a lossless format. Once an image's resolution is reduced, details are lost. If a display's resolution exceeds an image's resolution, the image will look blurry.

Resolution is measured in dots per inch (DPI) for printers or square pixels per inch (PPI) for computer displays, measured along the horizontal axis. Alternatives using SI metric units are to specify pixels per centimeter (pixels·cm^{-1}) or the pixel width in micrometers (µm).

Because the number of pixels in a scalable image is fixed, however, an expression of resolution is meaningless

FIGURE 16. Visible light photograph showing magnetic particle indication under ultraviolet lamp: (a) high resolution, at 40 pixels·cm^{-1} (100 PPI); (b) low resolution, at 10 pixels·cm^{-1} (25 PPI). Lower resolution blurs details. Lowering resolution sometimes enhances contrast or increases hue intensity — side effects better achieved by image processes that do not sacrifice detail.

(a)

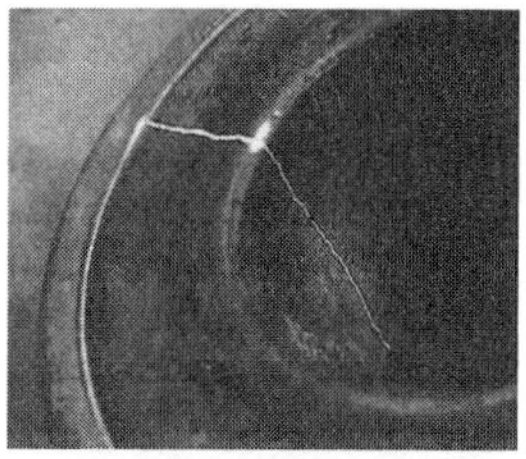

(b)

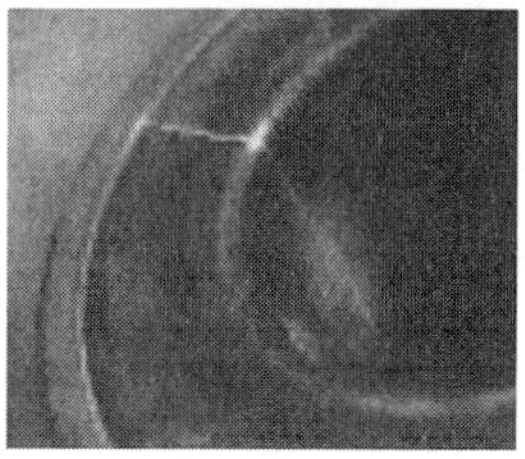

unless the viewed image size is also specified. If a given digital image is reduced to half its width, for instance, the number of pixels per unit of width and the resolution are doubled (Fig. 15). The data in the image file may remain the same, but smaller size makes details harder to see. Someone processing digital images must make decisions balancing image size versus resolution.

Brightness

Brightness is the common term for the luminance of light emitted by each pixel. Usually, one control setting adjusts brightness to affect all pixels in an image simultaneously. This parameter can be adjusted to compensate for poor visibility on overcast days. Too little brightness can make a picture dark as night, and too much brightness can make it washed out so no features stand out.

Saturation. In a color image, saturation is the intensity of a hue. A hue of zero saturation is black or gray. If all three hues have zero saturation, the image is colorless, black.

Contrast. The setting of contrast controls the degree of difference between light intensities at different points in the image. Like brightness, contrast is adjusted to affect all pixels in an image at once. As contrast is increased, for example, a dark indication becomes easier to see on a bright background and a bright indication becomes more visible on a dark background.

Contrast and brightness interact closely, and the inspector must sometimes adjust them to find the correct settings for each photograph. Figure 17 shows the effect of brightness and contrast on an image.

Image Integrity

Inspectors may be tempted to enhance images to justify accept/reject decisions. This temptation should be treated with caution. The inspector may consider several rules of thumb.

FIGURE 17. Photograph of cracked rod: (a) high contrast, high brightness; (b) high contrast, low brightness; (c) low contrast, high brightness; (d) low contrast, low brightness.

1. Local features including indications must not be exaggerated or obscured using image processing tools having descriptors such as *pencil, eraser, flood, spray* or *smudge*. These alterations of an image could be considered falsification of test results.
2. Image settings such as zoom, brightness and contrast may be freely adjusted, as photographers have done for generations.
3. A photograph of the reference standard should be included in the archive and used to evaluate modifications made to test images. How do the same modifications affect discontinuity visibility in the reference standard?

For some applications, a written statement on image processing may be a desirable part of an inspection procedure. Accept/reject decisions are usually made at the moment of inspection and in the presence of the test object. If some decisions are made later, that protocol can be documented in the procedure.

In some archives, a file's metadata, including its creation date, are part of its documentation. These metadata can be lost if an image is saved in another format.

Closing

In practice, the viewing quality of a digital photograph depends on several factors.

1. The vision acuity of the viewing human can be impaired by medical conditions such as myopia and color blindness.
2. The visibility of a shot is affected by things such as light and camera position.
3. The useful detail in an image is affected by its size and by its quality in terms of features such as contrast and resolution.

Technology has given the inspector options beyond the scope of the present discussion: (1) light measurement and illumination, (2) optical gaging and range finding, (3) photography sensitive to other wavelengths, including ultraviolet and infrared, and (4) video documentation of inspection, more for procedures than for indications. These and other options depend on hardware. The present discussion also does not consider marine environments or extremes of altitude or temperature.

Most inspectors already use a computer for communication, for research and for writing procedures. With the investment of a few hundred dollars for a camera and software, an inspector can add digital photography to the inspection options. With practice and planning, digital photography can be a valuable tool for inspection and quality control.

PART 3. Video[10,11]

The considerations discussed above for still images pertain also to motion pictures, or movies. The term *video* applies specifically to movie signals transmitted electronically. There have been three different platforms for moving pictures.

1. Film projection was used for commercial cinema in the twentieth century. Each frame was in effect an individual color slide, and they were sequenced on spools, or reels. A long movie consisted of thousands of frames and required several reels. Film of smaller width, and hence less resolution, was widely used at home and by industry. Some film was in black and white.
2. Analog video was used for television. The moving images were converted electronically into a series of horizontal lines that scanned across the screen successively in a raster pattern. Analog television can be in color or in black and white.
3. In digital displays, a movie consists of a series of still images, or frames, viewed in succession to recreate the appearance of motion (Fig. 18). Digital video can be viewed on a computer display or on a digital television screen.

Analog versus Digital[10]

An analog signal is any continuous variable signal and consists of a continuum of values (Table 3). It differs from a digital signal in that small fluctuations in the signal are meaningful. The primary disadvantage of analog signaling is that, if any signal acquires random noise as it is copied or transmitted, these random variations become dominant. Electrically these losses are lessened by shielding, good connections including coaxial and twisted pair cables.

A digital system uses a series of discrete values rather than a continuum of values. Waves are encoded as digital signals in a sequence of ones and zeros in the binary language of computing. Each selected individual point on the line is an integral single value. When reproduced on a video screen, a large number of single values appear as a stream of values representing the original wave.

It should be pointed out that videotape presents its analog signal as individual frames that appear as a succession of discrete image signals. These frames, however, are an artifact of the display medium. What appears to be a single analog frame is built out of adjacent lines in a raster arrangement. The signal input to the video recorder is continuous, the raster lines succeed each other more quickly than the human eye can track and the whole screen display is a composite of these line signals.

A great variety of analog devices persist in the twenty-first century: film cameras, microphones, speakers, luminaires (lamps), switches, machinery controls (for sorting in assembly lines, for example), cathode ray tubes and closed circuit monitors. In electronics, a digital-to-analog converter is needed to convert a digital, usually binary, code to an analog signal, usually an electric current or voltage. Signals are converted with switches or a network of resistors or current (electricity) sources. An analog-to-digital converter performs the reverse operation. Many digital devices,

FIGURE 18. Falling object in video: (a) 30 frames per second; (b) 25 frames per second.

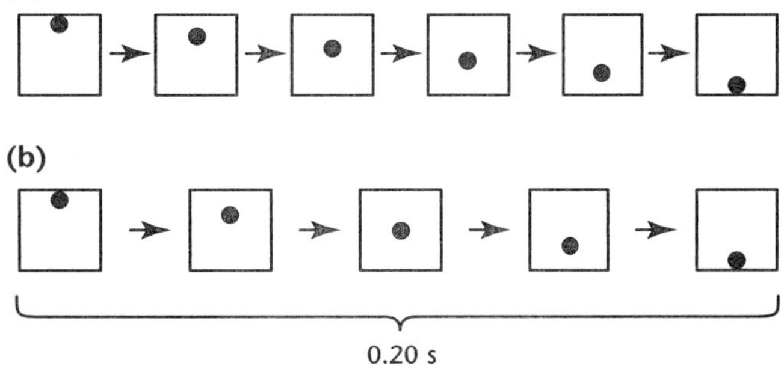

TABLE 3. Analog versus digital formatting.

Format	Sampling Interval
Analog	continuous stream of data
Digital with low sampling	values for discrete moments
Digital with higher sampling	discrete values in briefer intervals

including household appliances, incorporate circuitry to convert signals from a digital interface to analog performance, or vice versa, automatically.

Discontinuity Size Visibility[11]

In the twentieth century, the resolution of an analog television system was expressed as the number of lines in the picture. The electron beam produces a picture by drawing repeated lines of varying brightness across the tube. In analog video broadcasting, a signal with 525 lines has been used. About 480 lines actually form the picture and the balance are used to return the beam from the bottom to the top of the screen. There is also a kind of resolution in the horizontal direction, because television monitors are designed to have equivalent horizontal and vertical resolution. Closed circuit television systems used for visual nondestructive tests sometimes had resolution of 500 lines, higher than consumer broadcast systems of about 200 lines.

Usually, a video system cannot resolve narrower than a pixel. The smallest detail that can be resolved is predicted without magnification. If a lens with 2:1 magnification is used, the requirements on the electronic system are not so high and the size of the detail that can be resolved is smaller. In practice, it is best to magnify by moving the camera closer to the test object than merely to plug the video output into a larger screen. Also, because of considerations such as orientation and lighting, indications may be detected that are narrower than a resolution calculation would predict. Moreover, resolution is a characteristic of an entire system — the recording medium, the video camera and the monitor. The best way to quantify the resolution of a system is to use vision acuity charts and test objects with known discontinuities.

Photoelectric Sensors[11]

The photoelectric effect is the emission of electrons from some materials in the presence of electromagnetic radiation. Photoelectric sensors tuned to radiation were used for television cameras in the twentieth century. They converted information in the form of light into electrical signals that could then be amplified or processed to let the observer gather and interpret the test data. The information gathered may be from the visible region or in a form invisible to the eye, such as X-ray, ultraviolet or infrared radiation. There are two broad classifications of photoelectric devices: (1) detectors or measuring devices and (2) a two-dimensional imagers of visual data over an area.

A major category of photoelectric sensors, solid state image amplifiers, is discussed below.

Photoemissive devices include photoemissive cells and photomultipliers, photoconductive cells, photovoltaic cells and various devices that measure radiant energy directly, such as bolometers (thermal radiation detectors) and thermocouples. In the second category are various image converter tubes, image amplifier screens and television pickup tubes. Photoemissive devices feature materials that emit electrons under the influence of electromagnetic radiation, such as light. These electrons are then drawn away from the emitting surface by an electric field and used as the signal current, which may actuate relays to be amplified electronically. The emission of electrons in response to light is explained by the quantum theory of radiation. A potential barrier at the surface is the total energy an electron must have to escape. When its energy exceeds this threshold, an electron is emitted.

The photoelectric cell or multiplier phototube is used in many ways in industrial nondestructive testing. One of the most common uses is the measurement of radiant flux. In many respects, the phototube exceeds the capabilities of the human eye. It can detect not only radiation invisible to the eye but can also accurately measure quantities of light without reference to a standard, as required in a visual photometer. The familiar photoelectric exposure meter is no more than a barrier layer photovoltaic cell and a sensitive meter.

In addition to measuring light flux, photoelectric devices permit measurements of reflectance and transmission of materials, comparisons of two or more sources of light and (with the aid of filters or some type of spectrometer) colorimetric measurements. Phototubes find a natural application in spectrophotometry. Next to the measurement of radiant flux, perhaps the most widespread use of photoelectric devices is in monitoring and control applications: street lamps, smoke detectors, door openers and safety interlocks on punch presses. They inspect bottles after washing and detect foreign matter in filled soft drink bottles. They sort products such as beans, peas and coffee and actuate mechanisms for rejecting discolored products.

Many materials (primarily metals) permit photoemission. By far the most important such materials are compounds and alloys of the alkali metals, principally cesium. Of these, the most widely used are cesium antimony alloys and cesium oxygen silver compounds. The cesium antimony emitters are true alloys, while the cesium oxygen silver emitters are complex surfaces made from layers of cesium, cesium oxide and silver oxide in varying proportions on a silver base. Cesium antimony surfaces may have a quantum efficiency of 10 to 30 percent. Photoemitters of this type have responses that are selective to wavelength of illumination, so some respond to infrared and some to visible radiation or some other waveband. This selectivity varies greatly with processing.

Solid State Image Amplifier[11]

The solid state image amplifier is a sandwich of a photoconductive layer and an electroluminescent layer. The electroluminescent layer is composed of a material that emits light in response to an applied voltage. The photoconductor and the electroluminescent material are essentially in series across a suitable alternating current voltage supply. In darkness, the photoconductor is highly resistive and passes no current. When light falls on it, it becomes conductive and current flows through the sandwich, causing the luminescent material to emit light in the illuminated regions. Light amplifications up to 1000× have been obtained.

A modification of this light amplifier is the amplifying fluoroscope. In this device, the photoconductive material responds directly to X-ray radiation and, by the same principle as the light amplifier, converts it to visible light. An amplification of 100× over the output of a conventional fluoroscope has been obtained.

Charge Coupled Device[10]

Digital cameras are used to capture still images. The image can then be saved to memory and transferred to a computer. From the computer, the image can be manipulated, shared, stored, printed and much more.

Most digital cameras use a charge coupled device (CCD) as an image sensor. The charge coupled device is a collection of photosites, light sensitive diodes that convert light into electrical charge (photons into electrons). The brighter the light that hits a single photosite, the greater the electrical charge that will be accumulated. This electrical charge is then transported across the chip and read in one corner of the array. Each pixel's value is then turned into a digital value by an analog-to-digital converter. The charge coupled devices use a special manufacturing process to create the ability to transport to the chip without distortion. This process leads to sensors of high quality in terms of fidelity and light sensitivity.

The charge coupled device imager is a solid state silicon chip similar in structure to a photovoltaic cell but more complex. The charge coupled device is widely used for digital memory and analog signal processing, as well as optical imaging. A charge coupled device camera system consists of a lens system, a charge coupled device imager and processing electronics. The device includes a monolithic array of many loosely spaced, light sensitive, metal oxide semiconductor capacitors that transfer an analog signal charge from one capacitor to the next. Each capacitor corresponds to a picture element, or pixel, in a display.

The capacitors that collect the electrical charge are colorblind: they track only the total intensity of the light that hits the surface. To get a full color image, filters are needed. These filters allow the light to be seen in the three primary colors, which are then added together to attain a full color image in the full spectrum of colors. Each photosite collects varying amounts of electrical charge and therefore represent brighter and darker sections of the image. The electrical charge in each photosite is analog information, measured and digitized into binary form.

Figure 19 shows in detail how each pixel works. The lens focuses the image of an object on the surface of the charge coupled device, where the light is converted to electrons. This effect is photoelectric, the basis of common devices such as automatic night lights and certain types of paper copiers. The pixels are separate from each other, and the number of electrons produced in each one is proportional to the intensity of light focused on that particular pixel. This differential charging contains information needed to form an image.

FIGURE 19. Activation of one pixel from charge coupled device.

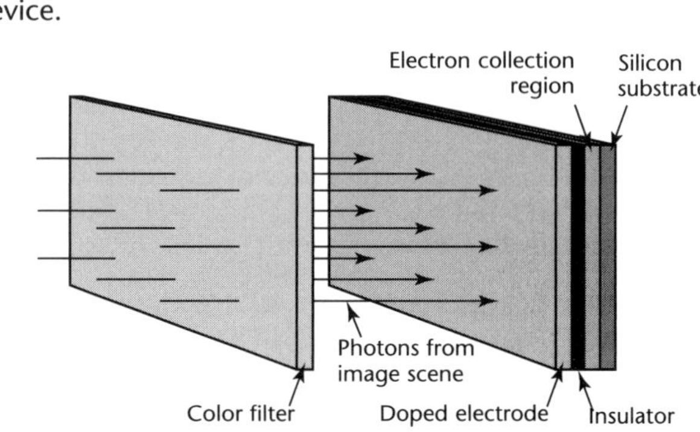

The amount of detail that can be captured is called the resolution and is measured in pixels. The more pixels there are, the more detail that can be captured. The more detail there is, the more a picture can be enlarged without becoming grainy and looking out of focus.

Some typical resolutions include the following: 256 × 256 pixels, resolution is so low that the picture quality is almost always unacceptable, 65 000 total pixels; 640 × 480 pixels, 307 000 total pixels; 1216 × 912 pixels, 1 109 000 total pixels; 1600 × 1200 pixels, almost 2 million total pixels; 1920 × 1080 pixels, 2 073 600 pixels is the pixel count for high definition television displays.

Unlike common photovoltaic cells, charge coupled device pixels feature a storage cell in each element that collects and stores the electrons as they are produced. The image of the object is then stored in the form of electrons on the charge coupled device, but is of no use unless the location and number of electrons in each pixel can be identified.

At this point, the charge coupling effect of the charge coupled device comes into play. Finite amounts of electrical charge (that is, packets of electrons) are transferred from one pixel site to the next, just as water is moved from one bucket to the next in a bucket brigade (Fig. 20). The force required to move these electrons is produced in the pixels by means of a voltage that alternately turns on and off. As this clock voltage alternates, the electrons move from one site to the other, and after the last pixel the electrons are converted to a voltage proportional to their number, this voltage is amplified, filtered, and then applied to the input of a video monitor.

Charge coupled device image sensor systems produce a brighter, higher resolution image than do conventional fiberscopes. Because of the charge coupled device's diminutive size, the silicon chip can be placed within the tip of a small diameter probe capable of penetrating the smallest apertures. Its advanced microelectronic technology enables the charge coupled device to function as a miniature television camera able to record and display images with great clarity on a video monitor. The device's ability to generate, detect, and shift distinct packets of electrons makes charge coupling suitable for many different information processing applications, including image sensing in video camera technology.

FIGURE 20. Electron transfer in charge coupled devices — analogy of bucket brigade.

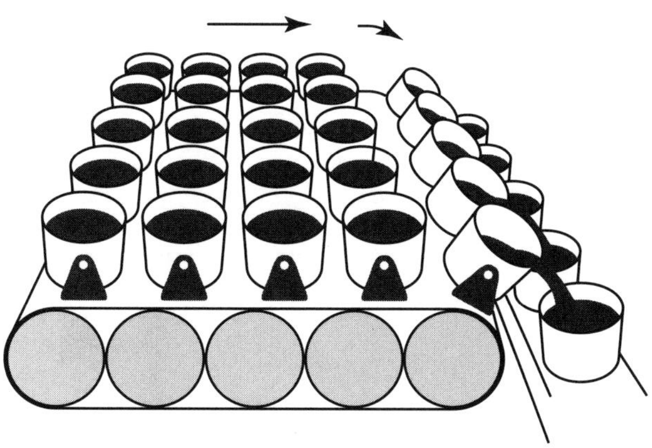

Digital File Formats

Codecs

A codec is software that en*co*des or *dec*odes a data stream for digital processing. Video codecs *co*mpress or *dec*ompress video files for transmission, storage or viewing. Most codecs are lossy, sacrificing data for the sake of efficiency. Many codecs are proprietary, designed to work with particular commercial software or hardware products; compatability problems can arise among versions, platforms and competing software.

A codec determines the format of the video file created with it, so the visual inspector selecting a codec for visual testing purposes must consider what format is desired for short term evaluation of files and for their long term archiving. The selection of codec also affects the selection of hardware and vice versa.

Codecs define the video settings such as the frame rate and size. There are many video codecs and formats. Some are listed in Table 4.

Moving Pictures Experts Group (MPEG)

The Moving Pictures Experts Group is a body in the International Organization for Standardization that develops consensus standards for the encoding of audiovisual information in digital formats. Codecs conforming to their standards use high levels of compression to achieve small file sizes; they format and read digital video files with extensions such as *MPG, MOV* and *AVI.*

MPEG-1. This codec is available in all personal computers in the twenty-first century.[17]

MPEG-2. A refined version of MPEG-1, MPEG-2 allows networking and multiple channels and is used on digital video disks and the internet.[18]

MPEG-4. This codec offers more efficient compression and is used in digital television and interactive media. MPEG-4 facilitates protection of intellectual property.[19]

MPEG-7. This codec specifies standardized metadata to enable search, tagging, cataloging and indexing.[20]

MPEG-21. This codec standardizes delivery across multiple platforms.[21]

MPEG-A. This codec integrates the features of its predecessors.[22]

Connectors

Table 5 lists several kinds of media for data transfer and storage.

Coaxial Cables[10]

Coaxial cable is used for transmission of analog or digital signals. In coaxial cables, a center conductor is surrounded by heavy dielectric insulation that is braided or foil shielded. The center-to-shield distance is controlled so that the impedance of the coaxial line matches the design load of the video circuitry feeding the signal. This matching minimizes losses and enables maximum power transfer. Most video coaxial cable has an impedance of 75 Ω.

Coaxial cables are used for video signals and can serve as control cables, actuating remote equipment.

TABLE 4. Digital formats and codecs of video files.

Designation	Format
Formats	
AVI	Audio Video Interleave
FLV	Shockwave Flash® file extension
MPEG	Moving Picture Experts Group, a standards body
MPEG 1	video format specification for compressing NTSC input
MPEG 2	video specification, compatible with analog interlacing
MPEG 3	video format specification
MPEG 4	video format specification
MP3	video format from MPEG 3
MP4	video format from MPEG 4
MOV	QuickTime® file extension
WME	Windows® Media Encoder
WMV	Windows® Media Video
SWF	Shockwave Flash®
VOB	video object, container format in DVD-Video media
WMF	Windows® metafile, native to Microsoft® software
Disk formats	
BD	Blu-Ray disk
CD	compact disk
DVD	digital video disk
ISO	extension for a file container that boots as virtual video disk

Ethernet

Ethernet cables are made of copper or fiber optic filaments and widely used for land based transmission of digital signals.[23] Ethernet was introduced as an upgrade to coaxial cable.

Thumb Flash Drives[10]

Thumb flash drives are convenient for plugging into a laptop for transfer of data or image files. In 2009, they use a universal serial bus connector and can be hung around the inspector's neck on a lanyard, attached to a key ring or kept in a pocket of the carrying case for a laptop computer. Inexpensive models in 2010 hold several gigabytes of data.

Universal Serial Bus (USB)

The universal serial bus is widely used for connected computer components. There are several generations and configurations of the bus. Four configurations developed by year 2010 are discussed below.

1. The USB 1.0 has a slow transfer speed of 1.5 megabytes per second and does not allow for streaming video. The USB 1.0 has been superseded by later designs.
2. Called *full speed,* the USB 1.1 has a transfer speed of 12 megabytes per second and allows for video transfer with an audio track. It has been widely used for most accessory devices in the first decade of the twenty-first century.
3. Called *high speed,* the USB 2.0 has a transfer speed of 480 megabytes per second and requires a correspondingly fast processor. It is backwardly compatible with USB 1.1 and uses the same connection.
4. Called *super speed,* the USB 3.0 has a transfer speed of 4.8 gigabytes per second and requires a correspondingly fast processor. It is backwardly compatible with most earlier USB devices and uses the same connection.

TABLE 5. Media for data storage.

Medium	Portability	Capacity
Floppy disk	poor	poor
Compact disk (CD)	good	good for images; poor for video
Digital video disk (DVD)	good	good
Blu-Ray™ disk (BD)	good	excellent
Flash™ drive	excellent	good for images; poor for video
External drive, cabled	awkward	excellent
External drive, wireless	connectivity limits	excellent

IEEE 1394

The standard IEEE 1394[24] defines a cable and connector similar to the universal serial bus. Its popular trade names include Firewire® and Lynx®. The IEEE 1394 connector preceded the universal serial bus and was intended to have similar goals. Using twisted pair wiring to move the data, IEEE 1394 has a serial bus that can handle over 400 megabits per second and over 60 devices per bus. IEEE 1394 supports devices that stream data in real time, including video. IEEE 1394 connections are inherently faster than USB 1.0 and so have been popular for fast transfer of large data streams, as in video recording.

Analog Video Connectors

Analog video connectors are used for transmitting signals to or from an analog device such as a telephone cable, television screen or video cassette player.

1. Connections referred to as *composite video* have one connector. Supplementary connectors may carry audio or control signals.
2. Connections referred to as *component video* carry analog signals and terminate in two or more plugs or sockets (sometimes colored red, green, blue and black).
3. Connections referred to as *separate video* (or *S video*) split the color signal among three or more wires in one cable and transmit images inferior to those provided by component video.

Visual inspectors may have to use analog connectors when working with video players. Analog videotapes may have been archived to document the service history of pressure vessels in a power plant, for example. The analog signals may be converted from analog to digital files through a modem.

Video Displays

Digital screen displays are available in a wide variety of sizes. Some popular sizes, measured in vertical pixels by horizontal pixels, are listed in Table 6.

Liquid Crystal Display[10]

Liquid crystal displays (LCDs) are used for color displays in many television screens, some computer monitors and many portable devices. Liquid crystal allows the display to be much thinner than cathode ray tube technology and consumes much less power.

The images are produced on flat surfaces by shining light through liquid crystals and colored filters. The liquid crystal display allows a flat display device made up of pixels in rows and columns to display an image. The individual pixels may emit colors or be monochromatic. The low levels of electric power produces a lit pixel by backlighting the screen, passing light through a layer of liquid crystal between two polarizing filters and the transparent electrodes between them. When the electrodes relax, the liquid crystal is without electrical charge and dims. Because the axes of polarity are perpendicular to each other, the light will not pass through. When the electrical charge is applied to the pixel, the molecules align themselves in helixes so that the light may pass from one side with vertical filters to the other side with horizontal filters. Hence the polarized filters oriented perpendicular to one another no longer block passage of the light once the liquid crystals have gained a helical twist going from horizontal to vertical. Uncharged crystals will be completely untwisted and the polarized perpendicular filters will block all light transmission. The more the crystal is charged, the more the helical twist will occur and allow more light to pass through the transparent crystal and more intensely light the pixel.

Plasma Display

Plasma displays operate similarly to liquid crystal displays. Electric charges stimulate minuscule packets of gas on the viewing surface so that they glow with desired color and intensity and together form a two-dimensional image.

Light Emitting Diode

A light emitting diode (LED) is a kind of transistor. Two-dimensional arrays of these transistors are actuated by the video signal and emit light of desired frequencies and intensities.

TABLE 6. Dimensions (in pixels) of some fixed grid displays. These specifications are sometimes called *resolution*, but true resolution depends also on physical dimensions of display.

Arrays (pixels)	Total Pixels	Aspect Ratio	Application
——	——	3 to 2	analog television
320 × 200	64 000	8 to 5	thumbnails and previews
640 × 480	307 000	4 to 3	portable devices
1024 × 768	786 432	4 to 3	extended graphics resolution
1600 × 1200	1 920 000	4 to 3	television
1920 × 1080	2 073 600	16 to 9	high definition television

Frame Rate

In digital video, moving pictures are produced by successive display of what are essentially still images, each image corresponding to a frame. Figure 18 uses the example of a falling object to show the difference between two frame rates. Capture rates of 60 frames per second are common in digital video cameras. Human perception performs well in viewing frames rates over 20 frames per second. In the second half of the twentieth century, most movies from digital files are each encoded to play at a particular rate, in synchrony with the computer's integral clock. Personal computers can display video recordings at any frame rate the visual inspector is likely to encounter.

Analog Image Transmission and Recording[10,11]

The following brief discussion of analog technology is included for visual inspectors who must work with quality assurance archives of service history. Fuller discussion is available elsewhere.[11]

Cathode Ray Tubes

Cathode ray tubes (CRTs) were found in most computer monitors and television sets in the second half of the twentieth century. The cathode is a heated filament inside a glass tube. The ray is a stream of electrons that pour off the heated cathode in a vacuum. Because the anode is positive, it attracts the electrons pouring off of the cathode. In cathode ray tubes, the radiation is a beam of electrons. In a black and white screen, there is only one phosphor that glows white when struck by the beam of electrons. In a color screen, there are three types of phosphors — red, green and blue — and three different electron beams to illuminate the colors together.

Video Tape Technology

The main task of video technology is to convert an optical image of a dynamic scene into an electrical signal that can be transmitted in real time to another location, and converted back without delay into an optical image that faithfully reproduces the original scene. Sound, synchronized to the scene, commonly accompanies the video image.

A video camera converts the two-dimensional optical image into a series of electrical impulses, the video signal. The video signal is converted by the monitor back into a two-dimensional picture that reproduces the dynamic scene of the original image in real time.

To dissect the optical image so that it can be represented as a series of electrical signals, narrow strips of the image are scanned. In principle, the image can be scanned by any electrooptical transducer that rapidly converts light intensities into electrical voltages or currents.

Line Scanning

In displaying a video image, analog video uses the raster scanning technique whereby electrons are beamed onto the phosphor coating on the screen a line at a time from left to right starting at the top left corner. At the end of the line, the beam is turned off and moved back to the left and down one line (Fig. 21), When the bottom right corner is reached, the gun is returned to the top left corner.

Interlacing illuminates a screen by displaying all odd lines in the frame first and then all even lines (262.5 scan lines each) which are combined and interlaced to produce a complete frame of 525 lines. The 2:1 standard interlace ratio means that even fields of horizontal scan fit neatly into odd fields of horizontal scan.

Figure 21. Electron beam scan path on analog video screen.

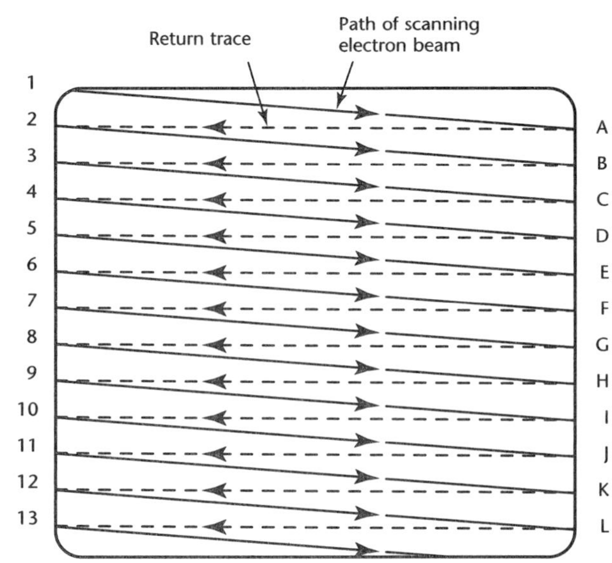

Analog Video Systems

Before the move to digital video in 2009, three systems were used worldwide for analog video: the National Television System Committee (NTSC), the phase alternating line (PAL) system and the SECAM (*séquentiel couleur à mémoire,* "sequential color with memory") system.

1. NTSC broadcasts are fully compatible with the video home system (VHS) standard for video tape recording and playback. In the United States, electric power is provided at 120 V at 60 Hz. The National Television System Committee developed the television standards in the United States — broadcasts at 60 half frames per second of 262.5 lines interlaced twice (odd and even) to make 30 frames per second of image, or 525 horizontal lines of resolution with 484 visible lines of image at 30 frames and 60 fields per second interlaced, the rest being used for other information such as synchronizing, data and captioning. The NTSC system has been used in North America and parts of Asia.
2. PAL broadcast signals are fully compatible with video tape standards of the same name. PAL is the standard for color television that was developed in Germany; it has been used in many nations of Europe, Africa and Asia. Electricity in Europe typically operates at 220 V and 50 Hz. German broadcasts have been at 25 half frames per second of 312.5 interlaced lines twice (odd and even) to make 50 frames per second image, or 625 horizontal lines of resolution with 576 visible lines of image at 25 frames/50 fields per second interlaced, the rest being used for other information such as synchronizing data and captioning. Two frames equal one field of image.
3. The SECAM system antedates the other two; it has been used in France and parts of Africa, Asia and Eastern Europe.

Figure 18 shows the difference between NTSC and PAL formats in terms of frame rate, NTSC being 25 and PAL being 30 frames per second. The two formats were incompatible in the years of analog video tape. Personal computers can display digitized video recordings at either frame rate.

The analog video is not a compilation of successive still image as in digital video but rather is a series of cycles in a raster signal. What appears to be a single image when a video tape is paused is in fact an image that the video player has constructed from successive lines of the continuous raster signal. Discussions of video electronics often refer to frame rate in hertz (Hz) — that is, cycles per second. The two measures are equivalent; 1 frame per second equals 1 Hz.

Closing

Digital technology has revolutionized the practice of visual testing. Personal computers are used for processing, archiving and transmitting images. Remote and portable devices integrate microprocessors for recording and transmission of test images on site. And advances in battery life, hard disk memory and processor speed have made possible the acquisition of test documentation, including video records, impractical to obtain in the twentieth century.

References

1. Pasquazzi, D.[A.] and W.R. Tweddell. Chapter 11, "Other Applications of Visual and Optical Tests": Part 2, "Applications of Photography in Visual Testing." *Nondestructive Testing Handbook,* second edition: Vol. 8, *Visual and Optical Testing.* Columbus, OH: American Society for Nondestructive Testing (1993): p 313-319, 336.
2. Schmidt, J.T. Section 11, "Recording of Magnetic Particle Test Indications." *Nondestructive Testing Handbook,* second edition: Vol. 6, *Magnetic Particle Testing.* Columbus, OH: American Society for Nondestructive Testing (1989): p 271-290.
3. Schmidt, J.T. "Color Photography of Magnetic Particle and Penetrant Indications." *Materials Evaluation.* Vol. 31, No. 3. Columbus, OH: American Society for Nondestructive Testing (March 1973): p 39-42.
4. Mosher, T.A. "Ultraviolet Photography: A System for the Nonprofessional Photographer." *Materials Evaluation.* Vol. 45, No. 7. Columbus, OH: American Society for Nondestructive Testing (July 1987): p 778-779, 781.
5. Chapter 9, "Recording of Magnetic Particle Indications": Part 5, "Digital Photography and Image Archiving." *Nondestructive Testing Handbook*, third edition: Vol. 8, *Magnetic Testing.* Columbus, OH: American Society for Nondestructive Testing (2008): p 243-250, 257.
6. Moore, P.O. "Digital Photography and Image Archiving." *Materials Evaluation.* Vol. 66, No. 7. Columbus, OH: American Society for Nondestructive Testing (July 2008): p 717-722.
7. Stephens, P. "Liquid Penetrant Panel Calibration." *Materials Evaluation.* Vol. 65, No. 10. Columbus, OH: American Society for Nondestructive Testing (October 2007): p 1016-1018.
8. Ang, T. *Advanced Digital Photography: Techniques and Tips for Creating Professional-Quality Images,* revised edition. Garden City, NY: Amphoto Books (2007).
9. Kelby, S. *The Digital Photography Book.* Berkeley, CA: Peachpit (2006).
10. *Nondestructive Testing: Remote Visual Inspection* (training modules). Skaneateles, NY: General Electric Corporation, General Electric Inspection Technologies [2005].
11. Generazio, E.R., D.J. Roth and V. Torrey. Chapter 5, "Other Instrumentation and Electronic Aids for Visual Testing": Part 2, "Video Technology." *Nondestructive Testing Handbook,* second edition: Vol. 8, *Visual and Optical Testing.* Columbus, OH: American Society for Nondestructive Testing (1993): p 131-147, 154.
12. Sayler, G.[C.] *ASNT Level III Study Guide: Visual and Optical Testing Method.* Columbus, OH: American Society for Nondestructive Testing (1998, revised 2006).
13. Mikhail, E.M., J.S. Bethel and J.C. McGlone. *Introduction to Modern Photogrammetry.* New York, NY: Wiley (2001).
14. *Manual of Photogrammetry,* fifth edition. Bethesda, MD: American Society for Photogrammetry and Remote Sensing (2004).
15. Luhmann, T., S. Robson, S. Kyle and I. Harley. *Close Range Photogrammetry: Principles, Methods and Applications.* Dunbeath, Caithness, United Kingdom: Whittles (2006).
16. Linder, D. *Digital Photogrammetry: A Practical Course.* New York, NY: Springer (2009).
17. ISO/IEC 11172, *Information Technology — Coding of Moving Pictures and Associated Audio for Digital Storage Media at up to about 1.5 Mbit/s* [MPEG-1]. Geneva, Switzerland: International Organization for Standardization (1993, 2008).
18. ISO/IEC 13818, *Information Technology — Generic Coding of Moving Pictures and Associated Audio Information* [MPEG-2]. Geneva, Switzerland: International Organization for Standardization (1996, 2009).
19. ISO/IEC 14496, *Information Technology — Coding of Audio-Visual Objects* [MPEG-4]. Geneva, Switzerland: International Organization for Standardization (2004, 2009).

20. ISO/IEC TR 15938, *Information Technology — Multimedia Content Description Interface* [MPEG-7]. Geneva, Switzerland: International Organization for Standardization (2004, 2009).
21. ISO/IEC 21000, *Information Technology — Multimedia Framework* [MPEG-21]. Geneva, Switzerland: International Organization for Standardization (1996, 2009).
22. ISO/IEC 23000, *Information Technology — Multimedia Application Format* [MPEG-A]. Geneva, Switzerland: International Organization for Standardization (2007, 2009).
23. IEEE 802, *Information Technology, Telecommunications and Information Exchange between Systems — Local and Metropolitan Area Networks — Specific Requirements*. New York, NY: Institute of Electrical and Electronic Engineers (2009).
24. IEEE 1394, *A High-Performance Serial Bus*. New York, NY: Institute of Electrical and Electronic Engineers (2008).

Bibliography

Still Photography

ANSI/I3A IT10, *Photography — Digital Still Cameras*. Washington, DC: American National Standards Institute (2004).

ISO DIS 12232, *Photography — Digital Still Cameras — Determination of Exposure Index, ISO Speed Ratings, Standard Output Sensitivity, and Recommended Exposure Index*. Geneva, Switzerland: International Organization for Standardization (2004).

ISO/IEC 10918, *Information Technology — Digital Compression and Coding of Continuous-Tone Still Images*. Geneva, Switzerland: International Organization for Standardization (2005).

ISO TS 22028, *Photography and Graphic Technology — Extended Colour Encodings for Digital Image Storage, Manipulation and Interchange*. Geneva, Switzerland: International Organization for Standardization (2006).

Moving Pictures

Kingslake, R. SPIE Monograph PM06, *Optics in Photography*. Bellingham, WA: International Society for Optical Engineering (Society of Photo-Optical Instrumentation Engineers) (1992).

Ray, S.E. *Applied Photographic Optics,* third edition. Burlington, MA: Elsevier, Focal Press (2002).

Robin, M. *Little Digital Video Book,* second edition. Berkeley, CA: Peachpit (2008).

Underdahl, K. *Digital Video for Dummies,* fourth edition. Indianapolis, IN: Wiley, For Dummies (2006)

SMPTE 2019, *Mapping VC-3 Coding Units into the MXF Generic Container*. White Plains, NY: Society of Motion Picture and Television Engineers (2009).

5

CHAPTER

Direct Visual Testing

Gregory C. Sayler, MD Helicopters, Mesa, Arizona

Douglas G. Krauss, Huddleston Technical Services, Redstone Arsenal, Huntsville, Alabama (Part 5)

PART 1. Circumstances of Viewing

Direct visual testing is a technique of the visual testing method of nondestructive testing. The American Society for Nondestructive Testing divides nondestructive testing into methods and subdivides methods into techniques. Direct viewing is a technique of visual testing in the same way that bubble testing is a technique of leak testing and gamma radiography is a technique of the radiographic test method.

Direct visual testing differs from indirect techniques, sometimes called *remote* visual testing, because the direct inspector is in the presence of the test object and has an unmediated view of the test surface, even if he looks through a device such as a magnifier or camera. In remote, or indirect, techniques, the inspector uses a borescope or propelled camera to view surfaces otherwise inaccessible. The distinction is fine, for the inspector may use a "remote" apparatus on a test object held in his hand, and an inspection may be viewed "directly" through a system of lenses and mirrors.

Observer's Attitude[1]

A complete representation of the visual field probably is not present in the brain at any one time. The brain must contain electrochemical activity representing some major aspects of a scene, but such a picture typically does not correspond to how the observer describes the scene. This discrepancy occurs because the observer adds experience and prejudices that are not themselves part of the visual field. Such sensory experience may reflect physical reality or may not.

Sensory data entering through the eye are irretrievably transformed by their contexts — an image on the retina is perceived differently if its background or context changes. Perceptually, the image might be a dark patch in a bright background that can, in turn, appear to be a white patch if displayed against a dark background. No single sensation corresponds uniquely to the original retinal area of excitation.

The context of a viewed object can affect perception and, in addition, the intention of the viewer may also affect perception. The number of visible objects in a scene far exceeds the typical description of the scene. And a great deal of information is potentially available to the observer immediately after viewing. If an observer has the intention of looking for certain aspects of a scene, only certain visual information enters the awareness, yet the total picture is certainly imaged on the retina. If a scene or an object is viewed a second time, many new characteristics can be discerned. This new information directly influences perception of the object, yet such information might not be available to the viewer without a second viewing.

Vision is selective in many common situations. An individual can walk into a room full of people and notice only the face of an expected friend. The same individual can walk right by another friend without recognition if the encounter is unexpected. Vision is strongly selective and guided almost entirely by what the observer wants and does not want to see. Any additional details beyond the very broadest have been built up by successive viewing. Both the details and the broad image are retained as long as they are needed, and then they are quickly erased.

The optical image on the retina is constantly changing and moving as the eye moves rapidly from one point to another — the sensing rods and cones are stimulated in ways that vary widely from one moment to the next. The mental image is stationary for stationary objects regardless of the motion of the optical image or, for that matter, the motion of the observer's head. It is very difficult to determine how a unique configuration of brain activity can result from a particular set of sensory experiences. A unique visual configuration must be a many-to-one relationship requiring complex interpretation. If an observer does not apply experience and the intellect, it is likely that a nondestructive visual test will be inadequate.

Viewing Angle[1-3]

Eye muscles may manipulate the eye to align the image on the lens axis. Different sensors in the retina receive images of different objects in the field of view (Fig. 1), and different banks of sensors basically require different stimuli to best perform their functions. Also, light rays entering the lens at angles not parallel to the lens axis are refracted to a greater degree. The angle changes the quality and quantity of the light energy reaching the retina. Even the color and contrast ratios vary and affect depth perception.

The included angle of five minutes of arc is commonly cited as optimum; it is the average in which an individual sees a sharp image. There are other angles to be considered when discussing visual testing. The angle of peripheral vision is not a primary consideration when performing detailed visual tests. It is of value under certain inspection conditions: (1) when surveying large areas for a discontinuity indication that (2) has a high contrast ratio with the background and (3) is observed to one side of the normal lens axis. The inspector's attention is drawn to this area and it can then be scrutinized by focusing the eyes on the normal plane of the lens axis.

The angle of view is very important during visual testing. The viewer should in all cases attempt to observe the target on the center axis of the eye. Figure 2 shows how the eye perceives an object from several angles and how the object appears to change or move with a change in viewing angle.

The angle of view should vary ideally not more than 45 degrees from normal, and a recommended viewing distance and angle for visual testing is to have the eye within 600 mm (24 in.) of the object and positioned at an angle not less than 30 degrees to the inspection surface as shown in Fig. 3.

The same principle applies to objects being viewed through accessories such as mirrors or borescopes. The field of view should be maintained much in the same way that it is when viewed directly. If the examination surface is immovable and situated so that the eye cannot be placed within this region, suitable visual aids, such as mirrors, must be used.

FIGURE 1. Peripheral vision.

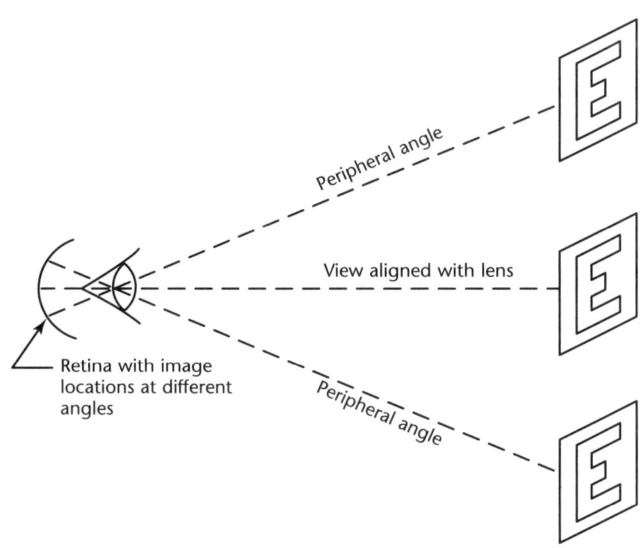

FIGURE 2. Eye position affects apparent size and location of object.

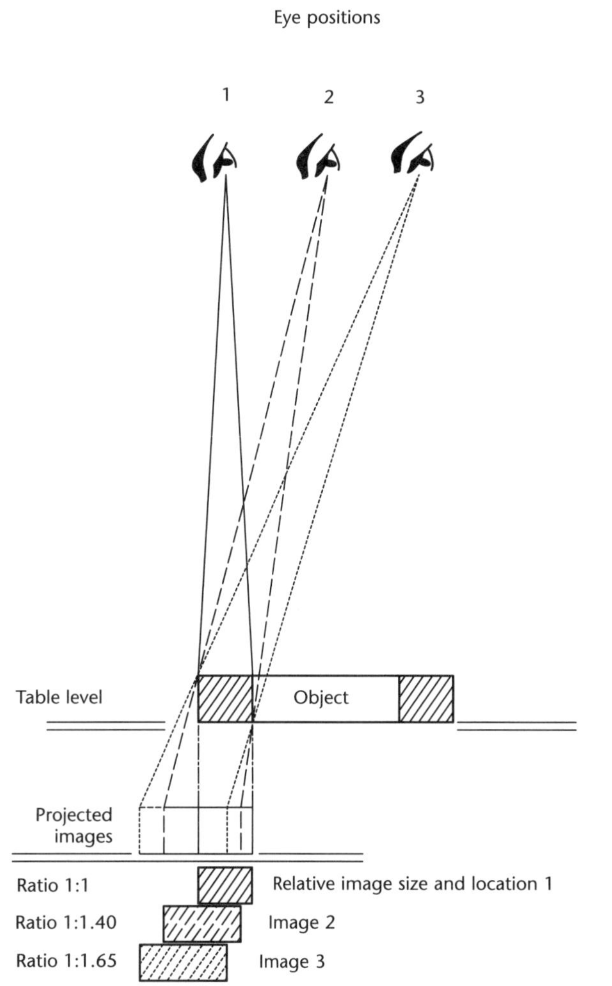

FIGURE 3. Range of viewing angle.

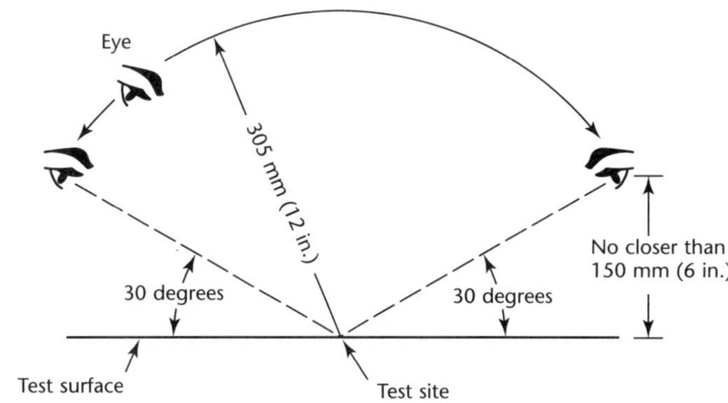

Pitfalls Related to Viewing Angle

Posture affects the way an object is observed. Appropriate posture and viewing angle minimize fatigue, eyestrain and distraction. The viewer's posture should make it easy to maintain the optimum view on the axis of the lens.

On reflective backgrounds, the viewing angle should be off normal but not beyond 45 degrees. This angle is maintained so that the light reflected off the surface is not directed toward the eyes, reducing the contrast image of the surface itself. It also allows the evaluation of discontinuities without distorting their size, color or location. The angle is important when using optical devices to view surfaces not available to direct line of sight.

Elevated objects are hard to test visually. Without adequate access, a surface beyond the reach of the inspector cannot be thoroughly inspected except in cases such as object verification or approximate location — to answer a question such as "Have the girders been painted yet?"

Mirrors[2]

Mirrors are common inspection aids. Easy to use, mirrors make inspection possible inside pipes and apertures and inside or behind objects obstructing the inspector's view.

Inspection mirrors come in a variety of shapes (Fig. 4). The head, holding the reflective surface, can be round or rectangular and range in size from 20 to 100 mm (0.8 to 4.0 in.). The handle, or stem, includes a scored surface or rubber sheath for easy handling. Smaller mirrors resemble those used by dentists and usually have a 150 mm (6 in.) long handle with a round head. The diameter of the handle can range from about 3 to 25 mm (0.1 to 1.0 in.).

Inspection mirrors frequently have telescoping handles that can almost double the handle length. Most have a double ball joint between the mirror and handle that lets the mirror swivel to any convenient angle.

The illumination of the area being inspected should be the same as that specified for the rest of the inspection.

Flashlights or other small portable light sources can provide adequate illumination, but strong direct lighting can cause contrastive shadows and reflected glare.

Many industrial environments are hard on the mirror's glass surface. A scratched mirror is a hindrance during the inspection. The mirror or its head should be replaced when marred.

Some specifications (in nuclear and aerospace industries) forbid the use of mirrors.

Light Manipulation by Mirrors

Mirrors change the direction of light by reflection and can be flat, convex or concave.

1. Flat or plane mirrors are arranged alone or in series to transmit light or an image.
2. Convex mirrors provide an enlarged field of view of the reflected image.
3. Concave mirrors each have a reflecting focal point. If a light is projected onto a spherical mirror normal to the curve of the surface, the light will be focused slightly in front of the mirror. If a point light source is placed at its focal point, the light will be reflected from the mirror so that it is parallel to the normal of the curve. A concave mirror can also be used for image enlargement. An image that is small compared to the width of the mirror will be reflected back in a diverging and optically reversed image.

FIGURE 4. Examples of inspection mirrors: (a) 32 mm (1.25 in.) diameter mirror, 375 mm (17 in.) long, with pocket clip and swiveling neck; (b) 60 mm (2.25 in.) diameter mirror with swiveling neck and length telescoping from 250 to 350 mm (10 to 14 in.); (c) 54 × 89 mm (2.13 × 3.5 in.) mirror with swiveling neck and length telescoping from 286 to 387 mm (11.25 to 15.25 in.).

(a)

(b)

(c)

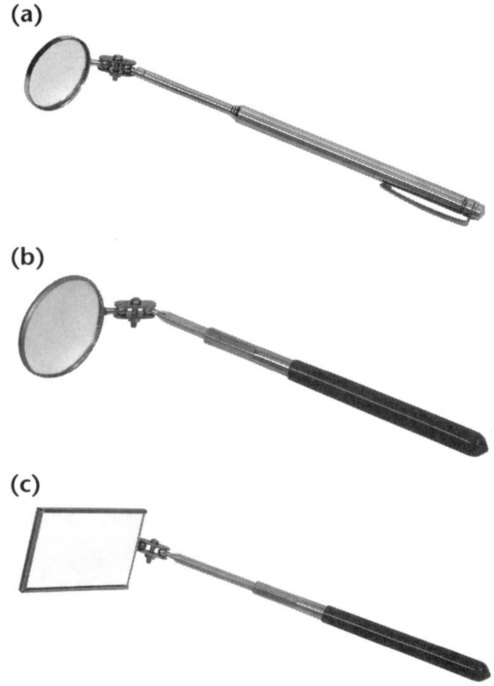

Interpretation of Mirror Images

Several precautions must be remembered in interpreting mirror images.

1. Curved mirrors can distort the apparent shape and size of an object. Concave mirrors make objects look smaller or farther away; convex mirrors (sometimes called *fish eye* mirrors) make objects look larger or closer to the viewer. Glass and plastic may be polarized to achieve these effects with a flat surface.
2. The inspection distance is equal to the distance from the area being inspected to the mirror plus the distance from the mirror to the inspector's eye. When using a mirror, the inspection angle is something other than normal to the surface being inspected. These factors can cause indications to be interpreted as smaller than they are. Measuring the size of the indication at the reflection in the mirror is only appropriate when pinpoint accuracy is unimportant or when the mirror is close to the surface being inspected. A set of pliable wires of known diameters is valuable when measuring indications in difficult-to-reach areas.
3. A mirror image is reversed, so an object on the right appears on the left and one on the left appears right. This reversal may affect documentation — photographs or descriptions in inspection reports.

Part 2. Illumination[2]

Visual Inspection Lighting

The purpose of lighting in a visual inspection area is to provide adequate contrast so that relevant objects or discontinuities are detected. Contrast detection is the most basic of visual tasks. It is a property of the difference between an object and its background of either luminance or color.

Luminance contrast is the difference in intensity of reflected light between the discontinuity and its background. Luminance contrast as related to optometry is discussed in this volume's chapter on vision acuity.

The contrast value is constant for any value of luminance but, like reflection, varies with the position of the subject (observer) and object. The probability of detection increases as the relative contrast value increases. Figure 5 illustrates this as a percentage of accuracy and relates it to the probability of detection.

Chromatic contrast is the difference in hue and saturation between an object and its background. Chromatic contrast produces visibilities less than 20 percent of the detectability of luminance contrast. Chromatic contrast may augment or detract from the achromatic luminance contrast, depending on the perceptive ability of the eye to detect the colors involved.

Figure 5. Probability of detection versus target contrast ratio.

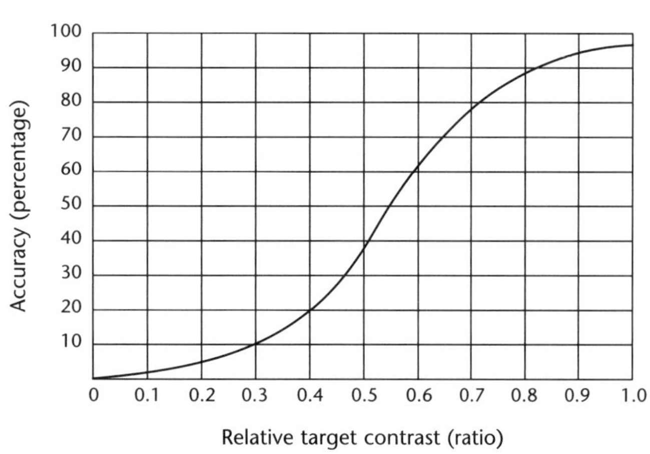

Quality of lighting or illumination in the inspection area refers to the distribution of the light sources in the area and implies that these devices aid visual performance and comfort. Quality illumination is composed of area lighting and specific test lighting.

To avoid inspector eye fatigue and to enhance the probability of detection due to size, the luminance ratios of the inspection area should be controlled. Table 1 lists the maximum recommended luminance ratio between the test object and the environment and for different areas of the environment. This table recommends maximum luminance ratios for areas where reflectances in the work area can be controlled when control of the surrounding area is limited.

Disability glare reduces visibility and visual performance. Glare is caused by light sources or reflections from light sources in the field of view. Discomfort glare produces visual discomfort. To reduce glare, it may be necessary to take one of more of the following steps: (1) decrease the intensity of the light source, (2) reduce the area of the light source, (3) increase the angle between the light source and the field of view or (4) decrease reflections by using a light source with a larger area and lower luminance.

Illuminance is very important in visual tasks. The Illuminating Engineering Society recommends the illuminance levels listed in Table 2 for the performance of visual tasks.

The spectral distribution of lighting can enhance or subdue a color. To enhance a color, the light source should be rich in the color. In most visual inspection situations, the color quality and spectral distribution of the light sources have no effect on vision acuity. When color discrimination or color

Table 1. Maximum recommended luminance ratios.

Ratio	Environment
3 to 1	between test surface and adjacent, darker surfaces
1 to 3	between test surface and adjacent, lighter surfaces
20 to 1	between test surface and distant, darker surfaces
1 to 20	between test surface and distant, lighter surfaces

matching is part of the inspection process, the color of light should be chosen carefully. Generally, the light color and spectral distribution should approximate sunlight.

When performing visual inspections under field conditions, it can be difficult to get enough light. Shadows in the inspection area can be controlled by a horizontal light source with a large surface area or by reflection from luminant walls. Harsh shadows generally cause eye strain, but some shadow effect can accentuate contrast and aid in the detection of depth and form. It is important to weigh the effects of illuminance on the probability of detecting a significant indication.

Area Lighting

Area or room lighting can be direct, semidirect or diffuse. In room lighting, ceiling lamps direct 90 to 100 percent of the output downward. This type of lighting produces high illuminance in the vertical direction but also can produce shadows, glare and veiling reflections. Highly reflective walls are generally recommended to control these effects.

Semidirect lighting directs 60 to 90 percent of the total light output downward with the balance of the light directed upward. This type of lighting eliminates ceiling shadows, but bright ceilings should be avoided.

Ambient lighting is diffuse, directing light upward, downward and sideways. This type of lighting generally provides good brightness relationships throughout the room and produces horizontal illuminance to soften shadows. Glare is minimal with this type of lighting.

Task Lighting

Task lighting may be classified based on its layout or orientation to the task to be performed. General task lighting provides approximately uniform illuminance over a broad area. It is generally produced by the area or room lighting.

Localized general lighting provides uniform general lighting and is oriented to provide optimum illumination in the inspection area. Commonly, ceiling lights are positioned close to the inspection area.

Local lighting includes desk lamps and other portable lighting equipment. Local lighting produces very high illuminance and is useful when shadow formation produces enhanced contrast. Local lighting can cause direct and reflected glare. To prevent eye fatigue due to constant eye adaptation, the general room lighting should provide at least 20 to 30 percent of the total illumination.

Flashlights, called *torches* in the United Kingdom, are common, inexpensive and useful (Fig. 6). They let the inspector direct light onto shaded and interior surfaces.

Table 2. Iluminance levels for visual tasks.[2]

Activity	Illuminance (lx)
General Lighting	
Occasional visual tasks	100 to 200
Tasks with high contrast or large size	200 to 500
Task Lighting	
Tasks with medium contrast or small size	500 to 1000
Tasks with low contrast or very small size	1000 to 2000

Figure 6. Flash lights: (a) household; (b) high power; (c) integrated in helmet.

(a)

(b)

(c)

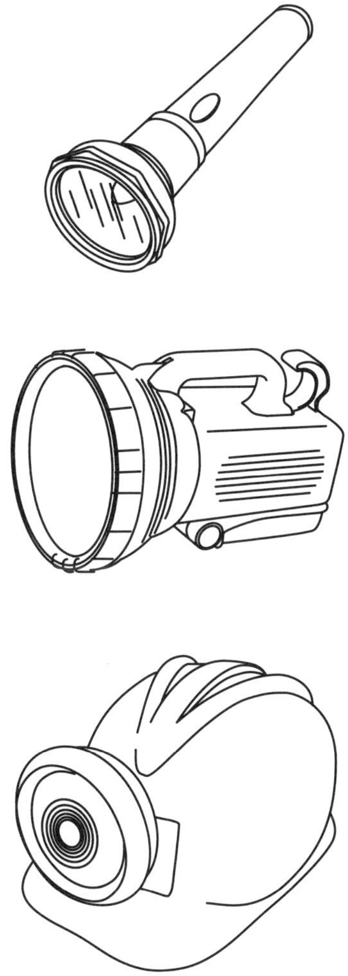

Lighting for Automated Inspection

When choosing lighting for automated inspection and machine vision applications, the goal is to optimize the contrast between a potential discontinuity and its surroundings by providing stable illumination that reduces signal processing complexity. Lighting that takes advantage of the spectral sensitivity of the light sensor and the optical characteristics of the test material are recommended.

The spectral sensitivity of sensors is different from that of the human eye. This difference allows for light sources ineffective during direct visual testing or light harmful to the eye: polarized, monochromatic and coherent (laser) light. Sensors can be optimized for self-radiating materials such as high temperature metals or glass. Most light sensors cannot adapt to differing levels of illumination as easily as the human eye, which makes consistent illumination critical in automated inspection.

When determining the lighting geometry and the type of light source for an inspection, the reflective, absorptive and transmissive characteristics of the test material are the primary factors.

1. Reflective characteristics are attributes of a diffuse or specular surface and of materials that reflect either selectively or nonselectively. Spectrally selective reflectors reflect uneven amounts of the incident light's initial wavelength distribution and absorb or transmit the balance. The nonuniform reflective characteristics of a spectrally selective reflector may be illuminated by a monochromatic source.
2. Absorptive characteristics can be selective or nonselective. Nonselective materials appear black or gray. Selective materials absorb some wavelengths more than others and so have a distinct color.
3. Transparent and translucent materials have transmissive characteristics. Transparent materials transmit the light spectrum with no apparent scatter. Translucent materials transmit a large percentage of the light, but some light scatters by diffusion in the material. Opaque materials transmit no light — all of the spectrum is absorbed or reflected.

The placement of the light source or sources and the amount of direct or diffuse light produced by each light source are critical for accurate inspections. The type of illuminator controls the amount of direct and diffuse light produced. Types of illuminators include condenser reflectors, spot projectors, diffuse light sources and collimators (Fig. 7).

1. A condenser projector has a lens that focuses a cone of light on a target surface.
2. A spot projector has a reflective surface designed to produce a gradually expanding cone of light.
3. Diffuse surface light sources use a translucent material to scatter the incident light.
4. Collimators use a small hole to produce a direct cone of light that expands very little.

Various illumination geometries are used with microscopes and machine vision. Translucent test objects may be

FIGURE 7. Types of light sources: (a) condenser reflector; (b) spot projector; (c) diffuse light source; (d) collimated light source.

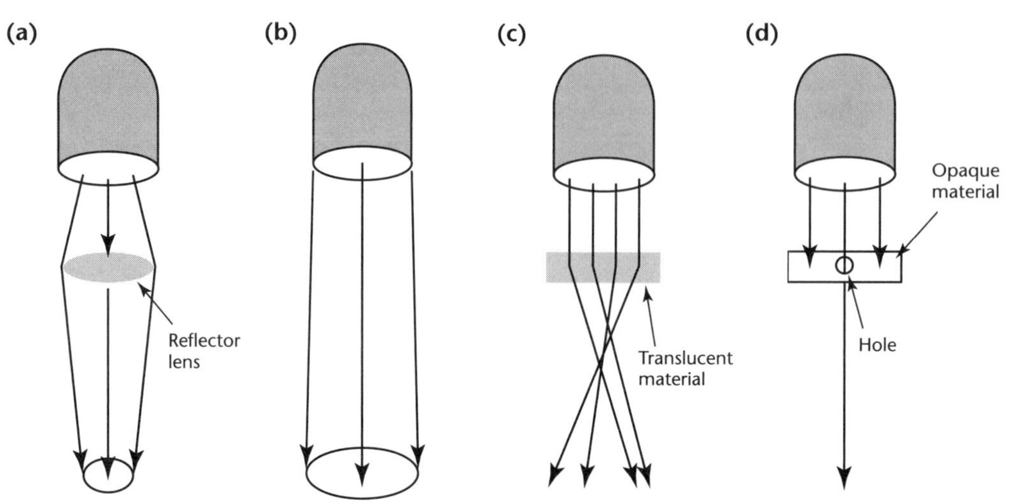

illuminated from the side opposite the detector, for instance. Diffuse front illumination (Fig. 8) floods the area of interest with as much light as possible and minimizes shadow formation. It typically is used on opaque materials where there are prismatic reflections or where the contrast values are very high. A directional light source makes differences in the surface texture more apparent.

Light Sources

Light sources for visual and optical inspection may be divided into five categories: solar, incandescent, luminescent, polarized and coherent light.

Sunlight

The brightness of sunlight varies with weather and time of day.

Direct sunlight can be a health hazard, burning skin and eyes. Glare from bright sunlight can interfere with nondestructive testing by obscuring indications and by fatiguing the inspector.

Incandescent Light

Incandescence is the emission of light due to the thermal excitation of atoms or molecules. Sources of incandescent light include filament lamps, pyroluminescence (such as glow in furnaces or foundries), gas mantles and carbon arc lamps.

Luminescent Light

Luminescence results from the excitation of a single valence electron. Luminescent light is more monochromatic in nature than incandescent light sources. Sources of luminescent light include gaseous discharge lamps, lasers, light emitting diodes and fluorescent lamps.

Polarized Light

The vibrations of polarized light have been oriented to show preference. This means that the vector describing the direction of the lightwave form is constant in time. Linear polarization means that either the vector or the waveforms have been aligned so that both are in the same plane. Although polarized light can be produced directly, it is most commonly produced using a conventional light source and a polarizing light filter.

Polarizing filters are used to control the intensity, color and glare of light. The intensity of light is controlled by using a pair of linear polarizing filters that can be rotated. This two-filter arrangement is capable of a smooth range of attenuation up to a factor of 100 000 to 1. This arrangement is used in some older photometers. Polarizing filters are also used in many glare reducing products such as sunglasses.

Some specialized optical techniques use polarized light because of its ability to produce uniform patterns of constructive and destructive interference of the light waves. This allows the characteristics of many products to be measured by assessing the interference patterns when polarized light is transmitted through or reflected from a test object. Techniques using polarized light include moiré fringe and birefringence techniques.

Coherent Light

Coherent light, such as light from a laser, is visible radiant energy or light with a high degree of phase coherence. Phase coherence requires both frequency coherence and the spatial coherence of polarized light. Phase coherence requires the individual waves of light to be of the same frequency and the sinusoidal curve of the waves to be aligned.

The light produced by most light sources has a broad spectrum and produces a diverging luminous area, but the rays of a laser or phased light are aligned in parallel. Phased light is produced by a ruby laser in a manner similar to the production of fluorescent light. The difference is that in the coherent light producing material, the excited electrons pause momentarily in an orbit with more energy than the ground state but less energy than the most excited state. This intermediate stage is known as the *metastable* state. The photons of light energy are released from the metastable state simultaneously producing spatial or time phased coherence.

FIGURE 8. Diffuse front illumination.

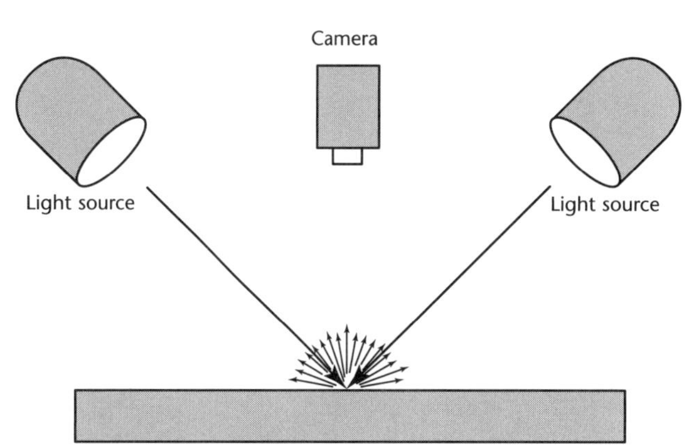

Photometry

The light used for visual testing is a nondestructive testing parameter that may be specified in a contract or written procedure. Hand held photometers provide digital measurements of ambient or direct light in lux. Photographers call them *light meters*. Photometry is discussed in this volume's chapter on light.

PART 3. Magnification[2,4,5]

Magnification is commonly achieved by the refracting of light through curved lenses. Refraction and basic lens types are discussed in this volume's chapter on the physics of light.

Devices for magnification used in visual and optical inspections range in magnifying power from 1.5× to over 2000×. Magnifiers are described by interrelated factors: (1) magnification power, (2) field of view, (3) working distance, (4) eye relief, (5) depth of field, (6) chromatic correction and (7) binocular or monocular vision.

1. The magnifying ability of a lens depends on the amount of curvature in the converging lens. Greater curvature produces a greater angle of refraction, reducing the focal length and so increasing magnification. The power of a magnifier describes the amount of enlargement compared to viewing the object 250 mm (10 in.) from the eye. An object 25 mm (1 in.) from the eye would appear to be ten times as large as the same object viewed at 250 mm (10 in.). The magnifier that produces enough refraction to allow the eye to focus at this distance has 10× magnification (10 power) and a focal length of 25 mm (1 in.).
2. Field coverage of conventional magnifiers ranges from 90 mm (3.5 in.) down to 0.15 mm (0.006 in.) wide.
3. The working distance of the magnifier is the same as the focal length. The distance from the magnifier to the object being inspected is adjusted to produce good focus. Close working distances accompany higher magnification. When choosing the magnification and working distance, the amount of room needed to perform the inspection and the tools to be used should be considered.
4. The eye relief is the distance from the inspector's eye to the device's lens.
5. The depth of field is the maximum range of distances in focus at the same time without a change in the viewing position. The field of view decreases as the magnification increases. A large depth of field eliminates the need for constant adjustments to the viewing position to maintain focus.
6. Distracting rainbows and colored bands are seen when different wavelengths of light refract in different directions. Chromatic correction mitigates this spectral divergence, often by using two lenses sandwiched together (a doublet), two layers with different indices of refraction.
7. Binoculars are used for stereoscopic viewing from a distance. Stereoscopic microscopes are used to inspect miniature assemblies such as printed circuit boards. Nearly all other magnifiers are monocular.

These magnifier attributes are interrelated. A high power magnifier, for example, has a short working distance, a small field of view and cannot be used for binocular observation; a low power magnifier, such as a rectangular reader lens, has a long working distance, a large field of view and can looked through by both eyes. To attain chromatic correction (to eliminate color fringing), the high power lens typically contains a cemented doublet or triplet of different optical glasses whereas the low power reader lens is sufficiently achromatic as a simple lens.

There are many variations of these characteristics. Commercial magnifiers can be 30× or higher in power, and there are many special mountings for particular applications.

The ideal magnifier has a large field of view, a large depth of field and an image without refractive or color aberrations, but in practice all magnifiers present a compromise. Generally, using as little magnification as possible will cause the fewest technical problems and provide the simplest inspection.

Magnifiers

Resolving powers range from 0.05 mm (0.002 in.) to 0.2 µm (8×10^{-6} in.). Powers of magnification refer to enlargement in one dimension only. A two-dimensional image magnified × 2, for example, doubles in width and in height while its area quadruples.

Magnifying Lenses

A test surface is viewed through a lens to obtain a magnification as great as desired.

The distance from lens to object is adjusted until the object is in the lens's depth of field and is in focus. The simplest form of a microscope is a single converging lens, often referred to as a simple magnifier. Magnification *M* of a single lens is determined by Eq. 1:

$$(1) \quad M = \frac{250}{f_{mm}} = \frac{10}{f_{in}}$$

In this equation, f_{mm} is the focal length of the lens in millimeters, f_{in} is the focal length of the lens in inches and 250 (or 10) is a constant that represents the average minimum distance at which objects can be distinctly seen by the unaided eye.

Using the equation, a lens with a focal length of 125 mm (5 in.) has a magnification of two widths or is said to be a two-power (2×) lens. For a simple magnifier, the focal length and working distance are about the same. For example, suppose that a component must be inspected without moving it and that a magnifier cannot be placed nearer than 75 mm (3 in.). A lens with a working distance (focal length) of at least 75 mm (3 in.) is required. From Eq. 1, that is shown to be a 3× or a three power lens.

The field of view is the area seen through the magnifier. With a simple magnifier, the diameter of the field of view is less than its focal length. Selection of a magnifier with the proper field of view is important. For example, if the test object is large, it takes too much time to use a 20× magnifier, with a field of view slightly greater than 10 mm (0.37 in.). The proper procedure is to use a low power magnifier first, marking questionable areas and then to inspect questioned areas with a higher powered magnifier.

Depth of field is the distance a magnifier can be moved toward or away from a subject with the subject remaining in good focus (sharply defined). At other distances, the subject is out of focus and not sharply defined. Depth of field varies with the power of the lens and is comparatively greater in lower power magnifiers, decreasing as the power of the lens increases.

Conventional Magnifiers and Readers

Table 3 shows the characteristics of a few typical magnifiers. These values are approximations because eye accommodation can cause each of the values to vary. Except for the reader lens, all magnifiers are used with the eye fairly close to the magnifier, giving the largest field of view. The reader lens is used binocularly and is normally held some distance away from the eyes.

Single lens magnifiers provide between 1.5× and 10× magnification. They may be handheld, mounted on flexible or rigid stands or mounted with a headband to allow free hands during viewing. In low cost magnifiers, the distortion caused by highly refracting lenses is controlled by decreasing the diameter of the lens. This decrease reduces the field of view.

For magnification above 7× to 10×, it is common for magnifiers to use more than one lens to provide for the increased magnification while limiting the spherical and chromatic aberrations produced by a highly refractive lens. Magnifiers with two or three lenses are referred to as doublets or triplets; these lens types are illustrated in this volume's chapter on light. Another type of lens, the coddington magnifier, has a thick groove cut into the outer diameter of a thick lens to eliminate spherical aberration (Fig. 9).

A loupe (or lupe) is a single or double lens magnifier where the lenses are held at the recommended working distance with a transparent cylinder. The clear spacing cylinder allows the use of ambient lighting to illuminate the test piece. Illuminating loupes have a battery powered light source for illumination. Loupes frequently include a contact reticle. Jewelers' loupes generally use the cylinder for purposes of eye relief rather than working distance. They are inserted into the eye socket and held in place with the facial muscles around the eye or attached to the frame of an eyeglass.

TABLE 3. Characteristics of typical magnifiers.

Magnifier	Working Field of View mm	Working Field of View (in.)	Power	Resolving Distance mm	Resolving Distance (in.)	Power µm	Power (in.)
Reader lens	90 × 40	(3.5 × 1.5)	1.5×	100	(4)	50	(0.002)
Eyeglass loupe	60	(2.375)	2×	90	(3.5)	40	(0.0015)
Doublet magnifier	60	(2.375)	3.5×	75	(3)	25	(0.001)
Coddington magnifier	19	(0.75)	7×	25	(1)	10	(0.0004)
Triplet magnifier	22	(0.875)	10×	20	(0.75)	7.5	(0.0003)

Because of its large diameter, the 3.5× doublet magnifier has a field as large as that of the 2× loupe. The double convex lens of the doublet magnifier with its central iris has a comparatively small field. The triplet is a three-element design having excellent optical correction for field coverage and reduction of color fringing. Its resolving power is the limit of detection for fine structures. In comparison, the doublet magnifier can barely differentiate two points 25 µm (0.001 in.) apart.

Surface Comparators

The surface comparator is a magnifier that provides a means for comparing a test surface against a standard surface finish. The observer views the two surfaces side by side, as shown in Fig. 10. The surface comparator uses a small battery powered light source, a semitransparent beam divider and a 10× triplet.

The light is divided between the reference surface and the standard surface. Flat and shiny surfaces reflect the filament image directly into the pupil of the eye so that these parts look bright. Sloping or rough surfaces reflect the light away from the pupil and such areas appear dark. This form of illumination sharply delineates surface pattern characteristics. The resolving power is about 7.5 µm (0.0003 in.). The field of view is about 1 mm (0.4 in.) diameter.

Measuring Magnifier

A measuring magnifier, or contact reticle (Fig. 11), has a graduated measuring scale in a transparent layer held against the test object to measure tiny details on its flat surfaces. Measuring magnifiers have one-, two- or three-lens magnifiers and a transparent housing that lets light fall on the measured surface. Contact reticles are available in a variety of scales and measuring systems including linear measurement, angular measurement, grid patterns and thread gages. Magnification is often between 6× and 20×; the diameter of the field of view is often about 25 mm

FIGURE 9. Cross sections: (a) triplet magnifier; (b) coddington magnifier.

(a)

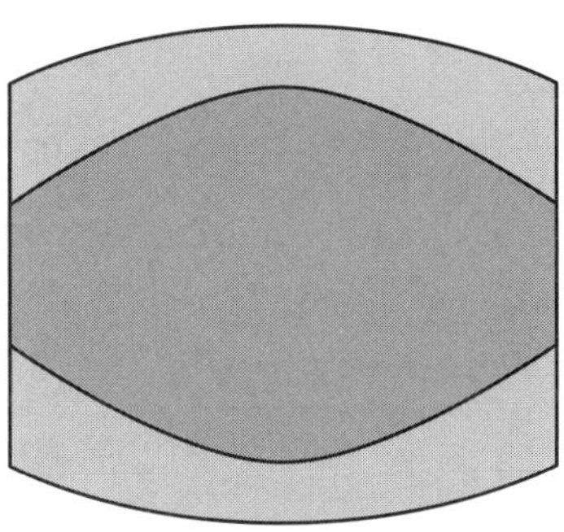

(b)

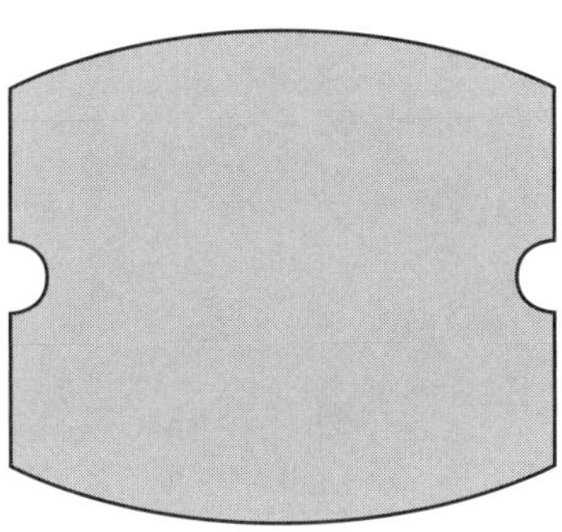

FIGURE 10. Surface comparator: (a) schematic; (b) two surfaces magnified for comparison.

(a)

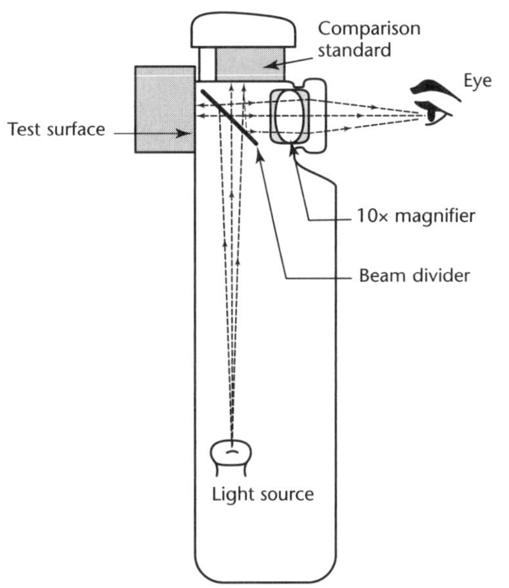

(b)

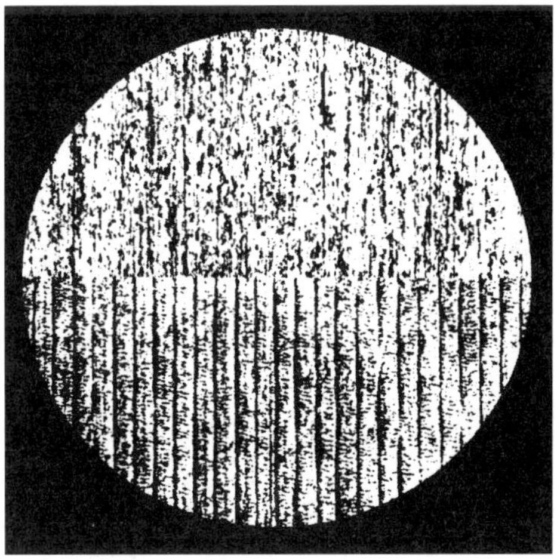

(1 in.); the resolving power is often about 1 mm (4 × 10^{-5} in.).

Illuminated Magnifiers

Illuminated magnifiers range from large circular reader lenses, equipped with fluorescent lighting and an adjustable stand, to a small battery powered 10× magnifier shaped like a pencil. Some illuminated magnifiers can be obtained in either a battery powered model or equipped for 115 V line operation. Such triplet magnifiers give about a 50 mm (2 in.) field of view. Resolving power is about 1.5 µm (6 × 10^{-5} in.).

Microscopes[2,4,5]

A magnifying glass magnifies what is visible to the unaided eye but is too small to evaluate; a microscope makes details visible that otherwise would remain invisible to the human eye.

In its simplest form, a microscope is a single biconvex lens in a housing adjustable for focus. Many forms of illumination are available, including bright field, dark field, oblique, polarized, phase contrast and interference.

Microscopy is usually destructive: a sample scraping removed from the surface of an object is discarded after analysis. There are many varieties of microscopes, however, and some can be used to examine very small products, such as printed circuits, or can be used to examine an area of a test object without sectioning or scraping it.

FIGURE 11. Typical measuring scales and reticles of measuring magnifier.

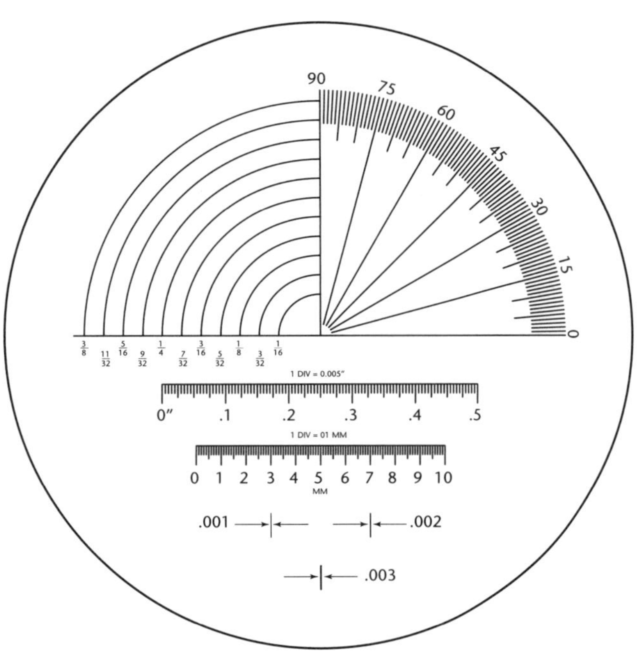

Low Power Microscopes

When magnifications above 10× are required, the short working distance of the magnifier becomes a problem and a low power compound microscope is preferred. Typical resolving power is about 7.5 µm (3 × 10^{-4} in.). At magnifications above 10×, low power microscopes are more convenient than simple magnifiers. Low power microscopes for shop or other stationary use include a large variety of devices ranging in complexity from simple to wide field stereoscopic microscopes. They use lensing systems that increase in complexity with the amount of magnification. At low and medium levels of magnification, a single lensing system can be used. At higher levels, compound lenses with two sets of lenses are common.

Field, pocket, pen or measuring microscopes are all terms used to describe small handheld magnifiers that have magnifying powers between 10× and 50× and more complex lens arrangements than simple magnifiers. These are powerful for field inspections but are difficult to use because of limited stability, small depths of field, rudimentary focusing systems, narrow fields of view and the need for intense illumination.

Stereo Microscopes (Medium Power)

The wide field stereoscopic microscope is very complex. It is basically two erect image microscopes, one for each eye, comprising two objectives, two erecting prisms, two inclination prisms and two eyepieces. The useful power range of the stereomicroscope is limited to 100×. The resolving power is about 5 µm (2 × 10^{-4} in.). Field coverage is inverse to its power: at 10× field coverage, the field is about 25 mm (1 in.). The instrument provides binocular vision, which makes possible its prolonged use for visual testing. As with low power magnifiers, manual manipulation during observation is practical. The stereoscopic microscope provides a true view of depth, so that test objects may be inspected in three dimensions.

Shop Microscope (Medium Power)

The shop microscope is similar to a wide field tube. It has a power of 40× and contains a built-in light source that may be operated from a battery or 115 V line current. The shop microscope contains a scale permitting direct measurement on the object plane to 0.025 mm (0.001 in.) or estimates to 6 µm (2.5×10^{-4} in.), over a scale length of 4 mm (0.15 in.). The field of view is 5 mm (0.22 in.) and the resolving power is about 3.3 µm (1.3×10^{-4} in.). The instrument weighs little, only 500 g (18 oz) including batteries.

Applications of the shop microscope include (1) on-site tests of plated, painted or polished surfaces, (2) detection of cracks, blowholes and other discontinuities and (3) measurement checking wear in mechanical components. Welding of machine tool frames, piping, structural members, pressure vessels, jigs and fixtures, can be quickly inspected.

In finishing and electroplating operations, surface tests with the shop microscope can detect cracks, blister, irregular deposits, pitting and poor quality buffing or polishing. It can reveal slag inclusions and poor surfacing of base metals before plating. On painted surfaces it permits quick and accurate evaluation of quality, uniformity and pigment distribution. In the graphic arts, it is used to check halftones for size, shape and distribution of dots.

Textile mills use shop microscopes for identification of fiber textures, distribution of coloring matter and test of weave, twist and other general characteristics. Fabric finishes, markings, lusters and dye transfers can be inspected for penetration and quality. In the paper industry, the shop microscope is used to check fiber uniformity, evenness of coating and wear of Fourdrinier wires.

Laboratory Microscope

Laboratory microscopes normally range in power from 100× to 2000× and over. They commonly use compound lensing systems with two or more separate sets of lenses. It is becoming more common for these microscopes to include provisions for a still or video camera. A video camera allows the image to be viewed on a monitor and provides an excellent means of documentation. Adapters are available to adapt conventional microscopes for use with a video or still camera.

Most laboratory microscopes were developed for medical use and rely on the transmission of light through the object, so they are limited to translucent objects.

Metallurgical Microscope

The metallurgical microscope is similar to a laboratory microscope with the addition of top or vertical illumination to permit viewing of opaque materials. The vertical illuminator, directly above the objective lens, is a semireflecting, thin, transparent plate. It directs light down through the objective lens onto the test object. The microscope is normally equipped with a built-in light source and has field and aperture iris controls in the illuminating arm. Because thick preparations are common in opaque test objects, the stage may be focused. Focusing permits an intense external light source without upsetting the illumination centering.

Although this microscope finds its principal applications in metallurgy, it can be used on almost any opaque material having a reasonably high reflectivity. When test objects are dark by nature (dark plastics, paints, minerals) or have excessive light scattering (fabrics, paper, wood, or biological specimens), a form of incident dark field illumination is superior to regular vertical illumination.

Metallographic Microscope

Microscopes designed to provide images of opaque objects have provisions for vertical illumination — that is, by reflected light rather than light transmitted through the test object from the other side. A metallographic microscope is a common example of this type of microscope. A metallographic microscope is a metallurgical microscope designed to integrate a camera for photographic documentation. The metallographic microscope's light source may have filters to alter the spectrum or polarize light to reduce glare.

Common metallographic techniques include dark field illumination, polarized light, phase contrast and differential interference. These techniques are increasingly being applied to the analysis of the microstructures of many nonmetallic and composite materials.

Polarizing Microscope

The addition of two polarizing elements and a circular stage converts a laboratory microscope into an elementary polarizing microscope. A polarizing element is a device that restricts light vibration to a single plane. This form of light is useful for studying most materials with directional optical properties, including fibers, crystals, sheet plastic and materials under strain. As such materials are rotated between crossed polarizers on the microscope stage, they change color and intensity in a way that is related to their directional properties.

Interference Microscope

The interference microscope is a tool using the wavelength of light as a unit of measure for surface contour and other characteristics. In one form of interference microscope, the stage is inverted and the test object is placed face downward. The image appears as a contour map, with a separation of one half-wave or about 0.25 µm (1×10^{-5} in.) between contour lines. Extremely precise measurements can be made with such equipment.

Applications of the interference microscope include the measurement, testing and control of very fine finishes, including highly polished or glossy finished surfaces, where the degree of surface roughness is within a few wavelengths of light. With coarser surfaces, the contour lines are close together and interpretation is difficult. An advantage of the interference microscope is that the test object is not moved manually during inspection.

PART 4. Surface Characteristics[2]

Surface characteristics that can be visually tested include surface texture (roughness or waviness), color (including gloss) and cleanliness. Cleanliness is a prerequisite for certain processes such as painting and liquid penetrant testing. Surfaces may also be visually tested closely to identify discontinuities and damage mechanisms, such as porosity, blistering, corrosion, flaking and fatigue. The visual inspector must flag apparent cracks for evaluation by other methods.

The variation from a specified nominal surface condition is controlled by dimensional tolerances and surface roughness specifications. Surface texture is usually measured in accordance with published standards.[6-11]

Surface Texture

The surface features of any object, regardless of scale, share three independent characteristics: form, roughness and waviness.

1. Variations in form or profile are typically controlled by the dimensional or geometric tolerance specifications. Surface profiles for larger tolerances, ±0.03 mm (±0.001 in.) and greater, are typically measured using standard dimensional equipment or optical comparators and should be measured with respect to a datum surface.
2. Surface roughness is critical wherever friction must be reduced, as in propellers, ball bearings or prosthetic joints. When measuring and reporting surface roughness, it is essential to choose a measurement parameter that assigns a numerical value to the characteristics that must be controlled. If only one parameter is chosen, there is a risk that some unimportant aspect of the surface may be overcontrolled, adding unnecessary cost to assure adequate control of the key characteristic. For example, the surfaces in Fig. 12 have the same roughness average R_a but different roughness peak R_p and roughness depth R_Y.
3. When the object of controlling surface roughness is to control waviness and roughness, the profile variation may be removed by using a line that describes the average surface. This line can then be applied as a profile filter.

Surface roughness may be measured with photogrammetry or profilometry. In photogrammetry, surface details are measured by comparing two photographic images offset from each other. In profilometry, points on the surface are successively measured and a series of optical or mechanical measurements are compiled.

Average roughness R_a is the average distance of the profile to the mean line. When controlling the height of individual peaks and valleys is important, measurements of maximum peak-to-valley height R_{max} and mean peak-to-valley

FIGURE 12. Comparison of profiles for average roughness R_a: (a) R_a = 2.4 µm; (b) R_a = 2.5 µm; (c) R_a = 2.4 µm. Surfaces of Figs. 12a and 12c have same average but different profiles.

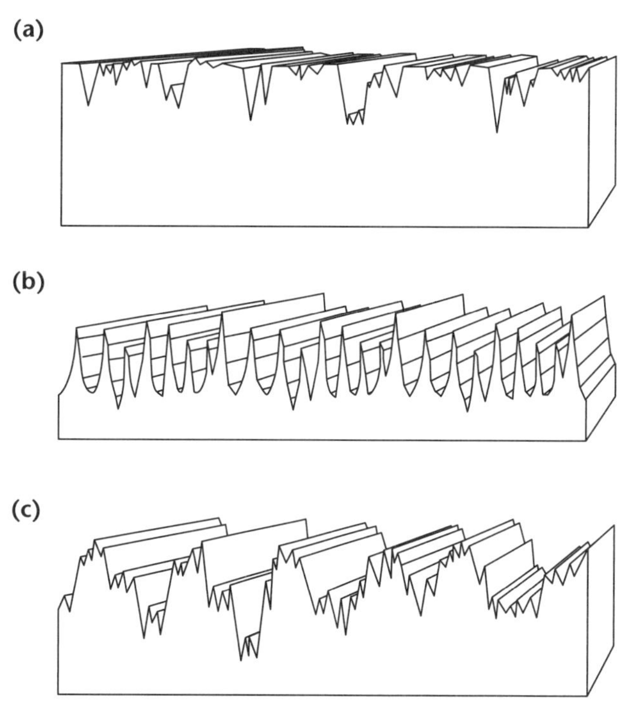

height R_z can control the total variation without adding the cost of an overly tight R_a requirement.

Typically, R_z is determined by dividing the measured length into five equal lengths and using the formula:

$$(2) \quad R_z = \frac{1}{5} \sum_{i=1}^{5} Z_i$$

where Z_i is the waviness variation.

The total waviness parameter is used to control the larger horizontal scale variation, waviness. Surface waviness and surface roughness measurements are similar and differ primarily in scale. In addition, the direction of the surface roughness should be considered. Isotropic surfaces are random. They are typically produced by processes such as casting or any kind of abrasive blasting. Anisotropic surfaces have a periodic irregularity, usually in one direction. They are produced by most machining operations. The measured value of anisotropic surfaces will change greatly with the direction of the measurement.

Table 4 lists the surface roughness produced by some manufacturing operations.

Color and Gloss

Color

The color of many materials is a significant part of commercial value. In cold rolled and surface treated steels, too, the surface colors are an important consideration in quality control.[12]

Colored light is reflected from a test object and perceived by an instrument or human observer. Color is a characteristic of light, and each hue corresponds to a waveband of electromagnetic radiation. The study of color is a branch of optics, which is a branch of physics and is discussed in this volume's chapter on light.

Color order systems describe color in terms of hue, value and saturation.

1. Hue, or chromaticity, indicates the value that separates the color in terms of its primary color constituents or its mix of primary constituents. The hue describes the color's redness, blueness and so forth.
2. Value describes the lightness or darkness of the color. Colors with a large portion of white or lightness have a high value. Dark colors have a low value.
3. Saturation (chroma) measures the distance from the corresponding neutral color. It is often referred to as the *color intensity.*

Human perception of color varies with the illumination and with the perceptive abilities of the observer. The communication of color requirements and the assessment of color against requirements can be imprecise. Color requirements can be effectively communicated by visual comparison to reference standards, a color order system or a color collection. Color order systems place colors into an orderly three-dimensional arrangement with a standard nomenclature to describe each color in the system.

1. Based on the three sets of opposing colors, the Natural Color System® is a proprietary system available from the Scandinavian Color Institute AB, Stockholm, Sweden. The chromatic colors are arranged in a circle with yellow, red, blue and green spaced at 90 degree intervals. The opposing monochromatic colors of black and white are represented by the central axis. The hue is described by its resemblance to the primary colors on a scale of 0 to 100 where the sum of the resemblances adds up to 100. Because red opposes green and blue opposes yellow, only two chromatic colors may be used at one time. The value of the other two chromatic colors is always zero.

TABLE 4. Surface roughness produced by some manufacturing operations.[2]

Process	Approximate Range from Surface	
	µm	(10^{-6} in.)
Casting and Forging		
Sand casting	30 to 10	(1100 to 400)
Hot rolling	30 to 10	(1100 to 400)
Forging	15 to 2.5	(600 to 100)
Permanent mold casting	4 to 1.5	(150 to 50)
Investment casting	4 to 1.5	(150 to 50)
Extruding	4 to 0.5	(150 to 25)
Cold rolling, drawing	4 to 0.5	(150 to 25)
Die casting	2 to 0.5	(75 to 25)
Finishing		
Flame cutting	30 to 10	(1100 to 400)
Sawing	30 to 1.5	(1100 to 50)
Planing, shaping	15 to 1.5	(600 to 50)
Drilling	7.5 to 0.5	(300 to 50
Chemical milling	7.5 to 0.5	(300 to 50)
Milling	7.5 to 0.5	(300 to 25)
Broaching, reaming	4 to 0.8	(150 to 25)
Grinding	2 to 0.2	(75 to 5)
Polishing	0.5 to 0.2	(35 to 5)
Superfinishing	0.1	(4)

2. Introduced by Albert H. Munsell around 1910, the munsell color system is similar in concept to the Natural Color System but uses ten hues equally spaced around the outside of a circle. The munsell color system is a common system used in the United States and Japan. Color trees are available in vertical slices that show the variation of chroma and value for a specified hue. Horizontal slices are also available that show the variation in hue and chroma from a given point on the neutral axis.
3 International Commission on Illumination (CIE) has promulgated a color system as an international standard.[13] When measured quantitatively, the reflected light of an object is measured in terms of its ability to create a response in three ideal color sensors that represent an ideal color observer. This measurement may be expressed in terms of the tristimulus value of the color or the chromaticity coordinates. The tristimulus value is calculated by multiplying the spectral power distribution by a reflectance factor. The tristimulus primaries specified by the International Commission on Illumination (CIE) are imaginary values of light and cannot be represented by an actual color stimulus.
4. Color collections are produced by many government and private sources to demonstrate fabric colors, paint colors or ceramic colors. The color collection is an excellent way to specify the exact color requirements for a specific product where the properties of gloss and texture are similar. They can be very difficult to correlate to other product forms or to other color collections or color scales.

These various color systems may be used to specify a desired color in a contract.

Color may also be measured with a colorimeter. A colorimeter is a photometer that attempts to duplicate the perception of color by a human observer: three colored filters simulate the effects of a light source and observer. They measure the tristimulus values or the color coordinate value of the reflected light.

Color measuring devices must be properly calibrated and maintained because they are affected by variations in temperature and humidity. These devices are simple and inexpensive but are limited to single combinations of light source and observer variables. Many digital cameras integrate colorimetric functions.

Gloss

Gloss, or reflectance, is a measure of the specular or mirror reflectance of a surface. Reflectance is measured by comparing the reflected light of the test piece to the reflection produced by a perfectly diffusing, perfectly reflective plane surface that is illuminated and viewed at the same angle as the test piece. The reflectance standard is usually a metal plate coated with magnesium oxide with a reflectance of about 98 percent.

The illumination angle and the angle of the light receiver must be equal when measured from the surface normal. The angle chosen for the measurement is based on the relative glossiness to best match the visual perception. Most measurements are made at 60 degrees from the surface normal. Very high gloss surfaces, such as automotive paints, are measured at 20 degrees; measurements of flat finishes are made at up to 85 degrees.

Part 5. Dimensional Measurement

Inspection methods that measure mass or length are often excluded from nondestructive testing because, although mass and length are material properties, the methods do not seek discontinuities. Nevertheless, the visual inspector is often given the task of measuring test objects for various purposes.

1. In receiving inspection, the visual inspector may confirm that components received were the components ordered. For instance, does the tubing have the needed diameter?
2. In assembly, the visual inspector may confirm that the correct parts are being used. For instance, are bolts of the correct size being installed?
3. In machining, the visual inspector may confirm that the operation is performed to specification. For instance, are rivets placed far enough from the edge of a plate?
4. In finishing, the visual inspector may confirm that the surface is treated to specification. For instance, has sheet metal been burnished or coated?
5. In nondestructive testing, the visual inspector needs to describe a visual indication. How long is the crack? How extensive is the blistering on the boiler's surface?

In many industries, the direct visual testing of bolts requires steel rulers, micrometers, vernier calipers, depth micrometers, thread gages and magnifying glasses.

Rulers and Tape Measures

Rulers and tape measures are familiar and easy to use. In the United States, these tools almost always show units of the English system (inches) on one side and of the International System (centimeters, millimeters) on the other side. Increasingly, the International System is specified for goods and services.

Calipers[3]

Calipers are used to obtain accurate linear measurements. Calipers come in a wide variety of sizes and configurations for measuring length, width, height, diameter and depth, and they can be either direct reading or indirect reading.

1. Indirect reading, or transfer type, calipers (Fig. 13) are used to transfer the dimension of an item from the item to a steel rule. For example, the measurement of an outside diameter is made by adjusting the caliper so that both legs lightly touch the widest portion of the item. This distance is then transferred to a steel rule to obtain the measurement. If performed properly, this type of measurement is accurate to 0.04 mm (0.016 in.).

Figure 13: Indirect calipers: (a) for outside measurement; (b) for inside measurement.

(a)

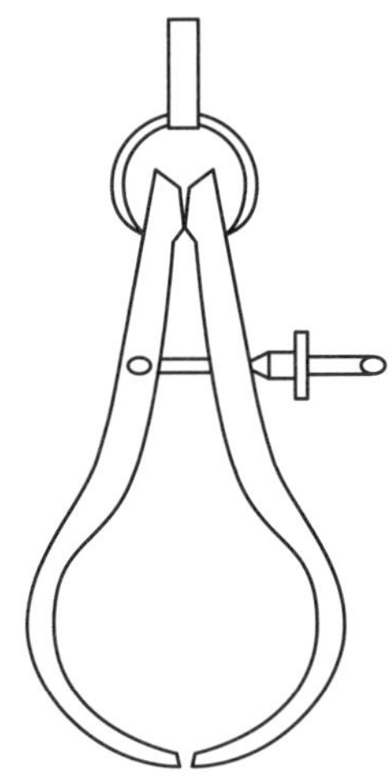

(b)

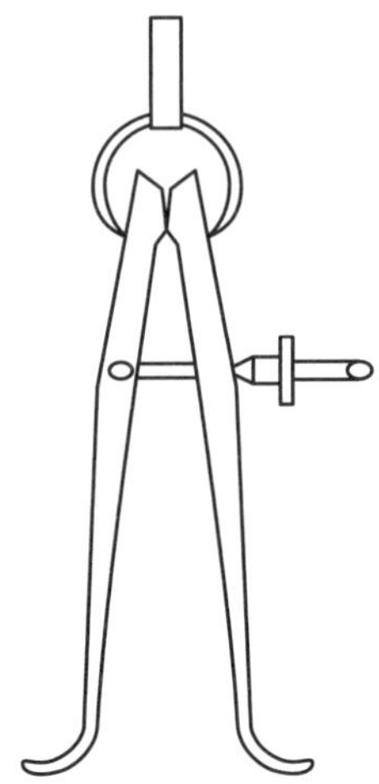

2. Direct reading calipers vary. A direct reading caliper can be simple, a rule with jaws for coarse measurements; or it could be of the vernier, dial, or electronic digital type, which are used for very accurate measurements (Fig. 14). All types of direct reading calipers consist of a fixed jaw on a beam along which a moveable jaw slides. The measurements are taken with the item between the jaws of the instrument.

Dial and electronic calipers are simple to use and read. Electronic calipers are the easiest to read because the actual measurement is displayed on a digital readout. The dial caliper may require some interpretation if the beam scale and the rotating indicator on the dial are graduated in different increments.

Vernier calipers are more difficult to use because the scale requires care during interpretation. The vernier consists of the fixed main scale, etched into the beam and the sliding vernier scale attached to the moveable jaw. To take a measurement with the vernier caliper, open the jaws larger than the maximum dimension of the item to be measured and slowly close the jaws around the item until light contact is made. For the greatest accuracy, the item must make even contact all along the thickness of the jaw faces.

Gages[3]

Some gages commonly used during visual testing include mechanical gages, weld fillet gages and cambridge gages.

Mechanical Gages (Micrometers)

Mechanical gages perform extremely precise measurements of linear dimensions. Mechanical gages are available in a wide variety of configurations for inside and outside measurements of flat, curved, threaded and cylindrical dimensions. The mechanical gage is a caliper that operates by determining how far the end of a screw travels in one complete revolution.

FIGURE 14: Direct reading caliper.

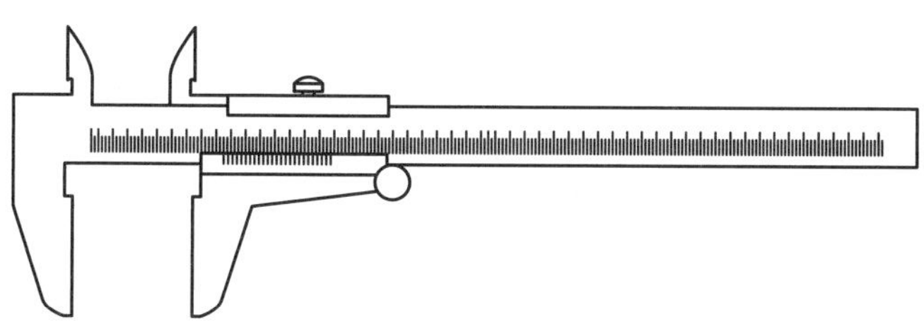

To better understand the operation of the mechanical gage, it is useful to know the components of the instrument. This includes the frame, anvil, spindle, barrel, thimble, screw, ratchet and clamp ring (Fig. 15). Measurement with the mechanical gage occurs between the anvil and spindle. As the thimble screw is rotated counterclockwise, the spindle and anvil separate and the barrel graduations are revealed in succession.

To make a measurement, rotate the thimble counterclockwise until the spindle is far enough away from the anvil to allow the test object to fit between them. With the part between the anvil and spindle, slowly turn the thimble clockwise to obtain contact between the anvil, part and spindle. A gentle pressure is all that is necessary to make an accurate measurement. Pressure can be applied in fine increments with a ratchet. Too much pressure could distort the frame and reduce accuracy; too little pressure could result in improper contact with the part, producing an inaccurate measurement. The part should be able to be rotated about the spindle axis with the feeling of a slight drag. Barrel graduations are revealed by the outward travel of the thimble. To read the dimension, note the largest major division uncovered, the graduation closest to the thimble and the thimble division aligned with the barrel reference line.

Weld Gages[3]

A common tool used in visual examination of weldments is the weld fillet gage. This simple, easy-to-use device measures leg lengths and determines if there is sufficient throat in weld fillets. This gage is basically a comparator — the acceptable size is etched into the gage and arcs are cut into the gage to allow space for the weld bead. The gage is placed square against the welded components and the actual weld is compared to the

FIGURE 15. Mechanical gage, or micrometer.

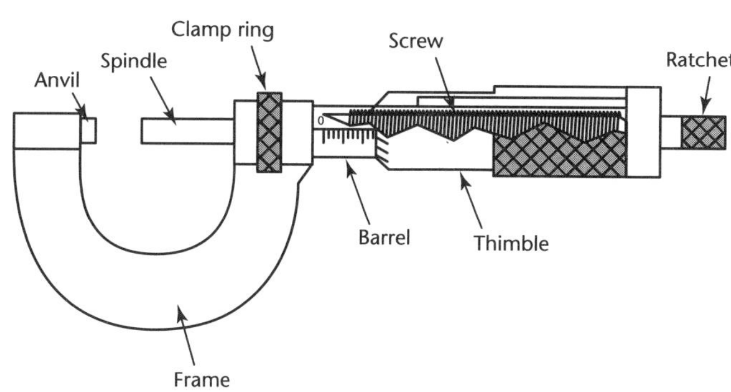

standards of the gage (Fig. 16). This type of gage offers a quick and precise means of measuring concave and convex fillet welds from 3 mm (0.13 in.) to 25 mm (1.0 in.).

Weld gages (Fig. 17) designed to measure offset displacement can be used to measure the size of fillet welds, the actual throat size of convex and concave fillet welds, reinforcement of butt welds and root openings of 8 mm (0.3 in.) and 3 mm (0.13 in.)

Another more versatile device used for weld inspection is the welding gage, commonly referred to as the cambridge gage. The Welding Institute, Cambridge, United Kingdom, developed this versatile tool — hence the name. With this device, joint preparation angles, joint misalignment, weld fillet size and depth measurements can be easily obtained. Figure 18 shows some typical applications.

This volume's chapter on electric power applications includes some discussion of weld gaging.

Tolerance Standards[2]

The physical features of an object (form, profile, orientation, location and size) must be controlled. The blueprint or drawing for the object must specify the attributes of each characteristic including tolerances. Requirements are specified and variations are assessed by using geometric dimensions and tolerances, uniformly communicated in a consensus standard. Dimensional tolerances are related to issues of surface roughness and waviness, discussed above, and are governed by some of the same standards.[6,7,11] Working to a consensus standard realizes the following benefits.

FIGURE 16. Fillet weld gage: (a) concave weld; (b) convex weld.

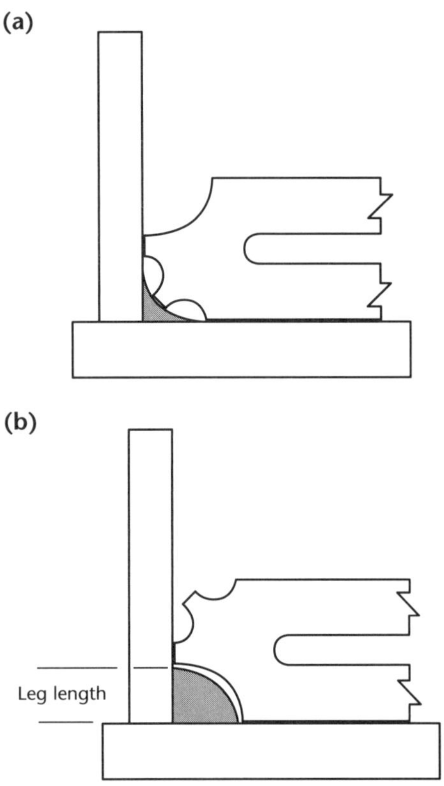

FIGURE 17. Displacement weld gaging: (a) leg length of fillet weld; (b) convexity of fillet weld; (c) concavity of fillet weld; (d) bead height of butt weld.

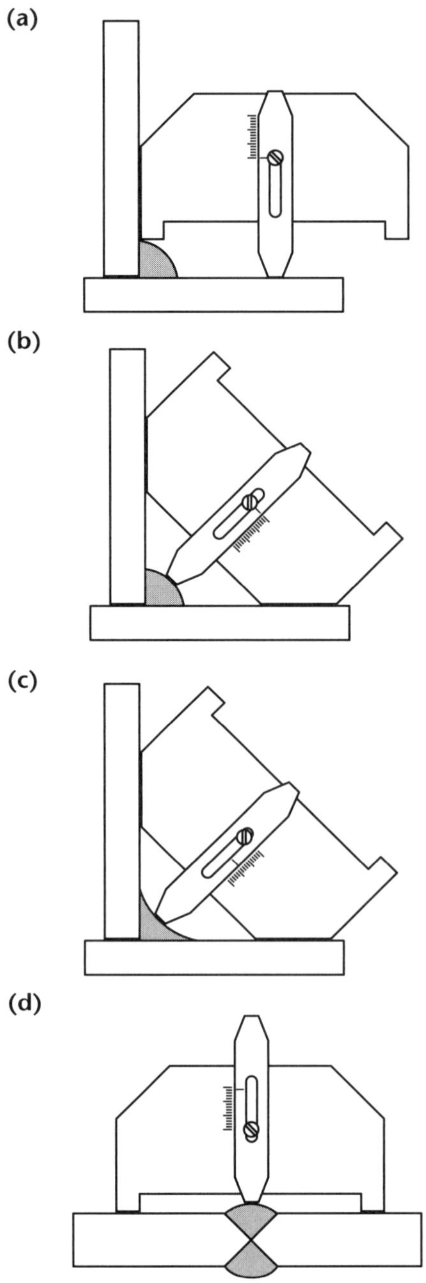

1. Computerized design tools, such as computer aided design (CAD) systems, are more feasible. The designer must have a complete understanding of inspection principles to properly apply geometric tolerancing principles. For designers to use points in space as construction points in their drawing can be a liability during manufacturing and inspection.

FIGURE 18. Cambridge gage: (a) zero to 60 degree angle of preparation; (b) excess weld metal in reinforcement; (c) pitting or depth of undercut; (d) fillet weld throat size; (e) high/low misalignment.

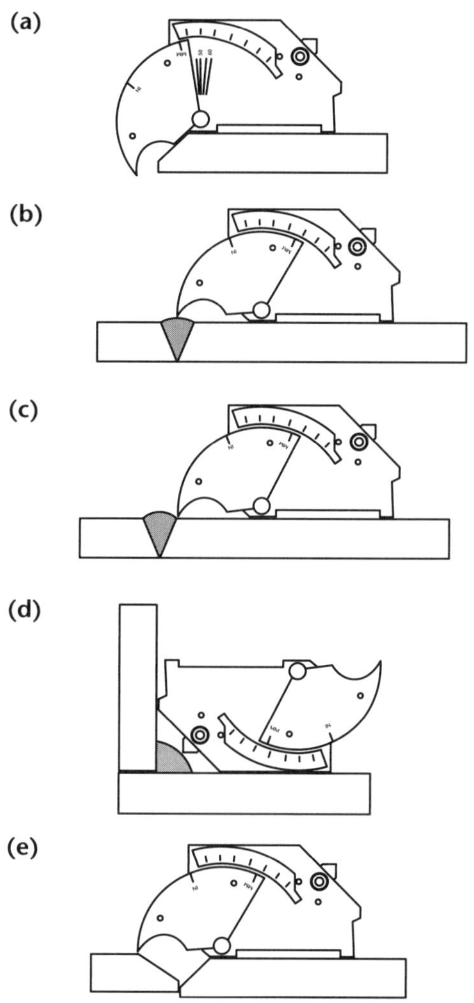

2. The part may be allowed to have the maximum possible tolerances while ensuring interchangeability and fit to other pieces. This will enhance the producibility and lower costs.
3. The specification of a consensus tolerance standard in the working procedures helps to answer concerns about quality and, in the event of material failure, about liability.
4. Controversy during manufacturing and inspection is reduced when the application of the design requirements is done consistently. Part tolerances can be established with regard to the part's functional surfaces. This consistency allows functional inspection gaging and manufacturing fixturing.

Consistency is generally accomplished in relation to datums and datum surfaces that serve as orientation or zero points. The datum plane or surface is the actual feature of the part used to establish the datum. A datum is a theoretically perfect point, axis, or plane derived from actual features. Any feature has variation. Datums are specified on the drawing and are controlled by establishing three separate datums to create an X,Y,Z coordinate system. Geometrically, three points establish a plane, two points establish a line and one point establishes a point in space. This arrangement is also called a 3,2,1 coordinate system.

Tolerances of orientation include requirements for perpendicularity, angularity or parallelism.

Tolerances of form include requirements for flatness, straightness, circularity and cylindricity. A feature has form only in terms of its relationship to itself. For example, the requirement for a flat surface means that each point on the surface must be within the specified band in terms of every other point of that surface. Any point on the surface is independent of any other surface.

Profile tolerances include the requirements for the profile of a line and the profile of a surface.

Basic dimensions are used to describe the theoretically exact size, profile, orientation, or location of a feature. They prevent ambiguity when describing the variation allowed in a feature. Size can be specified as a linear dimension, a diameter, a radius, an angle or any other quantity of size.

References

1. Bailey, W.[H.] Part 3, "Vision and Light." Section 1, "Fundamentals of Visual and Optical Testing." *Nondestructive Testing Handbook,* second edition: Vol. 8, *Visual and Optical Testing.* Columbus, OH: American Society for Nondestructive Testing (1993): p 9-21.
2. Sayler, G.[C.] *ASNT Level III Study Guide: Visual and Optical Testing Method.* Columbus, OH: American Society for Nondestructive Testing (1998, revised 2006).
3. Krauss, D.G. *ASNT Level II Study Guide: Visual and Optical Testing Method.* Columbus, OH: American Society for Nondestructive Testing (1998).
4. Chapter 4, "Basic Visual Aids and Accessories for Visual Testing": Part 3, "Magnifiers." *Nondestructive Testing Handbook,* second edition: Vol. 8, *Visual and Optical Testing.* Columbus, OH: American Society for Nondestructive Testing (1993): p 76-81.
5. Benford, J.R., L.E. Florry and A.A. Levin. Section 11, "Visual Inspection Equipment." *Nondestructive Testing Handbook,* first edition. Columbus, OH: American Society for Nondestructive Testing (1959).
6. ISO 2768, *General Tolerances.* Geneva, Switzerland: International Organization for Standardization (1989).
7. BS EN 22768-1, *General Tolerances.* Brussels, Belgium: European Committee for Standardization (1993).
8. ANSI B46.1, *Surface Texture (Surface Roughness, Waviness, and Lay).* New York, NY: American National Standards Institute (2000).
9. ANSI Y14.5M, *Dimensions and Tolerancing.* New York, NY: American National Standards Institute (2000). Superseded by ASME Y14.5M.
10. ISO 1302, *Geometrical Product Specifications (GPS) — Indication of Surface Texture in Technical Product Documentation.* Geneva, Switzerland: International Organization for Standardization (2002).
11. ASME Y14.5M, *Dimensions and Tolerancing.* New York, NY: American Society of Mechanical Engineers (2009). Supersedes ANSI Y14.5M.
12. Aizawa, H. and K. Miyagawa. Section 8, "Applications of Visual and Optical Tests in the Metals Industries": Part 2, "Visual and Optical Testing in the Steel Industry." *Nondestructive Testing Handbook,* second edition: Vol. 8, *Visual and Optical Testing.* Columbus, OH: American Society for Nondestructive Testing (1993): p 228-243.
13. CIE S 014-2; ISO 1164-2, *CIE Standard Illuminants for Colorimetry.* Wien, Österreich [Vienna, Austria]: Commission Internationale de l'Éclairage [International Commission on Illumination (CIE)] (2006).

Bibliography

Bureau of Naval Personnel. *Basic Optics and Optical Instruments.* New York, NY: Dover (1969).

Dunn, D. Chapter 3, "The Visual and Optical Testing Environment": Part 2, "Environmental Factors." *Nondestructive Testing Handbook,* second edition: Vol. 8, *Visual and Optical Testing.* Columbus, OH: American Society for Nondestructive Testing (1993): p 54-56.

Hecht, E. *Schaum's Outline of Theory and Problems of Optics.* New York, NY: McGraw-Hill (2007).

IESNA Lighting Handbook: Reference and Application, ninth edition. New York, NY: Illuminating Engineering Society of North America (2000).

ISO 3058, *Non-Destructive Testing — Aids to Visual Inspection — Selection of Low-Power Magnifiers.* Geneva, Switzerland: International Organization for Standardization (1998).

Johnson, B.K. *Optics and Optical Instruments: An Introduction,* third edition. New York, NY: Dover (1960).

Ness, S. Chapter 3, "The Visual and Optical Testing Environment": Part 3, "Physiological Factors." *Nondestructive Testing Handbook,* second edition: Vol. 8, *Visual and Optical Testing.* Columbus, OH: American Society for Nondestructive Testing (1993): p 57-62.

6

CHAPTER

Indirect Visual Testing

Michael W. Allgaier, Mistras Group, Princeton Junction, New Jersey

Thomas D. Britton, General Electric Sensing and Inspection Technologies, Skaneateles, New York (Part 3)

Trevor Liddell, General Electric Sensing and Inspection Technologies, Lewistown, Pennsylvania (Part 3)

Portions of Parts 1 and 2 are reprinted with permission from *Visual Examination Technologies* (EPRI learning modules), © [ca. 1982 and] 1996, the Electric Power Research Institute (EPRI), Charlotte, NC. ASNT has revised the text in 1993 and 2010, and deficiencies are not the responsibility of the Electric Power Research Institute.

Portions of Part 3 are reprinted with permission from *Nondestructive Testing: Remote Visual Inspection,* training modules © [2005] General Electric Corporation, Lewistown, PA. ASNT has revised the text in 2010, and deficiencies are not the responsibility of General Electric Corporation.

PART 1. Introduction to Indirect Visual Testing

Visual tests comprise five basic elements: the inspector, the test object, illumination, a recording method and usually an optical instrument. Each of these elements interacts with the others and affects the test results.

The human eye is an important component for performing visual nondestructive tests. However, there are situations where the eye does not have access to the test surface. In these cases mechanical and optical instruments can supplement the eye in a family of techniques called *indirect* visual testing. Indirect visual testing is sometimes called *remote* visual testing and is distinct from direct visual testing, where the inspector views the test surface with the naked eye and simple magnifiers and gages. Direct visual testing is discussed in a separate chapter.

Adjustable Focus[1]

Vision acuity affects the visual test, so it is important for borescopes to allow the eyepiece to be focused when used, just as binoculars are. Frequently, eyeglasses are inconvenient when the inspector is using a borescope: it is difficult to place the eye at the ideal distance from the eyepiece, and the view may be obscured by external glare and reflections. Rubber eyeshields on borescopes are designed to shut out external light but are not as effective when glasses are worn. For these reasons, it is critical that the inspector be able to adjust the instrument without wearing glasses to compensate for variations in vision acuity.

FIGURE 1. Objective distance (arrows) for forward and side viewing borescopes.

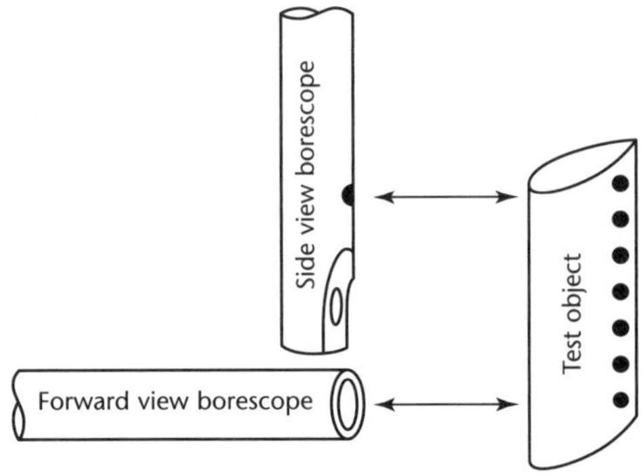

Effects of Test Object[1,2]

The test surface determines the specifications for (1) the instrument used during the visual test and (2) the required illumination. Objective distance, object size, discontinuity size, reflectivity, entry port size, object depth and direction of view are all critical aspects of the test object that affect the visual test.

1. Objective distance (Fig. 1) is important in determining the illumination source, as well as the required objective focal distance for the maximum power and magnification.
2. Object size, combined with distance, determines what lens angle or field of view is required to observe an entire test surface (Fig. 2).
3. Discontinuity size determines the magnification and resolution required for visual testing. For example, greater resolution is required to detect hairline cracks than to detect undercut (Fig. 3).

FIGURE 2. Arrows indicate portion of test object falling within field of view for forward viewing borescope.

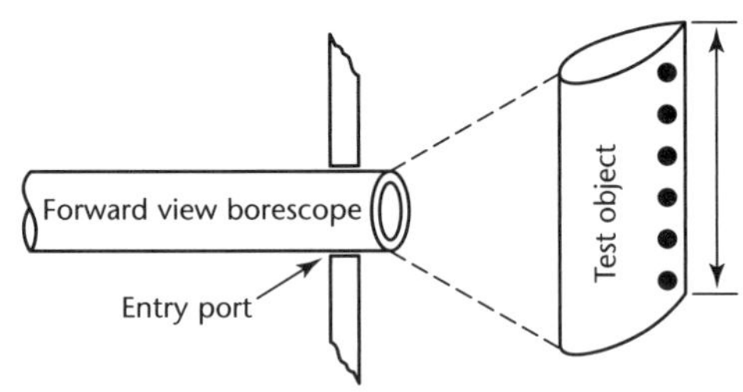

FIGURE 3. Discontinuity size affects resolution limits and magnification requirements.

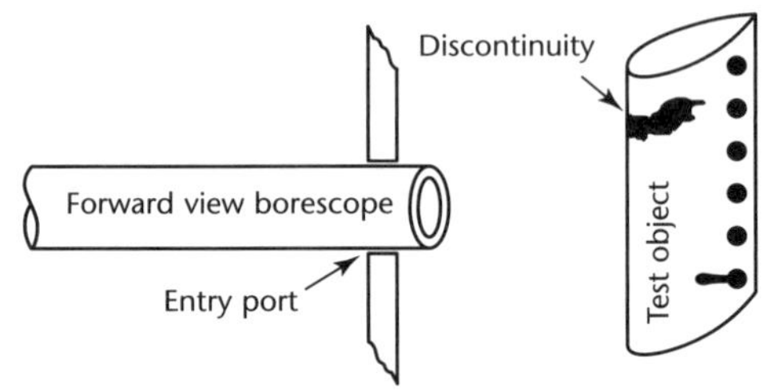

4. Reflectivity is another factor affecting illumination. Dark surfaces such as those coated with carbon deposits require higher levels of illumination than light surfaces do (Fig. 4).
5. Entry port size determines the maximum diameter of the instrument that can be used for the visual test (Fig. 5).
6. Object depth affects focusing. If portions of the object are in different planes, then the borescope must have sufficient focus adjustment or depth of field to visualize these different planes sharply (Fig. 6).
7. Direction of view determines positioning of the borescope, especially with rigid borescopes. Viewing direction also affects the required length of the borescope.

FIGURE 4. Reflectivity helps determine levels of illumination.

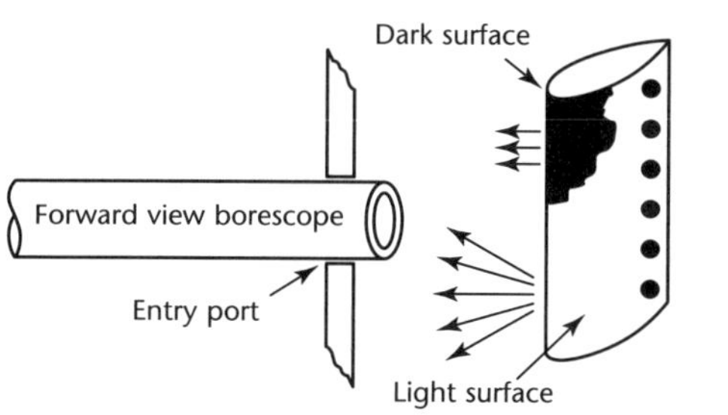

FIGURE 5. Entry port size (arrows) limits size of borescope.

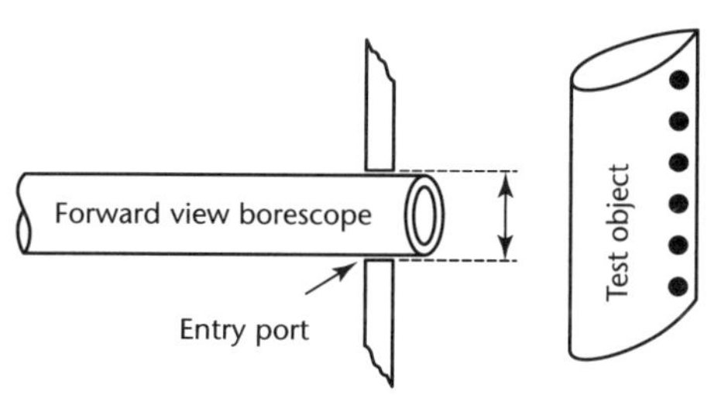

FIGURE 6. Object depth (arrows) is critical factor affecting focus.

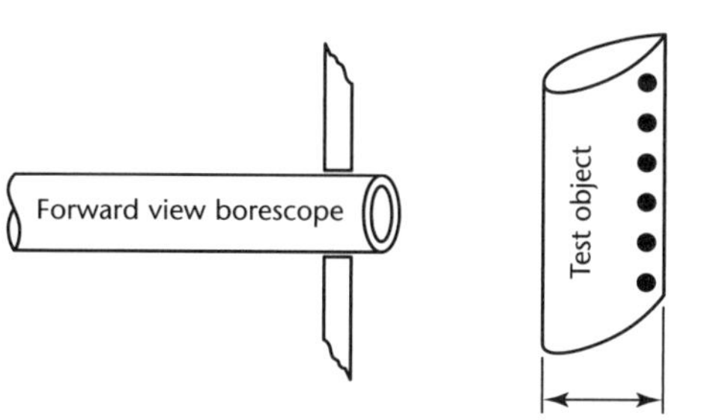

Some of the factors affecting visual tests with borescopes are in conflict and compromise is often needed. For example, a wide field of view reduces magnification but has greater depth of field (Fig. 7). A narrow field of view produces higher magnification but results in shallow depth of field. Interaction of these effects must be considered in determining the optimum setup for detection and evaluation of discontinuities in the test object.

Parts of Indirect Visual Test Instruments

Instruments for indirect visual testing typically have components for three functions.

1. A means of illumination is needed for most applications.
2. A viewing medium conveys or presents the image to the inspector's eye.
3. One or more control circuits let the inspector control a camera's or vehicle's motion.

An instrument widely used for indirect inspections is the borescope, a long instrument that may be rigid or flexible. The distal end has the objective lens directed at the test surface; the eyepiece, or ocular end, is close to the inspector and may include output to a video screen for viewing. Many borescopes also incorporate clamps or tweezers for object manipulation or retrieval.

FIGURE 7. Effects of viewing angle on other test parameters: (a) narrow angle with high magnification and shorter depth of field; (b) wide angle with low magnification and greater depth of field.

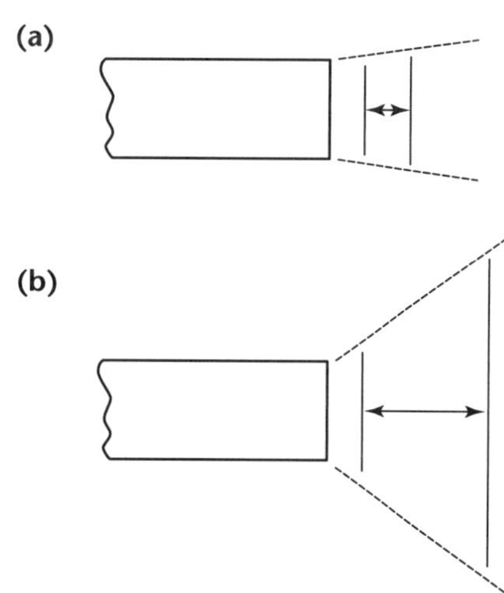

Illumination

In indirect visual tests, the inspector may work in a well lighted area. The test surface, however, is inside an object or on the far side of a barrier. Indirect visual tests can be classified according to the source of light that illuminates the test surface.

1. Light may be ambient in the chamber — if it is another room, for instance, or inside an operating furnace. Such applications might include hydroelectric penstock inspection, for example, but are unusual for most visual inspectors.
2. Light may be introduced by a lamp or flash unit at the distal (objective) end of the probing instrument. A lamp can be powered through a cable integral to a borecope. This design is problematic where there are hazardous fumes.
3. Light may be generated at the ocular (eyepiece) end of the instrument and conveyed to the test surface through fiber optic cable. This means of illumination is integral to the design of virtually all borescopes.
4. A camera based system may operate in darkness if the camera is sensitive to infrared or ultraviolet radiation at the wavelengths of interest. In condition monitoring, for example, an infrared camera may look for hot spots in an electric motor. Such applications are rare for most visual inspectors.

Image Transmission

There are various designs for indirect visual testing, and they can be classified according to the medium used to transmit the image from the objective lens to the viewing eyepiece or display.

Rigid Borescope. The light image may be transmitted through air inside the tube of a rigid borescope. This borescope is rigid so that the instrument's stiff body keeps the optical elements (lenses and possibly mirrors) aligned.

Fiber Optic Borescope. Light may be transmitted through a fiber optic borescope, having transparent (glass or plastic) fibers in a cable. A jargon term for such an instrument is *fiber scope*. The cable is thin, and the cable's flexibility lets it be threaded around interior corners to view surfaces difficult to access. At the objective end, the image may be captured by a camera and viewed on a video screen such as a computer display.

Camera Borescope. Light may be sensed by a small camera, converted to an electronic signal and sent through a cable to a video receiver. A single system called a *video probe* can offer a choice of still photography, motion video or both. The probe may be positioned in real time with a moving image; then a still image can be recorded to document the inspection.

Telemetric Instrument. In another sort of video probe, the video signal may be transmitted as radio signals to a receiver. Wireless transmission, however, opens up a range of possibilities. Such systems have been used for telemetry and robotics — the remote operation of vehicles such as submarines, aviation drones and outer space probes. Radio control can permit the nondestructive inspector to direct robotic inspections from a distance of many kilometers.

Visual test instruments in which a camera is mounted on a robot or crawler are not borescopes, and many of their applications, such as marine exploration,[3] are not nondestructive testing.

Applications of Indirect Techniques[4]

Machine Shops

Borescopes find applications in production machine shops, tool and die departments and in ferrous, nonferrous and alloy foundries. In production machine operations, borescopes of various sizes and angles of view are used to examine internal holes, cross bored holes, threads, internal surface finishes and various inaccessible areas encountered in machine and mechanical assembly operations. Specific examples are visual tests of machine gun barrels, rifle bores, cannon bores, machine equipment and hydraulic cylinders.

In tool and die shops, borescopes are used to examine internal finishes, threads, shoulders, recesses, dies, jigs, fixtures, fittings and the internal mating of mechanical parts. In foundries, borescopes are widely used for internal inspections to locate discontinuities, cracks, porosity and blowholes. Borescopes are also used for tests of many types of defense materials, including the internal surface finish of rocket heads, rocket head seats and guided missile components.

Electric Power Industry

In steam power plants, borescopes are used for visual tests of boiler tubes for pitting, corrosion, scaling or other discontinuities. Borescopes used for this type of work are usually made in 2 or 3 m (6 or 9 ft) sections. Each section is designed so that it can be attached to the preceding section, providing an instrument of any required length.

Other borescopes are used to examine turbine blades, generators, motors, pumps, condensers, control panels and other electrical or mechanical components without dismantling. In nuclear plants, borescopes offer the advantage that the inspector can be in a low radiation field while the distal, or sensor, end is in a high radiation field.

Photography of the interiors of large power plant furnaces during operation has been done since the 1940s using a periscope and camera. The periscope extends through the furnace wall and relays the optical image to the camera. A water cooled jacket protects the optical system and the camera from the furnace's high temperatures. With this equipment, still and motion picture studies have been made of the movement of the fuel bed and the action of the powdered fuel burner in furnaces operating at full load.

Petroleum and Chemical Industry

Borescopes are used for visual tests of high pressure catalytic cracking units, distillation equipment, fractionation units, hydrogenation equipment, pressure vessels, retorts, pumps and similar process equipment. Use of the borescope in the examination of such structures is doubly significant. Not only does it allow the examination of inaccessible areas without the lost time and expense incurred in dismantling, it avoids breakdown and the ensuing costly repair.

Visual tests of high pressure distillation units are used to determine the internal condition of tubes or headers. Evaporation tubes, fractionation units, reaction chambers, cylinders, retorts, furnaces, combustion chambers, heat exchangers, pressure vessels and many other types of chemical process equipment are inspected with borescopes or extension borescopes

Tank cars are inspected for internal rust, corrosion and the condition of outlet valves. Cylinders and drums can be examined for internal conditions such as corrosion, rust or other discontinuities.

Automotive Industry

Borescopes are widely used in the manufacturing and maintenance divisions of the automotive industry. Engine cylinders can be examined through spark plug holes without removing the cylinder head. The cylinder wall, valves and piston head can be visually tested for excess wear, carbon deposits and surface discontinuities. Crankcases and crankshafts are examined through wall plug openings without removing the crankcase. Transmissions and differentials are similarly inspected.

Borescopes are also useful for locating discontinuities such as cracks or blowholes in castings and forgings. Machined components such as cross bored holes can be examined for internal discontinuities. Borescopes are used to inspect cylinders for internal surface finish after honing. Tapped holes, shoulders or recesses also can be observed. Inaccessible areas of hydraulic systems, small pumps, motors and mechanical or electrical assemblies can be visually tested without dismantling the engine.

Aviation Industry

The use of borescopes for tests of airplane engines and other components without disassembly has resulted in substantial savings in costs and time. A borescope of 11 mm (0.44 in.) diameter by 380 mm (15 in.) working length can be used by maintenance and service departments for visual testing of engines through spark plug openings, without dismantling the engines. An excellent view of the cylinder wall, piston head, valves and valve seats is possible and several hundred hours of labor are saved for each engine test. Spare engines in storage can also be inspected for corrosion of cylinder wall surfaces.

Aircraft propeller blades are visually tested during manufacture. The entire welded seam of a blade can be inspected internally for cracks and other discontinuities. Propeller hubs, reverse pitch gearing mechanisms, hydraulic cylinders, landing gear mechanisms and electrical components also can be inspected with borescopes. Aircraft wing spars and struts are inspected for evidence of fatigue cracks and rivets and wing sections can be tested visually for corrosion. Borescopes used for tests of internal wing tank surfaces and wing corrugations subject to corrosion have saved airlines expense by reducing the time aircraft are out of service.

Infrastructure

Indirect visual testing is suited for the interrogation of channels, cavities, pockets, crevices and interstitial areas in all sorts of civil engineering structures. In interior areas, it can look for dampness and signs of corrosion, grout in masonry and places where brackets and braces have become detached from structures they are intended to support. Indirect techniques can explore air ducts, drains, cisterns, water lines, sewers, tunnels, crawl spaces and wall interiors.

Fiber optic borescopes have been used in the conservation of public statues (Fig. 8)[5,6] and to look inside a crypt (Fig. 9).[7]

FIGURE 8. Indirect visual testing of statues: (a) corrosion detection[5]; (b) checking for separation of weld joints in Statue of Liberty.[6]

(a)

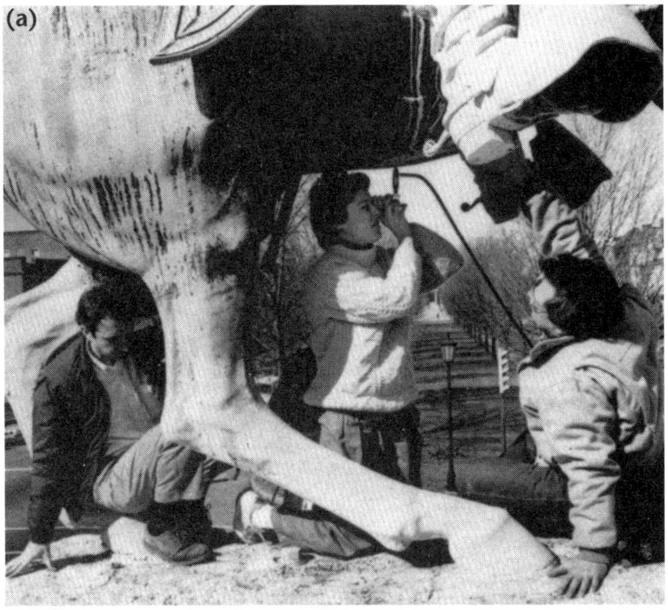

(b)

FIGURE 9. Indirect visual testing of crypt: (a) introduction of borescope from above; (b) view in borescope eyepiece.[7]

(a)

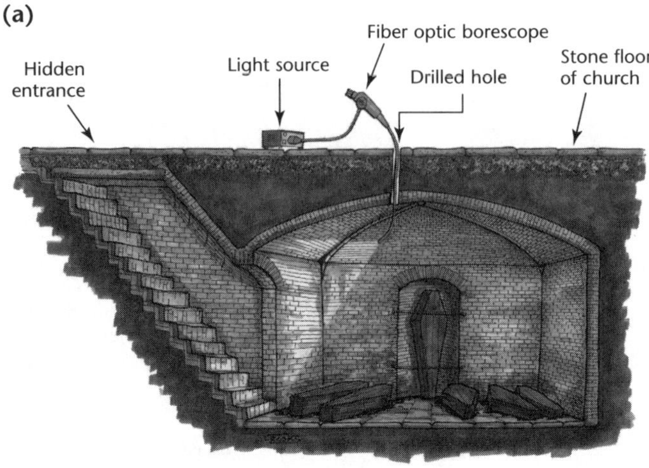

(b)

PART 2. Borescopy[4,8]

Rigid Borescopes

The rigid borescope (Fig. 10) was invented to inspect the bores of rifles and cannons. It was a thin telescope with a small lamp at the objective end for illumination. Most rigid borescopes now use a fiber optic light guide for illumination.

The image is brought to the eyepiece by an optical train consisting of an objective lens, sometimes a prism, relay lenses and an eyepiece lens. The image is not a real image but an aerial image: it is formed in the air between the lenses. This means that it is possible both to provide diopter correction for the observer and to control the objective focus with a single adjustment to the focusing ring at the eyepiece.

FIGURE 10. Lens system in representative rigid borescope.

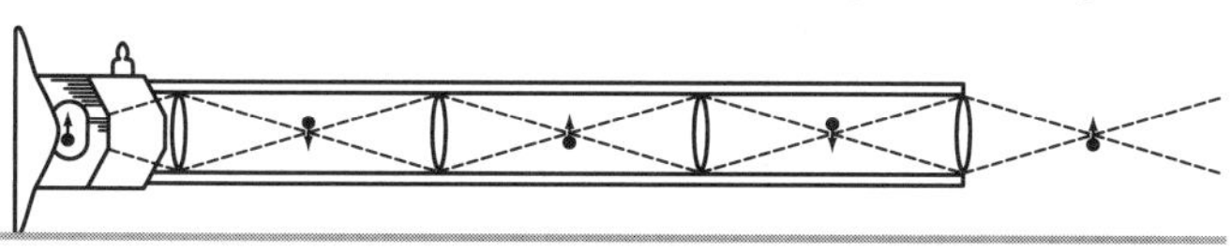

Focusing of Rigid Borescope

The focus control in a rigid borescope greatly expands the depth of field over nonfocusing or fixed focus designs. At the same time, focusing can help compensate for wide variations in eyesight among inspectors.

Figures 11 and 12 emphasize the importance of focus adjustment for expanding the depth of field. Figure 11 was taken at a variety of distances with fixed focus. Figure 12 was taken at the same distances as in Fig. 11 but with a variable focus, producing sharper images.

Need for Specifications

Because rigid borescopes lack flexibility and the ability to scan areas, specifications regarding length, direction of view and field of view become more critical for achieving a valid visual test. For example, the direction of view should always be specified in degrees rather than in letters or words such as north, up, forward or left. Tolerances should also be specified.

FIGURE 11. Borescope images with fixed focus (compare Fig. 12): (a) at 75 mm (3 in.); (b) at 200 mm (8 in.); (c) at 300 mm (12 in.).

(a)

(b)

(c)

FIGURE 12. Borescope images with variable focus (compare Fig. 11): (a) 75 mm (3 in.); (b) 200 mm (8 in.); (c) 300 mm (12 in.).

(a)

(b)

(c)

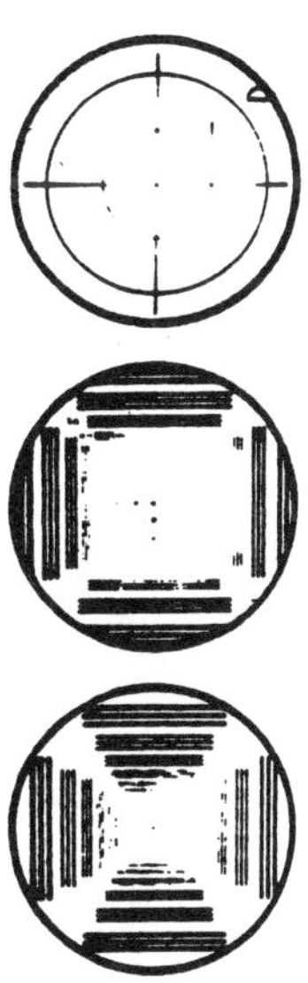

Some borescope manufacturers have considered the eyepiece to be zero degrees and therefore a direct view rigid borescope is 180 degrees. Other manufacturers start with the borescope tip as 0 degrees and then count back toward the eyepiece, making a direct view 0 degrees (Fig. 13).

Setup of Rigid Borescope

To find the direction and field of view during visual testing with a rigid borescope, place a protractor scale on a board or worktable. Position the borescope parallel to the zero line, with the lens directly over the center mark on the protractor. Remember that the optical center of a borescope is usually 25 to 50 mm (1 to 2 in.) behind the lens window.

By sighting through the borescope, stick pins into the board at the edge of the protractor to mark the center and both the left and right edges of the view field. This simple procedure gives both the direction of view and the field of view (Figs. 14 and 15).

FIGURE 13. Borescope direction of view: (a) forward; (b) side; (c) forward oblique; (d) retrospective.

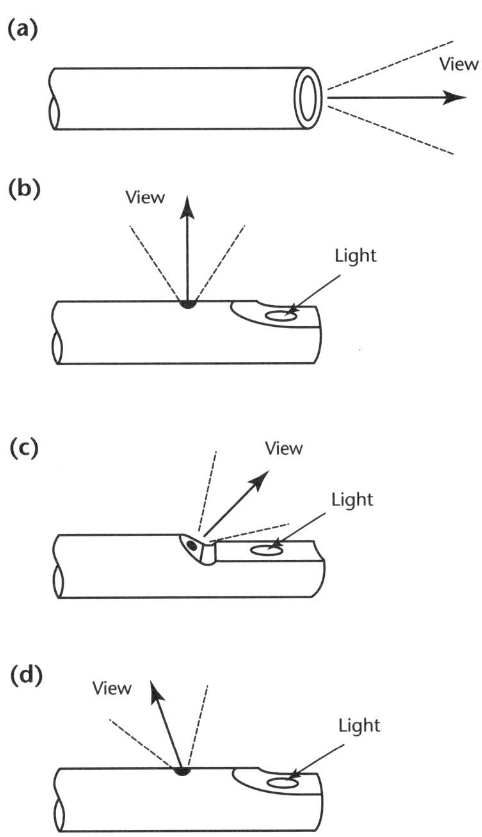

Miniature Borescope

One variation of the rigid borescope is called the *miniature borescope*. In this design, the relay lens train is replaced with a single, solid fiber. The fiber diffuses ions in a parabola from the center to the periphery of the housing, giving a graded index of refraction. Light passes through the fiber and forms an image at specific intervals. The aperture is so small that the lens has an infinite depth of field (like a pinhole camera) and no focusing mechanism is needed.

Accessories

Many accessories are available for rigid borescopes. Still and video cameras can be added to provide a permanent record of a visual test. Closed circuit television displays are common as well. Also available are attachments at the eyepiece permitting dual viewing or right angle viewing for increased accessibility.

Fiber Optic Borescopes

The industrial fiber optic borescope is a flexible, layered sheath protecting two fiber optic bundles, each comprising thousands of glass fibers. One bundle serves as the image guide, and the other bundle illuminates the test object.

Light travels in straight lines but optical glass fibers bend light by internal reflection to carry light around corners (Fig. 16). Such fibers are 9 to 30 µm (0.0004 to 0.0013 in.) in diameter, roughly one tenth of the thickness of a human hair.

A single fiber transmits very little light, but thousands of fibers may be bundled for transmission of light and images. To

FIGURE 14. Field of view for rigid borescope.

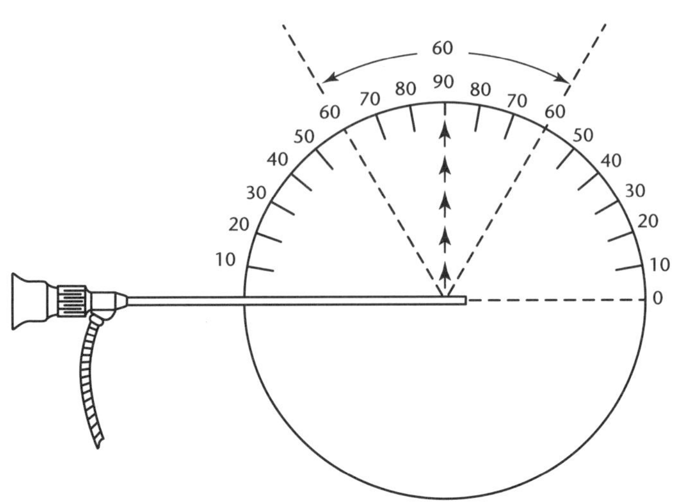

FIGURE 15. Distance to test surface versus field of view.

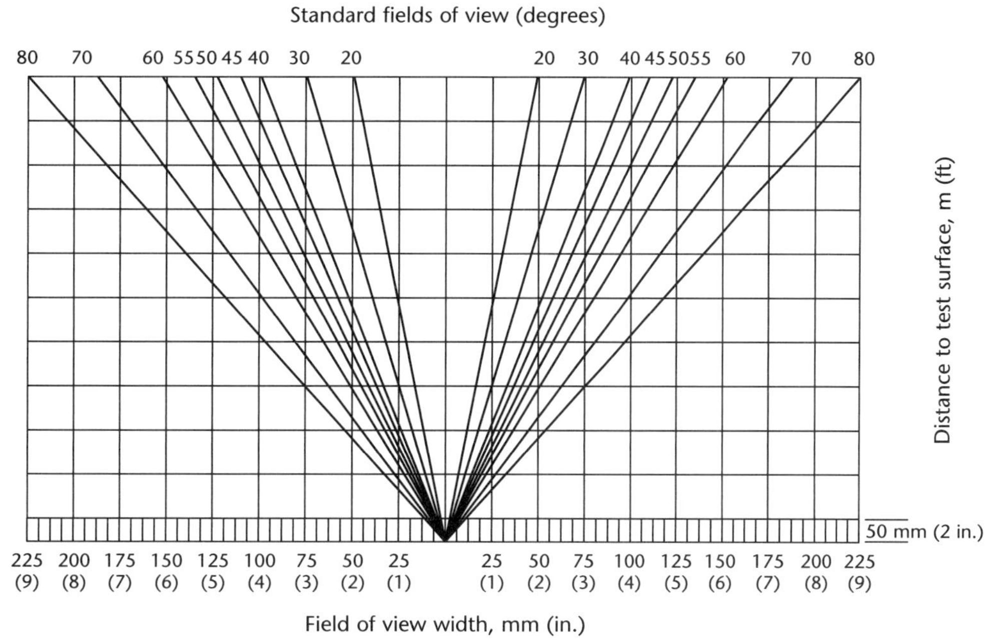

FIGURE 16. Internal reflection of light in optic fiber can be used to move light path in curve.

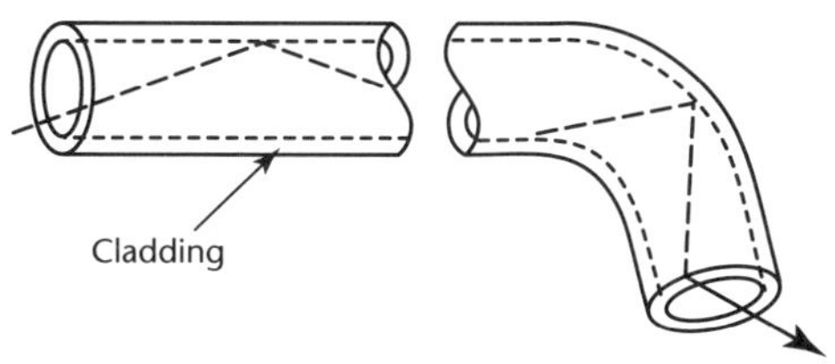

FIGURE 17. Light paths in fiber bundles: (a) uncoated fibers allow light to travel laterally through bundle and (b) coated fibers restrict light's path to original fiber.

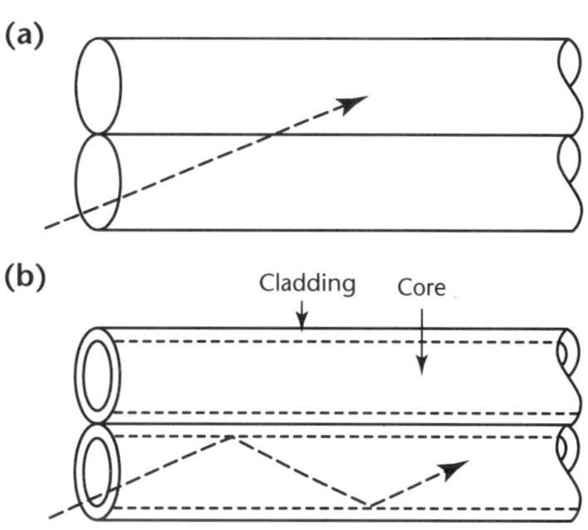

prevent the light from diffusing, each fiber consists of a central core of high quality optical glass coated with a thin layer of another glass with a different refractive index (Fig. 17). This cladding acts as a mirror — all light entering the end of the fiber is reflected internally as it travels (Fig. 16) and cannot escape by passing through the sides to an adjacent fiber in the bundle.

Although the light is effectively trapped within each fiber, not all of it emerges from the opposite end. Some of the light is absorbed by the fiber itself, and the amount of absorption depends on the length of the fiber and its optical quality. For example, plastic fiber can transmit light and is less expensive to produce than optical glass, but plastic is less efficient in its transmission and is unsuitable for fiber optic borescopes.

Fiber Image Guides

The fiber bundle used as an image guide (Fig. 18) carries the image formed by the objective lens at the distal end, or tip, of the borescope back to the ocular end, pr eyepiece. The image guide must be a coherent bundle: the individual fibers must be precisely aligned so that they are in identical relative positions at their terminations.

Image guide fibers range from 9 to 17 µm (0.0004 to 0.0007 in.) in diameter. Their size affects resolution, although the preciseness of alignment is far more important.

A real image forms on both highly polished faces of the image guide. Therefore, to focus a fiber optic borescope for different distances, the objective lens at the tip must be moved in or out, usually by remote control at the eyepiece section. A separate diopter adjustment at the eyepiece is necessary to compensate for differences in eyesight.

Fiber Light Guides

To illuminate the test object, another fiber bundle carries light from an external high intensity source. This light guide bundle is noncoherent (Fig. 19). These fibers are about 30 µm (0.0013 in.) in diameter, and the size of the bundle is determined by the diameter of the scope.

Fiber optic borescopes usually have a controllable bending section near the tip so that the inspector can direct the borescope during testing and can scan an area inside the test object. Fiber optic borescopes are made in a variety of diameters, some as small as 3.7 mm (0.15 in.), in lengths up to 10 m (30 ft), and with a choice of viewing directions.

FIGURE 18. Optical fiber bundle used as image guide.

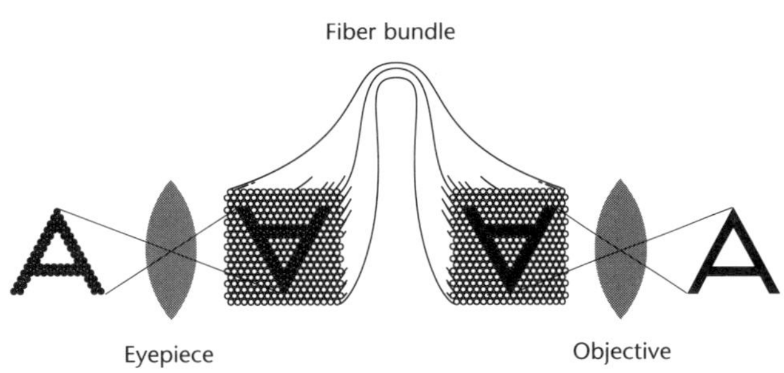

Borescope Optical Systems

Borescopes are precise optical devices containing a complex system of prisms, achromatic lenses and plain lenses that pass light to the observer with high efficiency. An integral light source at the objective end usually provides illumination of the test surface.

Angles of Vision

To meet a wide range of visual testing applications, borescopes are available in various diameters and working lengths to provide various angles of vision for special requirements. The most common vision angles are (1) right angle, (2) forward oblique, (3) direct and (4) retrospective (Fig. 13).

These types of vision are characterized by different angles of obliquity for the central ray of the visual field, with respect to the forward direction of the borescope axis (Table 1).

General Characteristics

Desirable properties of borescopic systems are large field of vision, no image distortion, accurate transmission of color values and adequate illumination.

The brightest images are obtained with borescopes of large diameter and short length. As the length of the borescope is increased, the image becomes less brilliant

FIGURE 19. Diagram of representative fiber optic borescope.

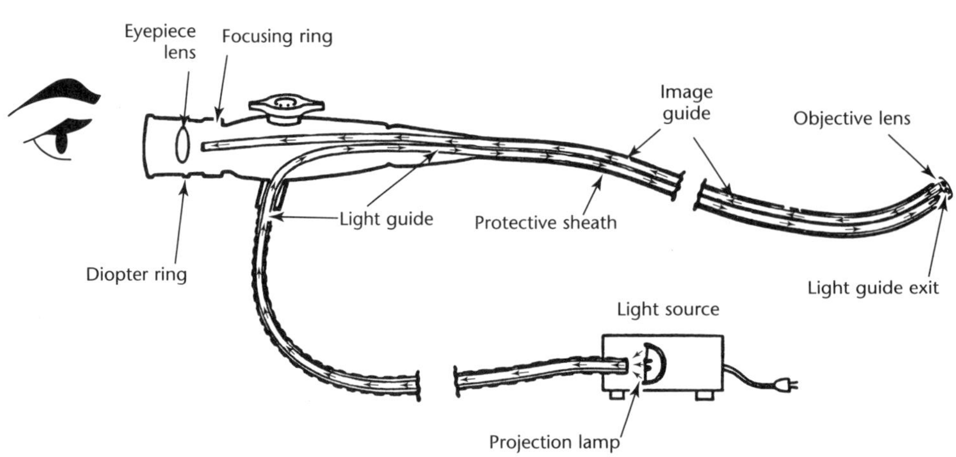

because of light losses from additional lenses required to transmit the image. To minimize such losses, lenses are typically coated with antireflecting layers to provide maximum light transmission.

Optical Components

The optical system of a borescope consists of an objective end, a middle lens system, correcting prisms and an ocular section (Figs. 10 and 19), angle of vision, the field of view and the amount of light gathered by the system.

The middle lenses conserve the light entering the system and conduct it through the borescope tube to the eye with a minimum loss in transmission. Design of the middle lenses has an important effect on the character of the image. For this reason, the middle lenses are achromatic, each lens being composed of two elements with specific curvatures and indexes of refraction. This design preserves sharpness of the image and true color values.

Depending on the length of the borescope, the image may need reversal or inversion or both, at the ocular end. This change is accomplished by a correcting prism within the ocular end for borescopes of small diameter and by erecting lenses for larger designs.

Depth of Focus, Field of View and Magnification

The depth of focus for a borescopic system is inversely related to the numerical aperture of its lens. Field of view, on the other hand, is relatively large, generally on the order of 50 degrees of angular field. This corresponds to a visual field of about 25 mm (1 in.) diameter at 25 mm (1 in.) from the objective lens. At different working distances, the diameter of the field of view varies with the working distance (Figs. 14 and 15).

The entrance pupil is that image of any of the lens apertures, imaged in the object space, which subtends the smallest angle at the object plane. Because the numerical aperture of borescope systems is usually very small compared with that of a microscope, the corresponding depth of focus is exceedingly large. This permits the use of fixed focus eyepieces in many small and moderately sized instruments. Focus and field of view are discussed in the chapter on photography.

The total magnification of borescopes varies with diameter and length but generally ranges from about 2× to 8× in use. Note that the linear magnification of a given borescope changes with working distance and is about inversely proportional to the object distance. A borescope with 2× magnification at 25 mm (1 in.) working distance therefore will magnify 4× at 13 mm (0.5 in.) distance. Magnification is discussed in the chapter on direct viewing.

Table 1. View orientations and angles of obliquity.

Orientation	Angle of Obliquity (degrees)	Angular Field (degrees)
Direct	0	45
Forward oblique	25	50
Forward	45	45
Right angle	90	50
Retrospective	135	45
Circumferential	0 and 90	45 to 15

Borescope Construction

A borescopic system usually consists of one or more borescopes having integral or attached illumination, additional sections or extensions, a battery handle, battery box or transformer power supply and extra lamps, all designed to fit in a portable case. If light is provided by a lamp at the objective end, insulated wires between the inner and outer tubes of the borescope serve as electrical connections. A contact ring permits rotation of the borescope through 360 degrees for scanning the object space without entangling the electrical cord. In most designs, however, illumination is provided by fiber optic light guides (Fig. 19).

Borescopes with diameters under 38 mm (1.5 in.) are usually made in sections, with focusing eyepieces, interchangeable objectives and high power integral lamps. This kind of borescope typically consists of an eyepiece or ocular section, a 1 or 2 m (3 or 6 ft) objective section, with 1, 2 or 3 m (3, 6 or 9 ft) extension sections. The extensions are threaded for fitting and ring contacts are incorporated in the junctions for electrical connections. Special optics can be added to increase magnification when the object is viewed at a distance.

Angulated borescopes are available with forward oblique, right angle or retrospective visual systems. These instruments usually consist of an objective section with provision for attaching an eyepiece at right angles to the objective section's axis. This permits inspection of shoulders or recesses in areas not accessible with standard borescopes. Eyepiece extensions at right angles to the axis of the borescope can be supplied, with provision to rotate the borescope with respect to the eyepiece extension, for scanning the object field.

Right Angle Borescopes

The right angle borescope can have a light source positioned ahead of the objective lens. The optical system provides vision at right angles to the axis of the borescope and covers a working field of about 25 mm (1 in.) diameter at 25 mm (1 in.) from the objective lens.

Applications of the right angle borescope are widespread. The instrument permits testing of inaccessible corners and internal surfaces. It is available in a wide range of lengths, in large diameters or for insertion into apertures as small as 2.3 mm (0.09 in.). It is the ideal instrument for visual tests of rifle and pistol barrels, walls of cylindrical or recessed holes and similar components.

Another application of the right angle borescope is inspection of the internal entrance of cross holes, where it may be critical to detect and remove burrs and similar irregularities that interfere with correct service. Drilled oil leads in castings can be visually inspected, immediately following the drilling operation, for blowholes or other discontinuities that cause rejection of the component. Right angle borescopes can be equipped with fixtures to provide fast routine tests of parts in production. The device's portability allows occasional tests to be made at any point in a machining cycle.

Direct Vision Borescope

The direct vision instrument provides a view directly forward with a typical visual area of about 19 mm (0.75 in.) at 25 mm (1 in.) distance from the objective lens. The light carrier is removable so that the two parts can be passed successively through a small opening.

Forward Oblique Borescopes

The forward oblique system is a design that permits the mounting of a light source at the end of the borescope yet also allows forward and oblique vision extending to an angle of about 55 degrees from the axis of the borescope.

A unique feature of this optical system is that, by rotating the borescope, the working area of the visual field is greatly enlarged.

Retrospective Borescope

The retrospective borescope has an integral light source mounted slightly to the rear of the objective lens. For a bore with an internal shoulder whose surfaces must be accurately tooled, the retrospective borescope provides a unique method of accurate visual inspection.

Panoramic Borescopes

Wide field borescopes have rotating objective prisms to provide fields of view up to 120 degrees. One application of wide field borescopes is the observation of models in wind tunnels under difficult operating conditions.

Special optical systems permit rapid panoramic scanning of internal surfaces. A mirror mounted in front of the objective lens system is rotated by turning a knob near the eyepiece. Swiveling in one plane covers the ranges of forward oblique, right angle and retrospective vision.

One form of panoramic borescope permits rapid scanning of the internal cylindrical surfaces of tubes or pipes. This instrument has a unique objective system that simultaneously covers a cylindrical strip 30 degrees wide around the entire 360 degrees with respect to the axis of the borescope. The diameter of this instrument is 25 mm (1 in.) and the working length is 1 m (3 ft) or larger.

Sectioned Borescopes

Borescopes under 38 mm (1.5 in.) diameter are often made in pieces, with the objective section 1 or 2 m (3 or 6 ft) in length. The additional sections are 1, 2 or 3 m (3, 6 or 9 ft) long with threaded connections. These sections may be added to form borescopes with lengths up to 15 m (45 ft) for diameters under 38 mm (1.5 in.).

Special Purpose Borescopes

Borescopes can be built to meet many special visual testing requirements. The factors affecting the need for custom designs include: (1) the length and position of the test area, (2) its distance from the entry port, (3) the diameter and location of the entry port and (4) inspector distance from the entry port.

Environmental conditions such as temperature, pressure, water immersion, chemical vapors or ionizing radiation are important design factors. The range of special applications is partly illustrated by the examples given below.

Miniature Borescopes. Miniature borescopes are made in diameters as small as 1.75 mm (0.07 in.), including the light source. They are useful because they can go into small holes. Inspection of microwave guide tubing is a typical application.

Periscopes. A large, multiangulated, periscope with a right angle eyepiece and a scanning prism at the objective end, with field of view 70 degrees in azimuth by 115 degrees in elevation, has been used for remote observation of otherwise inaccessible areas — cave exploration, for example.[8]

Indexing Borescope. Butt welds in pipes or tubing 200 mm (8 in.) in diameter or larger can be visually tested with a special 90 degree indexing borescope. The instrument is inserted in extended form through a small hole drilled next to the weld seam and is then indexed to the 90 degree position by rotation of a knob at the eyepiece. The objective head is then centered within the tube for viewing the weld. A second knob at the eyepiece rotates the objective head through 360 degrees for scanning the weld seam.

Reading Borescopes. Low power reading borescopes are used in plant or laboratory setups for viewing the scales of instruments such as cathetometers at inaccessible locations. The magnification is about 3× at 1 m (3 ft) distance.

Calibrated Borescopes. Calibrated borescopes are designed to meet specific test requirements. The external tubes of these instruments can be calibrated to indicate the depth of insertion during a test. Borescopes with calibrated reticles are used to determine angles or sizes of objects in the field when held at a predetermined working distance.

Ultraviolet Borescopes. Ultraviolet borescopes are used during fluorescent magnetic particle and fluorescent penetrant tests. These borescopes are equipped with ultraviolet lamps, filters and special transformers to provide the necessary wavelengths.

Harsh Environments. Waterproof and vaporproof borescopes are used for internal tests of liquid, gas or vapor environments. They are completely sealed and impervious to water or other types of liquid. Water cooled or gas cooled borescopes are used for tests of furnace cavities and jet engine test cells and for other high temperature applications.

Part 3. Camera Based Measurement[9]

Video Borescope

Relatively small openings can be accessed by using a charge coupled device sensor, light transmitted through a fiber bundle to the test surface (Fig. 20). The returning image is passed electronically from the charge coupled device through a wire to the image processor. From the processor, the reconstructed pixel image is viewed on a monitor, typically a liquid crystal display.

Small openings from 6 to 80 mm (0.25 to 3.0 in.) may be penetrated with the probe. Skillful manipulation can gain access to awkward locations not in the initial line of sight.

The total distance from the tip to the test surface is limited by the fiber bundle transmitting the light. If a the bundle is flexible enough to make sharp turns, then it may be too flexible to probe very long distances. Similarly, stiff bundles can probe farther but do not bend easily.

Small diameter tubes, pipe or openings into piping or tubing systems, pumps, valves, turbines, gear boxes or other components may be accessed if there are no multiple sharp turns for distances longer than several feet.

Video borecopes are often called *video probes* or *video scopes* to distinguish them from fiber optic borescopes, which they resemble.

Figure 20. Video probe: (a) tip; (b) attachments.

(a)

(b)

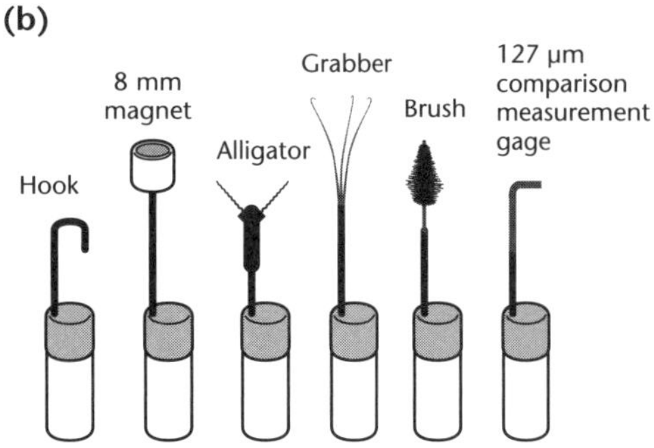

Characteristics Measured

In indirect visual testing, the test surface is typically the interior of a text object — for example, the inside surface of a pipe. To characterize or describe a feature, there are several ways that its shape, orientation or extent can be measured on that test surface.

1. A distance measurement is a linear measurement of objects on a flat surface perpendicular to the optical axis of the probe. (Note that the word *distance* here does not mean distance from the probe to the surface.)
2. A skew measurement is a linear measurement of an object on a surface at an angle that is skewed, or nonperpendicular, to the optical axis of the probe (Fig. 21).

Figure 21. Skew measurement is linear measurement of object on surface at skewed angle (nonperpendicular) to optical axis of probe.

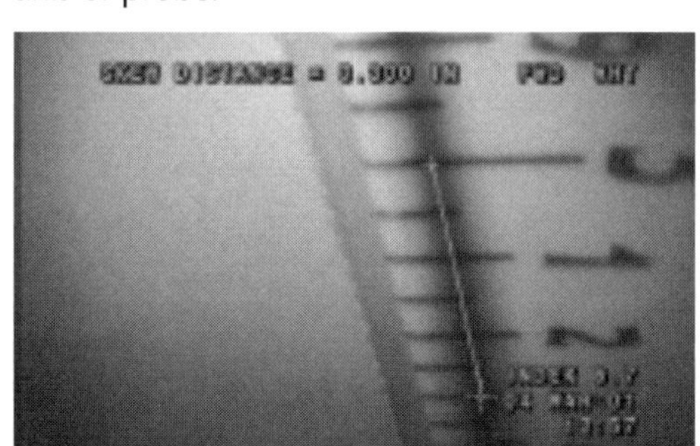

3. A depth measurement measures the height or depth of an object by determining the distances from the probe to nearer and farther surfaces and calculating the difference. The viewing angle should be perpendicular to the surface plane (Fig. 22) or the resulting measurement will be a diagonal distance rather than normal distance.
4. Area measurement calculates the surface area of a discontinuity or feature in a plane. The operator places multiple cursors in the image so that the dots when connected encompass the area of interest — a surface feature or anomaly (Fig. 23). Using image processing programming, the microprocessor calculates the surface area encompassed. Area measurement can be used to measure variations in material or surface conditions, for example, a patch of corrosion.
5. A circle gage (Fig. 24) is a quick and convenient way to check that a feature's diameter does not exceed a set amount in any direction. A circle is drawn and saved in the computer memory and may be superimposed on the test image during discontinuity assessment. A circle gage is like a reference object's image stored in the optical instrument's microprocessor. It can be used with the stereo or shadow techniques and is useful for quick accept/reject measurements.
6. Multiple short segments can be combined to measure the length of a nonlinear feature (Fig. 25). Segment lengths are calculated from point to point; the actual length depends on cursor placement and so may not follow the surface of the object.
7. Point-to-line measurement provides data in two dimensions, along an X axis and Y axis. The length of a line perpendicular to a line between two points is measured.

FIGURE 22. Depth measurement measures height or depth of object by determining distances from probe to upper and lower surfaces and calculating difference.

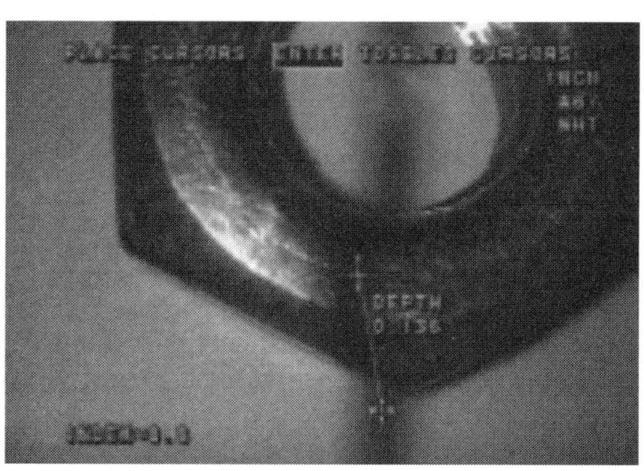

FIGURE 23. Cursors are connected to measure size or area of discontinuity on flat surface perpendicular to optical axis.

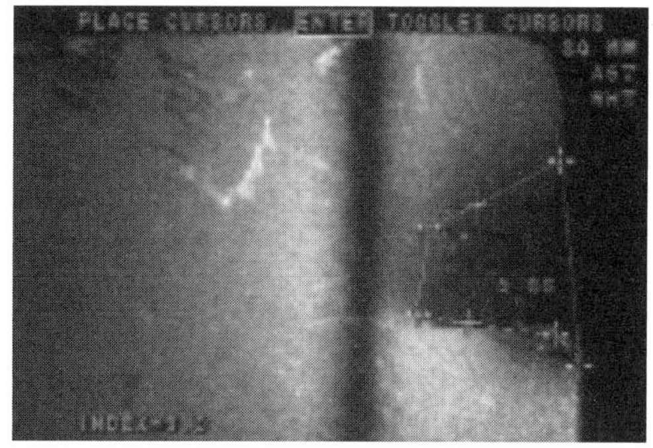

FIGURE 24. Circle gage technique makes circle of known size on image to compare to feature, based on known reference dimension.

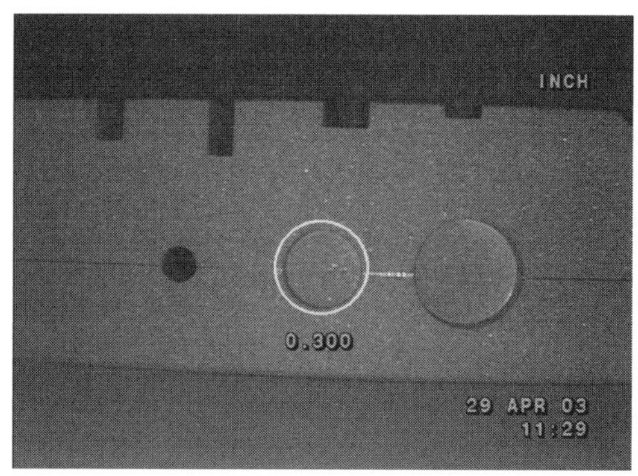

FIGURE 25. Multiple-segment technique is used to measure length of nonlinear feature by combining lengths of shorter segments.

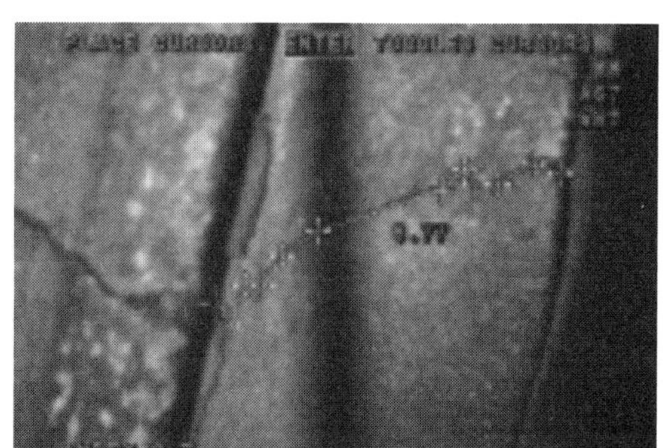

Note that commercially available systems do not display the distance from the probe to the surface. Integral software makes that calculation automatically and provides the inspector with data of interest — the size or extent of discontinuities or surface features.

Principles of Measurement

There are several designs that permit optical systems to measure objects and indications in the field of view. Three fundamental approaches may be used in combination: (1) triangulation, (2) comparison with a reference object (in the field of view or memory) and (3) pixel counting.

Triangulation

Indirect systems incorporate calibrated lens assemblies and image processing programs to perform measurements automatically, so the visual inspector does not need to learn trigonometry to measure indications. Understanding the principles behind the measurement, however, is valuable for purposes of test planning and interpretation.

Triangulation has been used for indirect measurements for centuries. The tangent of an angle is the ratio of the side opposite the angle divided by the side adjacent to the angle. The adjacent side is the inspector's position, and the opposite side is the distance to the object (Fig. 26). Using a calculator with a tangent function, the inspector can set up a ratio and solve for the distance to the object. For example, given a right triangle with an adjacent side of 10 m and a side angle of 70 degrees, what is the target distance X?

$$\text{Tan} = \frac{\text{Opposite side}}{\text{Adjacent side}} \tag{1}$$

$$\text{Tan } 70 = 2.7475 = \frac{X}{10 \text{ m}} \tag{2}$$

$$X = 27.5 \text{ m} \tag{3}$$

Accuracy

The accuracy of any measurement depends on the ability of an operator to operate the instrument and to recognize and capture a quality image. Accuracy is maximized by using the greatest magnification (filling as much of the screen as possible with the image), getting as close to perpendicular as possible and accurately placing the measurement cursors.

A reference block can be used to verify measurement accuracy. One such measurement verification block contains a side view chamber and one forward view chamber that contains 2.5400 mm ± 0.00508 mm (0.100 ± 0.0002 in.) test targets having calibration traceable to reference standards of the National Institute of Standards and Technology.

Techniques of Measurement

The three means of measurement — triangulation, comparison and pixel counting — are used in various combinations by three significant measurement techniques: stereo, shadow and comparison (Table 2).

Figure 26. Triangulation.

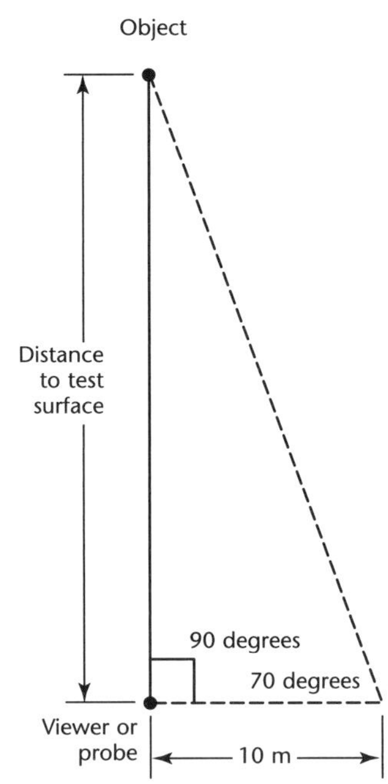

Table 2. Measurement techniques for indirect visual testing.

Measurement	Techniques: Stereo	Shadow	Comparison
Distance (length)	yes	yes	yes
Skew	—	yes	—
Depth	yes	yes	—
Area	yes	yes	yes
Circle gage	—	yes	yes
Multiple-segment	yes	yes	yes
Point to line	yes	yes	yes

Stereo Measurement

A stereoscopic image is a picture comprising two images of the same scene, taken from different viewpoints slightly displaced from each other, the viewpoints corresponding to the eyes right and left (Fig. 27). Some indirect visual test systems capture the twin images by offsetting one lens by a small, known distance. An inspector must view a stereoscopic image with both eyes to get stereo effect.

Parallax is the apparent motion of a relatively close object with respect to a more distant background as the location of the observer changes. This apparent displacement of an object is discussed also in this volume's section on photogrammetry. Stereoscopic parallax is caused by taking photographs of the same object but from different points of observation. Adjacent but overlapping photograph pairs can be used to measure depth and feature dimensions.

The difference in the angles formed by the lines of sight to two objects at different distances from the eyes is a factor in the visual perception of depth. Parallax is inversely proportional to an object's distance. An object near the inspector's nose has a much larger parallax than an object at arm's length

A stereo probe provides two views simultaneously on the left and right side of the screen. The farther something is from the observer, the less it appears to move in one view than in the other. The computer needs to recognize enough of a unique image in one view to recognize it in the adjacent view. The computer uses triangulation geometry to perform measurements. The shift is measured, and the distance is calculated. The shift in pixel distance between two cursors on an image plane from one stereo view and the other gives the ability to calculate distances.

There are five implementations of stereo measurement: length, point to line, area, multiple-segment length and circle gage.

The accuracy of stereo probe measurements depends on operator skill, image quality and magnification. The usable range for performing stereo measurements extends out to a target distance of about 30 mm (1.25 in.). For more accurate measurements, it is recommended that measurements be performed with a distance of 15 mm (0.64 in.) from tip to target.

Shadow Measurement

A known distance of a light source to an offset from the center hairline obstruction yields a known angle of the shadow being cast (Fig. 28). A shadow measurement is made with an image recalled from computer memory or frozen at the moment of inspection. The shift of the shadow cast across the face of the image permits trigonometric calculation of the distance to the surface on which the shadow lies. The points across the plane of the surface can be calculated by placing a cursor on different points of the shadow or points on a plane of the measured feature.

The shadow is cast by a gnomon in the optical instrument. (The term *gnomon* denotes the blade or column that casts a shadow on the face of a sundial; here, it denotes an artifact in the illumination system that casts a shadow on the test surface.) If the size of the gnomon is known, its shadow can be used to measure a feature on the test surface. In this case as in others, the calibrated system makes calculations and automatically provides measurements to the inspector.

Measurement can be made even if the objective is so close to a surface that the image is somewhat out of focus. In this case, features must be sharp enough so that the cursors can be placed accurately. There are various implementations of the

FIGURE 27. Stereo technique: (a) setup; (b) variation of parallax with distance of three different objects.

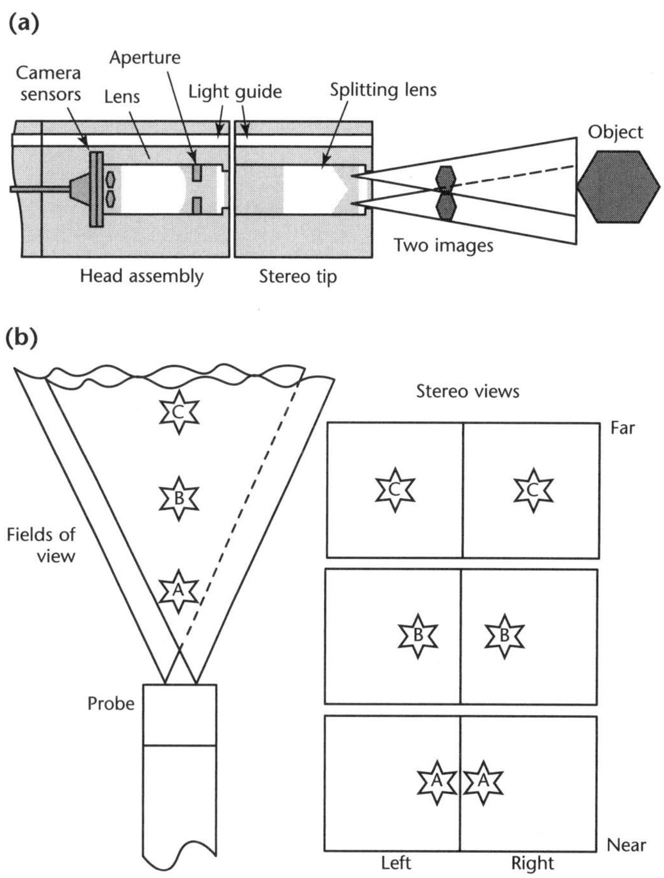

shadow technique, including distance, skew, depth, point-to-line length, multiple-segment length, area and circle gage.

The measurement surface must be nearly perpendicular to the probe view for an accurate measurement, except for skew measurements. The axis is perpendicular if the shadow appears in the viewer as an unbroken line that remains vertical even when the objective head is rotated. In this viewing condition, measurements can be taken anywhere in the frozen image.

In skew measurements, the shadow will appear sloped, nonvertical, indicating that the distance from the surface to the probe (and therefore magnification) varies. From a skewed viewing angle, measurements can be made only along the shadow, so the shadow must be aligned with the object to be measured before the image is captured and the measurement is made. A sloped shadow means the viewing tip is not perpendicular to the surface, so distance to the surface and magnification differ in different parts of the image.

In depth measurements, when the surface planes are perpendicular, the shadow will appear vertical in the image and remain vertical even when the probe is rotated. The shadow will break to the left to indicate a raised surface or to the right to indicate a recessed surface. Depth measurements can only be made along the shadow, so the shadow should be aligned vertically across the change in contour before freezing or capturing the image. When marking the depth or height to be measured, place the point cursors along the center of the shadow on the lower and upper surfaces near the depth break.

To ensure accuracy, the operator must place the cursor on the shadow image consistently.

FIGURE 28. In the shadow technique, the probe tip contains a charge coupled device imager with lens and fiber optic bundle carrying illumination from light source. The illumination lens has a gnomon on it that casts a shadow into field of view.

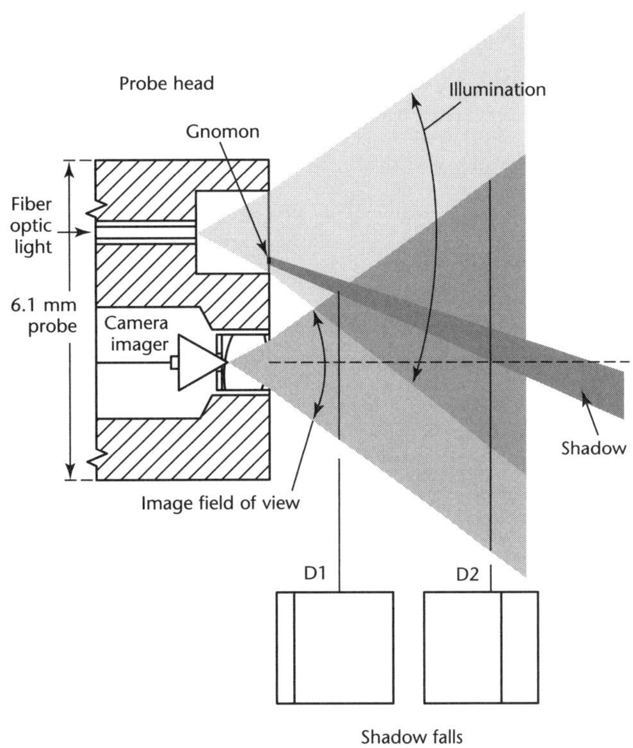

Comparison Measurement

Comparison measurements are made by using saved images. Measurement begins by measuring the dimension of a known reference object or feature (placed on the test surface with the probe or previously) to measure other objects in the same view and plane. These reference images must be at same focal distance to the test surface, and the measurement surface must be nearly perpendicular to the probe for accuracy. Pixels in the image are then counted to measure feature areas and dimensions.

The comparison technique is suited for measuring feature characteristics for their length, point-to-line distance, multiple-segment length, area and circle gage diameter. Although less accurate than shadow or stereo measurements, comparison measurements are useful if a shadow or stereo tip is unavailable, if the area or feature to be measured is too far for the shadow or stereo technique, if a shadow cannot be placed where needed, if the computer will not match two stereo images or if the approximate sizes of many items are to be checked quickly.

Measurement tips can be calibrated for specific probes and might not measure accurately with other probes. For optimum accuracy, stable seating of the measurement tips can be ensured by small O rings on the probe tip. The O ring should be checked each time the measurement tip is used, and it must be replaced if missing or excessively worn.

Depth Measurement

A depth measurement calculates the distance of a point above or below a plane surface. In the stereo technique, software requires the user to place three cursors on the image to define a plane surface and then to place a fourth cursor above or below that plane. The microprocessor will calculate the length of a perpendicular line between the fourth cursor and the defined plane.

In the shadow technique (Fig. 22), depth can only be measured along the shadow, so the shadow should be aligned vertically across the change in contour before freezing the image. When marking the depth or height to be measured, the point cursors should be placed along the center of the shadow on the lower and upper surfaces near the break. When the surface planes are perpendicular, the shadow will appear vertical in the image and remain vertical even when the probe is rotated. The viewing angle should be perpendicular to the surface plane to yield a true vertical depth.

Access to Particular Viewpoints

Several designs have the specific function of maneuvering the camera to where it can view an area of interest.

Pan, Tilt and Zoom Cameras

Cameras that can pan, tilt and zoom (Fig. 29) facilitate inspections by directing the camera on specific areas of interest. In the nuclear industry, these systems are used for inspection or surveillance activities in refueling, reactor vessel maintenance and inservice inspections. Loose parts, tools or debris dropped into power generation systems may be retrieved from hard-to-reach areas without disassembly of plant equipment.

The pan, tilt and zoom camera is suited for surveillance, large diameter piping, on-line applications, radiation environments, underwater applications, sumps, manifolds, and steam headers.

The pan, tilt and zoom camera is not suited for small diameter piping, heat exchangers or the space between the core shroud and the reactor vessel.

Panning, tilting and zooming permit the indirect viewing of vessels for cleanliness, corrosion, discontinuities, interior indications of cracks and many other features of the vessel surface. The collection of digital images of the inspection allows for historical comparisons of previous vessel inspections to identify and track changes of tanks or vessel profiles.

Push Cameras

A "push" camera is mounted on a rigid rod that the inspector inserts into an aperture or cavity; additional rods can be attached to extend the total length. A separate electronics cable controls the camera and carries images to the inspector or work station. A push camera can be adapted to various applications with specific cable lengths, camera diameters and lens options to perform a variety of inspections with one video system. Applications are found in the infrastructure, processing, power generation and water treatment industries. A single system inspects stainless tubing for weld condition or cleanliness, process piping, drain lines and heat exchangers. Camera heads can be swapped for different pipe openings and diameters.

The push camera design is unsuited for large diameter channels or open spaces where specific directional views are desired. Without pan, tilt and zoom capabilities, the view obtained yields minimum choice by the operator.

Fossil Power Generation. The push camera is suited for inspections of headers, turbines, boiler tubes, drain lines, heat exchangers, steam drums and other confined spaces. Varying lenses and camera heads provide flexibility for the specific situations.

Water Treatment. The push camera facilitates quick inspections of confined areas such as steam drums. One hand is used for camera operation; the other, for viewing and control functions up to 50 m (164 ft) away. A color camera provides color images to help identify water and steam related corrosion in pipe diameters from 31.8 mm (1.25 in.) to 406 mm (16 in.).

FIGURE 29. Camera head with pan, tilt and zoom capabilities.

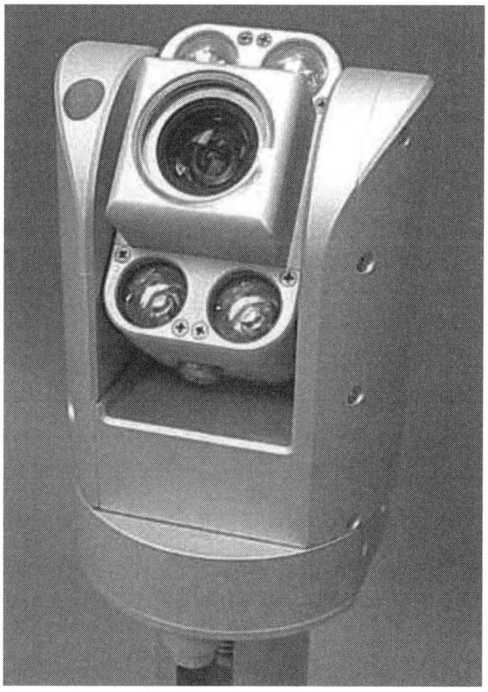

Infrastructure. The structural integrity of a building or the condition of a heavily trafficked bridge can be checked with accuracy and accessibility. Hand held operation and varying cable lengths allow for remote viewing in wastewater systems, buildings, bridges, underground piping, manholes and building airshafts.

Crawlers

Crawlers are useful for pipe inspections and entrances to hazardous spaces. Pipe crawler camera systems are commonly made of a tractor, lights, a camera (with or without pan and tilt), a cable reel, central control unit and various accessories and controls including auxiliary lights and centering devices. The typical combination is tractor, light, camera and cable (Fig. 30).

When space permits, robotic crawlers are available with options to mount a charge coupled device camera with two 35 W spot or flood lamps. The pan, tilt and zoom armatures of the camera can be mounted atop a robotic crawling platform or axially in front of a crawler.

Many robotic crawlers are portable and designed to inspect inside pipes with diameters ranging from 150 mm (6 in.) to 900 mm (36 in.). Crawlers can pass through restricted pipe, large offsets, and protruding pipe taps.

FIGURE 30. Video probe crawler: (a) side diagram; (b) application photograph.

(a)

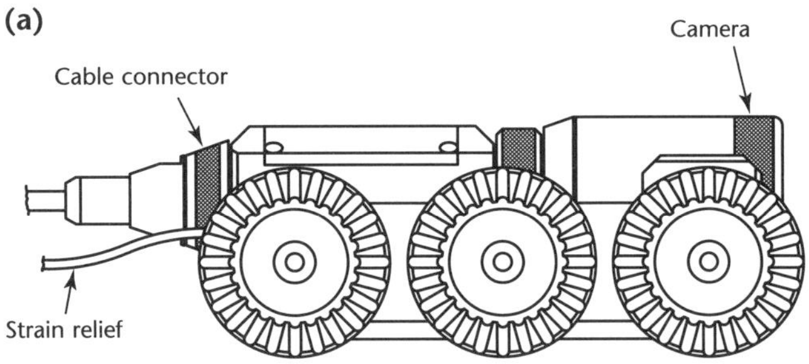

(b)

A crawler with 300 mm (12 in.) wheels and 13.6 kg (30 lb) of pull power can examine 200 mm (8 in.) and larger pipes and distances up to about 450 m (1500 ft). This arrangement allows operation in hazardous environments and 24× zoomed viewing up to 15 m (50 ft) from the camera.

Large tractors can carry the camera, lights, pull cable and pan and tilt mechanism — up to 9.5 kg (21 lb) — for distances up to 200 m (600 ft) long. As a rule, it is better to use the more stout system when space and configuration allow. The wheels can be removed to fit 250 mm (10 in.) pipe, but crawlers still need room to make a turn.

Adverse terrain can be navigated, and steep inclines can be negotiated. Some debris can be tolerated in the pipe or tank, and some bumps can be overcome, but a crawler needs clearance to get over an obstacle. After the crawler gets over debris in the outbound travel, it can be difficult to retrieve it over the same debris if the downstream or far side of an obstruction is not tapered for the return trip (Fig. 31).

Crawlers can explore enclosed spaces and channels such as air ducts and electrical conduits, lubrication oil lines, steam lines, steam headers and small pressure vessels and tanks. If waterproof, crawlers are suited for environments that may have liquid up to 100 mm (up to 4 in.) deep— sewer lines, concrete pipe storm lines and intake or discharge lines. A variety of wheel assemblies are available for retrieving loose parts and maneuvering through wreckage.

FIGURE 31. Obstruction must be tapered to permit travel over it: (a) obstruction with taper for travel to left but not to right; (b) minor obstruction.

(a)

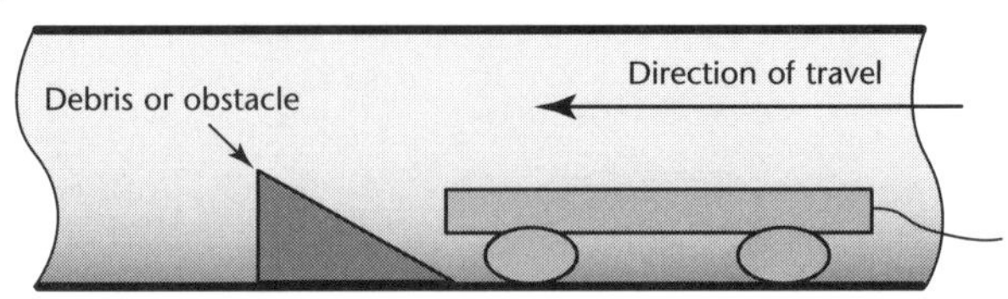

(b)

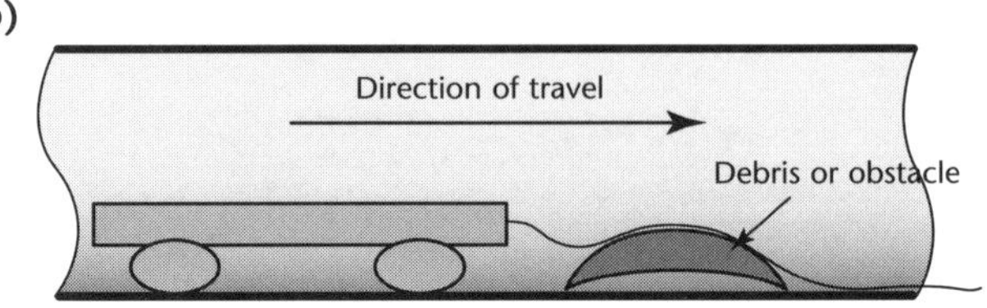

References

1. Section 4, "Basic Aids and Accessories for Visual Testing": Part 1, "Basic Visual Aids." *Nondestructive Testing Handbook,* second edition: Vol. 8, *Visual and Optical Testing.* Columbus, OH: American Society for Nondestructive Testing (1993): p 74-75.
2. EPRI *Visual Examination Technologies,* training modules. Charlotte, NC: Electric Power Research Institute (1996).
3. Froehlich, H.E. "Building the Personnel Sphere of the Deep-Submersible Alvin." *Materials Evaluation.* Vol. 48, No. 6. Columbus, OH: American Society for Nondestructive Testing (June 1990): p 785-786, 788-789, 791-793.
4. Section 4, "Basic Aids and Accessories for Visual Testing": Part 3, "Borescopes." *Nondestructive Testing Handbook,* second edition: Vol. 8, *Visual and Optical Testing.* Columbus, OH: American Society for Nondestructive Testing (1993): p 82-91.
5. "Borescopes Check for Corrosion Inside Statue." *Materials Evaluation.* Vol. 45, No. 6. Columbus, OH: American Society for Nondestructive Testing (June 1987): p 640, 716.
6. "The Statue of Liberty: The Restoration in Retrospect." *Materials Evaluation.* Vol. 44, No. 8. Columbus, OH: American Society for Nondestructive Testing (July 1986): p 891-895.
7. "Scope Turns Clock Back to 18th Century." *Materials Evaluation.* Vol. 43, No. 10. Columbus, OH: American Society for Nondestructive Testing (September 1985): p 1246-1247.
8. Benford, J.R., L.E. Flory and A.A. Levin. Section 10, "Visual Inspection Equipment." *Nondestructive Testing Handbook,* first edition. Vol. 1. Columbus, OH: American Society for Nondestructive Testing (1959).
9. *Nondestructive Testing: Remote Visual Inspection,* training modules. Lewistown, PA: General Electric Corporation [2005].

Bibliography

Ammann, F. *Industrial Endoscopy: Inspection[,] Quality Control [and] Non Destructive Testing.* Munich, Germany: Resch (1984).

Anderson, M.T., S.E. Cumblidge and S.R. Doctor. "An Assessment of Remote Visual Testing System Capabilities for the Detection of Service Induced Cracking." *Materials Evaluation.* Vol. 63, No. 9. Columbus, OH: American Society for Nondestructive Testing (September 2005): p 883-891.

"Borescope Speeds Weld Inspection on Stainless Steel Tubing." *Materials Evaluation.* Vol. 45, No. 4. Columbus, OH: American Society for Nondestructive Testing (April 1987): p 442-443.

"Borescopes Help Inspect Race Car Engines." *Materials Evaluation.* Vol. 48, No. 2. Columbus, OH: American Society for Nondestructive Testing (February 1990): p 290.

Boyd, C. and M. Lyman. "Video Borescope Technology Helps Aircraft Engine Maintenance." *Materials Evaluation.* Vol. 51, No. 5. Columbus, OH: American Society for Nondestructive Testing (May 1993): p 543.

Cabe, T. and R. Olmsted. Section 5, "Other Instrumentation and Electronic Aids for Visual Testing": Part 3, "Borescopes." *Nondestructive Testing Handbook,* second edition: Vol. 8, *Visual and Optical Testing.* Columbus, OH: American Society for Nondestructive Testing (1993): p 148-153.

"Coast Guard Uses Borescope in War against Drug Traffic." *Materials Evaluation.* Vol. 46, No. 9. Columbus, OH: American Society for Nondestructive Testing (August 1988): p 1168.

Davis, R.S. "Remote Visual Inspection in the Nuclear, Pipeline and Underwater Industries." *Materials Evaluation.* Vol. 48, No. 6. Columbus, OH: American Society for Nondestructive Testing (June 1990): p 797-798, 800-803.

Egger, E. Section 5, "Other Instrumentation and Electronic Aids for Visual Testing": Part 1, "Remote Positioning and Transport Systems." *Nondestructive Testing Handbook,* second edition: Vol. 8, *Visual and Optical Testing.* Columbus, OH: American Society for Nondestructive Testing (1993): p 128-130.

EPRI TM101PS, *Visual Examination Technologies:* Level 1, *Practical/Specific.* Charlotte, NC: Electric Power Research Institute (1996).

EPRI TM102CG, *Visual Examination Technologies:* Level 2, *General.* Charlotte, NC: Electric Power Research Institute (1996).

Generazio, E.R., D.J. Roth and V. Torrey. Section 5, "Other Instrumentation and Electronic Aids for Visual Testing": Part 2, "Video Technology." *Nondestructive Testing Handbook,* second edition: Vol. 8, *Visual and Optical Testing.* Columbus, OH: American Society for Nondestructive Testing (1993): p 131-147.

Lorenz, P.G. "Expanding Technology Adds Value to Testing, Visual and Remote Visual Inspection Procedures." *Materials Evaluation.* Vol. 55, No. 9. Columbus, OH: American Society for Nondestructive Testing (September 1997): p 1001-1004.

Lorenz, P.G. "Video Borescopes." *Materials Evaluation.* Vol. 61, No. 9. Columbus, OH: American Society for Nondestructive Testing (September 2003): p 982-985.

Lorenz, P.G. *The Science of Remote Visual Inspection (RVI): Technology, Applications, Equipment.* Lake Success, NY: Olympus (1990).

McGarry, G. "Penetrants and Fiber Optic Borescope Help Ensure Safety of Shuttle Training Aircraft." *Materials Evaluation.* Vol. 52, No. 10. Columbus, OH: American Society for Nondestructive Testing (October 1994): p 1180.

Pellegrino, B.A. "Remote Visual Testing for Internal Pressure Vessel Inspection." *Materials Evaluation.* Vol. 56, No. 5. Columbus, OH: American Society for Nondestructive Testing (May 1998): p 606-609.

Remote Visual Testing Training Course Manual. Flanders, NJ: Everest VIT (1988, 2000).

Samsonov, P. "Remote Visual Inspection for NDE in Power Plants." *Materials Evaluation.* Vol. 51, No. 6. Columbus, OH: American Society for Nondestructive Testing (June 1993): p 662-663.

Schaack, D.F. "Measuring Borescope Optical Quality." *Materials Evaluation.* Vol. 61, No. 2. Columbus, OH: American Society for Nondestructive Testing (February 2003): p 168-172.

Siddiqi, A.A. NASA SP-2002-4524, *Deep Space Chronicle: A Chronology of Deep Space and Planetary Probes 1958–2000.* Monograph in Aerospace History. No. 24. Washington, DC: National Aeronautics and Space Administration (2002).

7

CHAPTER

Machine Vision for Visual Testing

Zheng Liu, National Research Council Canada, Ottawa, Ontario, Canada

Hiroyuki Ukida, University of Tokushima, Tokushima-shi, Japan

Part 1. System Architecture of Machine Vision System

Machine vision is the application of computer vision to industry and manufacturing. It is a specialization within system engineering, which encompasses computer science, optics, mechanical engineering, and industrial automation. One definition of machine vision is "the use of devices for optical, noncontact sensing to automatically receive and interpret an image of a real scene in order to obtain information and/or control machines or processes."[1]

For nondestructive inspection, visual inspection is usually performed by an experienced inspector. However, a tedious and difficult task may cause the inspector to tire prematurely and degrade the quality of inspection. Repetitive and dangerous inspection demands machine vision to replace human inspection so that precise information can be extracted and interpreted consistently. With technological advances on computer, camera, illumination, and communication, widespread application of machine vision systems to nondestructive testing is foreseen.

The basic architecture of a personal computer based machine vision system is given in Fig. 1. The main components include light source, detector, optics, frame grabber and computer. Machine vision does not necessarily mean the use of a computer. Specialized image processing hardware is even capable of achieving a higher processing speed and can replace the computer.[1] The modern approaches may use a camera with the capability to interface directly with a personal computer, a system designed on an image processing board, or a vision engine that plugs into a personal computer.[2]

A smart camera is a self-contained and standalone unit with communication interfaces. A typical smart camera may consist of the following components: image sensor, image digitization circuit, memory, digital signal processor, communication interface, input/output ports and a built-in illumination device. An embedded vision computer, which is a standalone box with frame storage and intelligence, is intermediate to the personal computer based vision system and a smart camera.[2] The system differs from smart cameras, in that the camera is tethered to the unit rather than self-contained. Different system configurations have their own advantages for different applications. A personal computer based machine vision system has the greatest flexibility and capability of handling a wider range of applications.

Figure 1. Typical architecture of machine vision system.

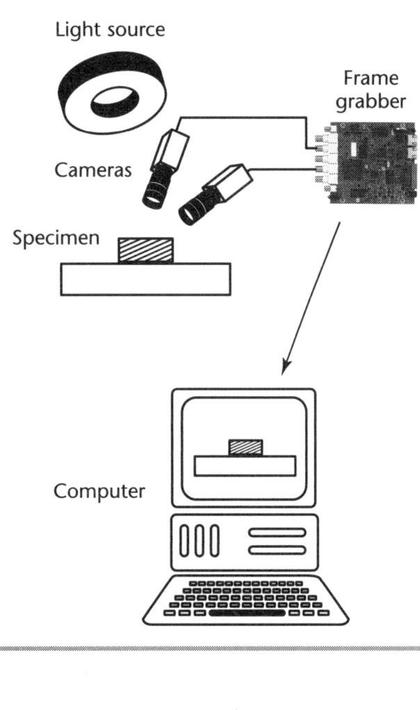

Figure 2. Four basic parameters for optics.

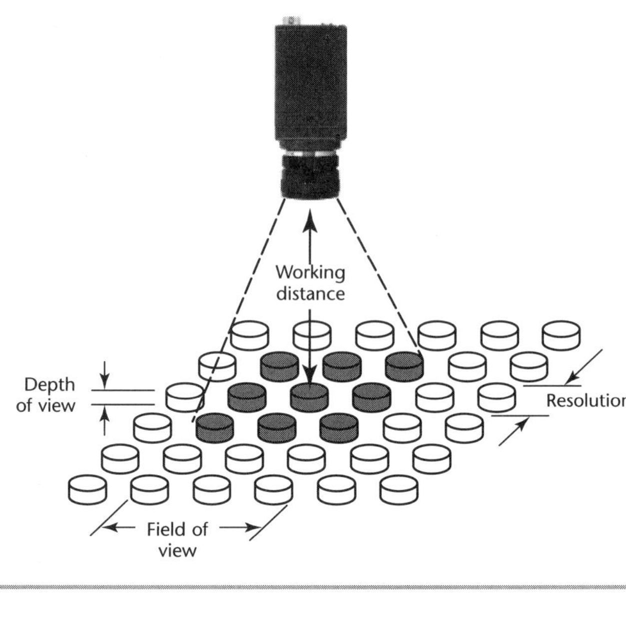

Optics and Lighting

Optics

Optics is one of the basic elements for a machine vision system. The optics creates an image such that there is a correspondence between object points and image points and contributes to the object enhancement.[3] However, the optics may also introduce distortions and aberrations. The optics includes lenses, mirrors, beam splitters, prisms, polarizers, color filters and gratings. Optics has three functions in a machine vision system: produce a two-dimensional image of the scene at the sensor; eliminate some of the undesired information from the scene image with various filters; transfer or modify the light before it arrives at the scene.[3]

There are four basic parameters for specifying the optics of a vision system: field of view, resolution, working distance and depth of field.[4] They are illustrated in Fig. 2. Field of view is the extent of the observable scene measured as an angle subtended from the focal point. The resolution of a system is the minimum size of a distinguishable feature of the object under inspection. The depth of field of a lens is its ability to maintain a desired resolution as the object is positioned closer to or farther from the lens focus.

The second element of a machine vision system is lighting. An application specific lighting or illumination can yield consistent appearance, which is essential to the success of a machine vision system. The lighting should maximize the contrast of features of interest while minimizing all other features.

Illumination

Illumination can be provided by one or more of the following techniques: front lighting, backlighting, coaxial lighting, structured illumination, strobed illumination or polarized light.

As illustrated in Fig. 3a, the bright field mode for front lighting uses any light source in the line of sight of the camera upon direct reflection from the test surface. Matte surfaces will appear darker than specular surfaces because the scattering of the matte surface returns less light to the camera. In contrast, sharp reflection returns more light. Dark field is any light source that is outside the line of sight of the camera upon direct reflection. In a dark field, light scattering from a matte surface will reach the camera and create a bright region. Similarly, a bright field for backlighting is any light source in the line of sight of the camera upon direct transmission from the source, whereas dark field is outside of the line of sight of the camera upon direct transmission (Fig. 3b).

Front lighting is the most convenient configuration of illumination for machine vision systems. A front lighting setup may be a single point source, a combination of lighting configurations, or may encompass the entire dome. Figure 4 shows the directional lighting, diffuse dome lighting, and oblique lighting. Light from one direction can create high contrast transitions to highlight an object's features. A diffuse dome light is one in which the light may scatter in many different directions. This makes it useful to light curved and specular surfaces. Oblique lighting is obtained by restricting the light to lower incoming angles. With this technique, flat specular surfaces reflect light away from the camera while small raised or depressed areas reflect light back.

In back lighting in Fig. 5a, the light source and camera are placed on opposite sides of the object under inspection. This arrangement creates dark silhouettes

FIGURE 3. Bright and dark field mode: (a) front light; (b) backlighting.

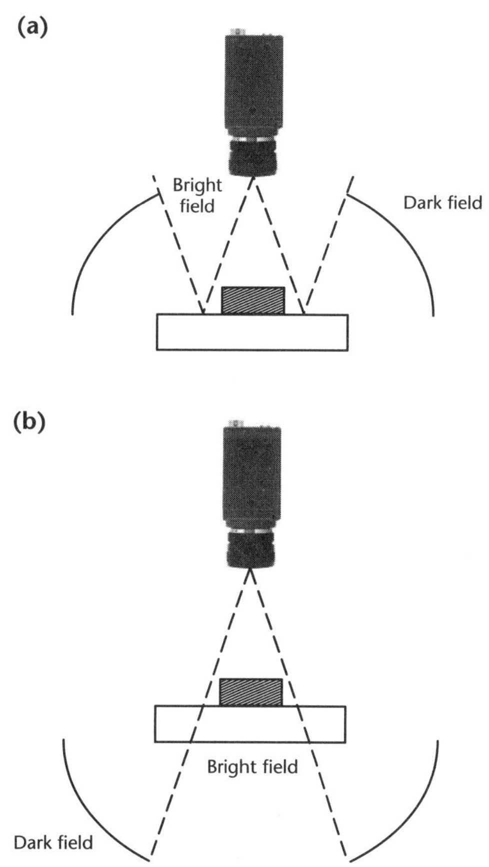

against a bright background. A beam splitter is used to create a coaxial or on-axis illumination as shown in Fig. 5b. On-axis or coaxial illumination is particularly effective for enhancing differentially angled, textured, or topographic features on a flat object.

Structured illumination is the projection of a light pattern at a known angle onto an object so that the dimensional information can be acquired.[5] The light pattern can be a plane, grid, or other complex shapes. The intersection of the pattern and an object can be translated into geometric information. Thus, the shape of an object can be determined.

Strobed illumination is also known as *strobe lighting*. A flash of light illuminates the test object momentarily.[6] Strobe lighting is applied in alarm systems, theatrical lighting and high visibility running lights.

Unpolarized light is an electromagnetic wave vibrating in various directions. Polarization limits such vibration to a single plane, which includes the optical axis. The techniques include transmission, reflection, refraction and scattering. Polarization techniques can improve the optical front portion of a machine vision system.

FIGURE 4. Some configurations of front lighting: (a) directional lighting; (b) diffuse dome lighting; (c) oblique lighting.

(a)

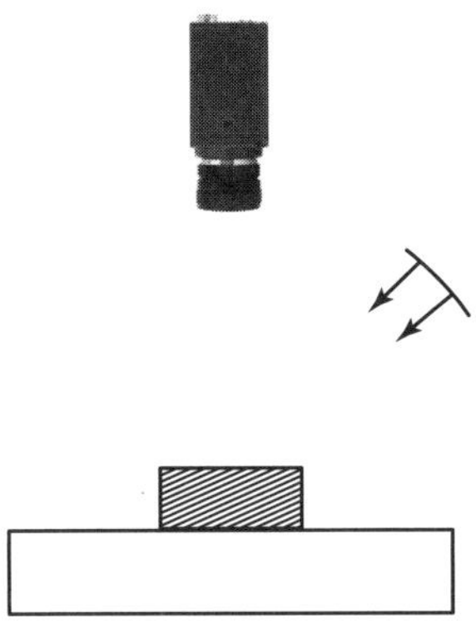

(b)

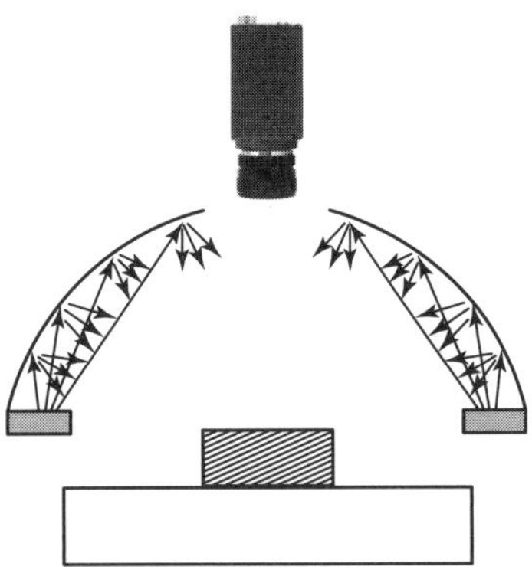

(c)

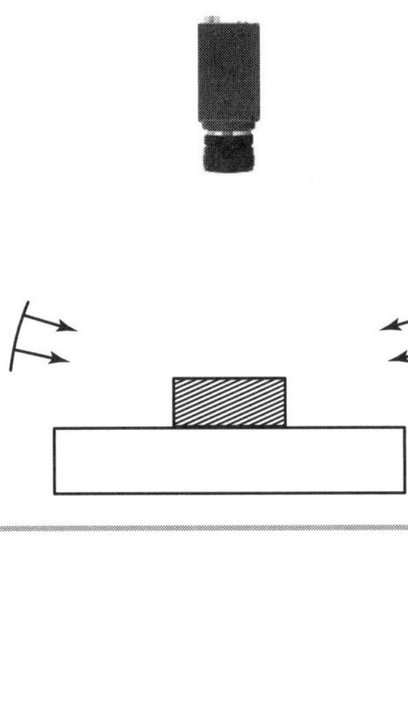

Light Shapes

The shapes of the light source include point, line, and area.[7] Compared to the environment where it stands, a point source can be approximated by an extremely small sphere. The point source has two types of models: a nearby point source and a point source at infinity.[7] The line source has the geometry of a line. A tubular fluorescent light bulb is an example. An area source is an area that radiates light. Area sources are often modeled as surface patches and the emitted radiance is independent of position and direction.

FIGURE 5. Illumination: (a) backlighting; (b) coaxial illumination.

(a)

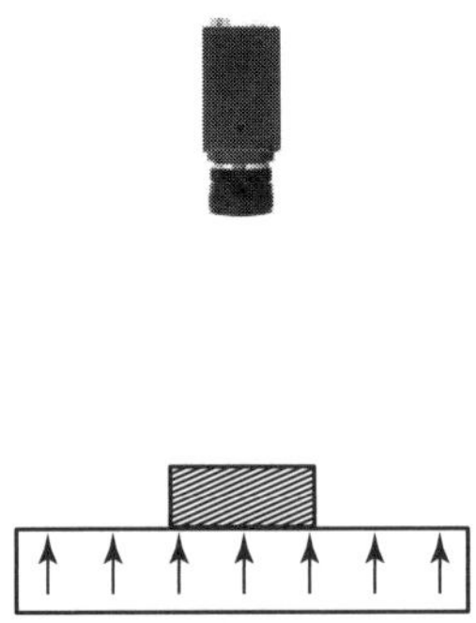

(b)

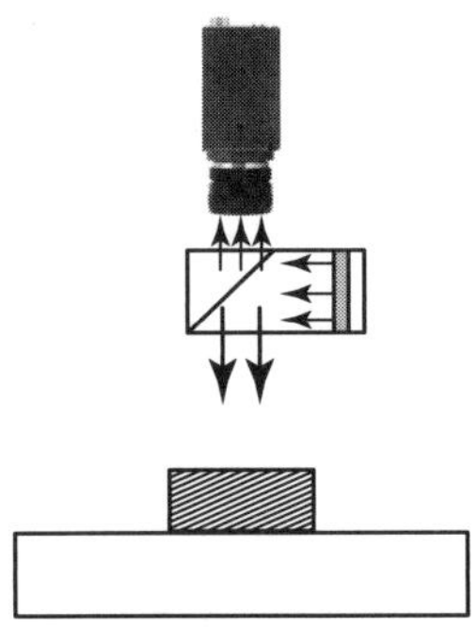

Lighting Arrangement

Machine vision commonly uses lighting sources of the following types: fluorescent, quartz halogen, light emitting diode, metal halide (mercury), xenon and sodium.[8] Table 1 gives a brief description of each type of lighting source.

The choice of a light source depends on the requirements for the brightness and spectrum content. Each type has its advantages and disadvantages. An inspection must find the suitable lighting solution for the application.

Cameras

A camera is used to capture images of still or moving objects. Traditional cameras capture light onto photographic film or photographic plate. Nowadays, digital cameras use the solid imaging device — that is, a charge coupled device or a complementary metal oxide semiconductor (CMOS) — to capture images, which can be stored in computer memory for later use.

Charge Coupled Device

A charge coupled device (CCD) is an electronic detector consisting of many square photosensitive pixels,[9] known as a *capacitor array* (photoactive region) or *transmission region*. An image is projected on the capacitor array through a lens. Each capacitor accumulates an electric charge proportional to the light intensity at that location. A control circuit can transfer the charge to a charge amplifier, which converts the charge into voltage, by a control circuit.

Silicon based charge coupled devices are monochromatic in nature. Three techniques are commonly used to extract color information for a given scene as shown in Fig. 6.[10] Color sequencing alternates optical filters with desired red green blue characteristics. A color image can also be created using integral color filter arrays (CFA) over the charge coupled device.[11] A color filter array registers the intensity of a single color at each pixel.[11] By interpolation with the color intensities of adjacent pixels, the intensity of a color at each pixel can be estimated.

A three–charge coupled device camera has three separate charge coupled devices and provides higher image quality than does a one–charge coupled device camera. In a three–charge coupled device camera, each charge coupled device takes a separate measurement of red, green and blue light for each pixel. Light is split by a trichroic prism assembly and a corresponding color filter is placed in each of the three imaging planes. A color image is obtained by synchronizing the outputs of the three charge coupled devices.

Complementary Metal Oxide Semiconductor

In a complementary metal oxide semiconductor camera, each pixel has its own charge-to-voltage conversion. Most functions, such as amplification, noise correction, and digitization circuits, are integrated into a chip, which can output digital bits.[12] Although less flexible, such design makes a complementary metal oxide semiconductor camera more reliable.

Charge Injection Devices

Charge injection device (CID) cameras have a detection mechanism similar to that of a charge coupled device. The difference lies in their readout system. The collected charge in a charge injection device camera does not transfer from site to site in the charge injection device array. Instead, a displacement current, which is proportional to the stored charge, is read when charge packets are shifted between capacitors within

TABLE 1. Brief description of lighting sources.

Lighting	Description
Fluorescent lighting	Illumination using electricity to excite mercury vapor to produce short wave ultraviolet radiation, which causes a phosphor to fluoresce, producing visible light.
Quartz halogen lamp	Incandescent light bulb with envelope made of quartz and with filament surrounded by halogen gas.
Light emitting diode (LED)	Semiconductor diode that emits light when electrical current is applied.
Metal halide lamp	Lamp that produces light by passing an electric arc through high pressure mixture of argon, mercury and various metal halides.
Xenon	Element used in arc and flash lamps. Xenon arc lamps use ionized xenon gas to produce bright white light; xenon flash lamps are electric glow discharge lamps that produce flashes of very intense, incoherent, full spectrum white light.
Sodium	Element used in some vapor lamps. Sodium gas discharge lamps use sodium in excited state to produce light. There are two types: low pressure and high pressure lamps.

individually selected pixels.[13] The displacement current is then amplified, converted to a voltage and fed outside as a digitized signal. To clear the array for new frame integration, the electrodes in each pixel are momentarily switched to be grounded.

Camera Calibration

The most common camera model is the pinhole model shown in Fig. 7.[14] The focal length *f* represents the distance between O and image center o. The three-dimensional reference frame XYZ is called the *camera frame.* The purpose of camera calibration is to estimate the intrinsic and extrinsic parameters of the camera model.

1. Intrinsic parameters link the pixel coordinates of an image point with the corresponding coordinates in the camera reference frame.
2. Extrinsic parameters define the location and orientation of the camera reference frame with respect to a known world reference frame.

In detail, the intrinsic parameters include the focal length *f*, the transformation between camera frame coordinates and pixel coordinates, and the geometric distortion introduced by the optics. The extrinsic parameters include a translation vector *T* and a rotation matrix *R*, which relate the coordinates of a point *P* in the world P_w and in camera frame P_c:[14]

FIGURE 6. Three techniques to extract color information: (a) color sequential capture; (b) integral color filter array; (c) three-chip color capture.

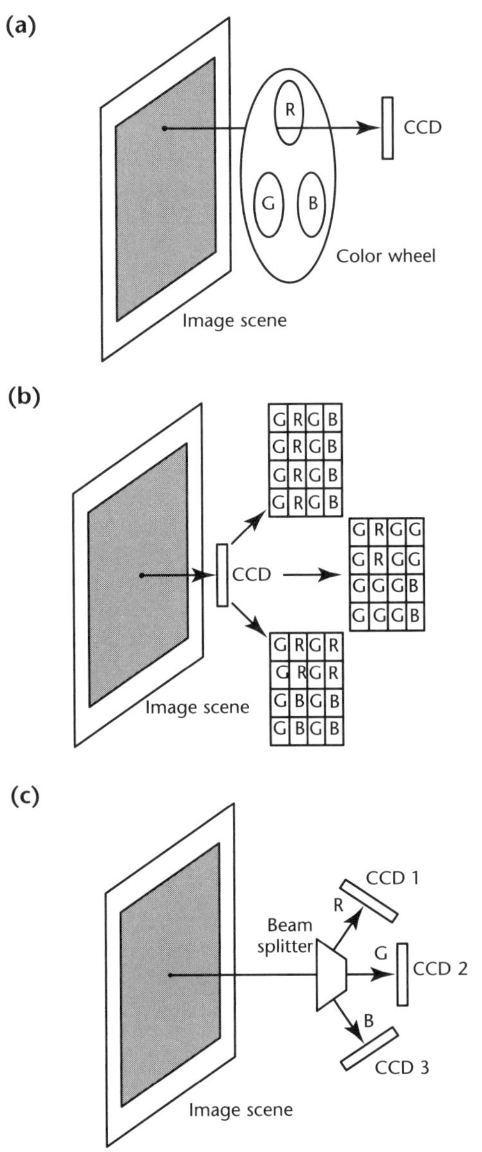

Legend

CCD = charge coupled device
B = blue
G = green
R = red

FIGURE 7. Pinhole camera: (a) model; (b) transformation between camera frame coordinates and pixel coordinates.[14]

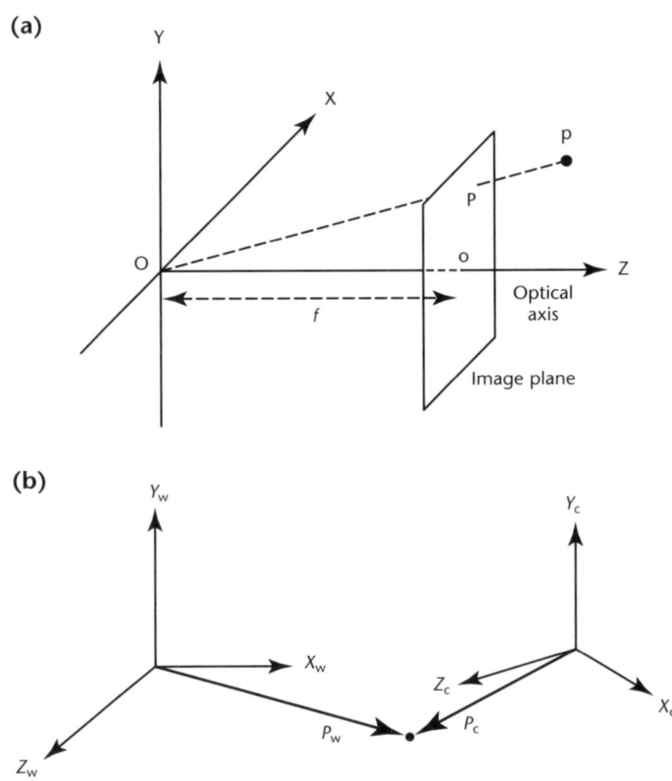

Legend

c = subscript denoting camera
f = focal length
O = center of projection
o = center of image
OZ = optical axis
w = subscript denoting world

$$P_c = R(P_w - T) \quad (1)$$

Numerous approaches are available to achieve this.[15,16]

Camera Interface

The camera interface is hardware that connects different image sensors and provides a standard output. A frame grabber is such a device that can capture individual frames from an analog or digital video stream and store and/or compress these frames in real time.

Analog frame grabbers accept and process analog video signals while digital ones deal with digital video streams. The physical interface standards include CameraLink®, USB (universal serial bus), GigE Vision®, and IEEE 1394.[17,18] Table 2 gives the basic information about these standards.[19-22]

The distinguishing difference between these interfaces is the speed.[18] Other differences include cable length, whether power is supplied over the cable, and the level of software support. Table 3 shows the major differences between these interfaces.[23]

Table 2. Summary of camera interfaces.

Interface Standard	Issuing Organization	Comment
Camera Link®	Automated Imaging Association, Ann Arbor, MI	serial communication protocol that extends base technology of Channel Link® for vision application[19]
USB	USB Implementers Forum	universal serial bus
GigE Vision®	Automated Imaging Association, Ann Arbor, MI	based on gigabit ethernet standard with fast data transfer, allowing standard, long, low cost cables[21]
IEEE 1394	IEEE, New York, NY	Interface standard for high speed communication and isochronous (real time) data transfer for high performance and time sensitive applications[22]

Table 3. Comparison of camera interfaces.

Category	Camera Link®	USB[a]	GigE Vision®	IEEE 1394[b]
Topology	master and slave	master and slave ("on the fly")	networked, peer to peer	peer to peer
Maximum bit rate[c]	2380 Mbps	480 Mbps	1000 Mbps	~400 ~800 Mbps
Isochronous mode	yes	yes	no	yes
Maximum sustained bit rate	2380 Mbps	432 Mbps	930 Mbps	~320 to ~640 Mbps
Cable distance (copper)	10 m	5 m	25 m	~4.5 to ~100 m
Bus power	none	up to 0.5 A	none	up to 1.5 A

a. USB = universal serial bus
b. IEEE = IEEE [formerly Institute of Electrical and Electronics Engineers], New York, NY
c. Mbps = 10^6 bits per second

Part 2. Algorithms and Software

Software has been developed to implement algorithms in a variety of machine vision tasks: image processing, image segmentation, geometric transformations, pattern recognition and nondestructive testing.[6,24-26]

Image Processing

Convolution Operator

In the image processing, the convolution operator is the most useful and basic operation. Figure 8 shows an example of this operation.

Let image $I(x,y)$ be an image to process. To apply the convolution operator, a small image $w(i,j)$, which includes weights multiplied by each pixel in $I(x,y)$, is prepared. $w(i,j)$ is also called the operator, mask or kernel. When the size of $w(i,j)$ is $M \times N$, the convolution operator in a coordinate (x,y) is described:

$$I'(x,y) = \sum_{i=-\frac{M}{2}}^{\frac{M}{2}} \sum_{j=-\frac{N}{2}}^{\frac{N}{2}} \left[w\left(i+\frac{M}{2}, j+\frac{N}{2}\right) \times I(x+j, y+j) \right] \tag{2}$$

In this calculation, $w(i,j)$ and $I(x,y)$ overlap the center of $w(i,j)$ and the coordinate (x,y) at the same position. Corresponding pixels are multiplied and summed. This calculation is the discrete version of the orthogonal transformation. Hence, the convolution operator is related to the fourier transformation in the frequency domain. The following sections describe image processing using the convolution operators.

Figure 8. Arrows depict movement of convolution operator.

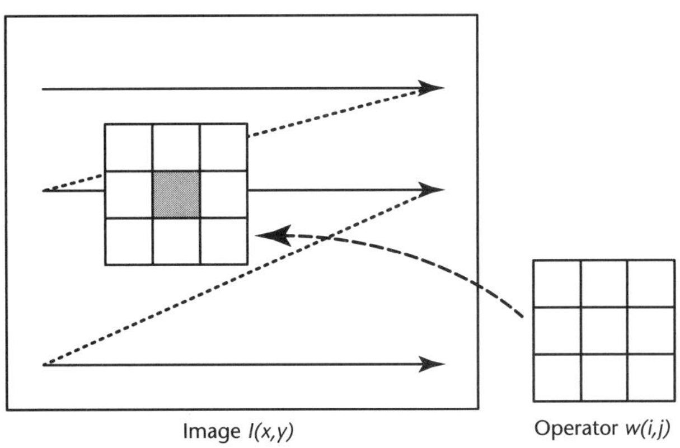

Edge Detection

The edge in the image is the part in which the pixel value (intensity or color) changes drastically. Therefore, parts of the edge in the image are estimated by calculating the differentiation of pixel values. However, image data cannot be expressed by mathematical functions, so the difference is usually used instead of the differentiation. Moreover, the convolution operator is used to calculate the difference. Figure 9 shows a simple operator of difference. Figure 9a is the difference along the X axis. The image data are spread on a two-dimensional plane; hence the operator along the Y axis (Fig. 9b) is also used.

Here, $I_x(x,y)$ and $I_y(x,y)$ are the results of differentiation along the X and Y axes respectively. The value $\sqrt{[I_x^2(x,y) + I_y^2(x,y)]}$ is the magnitude of the gradient, and the

Figure 9. Differential operator: (a) X direction; (b) Y direction.

(a)

0	0	0
0	1	–1
0	0	0

(b)

0	0	0
0	1	0
0	–1	0

value of $\tan^{-1}[I_y^2(x,y) + I_x^2(x,y)]$ is the direction of the edge. A simpler version of the magnitude of the gradient is $|I_x(x,y)| + |I_y(x,y)|$.

Moreover, there are other operators to calculate differences, such as roberts, prewitt and sobel operators (Fig. 10). The roberts operator can detect edges in the direction of a slant. The edges detected by prewitt and sobel operators tend to be thick.

Edge parts can be extracted by using the first order and second order differentiation. A typical operator is the laplacian operator. Three commonly used small kernels are given in Fig. 11. This operator expresses the magnitude of the edge, combining *x* and *y* directions. According to definitions, there are some variations. The position of edge is the zero origin point because large gradient values are found around the edge points.

The edge parts correspond to the high frequency parts of the image intensity. Therefore, the edge extraction process is the high pass filtering in the frequency domain. (This is proved mathematically.)

Noise Reduction

In image data, there are various noises of which the noise called *salt and pepper* is typical. Such noises are expressed by pixels in random positions, not fixed. The intensities of these noises differ from those of surrounding pixels, so the noise is conspicuous. The salt and pepper noise is caused by the flicker of the illumination and the variation in the performance of imaging sensor elements.

The operation called *smoothing* is a simple way to reduce such noises. This process is used to obtain the local averaged intensity value. Such a calculation can be performed by the convolution operation. Figure 12 shows a 3 × 3 smoothing operator. Smoothing blurs the image. The smoothing operation is a kind of low pass filtering in the frequency domain.

FIGURE 10. Operators for edge detection: (a) roberts operators; (b) prewitt operators; (c) sobel operators.

(a)

0	0	0
0	1	0
0	0	–1

0	0	0
0	0	1
0	–1	0

(b)

–1	0	1
–1	0	1
–1	0	1

–1	–1	–1
0	0	0
1	1	1

(c)

–1	0	1
–2	0	2
–1	0	1

–1	–2	–1
0	0	0
1	2	1

FIGURE 11. Laplacian operators: (a) laplacian 1; (b) laplacian 2; (c) laplacian 3.

(a)

0	1	0
1	–4	1
0	1	0

(b)

1	1	1
1	–8	1
1	1	1

(c)

–1	2	–1
2	–4	2
–1	2	–1

FIGURE 12. Smoothing operator.

1/9	1/9	1/9
1/9	1/9	1/9
1/9	1/9	1/9

An effective technique to remove noise is median filtering. In this technique, the pixels in a local region are sorted in terms of its intensity value, and the median is picked up as the new intensity for the pixel at the center position (x,y) (Fig. 13). If the noise is included in the local region, it will be arranged in the first or the last after sorting. Unlike the convolution operator, the median filter is nonlinear. Smoothing blurs the edges of an image, but the median filter does not.

Image Segmentation

Binarization

Binarization is used to identify pixels in the image, as being either the object or the background. From the assumption that the intensities of the pixels in the object are brighter (or darker) than the background, the binarization can be expressed as follows.

For an object with a higher intensity than the background, there are:

(3) $$I(x,y) \geq t \rightarrow B(x,y) = 1$$

(4) $$I(x,y) < t \rightarrow B(x,y) = 0$$

For an object that is darker than the background, there are:

(5) $$I(x,y) \leq t \rightarrow B(x,y) = 1$$

(6) $$I(x,y) > t \rightarrow B(x,y) = 0$$

where $I(x,y)$ is the original monochrome image and $B(x,y)$ is the binary image. $B(x,y) = 1$ denotes the object and 0 is the background. The boundary value t is the threshold that divides the object from the background.

The importance of the binarization is how to estimate the threshold value. Two popular techniques for estimating the threshold, *P* tile and discriminant analysis, are described below.

P Tile Technique

The *P* tile technique can estimate the threshold when the number of pixels (rate of areas) of the object and the background of the image are known. When this technique is applied to a monochrome image, the histogram (the frequency distribution) of the intensity in the image can be obtained. The area of the object occupies *P* percent in the histogram from the bright (dark) side. Therefore, the threshold is decided from the histogram (Fig. 14).

In the *P* tile technique, the rate of area *P* should be known in order to estimate the threshold. For example, in a document image there are only letters and the rate of black pixels (letters) is considered about 5 percent in the whole image, but if figures, tables and photographs are included in the image, it is difficult to define the rate *P*.

Discriminant Analysis

Discriminant analysis can estimate a threshold from only image data. This technique also uses the histogram of pixel intensity.

With this technique, the histogram is divided into two classes: dark and bright pixels. Using the between-class variance σ_{bc}^2 and the within-class variance σ_{wc}^2, the separation metric $\sigma_{bc}^2 (\sigma_{wc}^2)^{-1}$ is defined. The threshold is determined when the separation metric reaches its maximum value.

Assuming that the histogram is divided into two classes by a threshold t (Fig. 15), let ω_1 be the number of pixels, m_1 be the average and σ_1^2 be the variance in the dark pixel class. In the bright pixel class, ω_2, m_2 and σ_2^2 are also defined similarly while m and σ^2 show the average and variance in the whole image.

FIGURE 13. Process of median filtering.

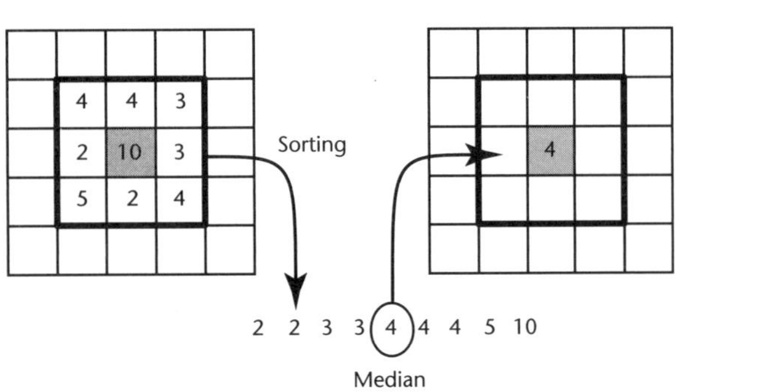

FIGURE 14. *P* tile technique.

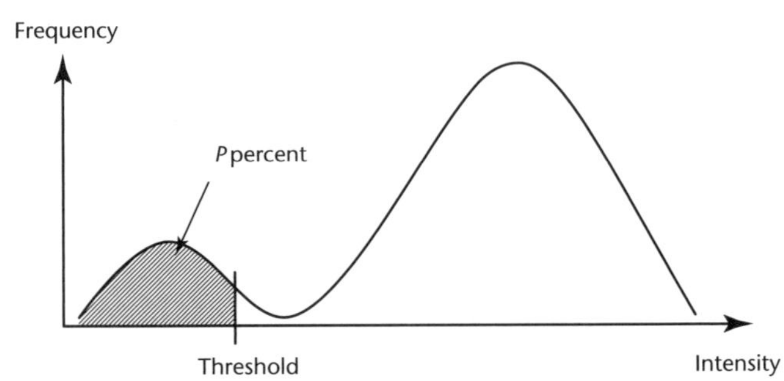

The within-class variance σ_{wc}^2 is expressed:

$$(7) \quad \sigma_{wc}^2 = \frac{\omega_1\sigma_1^2 + \omega_2\sigma_2^2}{\omega_1 + \omega_2}$$

The between-class variance (σ_{bc}^2) is:

$$(8) \quad \sigma_{bc}^2 = \frac{\omega_1(m_1 - m)^2 + \omega_2(m_2 - m)^2}{\omega_1 + \omega_2} = \frac{2\omega_1\omega_2(m_1 - m_2)^2}{(\omega_1 + \omega_2)^2}$$

On the other hand, there is a following relation between σ^2, σ_{wc}^2 and σ_{wc}^2:

$$(9) \quad \sigma^2 = \sigma_{wc}^2 + \sigma_{bc}^2$$

From the above equations, the separation metric is formulated:

$$(10) \quad \frac{\sigma_{bc}^2}{\sigma_{wc}^2} = \frac{\sigma_{bc}^2}{\sigma^2 - \sigma_{bc}^2}$$

Because σ^2 is fixed, when σ_{bc}^2 reaches its maximum value, the separation metric is maximized. Therefore, in the discriminant analysis technique, σ_{bc}^2 is calculated by changing the value of t and searching for the adequate threshold in case of maximum σ_{bc}^2.

By using discriminant analysis, the threshold is uniquely estimated for any monochrome images. Although only the case of binarization (two classes, black and white) is demonstrated, this technique can also be applied to estimate multiple thresholds.

FIGURE 15. Discriminant analysis technique.

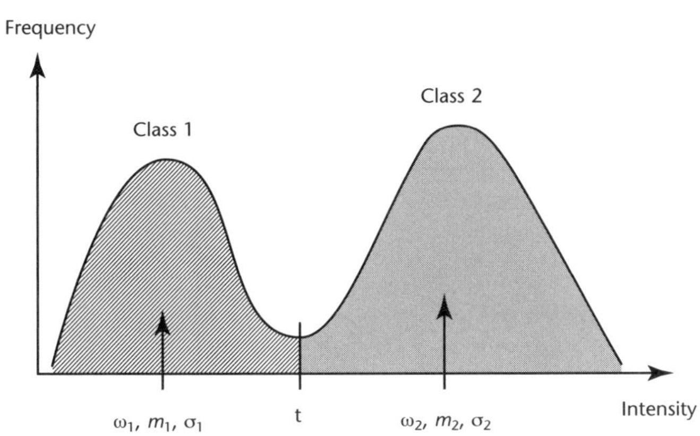

Mathematical Morphology

A logical process applied to a binary image, mathematical morphology can perform noise reduction, lacuna restoration and so on. Some practical techniques of this process are described below.

Dilation

Dilation expands the object region in a binary image X using the structuring element Y. The structuring element is a small binary image with various shapes. A pixel in the structuring element is defined as an origin of the element.

The operation of dilation is defined as:

$$(11) \quad X \oplus Y = \{z \mid z = x + y \text{ for } x \in X, y \in Y\}$$

This formulation is valid only when $\forall x \in X$ and when the origin of the structuring element y_{org} should be located in X. Examples of the dilation are shown in Fig. 16. When the structuring elements

FIGURE 16. Examples of dilation with different structuring elements: (a) 2 × 2 set; (b) four connected sets.

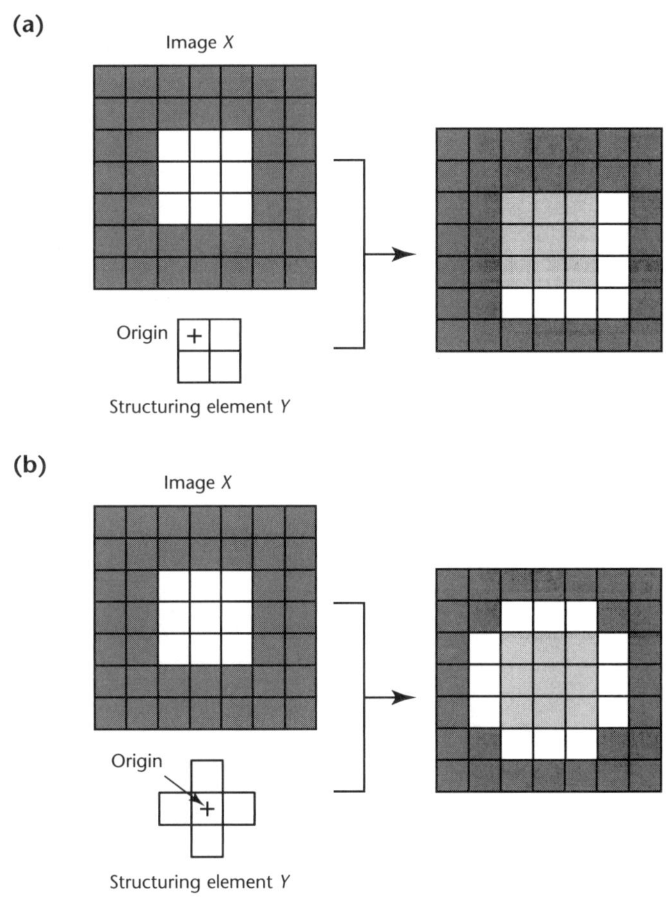

are different, the results are of different shape.

Erosion

Erosion is the operation opposite to dilation. Erosion shrinks the binary image according to the structuring element. The operation of erosion is defined as:

$$\textbf{(12)}\quad X \otimes Y = \left\{ z \,\middle|\, z + y \in X \text{ for } \forall y \in Y \right\}$$

Note that this formulation is valid only when z and y_{org} are at the same position. Examples of the erosion are shown in Fig. 17. As in the case of dilation, when the structuring elements are different, the resulting shapes are different.

Opening and Closing

The operations that combine dilation and erosion are called the opening and closing operations. They are defined as follows. The opening operation eliminates the isolated noise and small regions:

$$\textbf{(13)}\quad X \circ Y = \left(X \otimes Y \right) \oplus Y$$

The closing operation fills the holes:

$$\textbf{(14)}\quad X \bullet Y = \left(X \oplus Y \right) \otimes Y$$

FIGURE 17. Examples of erosion with different structuring elements: (a) 3 × 3 set; (b) four connected sets.

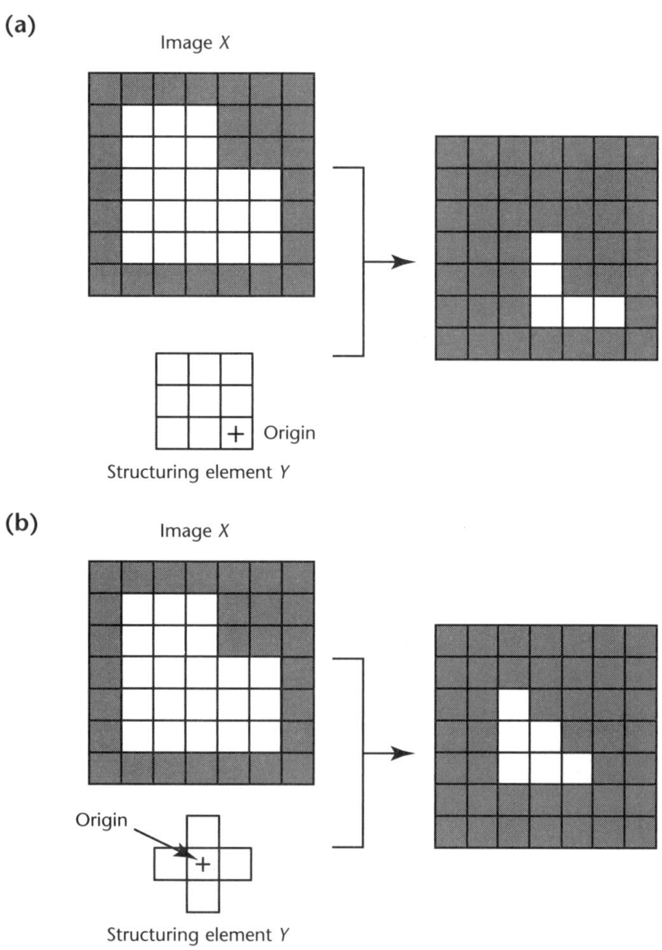

FIGURE 18. Geometric transformations: (a) original image; (b) scaling (1.2*x*, 0.6*y*): (c) translation; (d) rotation.

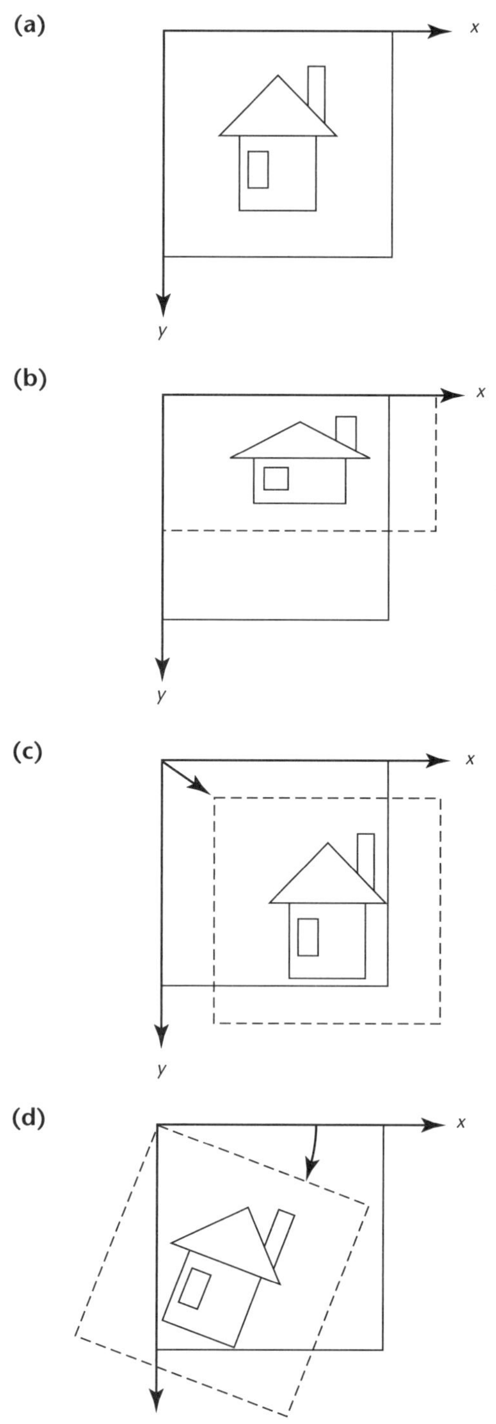

When the dilation is applied iteratively, the object region becomes large. When the erosion is iterated, the object region becomes small. The opening and closing operations can conserve the size of main parts of the image, even if the iterative operations are applied.

Geometric Transformation

The geometric transformation (for example, the object size is changed, the slant is corrected and so on) is a major technique in image processing. Many geometric transformations can be expressed as the affine transformation using a 3 × 3 matrix. Let (x,y) be the coordinate before transformation and (x',y') be the transformed coordinates. The affine transformation is formulated:

$$\textbf{(15)} \quad \begin{bmatrix} x' \\ y' \\ 1 \end{bmatrix} = \begin{bmatrix} a & b & c \\ d & e & f \\ 0 & 0 & 1 \end{bmatrix} \begin{bmatrix} x \\ y \\ 1 \end{bmatrix}$$

The following equations describe the geometric transformations of the original image in (Fig. 18a). Equation 16 gives scaling rates α and β respectively for X and Y axes (Fig. 18b):

$$\textbf{(16)} \quad \begin{bmatrix} x' \\ y' \\ 1 \end{bmatrix} = \begin{bmatrix} \alpha & 0 & 0 \\ 0 & \beta & 0 \\ 0 & 0 & 1 \end{bmatrix} \begin{bmatrix} x \\ y \\ 1 \end{bmatrix}$$

Equation 17 gives the amount of translation t_x,t_y (Fig. 18c):

$$\textbf{(17)} \quad \begin{bmatrix} x' \\ y' \\ 1 \end{bmatrix} = \begin{bmatrix} 1 & 0 & t_x \\ 0 & 1 & t_y \\ 0 & 0 & 1 \end{bmatrix} \begin{bmatrix} x \\ y \\ 1 \end{bmatrix}$$

Equation 18 gives rotation as an angle from the X axis around the origin (Fig. 18d):

$$\textbf{(18)} \quad \begin{bmatrix} x' \\ y' \\ 1 \end{bmatrix} = \begin{bmatrix} \cos\theta & -\sin\theta & 0 \\ \sin\theta & \cos\theta & 0 \\ 0 & 0 & 1 \end{bmatrix} \begin{bmatrix} x \\ y \\ 1 \end{bmatrix}$$

The skewing in Eq. 19 uses slants p,q from X or Y axis (Fig. 19):

$$\textbf{(19)} \quad \begin{bmatrix} x' \\ y' \\ 1 \end{bmatrix} = \begin{bmatrix} 1 & p & 0 \\ q & 1 & 0 \\ 0 & 0 & 1 \end{bmatrix} \begin{bmatrix} x \\ y \\ 1 \end{bmatrix}$$

Figure 20 depicts reflection in three directions — along X axis (Eq. 20), along Y axis (Eq. 21) and where $y = x$ (Eq. 22):

$$\textbf{(20)} \quad \begin{bmatrix} x' \\ y' \\ 1 \end{bmatrix} = \begin{bmatrix} 1 & 0 & 0 \\ 0 & -1 & 0 \\ 0 & 0 & 1 \end{bmatrix} \begin{bmatrix} x \\ y \\ 1 \end{bmatrix}$$

$$\textbf{(21)} \quad \begin{bmatrix} x' \\ y' \\ 1 \end{bmatrix} = \begin{bmatrix} -1 & 0 & 0 \\ 0 & 1 & 0 \\ 0 & 0 & 1 \end{bmatrix} \begin{bmatrix} x \\ y \\ 1 \end{bmatrix}$$

$$\textbf{(22)} \quad \begin{bmatrix} x' \\ y' \\ 1 \end{bmatrix} = \begin{bmatrix} 0 & 1 & 0 \\ 1 & 0 & 0 \\ 0 & 0 & 1 \end{bmatrix} \begin{bmatrix} x \\ y \\ 1 \end{bmatrix}$$

By multiplying several matrices, a composite transformation can be obtained. Because the transformation is a multiplication of multiple matrices, the change of the sequence of these matrices will give a different (transformation) result.

FIGURE 19. Skewed geometric transformations of image in Fig. 18a: (a) skewing ($p = \tan\theta_p$, $q = 0$); (b) skewing ($p = 0$, $q = \tan\theta_q$).

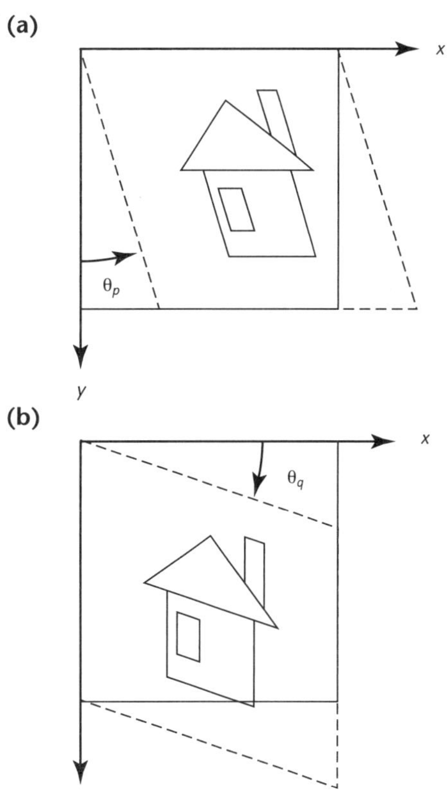

Pattern Recognition

Template Matching

Template matching is usually used to search for an object in an image. In this technique, a small image of an object called the *template* is prepared first. The next step is to match the template with the whole image. In this process, the similarity is estimated at each location of the image. As a result, the object in the template is considered to be where the similarity reaches the maximum value for the whole image (Fig. 21).

Let the size of template image $T(x,y)$ be $t_w \times t_h$. To calculate the similarity at (x,y) in the input image $I(x,y)$, Eqs. 23 to 25 are applied. Equation 23 expresses the sum of absolute difference (SAD):

$$R_{SAD}(x,y) = \sum_{i=0}^{t_w-1} \sum_{j=0}^{t_h-1} \left| I(x+i,y+j) - T(i,j) \right| \tag{23}$$

Equation 23 expresses the sum of squared difference (SSD):

$$R_{SSD}(x,y) = \sum_{i=0}^{t_w-1} \sum_{j=0}^{t_h-1} \left[I(x+i,y+j) - T(i,j) \right]^2 \tag{24}$$

Equation 24 expresses the (normalized cross correlation (NCC):

$$R_{NCC}(x,y) = \sum_{i=0}^{t_w-1} \sum_{j=0}^{t_h-1} \left[I(x+i,y+j) \times T(i,j) \right] \div \left[\sqrt{\sum_{i=0}^{t_w-1} \sum_{j=0}^{t_h-1} I(x+i,y+j)^2} \times \sqrt{\sum_{i=0}^{t_w-1} \sum_{j=0}^{t_h-1} T(i,j)^2} \right] \tag{25}$$

In Eqs. 23 to 25, the smaller the value of the sum of absolute difference and the sum of squared difference, the higher the similarity. For normalized cross

FIGURE 20. Reflected geometric transformations of image in Fig. 18a: (a) reflection (*y*); (b) reflection (*y* = *x*); (c) reflection (*x*).

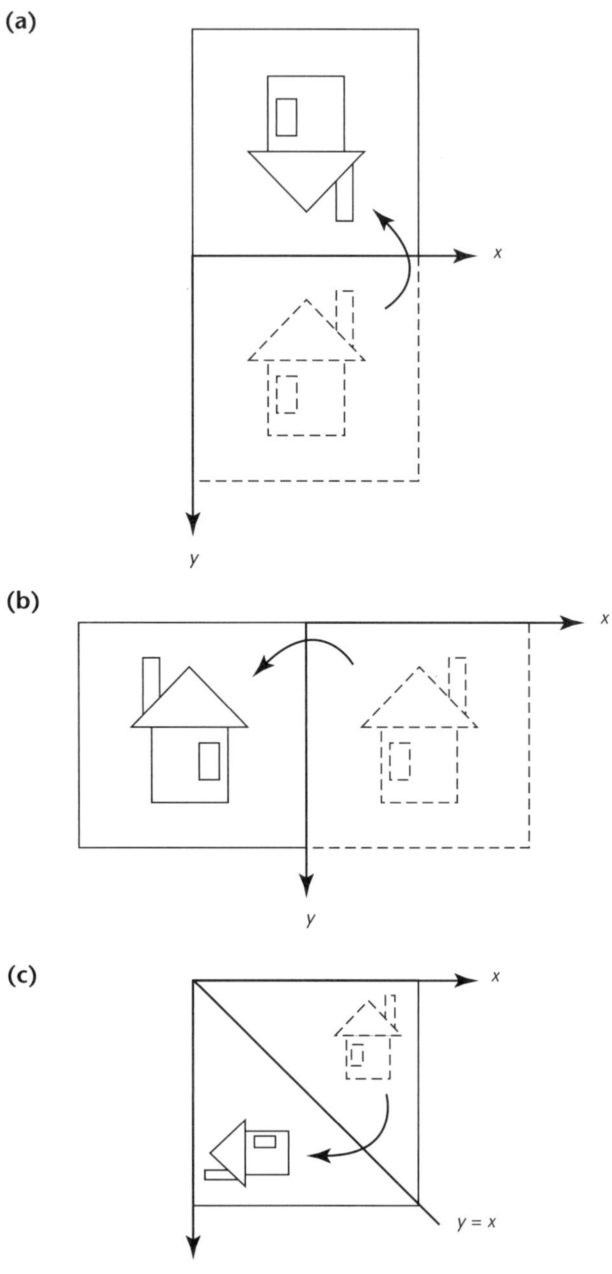

FIGURE 21. Template matching technique.

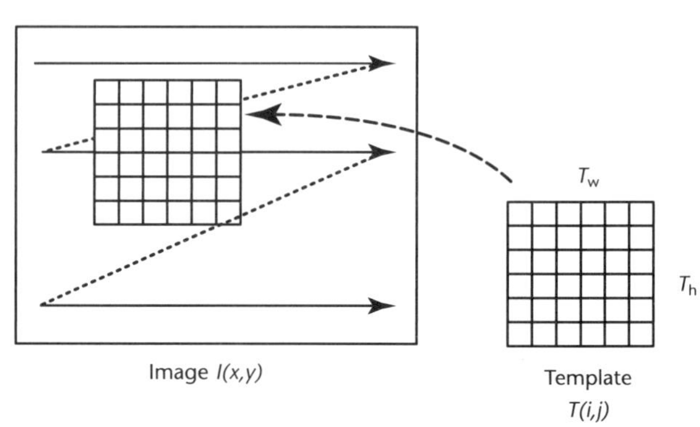

correlation, a higher value indicates a higher similarity. The similarity estimation is carried out over the whole input image and the position at which the highest similarity is obtained is the location of the object.

The computational costs of the sum of absolute difference and the sum of squared difference are small and they can be estimated rapidly. However, if the brightness (or gain) of the input image is different from that of the template image (in other words, the input and template images are of different illumination conditions), the similarity will be low. Therefore, an accurate match will not be achieved. The normalized cross correlation estimates the correlation between the input and template images, which is less likely to be affected by illumination changes, but comes with a higher computational cost.

So far as the template matching technique is concerned, the size and pose of the object in the template image and those of the corresponding patterns in the input image need to be the same. If not, the similarity becomes low and an accurate match will not be obtained. In this case, it is necessary to apply the geometric transformations, described in the previous section, to the template image. However, this step requires estimating the transformation parameters, so it is not efficient. When the size and pose of the object in the template image vary from those in the input image, it is still possible to do the matching with color information or other high dimensional features.

a-b Hough Transform

The template matching technique can be applied to any pattern of an object. If the object can be expressed by the mathematical models (such as line or circle), it can be searched more efficiently in the input image with a hough transform.

The line detection with hough transform is first described. A straight line in *x-y* space is modeled by parameters *a* and *b*:

$$y = ax + b \tag{26}$$

This formulation can be transformed as:

$$b = (-x)a + y \tag{27}$$

This formulation shows another straight line in *a-b* space. Here, the *a-b* space is called the *parameter space.*

As shown in Fig. 22a, when a line crosses two points (x_1,y_1) and (x_2,y_2) in *x-y* space, parameters $\hat{a}$ and b of this line can be expressed:

$$\left(\hat{a},\hat{b}\right) = \left(\frac{y_2 - y_1}{x_2 - x_1}, \frac{x_2 y_1 - y_2 x_1}{x_2 - x_1}\right) \tag{28}$$

On the other hand, points (x_1,y_1) and (x_2,y_2) in *x-y* space correspond to the following lines in the parameter space:

$$b = (-x_1)a + y_1 \tag{29}$$

$$b = (-x_2)a + y_2 \tag{30}$$

And the cross point of these lines is equal to $\hat{a},\hat{b}$ (Fig. 22b).

Therefore, points on a straight line in an image (*x-y* space) correspond to lines that cross a point in the parameter space. By estimating the crossing point of straight lines in the parameter space, we can obtain parameters (*a* and *b*) of a straight line in an image (*x-y* space).

However, there are many cross points of the straight lines in the parameter space, so it is difficult to estimate adequate points mathematically. To overcome this problem, the cross points are obtained by a voting process in the hough transform. In this technique, the

FIGURE 22. *a-b* hough transform: (a) *x-y* image space; (a) *a-b* parameter space.

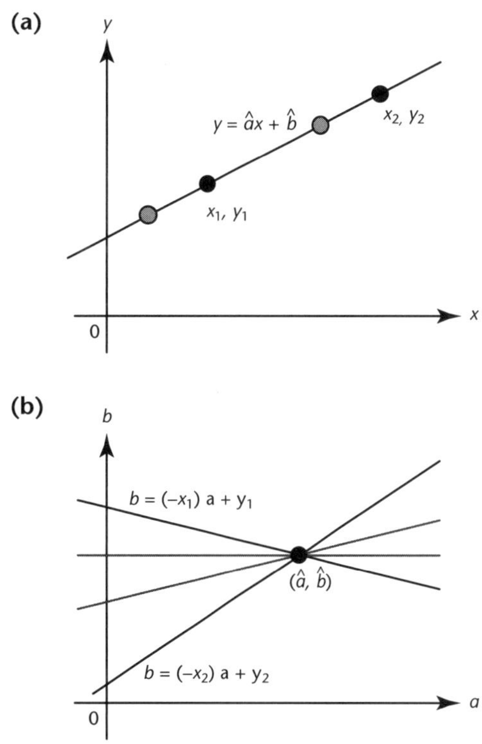

parameter space is divided into small rectangular regions along each axis. Each small rectangular region is called a *cell.* The cell works as a counter. If a line crosses a cell, the counter increases by one. Finally, the coordinates of the cell, which have maximum value of voting, are the estimated parameters of the line in the image. The voting process in the parameter space can be considered as a parameter estimation process for a line, which crosses (or is supported by) many points.

The least square technique can also be used to estimate line parameters. Basically, this technique estimates only one line. The hough transform can estimate multiple lines simultaneously by picking up cells whose voting number is beyond a threshold.

ρ-θ Hough Transform

Generally, the range of line parameters is from $-\infty$ to $+\infty$, which is the same as the range of the parameter space. The range of the image (in x-y space) is limited but the range of parameter a (slant of line) is from $-\infty$ to $+\infty$. It is difficult to prepare cells in such a range for computation. Therefore, the model of a straight line can be rewritten as:

$$(31)\quad \rho = x\cos\theta + y\sin\theta$$

where ρ is an assigned distance from the origin to the line and θ is an angle representing the altitude of the line. When the range of the image (x-y space) is limited to $0 \le x \le w$, $0 \le y \le h$, the ranges of ρ and θ are:

$$(32)\quad -\sqrt{w^2+h^2} \le \rho \le \sqrt{w^2+h^2}$$

$$(33)\quad 0 \le \theta \le \pi$$

For a coordinate, (x_1,y_1) on the image (x-y space), Eq. 31 becomes:

$$(34)\quad \rho = A\sin(\theta+\alpha)$$

where:

$$(35)\quad A = \sqrt{x_1^2 + y_1^2}$$

and where:

$$(36)\quad \alpha = \tan^{-1}\frac{x_1}{y_1}$$

It means that a point in x-y space corresponds to a sine wave in (ρ-θ) parameter space. By using the voting process along this trajectory, the cross points can be estimated (Fig. 23).

The ρ-θ hough transform has a limited parameter space, but the computational cost is still high because of the calculation of sine waves. To avoid such a problem, another technique is proposed. The γ-ω hough transform uses piecewise line segments in the parameter space to perform the voting process rapidly.[27] Moreover, because the coordinates of the image plane are discrete values, by considering the cell redivision in the parameter space, line parameters can be estimated more accurately.[28]

If the parameter space is expanded to three dimensions, circles can be detected in an image. Moreover, if high dimensional parameter space is considered, it is possible to detect various patterns. However, the higher the dimension of parameter space, the more the computational costs. In this case, huge numbers of cells are needed and it is difficult to perform any pattern detection. To overcome this problem, a generalized hough transform, which detects the position and pose of a pattern by the voting process, is an option.

Figure 23. ρ-θ hough transform: (a) x-y image space; (a) ρ-θ parameter space.

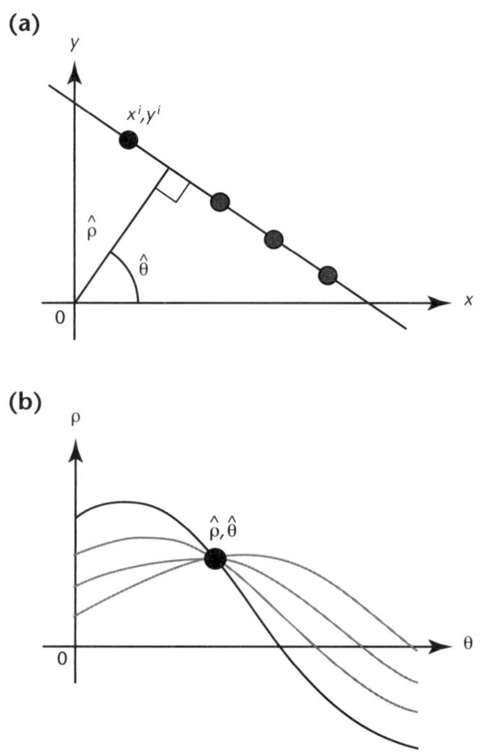

Machine Vision for Nondestructive Testing

Nondestructive testing is one of the important applications of machine vision. Nondestructive testing using radiation beyond the visible spectrum, such as infrared, ultraviolet and X-ray radiation, is described in other volumes of the *NDT Handbook* and in the introductory chapter of this volume. This chapter focuses on the use of visible light in a machine vision system. In nondestructive testing, the purpose of a machine vision system is to capture and characterize anomalies of the object under inspection, that is, to inspect for structural and surface quality.[29]

Three types of results may be obtained from a machine vision system. The first type is an enhanced image in which the discontinuities are highlighted or intuitively presented so that the inspector can easily make a subjective assessment. One example is the edge-of-light surface inspection technique, which uses the edge of light to highlight the surface slope or deformation.[30] Figure 24 shows the result of edge-of-light inspection of an aircraft lap joint. Figure 24a shows the lap joint; in the corresponding edge-of-light scan (Fig. 24b), bright and dark regions present the surface deformation. Such deformation implies the potential hidden corrosion between the two layers.

A similar technique, double-pass retroreflection surface inspection, has also been applied to the same application.[31,32] Figure 25 also shows the inspection result of aircraft lap joints. Figure 25a shows a picture of the specimen whereas Fig. 25b is the double-pass retroreflection image. These two techniques implement enhanced visual inspection through the design of a special machine vision system.

Image processing techniques can also achieve an enhanced image to facilitate the inspection. A three-dimensional stereoscopic visual system has been built to inspect aircraft skin.[33] Algorithms to enhance monoscopic and stereoscopic images were developed. A high frequency emphasis algorithm consists of two steps as illustrated in Fig. 26.[27] The live image passed through low pass and high pass filters. Then, a fraction of low frequency content was added back to the high pass filtered image. This algorithm emphasized the high frequency features while attenuating the background low pass information. Therefore, the potential surface flaws or cracks were highlighted.

For stereoscopic image enhancement (Fig. 26), the high frequency emphasis algorithm was applied to the left and right images of the stereoscopic image. An augmented stereoscopic, high frequency emphasis algorithm was implemented:

FIGURE 24. Enhanced surface inspection of aircraft lap joint: (a) aircraft lap joint; (b) image resulting from edge-of-light technique.

(a)

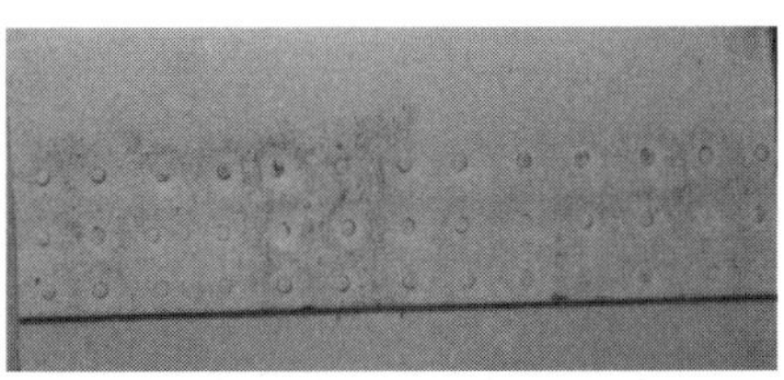

(b)

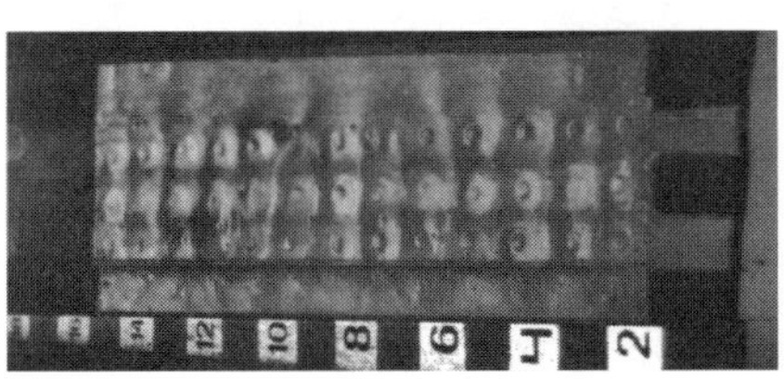

FIGURE 25. Enhanced surface inspection of aircraft lap joint: (a) aircraft lap joint; (b) image resulting from double-pass retroreflection technique.

(a)

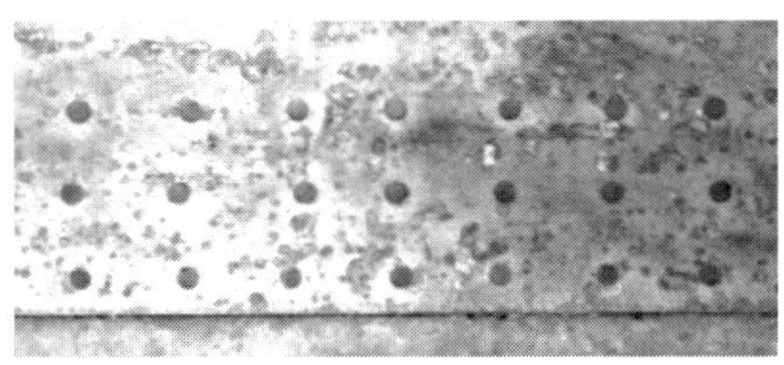

(b)

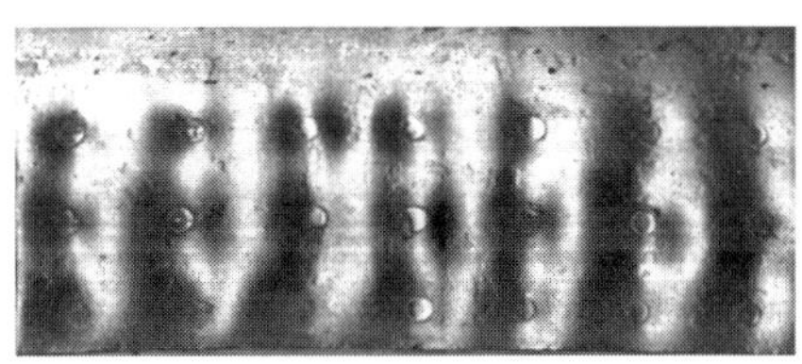

FIGURE 26. High frequency emphasis algorithm.[27]

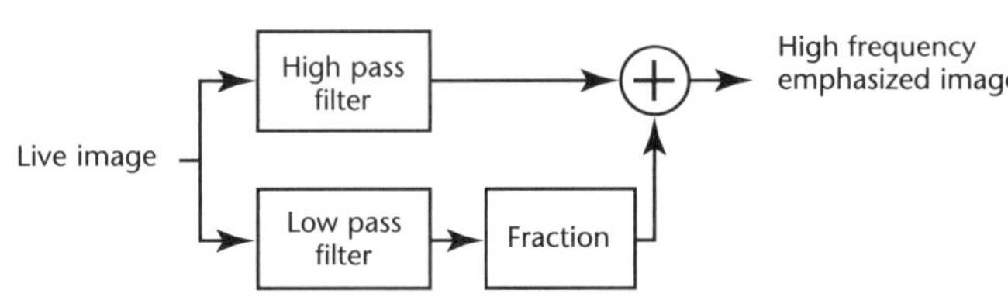

1. High frequency emphasis algorithms are applied to the left and right images.
2. To identify the features of interest, images are dynamically threshold filtered.
3. The original left and right images are overlaid with the depth offset desired for identified features.
4. The processed images are displayed stereoscopically on the screen. The eyewear of the inspector or operator can help highlight features of interest.

The second type of result is binary — that is, crack or noncrack. Binary results are useful for the inspection of a specific part, where a binary accept/reject decision may follow. As described in one study,[34] a crack detection algorithm shown in Fig. 27 was developed to identify the surface cracks on aircraft skin. Cracks frequently happen near rivets; therefore, the first step is to detect rivets by detecting the circular arcs in the image.

FIGURE 27. Surface crack detection algorithm.

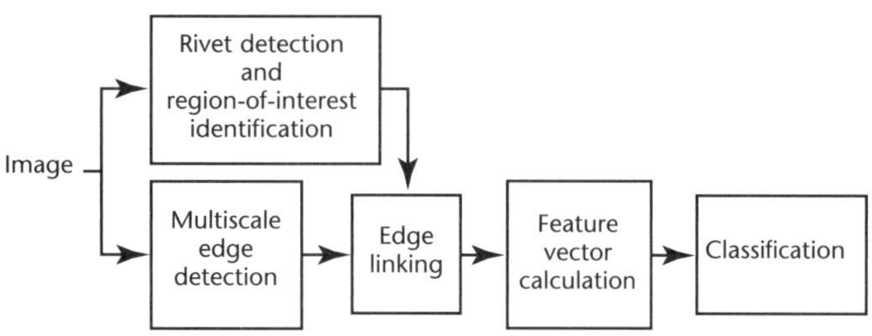

FIGURE 28. Neural network used for crack classification.

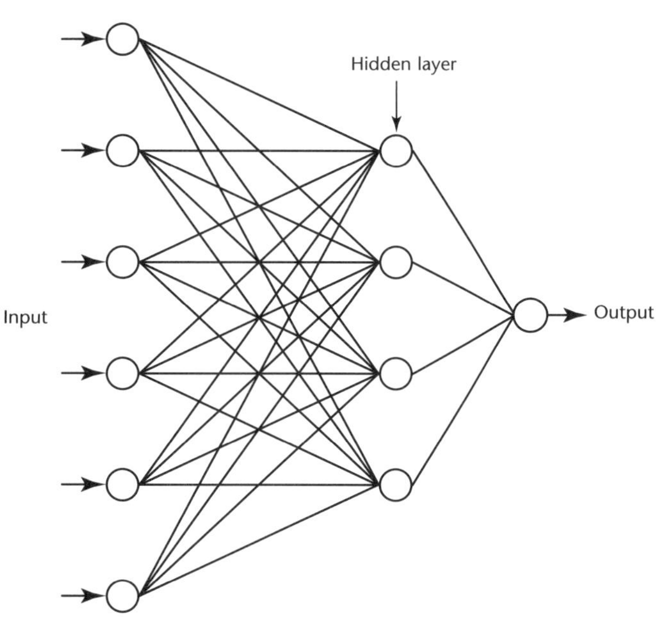

Once the edge maps of the rivets are detected, the region of interest can be determined with the centroid of the rivet.

Once the region of interest is identified, the multiscale edge detection is applied to the region of interest to generate a list of edges at different scales. This technique will help discriminate cracks from noncracks according to the size of a typical crack in comparison to other objects such as scratches and repair plates appearing on the surface. A coarse-to-fine edge linking process traced an edge from the coarse resolution (high scale) to a fine resolution (low scale). The propagation depth of all edges presented at scale one was found. Here, the propagation depth means the number of scales in which the edge appears. A feature vector for each edge in scale was generated so that the edges of cracks could be discriminated from those of noncracks. The feature vector includes the following: average wavelet magnitude of active pixels, which belong to the edges; the propagation depth number; average wavelet magnitudes of any linked edges in scale two and scale four; spins of sum W_X and sum W_Y, where W_X, W_Y are the wavelet coefficients in the x and y direction of an active pixel at scale one; and the number of active pixels.[34]

A neural network as shown in Fig. 28 was trained to classify the inputs — feature vectors of edges in the region of interest — into cracks and noncracks. The feature vectors used for the training may represent the cracks that need immediate repair. In this case, the classification result indicating a crack calls for further investigation of the corresponding region of interest for repairing. An accuracy rate of 71.5 percent and a false alarm rate 27 percent for the neural network based classification were reported.

The third type is more informative, which allows quantitative information

FIGURE 29. Pillowing deformation: (a) on aircraft lap joints; (b) on hidden, faying surface.

(a)

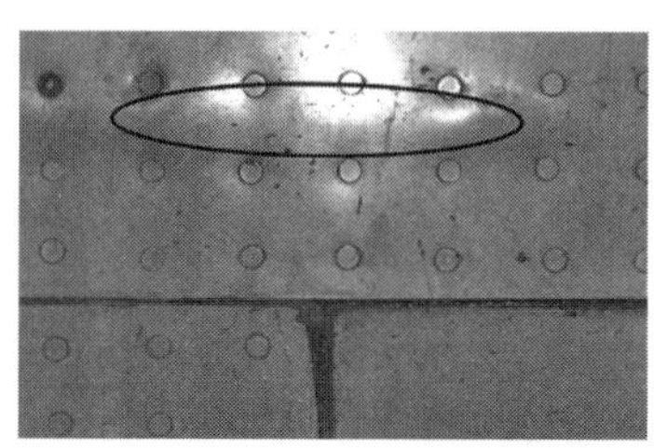

(b)

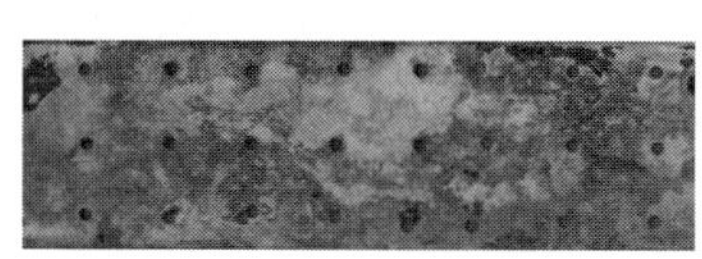

about the discontinuity to be derived. For the application of aircraft inspection, corrosion detection is crucial to the risk assessment of structural integrity. One type of corrosion occurs on the interior, hidden surface of aircraft lap joints if sealant and corrosion protection systems break down. Corroded product is of much higher volume than the original material and this will cause an expansion of the skins between rivets. This phenomenon is known as *pillowing*. An example is shown in Fig. 29. Figure 29a shows an example of pillowing on a lap joint whereas Fig. 29b shows the corroded area on the faying surface. Another type of corrosion happens to the surface, which can be detected by its suggestive texture captured by a machine vision system. In a procedure for surface corrosion detection,[34] the image was first decomposed into subimages with a discrete wavelet transform. Figure 30 shows a three-level decomposition, which consists of ten subimages. Let $W_j(k,l)$ be the wavelet coefficient at (k,l) in the subimage W_j. The original image was divided into nonoverlapping blocks each of 8 × 8 pixels. For each block $B(i)$, a ten-dimensional feature vector was created. The element is $E_j(i)(j = 1, \ldots, 10)$, the corresponding energy function in subimages and can be expressed:

$$E_j(i) = \sum_{(k,l)\in B(i)} w_j(k,l)^2 \quad (37)$$

Then, a nearest neighbor classifier was trained to classify the original image into corrosion and corrosion free regions. A detection rate of 95 percent of the test set was reported.[34] Once the original image is classified, postprocessing can be carried out to calculate the corrosion area. Therefore, the information about the size of the corroded area is available.

A more general procedure is shown in Fig. 31. The image is first preprocessed for enhancement and noise removal so that the features of targeted objects (discontinuities) are highlighted. The discrimination of different objects will be achieved in a feature space. Extraction of image features can be done in the spatial domain and frequency domain. There are numerous approaches available for this purpose. Some of these techniques have

FIGURE 30. Wavelet decomposition of image: (a) three-level decomposition into ten images; (b) procedure for classification.

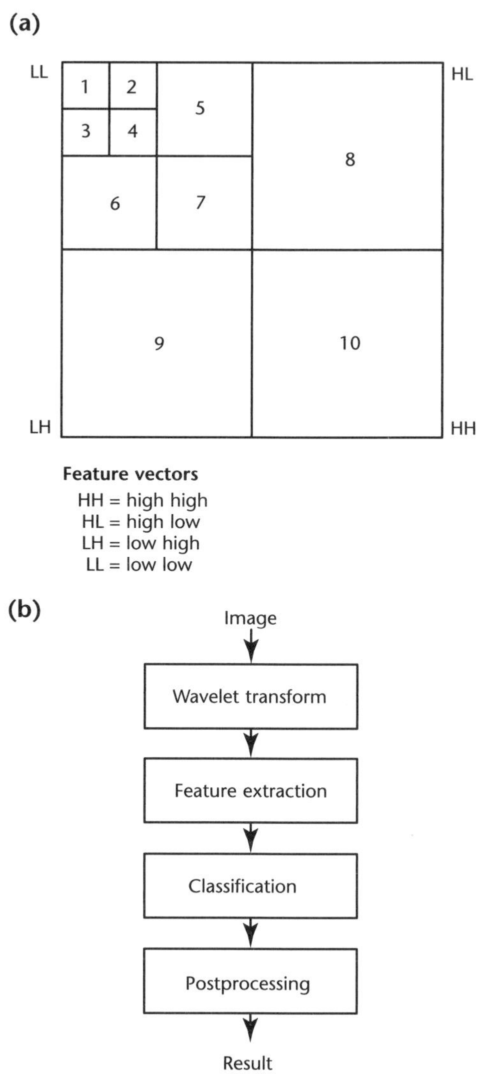

FIGURE 31. Wavelet decomposition of image and procedure for classification.

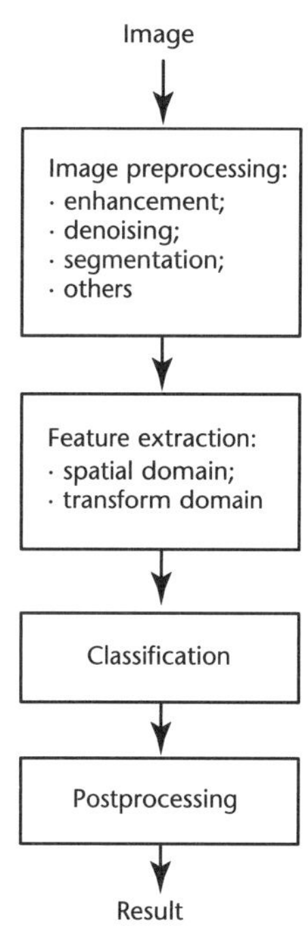

been described in the previous section of this chapter. Sometimes, postprocessing is needed to further characterize the classified results as described in the example above. The measurement results can also be compared with calibrated samples for quantitative analysis. Such comparison can also be done in the feature space.

Conclusion

This chapter provides a general description of machine vision techniques for nondestructive testing. Both the system architecture and algorithm implementation for machine vision are described. A good understanding of the application's requirements is essential to the success of a machine vision system. The technical advances in machine vision make it applicable to varied nondestructive test applications. The capability of a machine vision system can be further expanded and enhanced by incorporating multiple image modalities or other nondestructive test techniques, which may provide complementary information.

References

1. Batchelor, B.G. and P. F. Whelan. *Intelligent Vision Systems for Industry.* Bruce G. Batchelor, Cardiff, United Kingdom; Paul F. Whelan, Dublin, Republic of Ireland (2002).
2. Zuech, N. "Smart Cameras vs. PC-Based Machine Vision Systems." *Machine Vision Online.* Ann Arbor, MI: Automated Imaging Association (October 2002).
3. Zuech, N. "Optics in Machine Vision Applications." *Machine Vision Online.* Ann Arbor, MI: Automated Imaging Association (August 2005).
4. Fales, G. "Ten Lens Specifications You Must Know for Machine-Vision Optics." *Test and Measurement World.* Web page. Waltham, MA: Reed Elsevier (27 October 2003).
5. "What is Structured Light?" Web page. Salem, NH: StockerYale (2009).
6. Casasent, D.[P.], Y.F. Cheu and D. Clark. Chapter 4: Part 4, "Machine Vision Technology." *Nondestructive Testing Handbook*, second edition: Vol. 8, *Visual and Optical Testing.* Columbus, Ohio: American Society of Nondestructive Testing (1993): p 92-107.
7. Forsyth, D.A. and J. Ponce. *Computer Vision: A Modern Approach.* Upper Saddle River, NJ: Prentice Hall (2002).
8. Martin, D. *Practical Guide to Machine Vision Lighting.* Web pages. Austin, TX: National Instruments Corporation (November 2008).
9. Hainaut, O.R. "Basic Image Processing." Web pages. Santiago, Chile: European Organisation for Astronomical Research in the Southern Hemisphere, European Southern Observatory (December 1996).
10. Users Manual MTD/PS-0218, *Kodak Image Sensors.* Revision 2.0. Rochester, NY: Eastman Kodak (July 2008).
11. Peterson, C. "How It Works: The Charged-Coupled Device, or CCD." *Journal of Young Investigators.* Vol. 3, No. 1. Durham, NC: Journal of Young Investigators, Incorporated (March 2001).
12. Litwiller, D. "CCD vs. CMOS: Facts and Fiction." *Photonics Spectra.* Vol. 35, No. 1. Pittsfield, MA: Laurin Publishing (January 2001): p 154-158.
13. *Charge Injection Device Research at RIT.* Web site. Rochester, NY: Rochester Institute of Technology, Center for Imaging Science (2009).
14. Trucco, E. and A. Verri. *Introductory Techniques for 3-D Computer Vision.* Upper Saddle River, NJ :Prentice Hall (1998).
15. Bouguet, J.-Y. *Camera Calibration Toolbox for Matlab.* Web pages. Pasadena, CA: California Institute of Technology (2009).
16. Wang, J., F. Shi, J. Zhang and Y. Liu. "A New Calibration Model of Camera Lens Distortion." *Pattern Recognition.* Vol. 41, No. 2. Amsterdam, Netherlands: Elsevier, for Pattern Recognition Society (February 2008): p 607-615.
17. IEEE 1394, *High-Performance Serial Bus.* New York, NY: IEEE (2008).
18. Wilson, A. "Camera Connections." *Vision Systems Design.* Tulsa, OK: PennWell Corporation (April 2008).
19. *Specifications of the Camera Link Interface Standard for Digital Cameras and Frame Grabbers.* Ann Arbor, MI: Automated Imaging Association (Annex D, 2007).
20. "Universal Serial Bus," Web site. Beaverton, OR: USB Implementers Forum (2008).
21. "GigE Vision®." Web page. *Machine Vision Online.* Ann Arbor, MI: Automated Imaging Association (2009).
22. *1394 Technology.* Web page. Southlake, TX: 1394 Trade Association (2008).
23. Sgro, J. "USB Advantages Offset Other Interfaces." *Vision Systems Design.* Tulsa, OK: PennWell Corporation (September 2003).
24. Jain, R., R. Kasturi and B.G. Schunck. *Machine Vision.* New York, NY: McGraw-Hill (1995).
25. Gonzalez, R.C. and R.E. Woods. *Digital Image Processing*: Upper Saddle River, NJ: Prentice Hall (2002).
26. Otsu, N. "A Threshold Selection Method from Gray-Level Histograms." SMC-9, *IEEE Transactions on Systems, Man, and Cybernetics.* Vol. 9, No. 1. New York, NY: IEEE (January 1979): p 62-66.

27. Wada, T., T. Fujii and T. Matsuyama. "γ - ω Hough Transform — Linearizing Voting Curves in an Unbiased ρ - θ Parameter Space" [in Japanese]. *IEICE D-II.* Vol. J75, No. 1. Minato-ku, Tokyo, Japan: Institute of Electronics, Information and Communication Engineers (1992): p 21-30.
28. T. Wada, M. Seki and T. Matsuyama, "High Precision γ - ω Hough Transform Algorithm to Detect Arbitrary Digital Lines" [in Japanese]. *IEICE D-II.* Vol. J77, No. 3. Minato-ku, Tokyo, Japan: Institute of Electronics, Information and Communication Engineers (1994): p 529-530.
29. Malamas, E.N., E.G.M. Petrakis, M.E. Zervakis, L. Petit and J.D. Legat. "A Survey on Industrial Vision Systems, Applications and Tools." *Image and Vision Computing.* Vol. 21, No. 2. Amsterdam, Netherlands: Elsevier (February 2003): p 171-188.
30. Komorowski, J.P. and D.S. Forsyth. "The Role of Enhanced Visual Inspection in the New Strategy for Corrosion Management." *Aircraft Engineering and Aerospace Technology.* Vol. 72, No. 1. Bingley, United Kingdom: Emerald Group Publishing (2000): p 5-13.
31. Komorowski, J.P., N.C. Bellinger, R.W. Gould , A. Marincak and R. Reynolds. "Quantification of Corrosion in Aircraft Structures with Double Pass Retroreflection." *Canadian Aeronautics and Space Journal.* Vol. 42, No. 2. Kanata, Ontario, Canada: Canadian Aeronautics and Space Institute (1996): p 76-82.
32. Komorowski, J.P., S. Krishnakumar, R.W. Gould, N.C. Bellinger, F. Karpala, O.L. Hageniers. "Double Pass Retroreflection for Corrosion Detection in Aircraft Structures." *Materials Evaluation.* Vol. 54, No. 1. Columbus, OH: American Society for Nondestructive Testing (January 1996): p 80-86.
33. Siegel, M. and P. Gunatilake. "Remote Inspection Technologies for Aircraft Skin Inspection." ETVSIM'97, *IEEE Workshop on Emergent Technologies and Virtual Systems for Instrumentation and Measurement* [Niagara Falls, Ontario, Canada, May 1997]. New York, NY: IEEE (1997): p 1-10.
34. Gunatilake, P., M.W. Siegel, A.G. Jordan and G.W. Podnar. "Image Understanding Algorithms for Remote Visual Inspection of Aircraft Surface." *Machine Vision Applications in Industrial Inspection* V [Orlando, FL, April 1997]. Proceedings Vol. 3029. Bellingham, WA: International Society for Optical Engineering (Society of Photo-Optical Instrumentation Engineers) (1997): p 2-13.

8

CHAPTER

Visual Testing of Metals

David R. Atkins, Packer Engineering, Naperville, Illinois (Parts 1 to 3)

Stanislav I. Rokhlin, The Ohio State University, Columbus, Ohio (Part 4)

Michael A. Urzendowski, Valero Energy, San Antonio, Texas (Parts 1 to 3)

Robert W. Warke, LeTourneau University, Longview, Texas (Parts 1 to 3)

Portions of Part 6 are reprinted with permission from Donald J. Wulpi, *Understanding How Components Fail,* © 1985, ASM International, Materials Park, OH. Reprinted with permission. Reference number[9] superscripted in headings indicates sections adapted. ASNT has revised the text in 1993 and 2010, and deficiencies are not the responsibility of ASM International.

PART 1. Metal Processing

Inspection for Discontinuities in Steel[1]

Surface Inspection

In the manufacturing processes, it is common practice to inspect the top surface of cut lengths as they are sheared to length from coils. In the case of coils, the outside surface of the coil is inspected during processing and rewinding.

Sheet steel in coils or cut lengths may contain surface imperfections that can be removed with a reasonable amount of metal finishing by the purchaser. Because the top side of a cut length or outside of a coil is ordinarily the inspected side, the opposite side may contain more surface imperfections. To minimize the amount of metal finishing, the inspected side can be used for the most critical surface of a fabricated part.

When it is not possible to use the inspected side for the most critical surface of a fabricated part, the producer should be notified. It is sometimes possible for a producer to inspect the bottom surface or reverse the inspected top side to the bottom side of a lift of cut lengths or inside surface of a coil.

Coils contain more surface imperfections than cut lengths because the producer does not have the same opportunity to sort portions containing such imperfections as is possible in the case of cut lengths.

Limitations of Inspection, Testing and Certification

When a purchaser's specifications stipulate that inspection and tests (except product analysis) for acceptance of the steel be made by an agent of the purchaser before shipment from the mill, the producer affords the purchaser's representative all reasonable facilities to assure that the steel is being furnished in accordance with the specification.

There are a number of intrinsic features of steel making and finishing processes that affect the properties or conditions of the finished products, and those effects cannot always be precisely known. Therefore, it is technically impossible to give unconditional certification of complete compliance with all prescribed requirements. That fact is manifest to those having a technical knowledge of the subject and those skilled in the manufacture of steel, and is recognized in applying a factor of safety. For example, the phenomenon of segregation causes variations in chemical composition, mechanical test results and soundness. Variations in manufacturing practice such as control of temperature, which cannot always be regulated with exactness, sometimes cause variations in mechanical properties in different parts of a batch of steel.

Because of these and other conditions which are present in steel mill operations, it is difficult to identify a practical test method to ensure the detection and rejection of every piece of steel that varies from the specified requirements with regard to dimensional tolerances, chemical composition, mechanical properties, surface or internal conditions.

Discontinuities[2]

Discontinuities are variations in the geometry or composition of an object. Such variations inherently affect the physical properties of the object and may in turn affect the object's service life. Not all discontinuities are defects. The definition of *defect* changes with the type of component, its construction, its materials, its use and the specifications or codes in force. It should be well understood that a discontinuity harmless in one object may be critical in another.

Detection of discontinuities is a process that largely depends on the discontinuity's physical characteristics and location — in the case of surface cracks, a critical parameter is the ratio of surface opening to crack depth. However, crack depth and width are not the only factors affecting detectability; length and orientation to the surface are also important.

To better detect and interpret visual discontinuity indications, it is necessary to know the basic material characteristics of the test object. Furthermore, it is also important to consider how the material is produced, what manufacturing processes are used to form the finished product and what discontinuities are typically initiated by the processing operations.

During the various stages of material processing, certain discontinuities can be expected. Typically, a discontinuity is categorized by its manufacturing stage: casting, forging, welding, processing and service. A discontinuity may be created at one stage and be detected in a later stage, when processing has changed its shape or accessibility (Fig. 1). The text that follows describes discontinuities that may originate from operations in each of the five stages. The listing is provided only for educational purposes and may not apply to all test objects.

FIGURE 1. Development of discontinuity in metal.

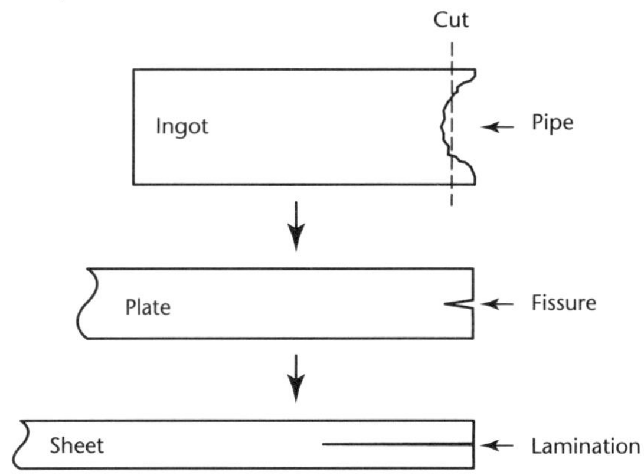

PART 2. Visual Testing of Cast Ingots[2]

Discontinuities in Castings

When metals are produced, molten metal solidifies into ingot form. During solidification, foreign materials and gas bubbles may be trapped in the ingot and form what is known as *inherent* discontinuities. Many of these are removed by cropping but a number of them can remain in the ingot. Such discontinuities then can be rolled, forged and sectioned along with the material in its subsequent processing operations.

Several inherent discontinuities occur commonly in metals (Table 1).

Cold Shut

A cold shut is initiated during the metal casting process. It occurs because of imperfect fusion between two streams of metal that have converged. Cold shuts may also be attributed to surging, sluggish molten metal, an interruption in pouring or any factor that prevents fusion where two molten surfaces meet.

This discontinuity produces cracks with smooth or rounded edges similar to seams (Fig. 2).

Pipe

During solidification, molten metal shrinks. In the case of a casting, there eventually can be insufficient molten metal for completely filling the top of the mold. Shrinkage occurs all over the casting as the metal cools. As a result, a cavity forms, usually in the shape of an inverted cone or cylinder (Fig. 3).

If this shrinkage cavity is not completely removed before rolling or forging into final shape, it becomes elongated and appears as voids called *pipe* in the finished product. Pipe can also result from extrusion, caused by the oxidized surface of a billet flowing inward toward the center of a bar at the back end. At the end of a billet, pipe usually appears as a small rounded cavity between the surfaces (Fig. 4).

Hot Tears

At the elevated temperature associated with solidification, cast materials are susceptible to hot tears. Segregation of low melting point impurities results in localized loss of ductility and strength. Lacking these, the cooling metal can tear and crack in the mold because of restraint from the mold. In addition, uneven cooling in thin sections or corners that adjoin heavier masses of metal can result in higher metal surface stresses that in turn produce hot tears. Hot tears occur especially at thick-to-thin transitions.

FIGURE 2. Indication of cold shut in casting, enhanced here with magnetic particles.

TABLE 1. Discontinuities in ferromagnetic castings.

Discontinuity	Location	Cause
Cold shuts	surface or subsurface	meeting of two streams of liquid metal that do not fuse
Hot tears	surface	adherence to core or mold during cooling
Inclusions	surface or subsurface	contaminants introduced during casting process
Pipe, shrinkage	subsurface	absence of molten metal during final solidification
Porosity	surface or subsurface	entrapped gases during solidification of metal
Segregation	surface or subsurface	localized differences in material composition

Hot tears appear on the surface as a ragged line of variable width and numerous branches. In some instances, the cracks are not detectable until after machining because the tearing can be subsurface.

Blowholes and Porosity

Gas porosities are rounded cavities (flattened, elongated or spherical) caused by the accumulation of gas bubbles in molten metal as it solidifies. A small percentage of these bubbles rise through the molten metal and escape. However, most are trapped at or near the surface of the ingot when solidification is complete. During rolling or forging of the ingot, some of these gas pockets are fused shut. The remaining pockets may appear as seams in the rolled ingot.

FIGURE 3. Longitudinal section of two ingots, showing typical pipe and porosity: (a) detectable; (b) severe.

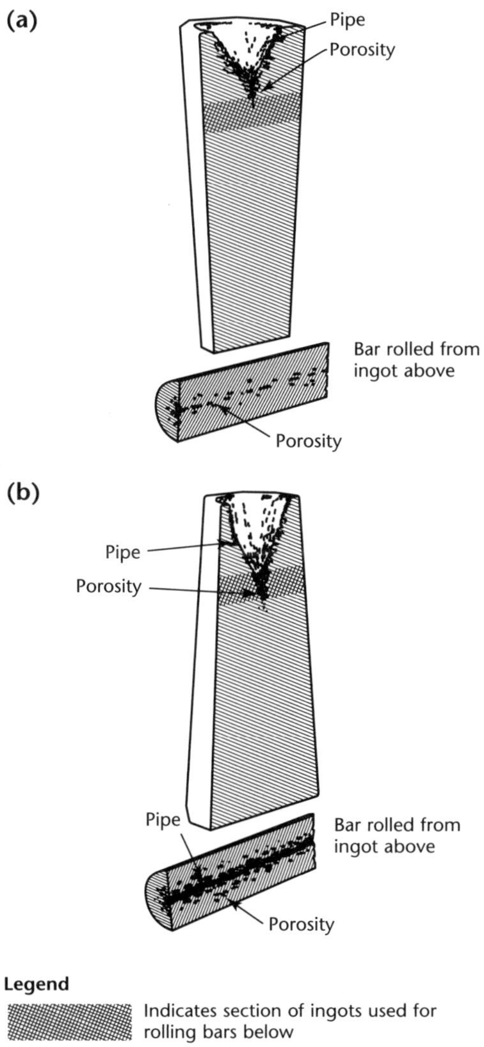

Blowholes are conical, wide at the surface and tapering internally. Deep blowholes not rolled shut may appear as laminations after becoming elongated in the rolling operation.

A surface condition resembling porosity is called *orange peel* because of its appearance (Fig. 5).

Nonmetallic Inclusions

Inclusions in ferrous alloys are usually oxides, sulfides or silicates either inherent in the base metal or introduced during the melting operation. These inclusions are caused by conversion of iron ore in the blast furnace. Dirty remelt, crucibles or rods or poor linings may introduce nonmetallic inclusions into the molten metal. Other contributing factors are poor pouring practice and inadequate gating design that can produce turbulence within the mold.

Nonmetallic inclusions in ingots can, after forging, become stress risers because of their shape, discontinuous nature and incompatibility with the surrounding material. In many applications, it is the presence of these inclusions that lowers the ability of a metal to withstand high impact, static or fatigue stresses. Moreover, the effect of inclusions depends on their size and shape, their resistance to deformation, their orientation relative to applied stress and the tensile strength of the material. Many inclusions can be of a more complex intermediate composition than their host materials and each grade and type of metal has its own characteristic inclusions.

FIGURE 4. Pipe lamination is separation midway between surfaces containing oxide inclusions: (a) surface view; (b) internal section.

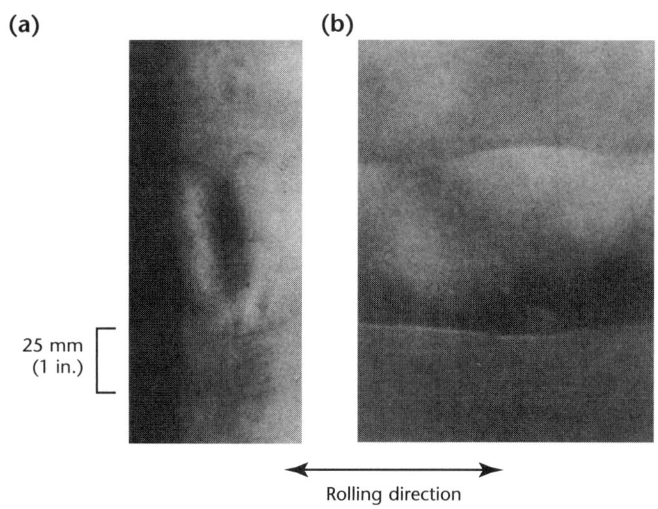

FIGURE 5. Orange peel is coarse grain condition that becomes evident during drawing: (a) drawn surface; (b) formed surface; (c) orange peel strain, pebbly surface condition that develops during drawing.

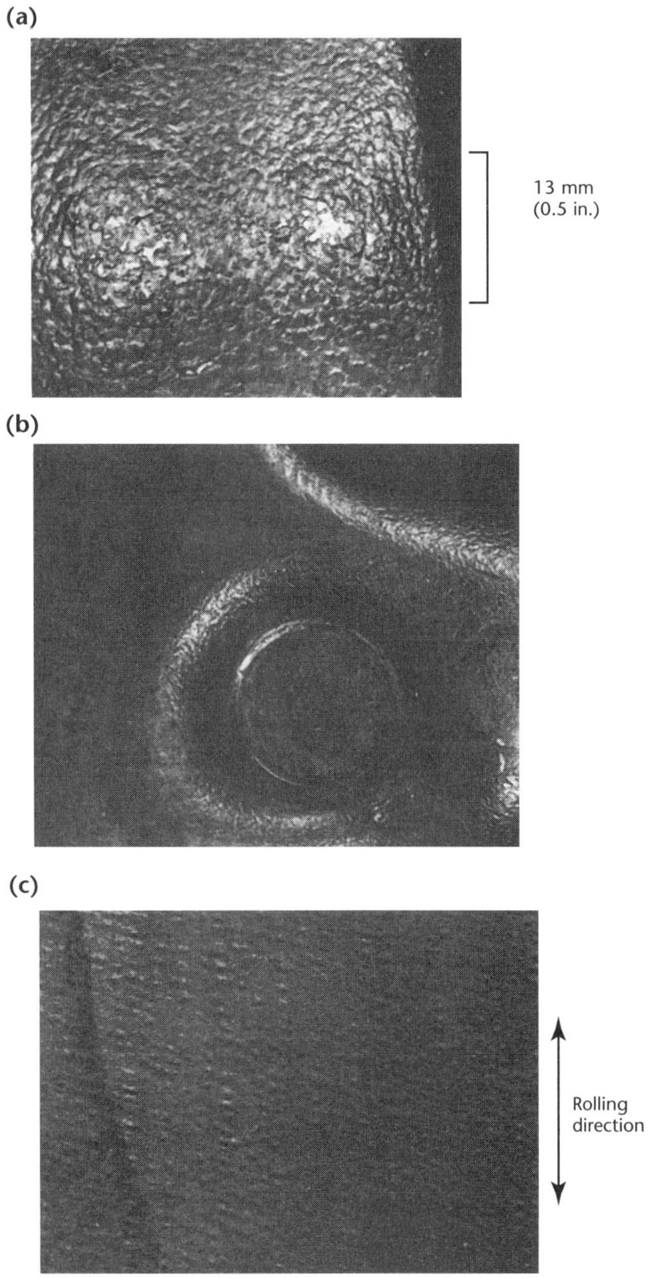

When steel is mechanically rolled or formed, typically, inclusions form plastically into elongated shapes and to appear in longitudinal sections as stringers or streaks. In transverse cross sections, the inclusion's shape is more globular or flat.

Segregation

Segregation is a localized difference in a material's chemical composition. During solidification of molten metal, certain elements may concentrate in limited areas, resulting in an uneven distribution of some of the alloying elements of the steel. Equalization of the compositional differences can be achieved by hot working (forging or rolling). However, segregation is sometimes carried into the wrought product.

When not detected, segregation can affect corrosion resistance, forging and welding characteristics, mechanical properties, fracture toughness and fatigue resistance. Furthermore, quench cracks, hardness variations and other discontinuities are likely to result during heat treating of materials that exhibit segregation of alloying elements.

Part 3. Visual Testing of Forgings and Rolled Metal[2]

Discontinuities in Forgings

Discontinuities that originate during hot or cold forming are said to be *primary processing discontinuities*. The processing of a wrought product by rolling, forging, casting or drawing may introduce specific discontinuities into the product and inherent discontinuities that were at one time undetectable or insignificant may propagate and become detrimental.

The following is a brief description of some primary processing discontinuities in metals (Table 2).

Seams

As an ingot is processed, surface discontinuities such as gas pockets, blowholes and cracks are rolled and drawn longitudinally. When these discontinuities exist, an underfill of material occurs during the rolling operation. Seams may also be initiated in the semifinishing and finishing mills because of faulty, poorly lubricated or oversized dies.

As a result of multiple passes during rolling operations, underfilled areas are rolled together to form a seam. The surfaces are typically oxidized and may be intermittently welded together to form very tight, usually straight cracks that vary in depth from the surface.

Laminations

Laminations are separations that are typically aligned parallel to the worked surface of a material. They may be the result of blowholes, internal fissures, pipe, inclusions, seams or segregations that are elongated and flattened during the rolling process. They can be surface or subsurface, are generally flat and extremely thin (Figs. 4 and 6).[3]

Laminations can be detected at an end or at a transverse cross section through a rolled plate.

Stringers

Stringers are predominantly found in bar stock. They originate by the flattening and lengthening of nonmetallic inclusions during rolling.

Stringers are typically subsurface, semicontinuous straight lines parallel to the length of the bar stock in straight rolls and spiral in angular rolls. The seams are oriented in the rolling direction.

Flash Line Tears

As the dies close in the final stage of the forging process, a small amount of metal is extruded between the dies. This extruded metal is called *flash* and must be removed by trimming.

If the trimming is not done or not done properly, cracks or tears can occur along the flash line (Fig. 7).

Discontinuities in Rolled Metal

Cooling Cracks

After bar stock is hot rolled, placed on a bed or cooling table and allowed to reach room temperature, cooling cracks may develop from uneven cooling. Such cracks

Table 2. Discontinuities in ferromagnetic forgings.

Discontinuity	Location	Cause
Bursts	surface or subsurface	forming processes at excessive or improper temperatures
Cooling cracks	surface	uneven cooling during cold drawing
Cupping	subsurface	internal stresses during cold drawing, forging or extrusion
Hydrogen flakes	subsurface	abundance of hydrogen during elevated temperatures
Laminations	subsurface	elongation and compression of inherent discontinuities during rolling
Laps	surface	material folded over and not fused
Seams	surface	elongation of unfused surface discontinuities in rolled products
Stringers	subsurface	elongation and compression of inherent discontinuities during forming

are typically longitudinal and usually vary in depth and length. Although often confused with seams, cooling cracks do not exhibit surface oxidation.

Cooling cracks tend to curve around the object shape and so are not necessarily straight.

FIGURE 6. Lamination in rolled plates: (a) visible from end; (b) visible in holes; (c) visible from side.

(a)

(b)

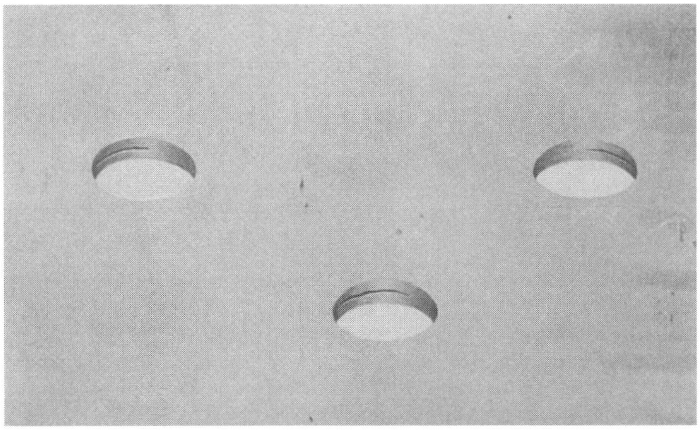

(c)

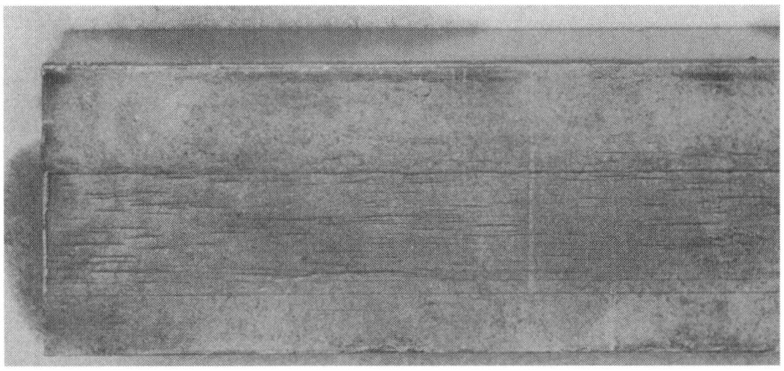

Forged and Rolled Laps

Forging laps are the result of metal being folded over, forming an area that is squeezed tight but not fused (Fig. 8). They are caused by faulty or misaligned dies, oversized blanks or improper handling of the metal in the die. Forging laps are usually open to the surface and are either parallel or at a small angle to the surface.

Rolled laps are a condition similar to a seam. Excessive material is squeezed out during a rolling pass, causing a sharp overfill or fin. When rotated for the following pass, the material is rolled back into the bar. Because of its heavily oxidized surface, the overfill cannot be fused to other material in the rolling operation. Rolling laps are usually straight or slightly curved from the longitudinal axis and are either parallel or at a small angle to the object surface.

Internal and External Bursts

Internal bursts are found in bars and forgings and result from excessive hot working temperatures. Discontinuities that exist before forming (porosity, pipe,

FIGURE 7. Flash lines and laps in forgings.

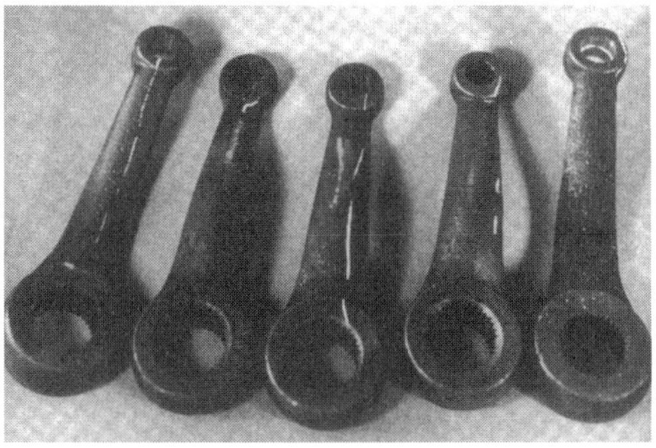

FIGURE 8. Wet fluorescent magnetic particle indication of forging lap in connecting rod.

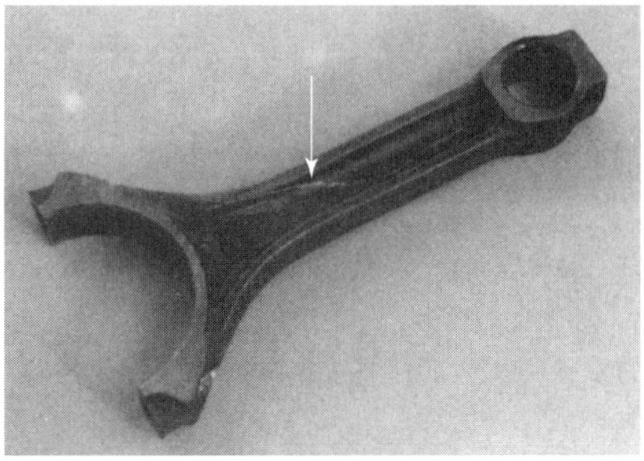

inclusions or segregation) are pulled apart because of the high tensile stresses developed during the forming operation.

Rolled and forged metals may also develop internal bursts (Fig. 9) when the equipment cannot work through the metal's cross section.

External bursts typically occur when the forming section is too severe or where sections are thin. External bursts may also be formed if rolling and forging equipment is not powerful enough: The outer layers of the metal are deformed more than the internal metal and the resulting stress causes an external burst. Forming during improper temperatures may also cause external bursts.

Hydrogen Flakes

Flakes are formed while cooling after the forging or rolling operations. Flakes are internal fissures attributed (1) to stresses produced by localized metallurgical transformations and (2) to hydrogen embrittlement, decreased solubility of hydrogen after rapid cooling.

Hydrogen is available in abundance during most manufacturing operations. When permitted, hydrogen dissipates freely at temperatures above 200 °C (390 °F), so that the solubility of hydrogen in material proportionally increases with increasing time and temperature. Hydrogen flakes are usually found deep in heavy steel forgings, are extremely thin and are aligned parallel with the grain.

FIGURE 9. Forging bursts: (a) internal; (b) external, with stretch mark indications.

(a)

(b)

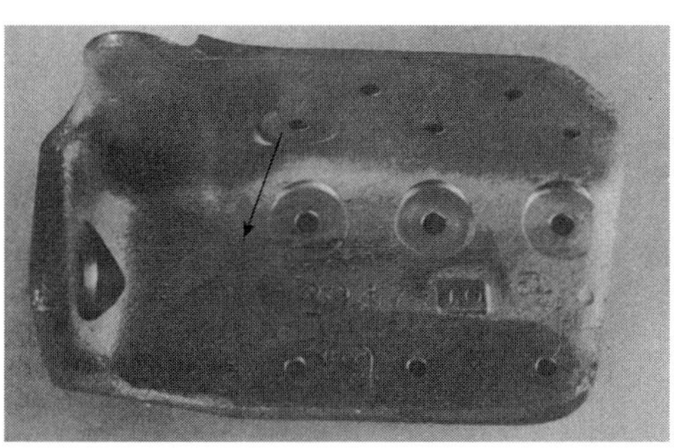

Rolling of Sheet Steel[1,4-6]

Many common surface discontinuities are visible to the unaided eye before fabrication. The following become evident only after forming: orange peel (Fig. 5), strain (Fig. 10), fluting (Fig. 11) and ghost lines (Fig. 12).

FIGURE 10. Stretcher strains are irregular surface patterns of ridges and valleys that develop during drawing.

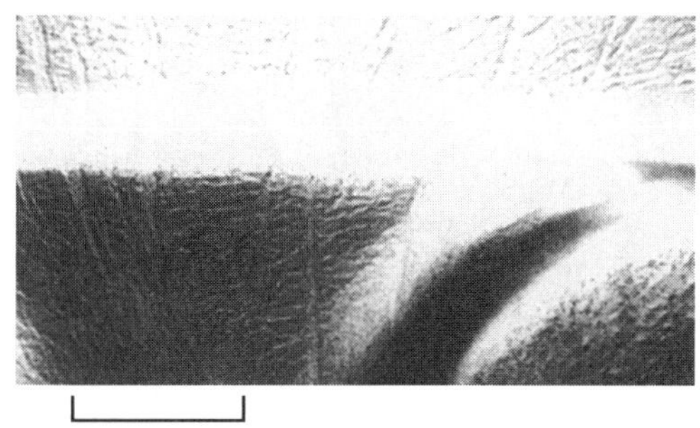

25mm
(1 in.)

FIGURE 11. Fluting is series of sharp parallel kinks or creases occurring in arc when sheet steel is formed cylindrically; photograph shown is from test specimen.

Figures 13 to 21 show visual indications that can be detected after the rolling operation.

FIGURE 12. Ghost lines are lineal irregularities in surface that develop in drawing. They are parallel to direction of rolling: (a) first example; (b) second example.

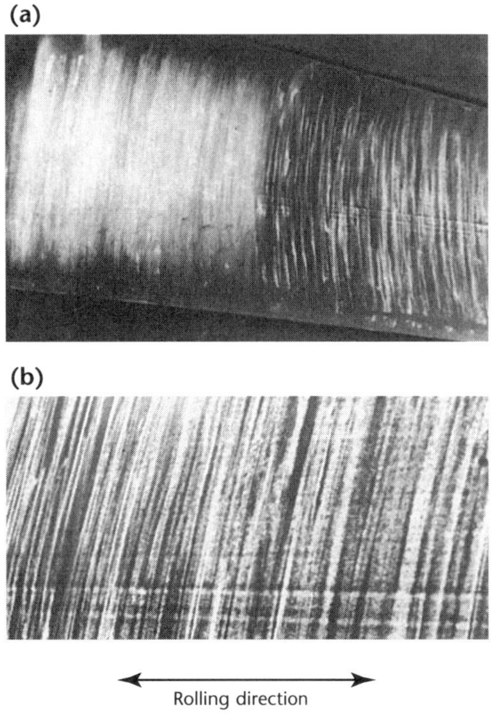

FIGURE 13. Coil breaks are creases or ridges that appear as parallel lines, transverse to direction of rolling and generally extending across width of sheet: (a) closeup; (b) appearance after flattening.

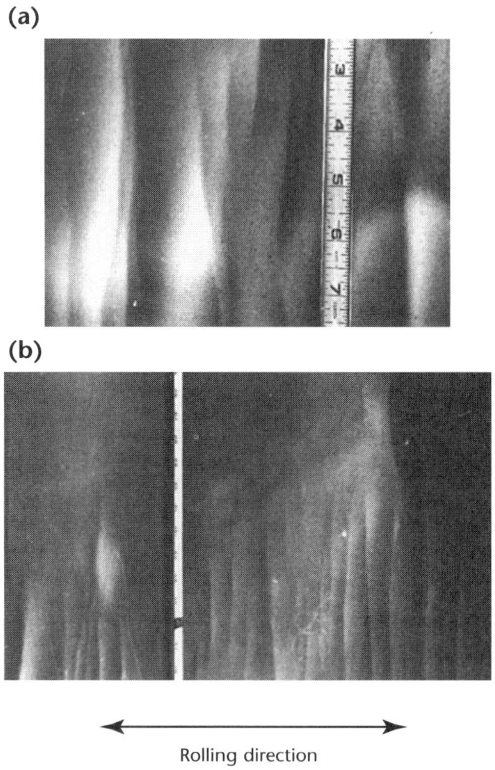

FIGURE 14. Coil weld is joint between two lengths of metal within coil. Coil welds are not necessarily visible in finished product.

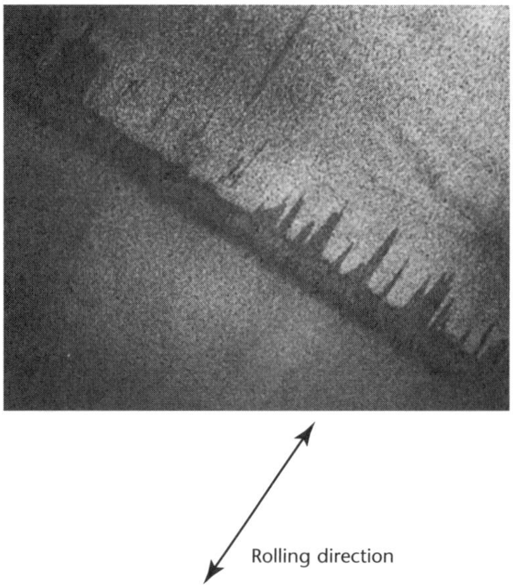

Figure 15. Edge breaks are short creases that extend in varying distances from side edge of temper rolled sheet.

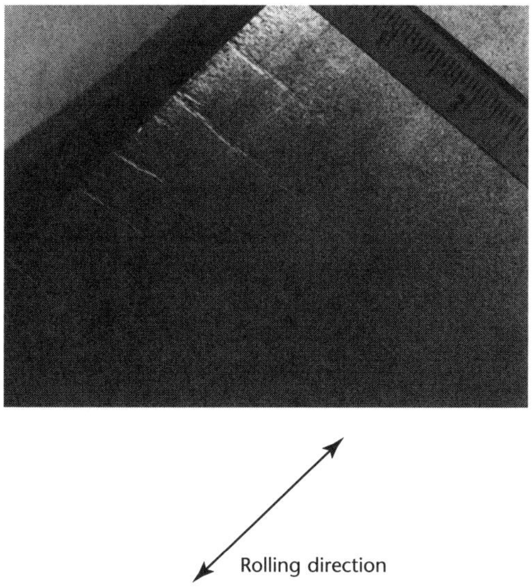

Figure 16. Floppers are lines or ridges diagonally transverse to direction of rolling and generally confined to midway between edges of coil as rolled. They are somewhat irregular and tend toward flat arc shape.

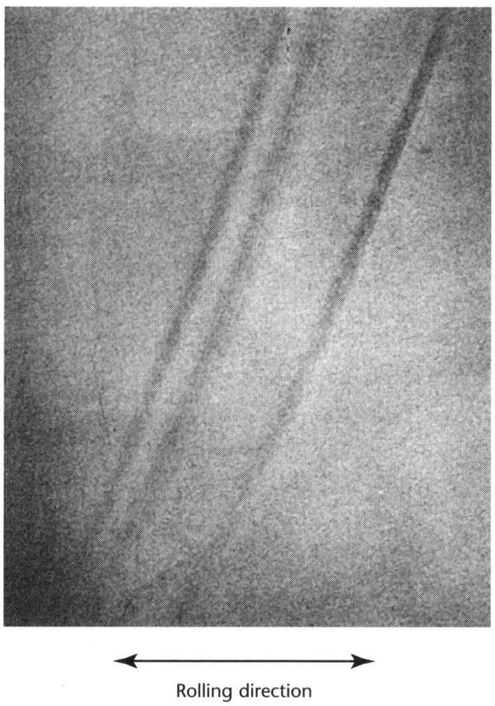

Figure 17. Friction digs are series of relatively short scratches variable in form and severity.

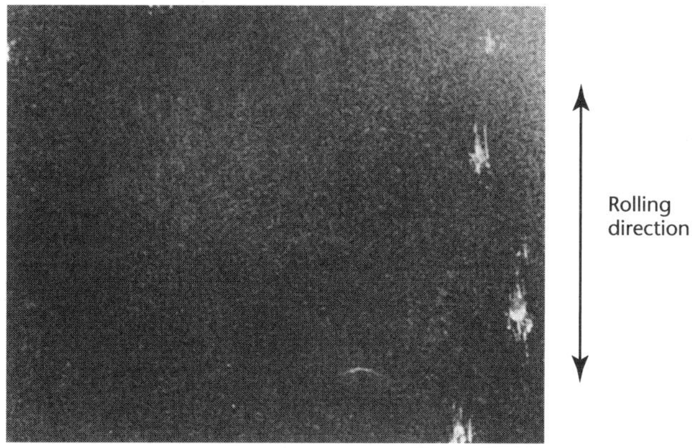

Figure 18. Pinchers are fernlike ripples or creases usually diagonal to rolling direction.

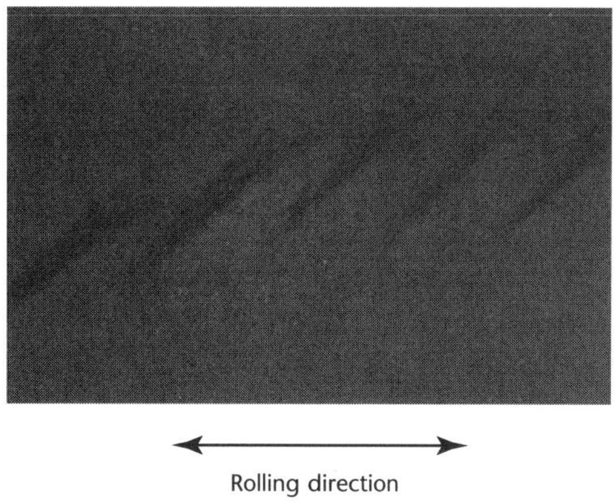

Figure 19. Rolled-in dirt is extraneous matter rolled into surface of sheet.

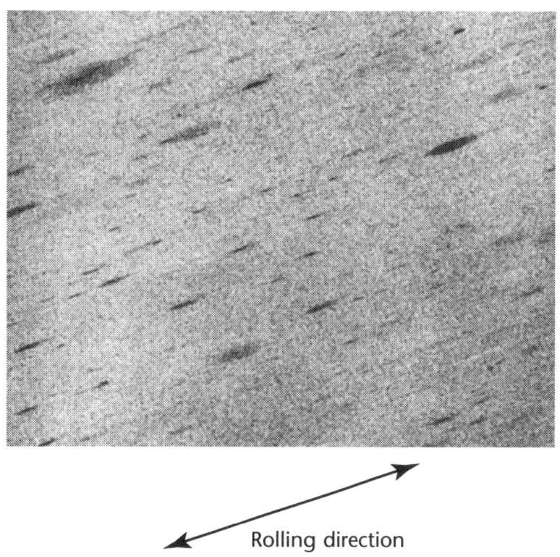

FIGURE 20. Slivers are surface ruptures somewhat similar in appearance to skin laminations but usually more prominent.

Rolling direction

FIGURE 21. Sticker breaks are arc shaped coil breaks usually near middle of sheet.

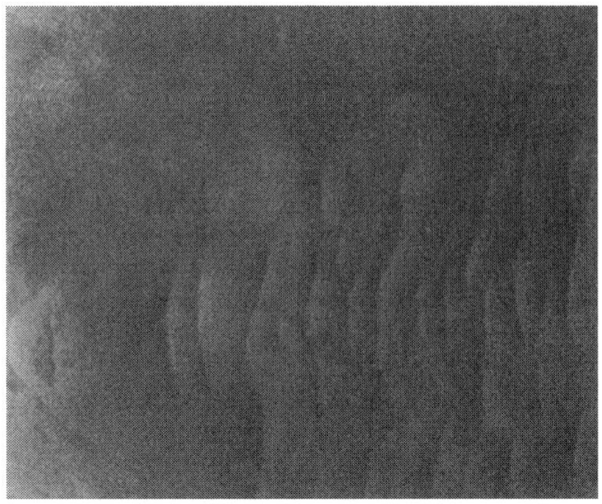

Rolling direction

PART 4. Visual Testing of Welds[2,7]

The discontinuities described below relate mainly to fusion welding; a few may also apply to resistance and solid state processes. The discussion covers discontinuities that lend themselves to detection by visual testing (Table 3).

Acceptance or rejection of a weldment, based on the detection of a particular discontinuity, is determined by the requirements of the designer and the applicable code. The *Structural Welding Code,* published by the American Welding Society, is specified for many diverse projects.[8]

Although many of the following discontinuities occur in several types of welded joints, the differences in joint geometry produce differences in the location and orientation of the discontinuities.

Porosity

Porosity is composed of cavities or pores that form when the weld metal solidifies. Gases result from chemical and physical reactions during welding. These gases dissolve in the metal while it is hot and then separate as it cools. The solubility of gases in metals can be very high at high temperatures. A certain amount of gas is always generated in standard welding but is usually not detectable. At times, however, excessive gas is produced and leads to the discontinuity called *porosity.*

Porosity is usually spherical but can occur in various shapes and sizes, depending on welding conditions. The pores usually lack sharp edges and corners.

1. The porosity may consist of single, isolated pores.
2. The distribution of porosity within the weld metal may be clustered, usually because of improper initiation or termination of the welding arc.
3. The porosity within the weld metal may be linear. Linear distribution can result from welding over a contaminant in a linear junction such as a corner or crevice.

Porosity can be distributed in ways related to welding condition and may be found anywhere throughout the weld metal. It will not be found in the unmelted heat affected zone.

Porosity often results from dirt, rust or moisture on the base or filler metal surface before welding and can be prevented by maintaining cleanliness and dryness. Other contributing factors include base metal composition (such as high sulfur content), high solidification rate and improper welding technique (such as excessive arc length or lack of shielding gas).

Most welds contain some porosity. All but the smallest surface pores should be visible to the unaided eye.

The restrictions on porosity in some welds may be rather lenient. Spherical

TABLE 3. Weld discontinuities.

Discontinuity	Location	Cause
Cold cracking	surface or subsurface	combination of atomic hydrogen, hardenable material and high residual stresses
Hot cracking	surface or subsurface	segregation during solidification (see *Liquation* and *Solidification*)
Inclusions, oxide	subsurface	mixing oxides on base metal surface into weld pool
Inclusions, slag	subsurface	improper cleaning of a previous weld pass
Inclusions, tungsten	subsurface	molten weld pool or filler metal comes in contact with tip of tungsten electrode
Lack of fusion	surface or subsurface	failure of weld metal to coalesce with base metal
Lack of penetration	surface or subsurface	inadequate penetration of weld joint root by the weld metal
Lamellar tearing	surface	delamination of the base material during solidification and cooling of weld metal
Liquation	surface or subsurface	segregation in heat affected zone of material
Porosity	surface or subsurface	vaporized constituents in molten weld metal are entrapped during solidification
Overlap	surface	insufficient amperage or travel speed
Solidification	surface or subsurface	dendritic segregation of low melting point constituents opening during solidification
Undercut	surface	oversized weld pool (related to excessive amperage, travel speed and electrode size)

porosity does not seriously weaken a welded joint. First, spherical porosity does not act as a stress riser, meaning that it does not behave like a sharp notch that would indeed weaken a joint. Secondly, the strength of weld metal is customarily greater than the nominal strength of the material being joined. (It is possible, however, for porosity to increase ductility.) Welding electrodes, with their special coatings, are designed to deposit a relatively high strength material, one with a very fine cast structure as a result of an extremely rapid cooling rate. Most weldments then can accommodate a fair amount of porosity, especially when the service conditions are predictable. An exception to this lenience is for severe linear and clustered porosity in highly stressed and fatigue sensitive joints.

Spherical Porosity

Uniformly distributed, spherical porosity indicates that the cause of the difficulty has been averaged over the entire weld.

Clustered spherical porosity occurs in zones where there has been a perturbation in the welding process, especially when welding has been stopped and started up again. In such cases the pores would be spherical.

Linear spherical porosity is confined to the earlier portions of the weld, often following a joint edge. Linear spherical porosity usually indicates contaminated material.

Blowholes

Often the surface discontinuities called *blowholes* are found where gas pockets have reached the surface of the weld pool but do not fully escape before solidification takes place. Blowholes should be removed before any subsequent weld passes are deposited because they are likely places for slag entrapment.

Long Porosity

Piping is a term applied to porosity of a teardrop shape; it usually has a linear distribution. One type of elongated pore is often called a *wormhole*.

Pipeline welding can have a linear type of porosity in the weld root. The root bead has a tunnel that follows the root contour.

Herring Bone Porosity

Herring bone porosity is a special case, occurring in automatic arc welding processes. It has a tear drop shape, similar to piping porosity, and is distributed in linear fashion with a high degree of regularity. Herring bone porosity is caused by contamination, generally the presence of air. In the gas metal arc process, a special atmosphere provided by the equipment helps avoid this discontinuity.

Inclusions

Inclusions are pieces of slag or other solid materials that have been trapped by weld metal. This form of discontinuity, like porosity, is consistently found by radiography. It can occur in nearly spherical, isolated cases but the tendency is for inclusions to accumulate linearly. Most inclusions consist of residual slag from the electrode coatings or fluxes in welding. The trapped material tends to occur in the welding direction and therefore normally appears in a line.

In building up a weld, a channel effect occurs on both sides of the bead. If through a fault in the process some slag is trapped, there could be two affected zones, corresponding to the weld bead width. When such parallel slag lines occur they are called *wagon tracks*.

In multiple-pass welding, the slag covering will remain on the as-welded work piece, just as it does on properly completed welds, and measures must be taken to ensure its removal. If this interbead slag is not completely removed before the next pass, inclusions can result.

All the welding processes involving flux can exhibit slag inclusions. Just as for porosity, the automatic processes, when they go awry, tend to exhibit long, regular discontinuities.

The limits on slag are more stringent than for porosity because the linear form of the inclusion can act as a stress riser. The inspector is usually directed by the applicable code in determining the allowable length and width of such inclusions.

An intermittent slag line is actually a single slag deposit with gaps. The accumulated length of the broken run plus the gap space may be used to determine the total discontinuity length, in some specifications.

Cracks

A crack may be defined as a split, exhibiting a sharp tip and typically a very narrow opening. Cracks are the most serious form of discontinuity because their sharp tip acts as a severe stress riser. For cyclic loading, cracks are always dangerous. The actual size of a crack may exceed its detected length. Further portions of cracks can be closed very tightly because of weld shrinkage stresses, and a tightly closed crack might not be seen in a visual test.

As in the case of incomplete fusion, cracks are ordinarily detectable only by ensuring that the beam direction is in line with the crack. When suspecting a crack, the visual inspector flags the area for assessment with other nondestructive test methods.

After the test object is rejected, repair and reinspection are options.

Hot cracks and cold cracks may not break the surface.

Cold Cracking

Cold cracking is also known as *underbead* or *delayed cracking*. It is a form of hydrogen induced cracking that appears in the heat affected zone or weld metal of low alloy and hardenable carbon steels. Cracking of this type may occur immediately on cooling or after a period of hours or even days. The principal factors contributing to cold cracking are (1) the presence of atomic hydrogen, (2) a hard martensitic microstructure in the heat affected zone and (3) high residual tensile stresses resulting from restraint.

Sources of atomic hydrogen include moisture in the electrode covering, shielding gas or base metal surface (including hydrated rust), as well as contamination of the filler or base metal by a hydrocarbon (oil or grease). Dissociation of water vapor or a hydrocarbon in the welding arc results in the rapid diffusion of atomic hydrogen into the molten weld pool and subsequently into the base metal's heat affected zone. If the zone's cooling rate is high enough and the steel is hardenable enough (a function of carbon and alloy content), a martensitic microstructure may form and the hydrogen atoms may then collect at internal discontinuities. Residual stresses caused by weld shrinkage, or externally applied tensile stresses, result in hydrogen induced cracks initiating at the hydrogen rich discontinuities.

Cold cracks produce sharply defined indications if they are open to the test object surface, as in the case of underbead cracks that extend to the weld toe. Weld metal cracks may be oriented in any direction and are often associated with nonmetallic inclusions. Subsurface indications are less pronounced or may be undetectable, depending on depth.

Hot Cracking

Hot cracking is a term applied to several varieties of weld metal and heat affected zone cracking, all of which occur at elevated temperatures. The following types are the most common hot cracks.

Solidification. Solidification cracking occurs near the solidification temperature of the weld metal and is caused by the presence of low melting point constituents, typically iron sulfides, that segregate to the weld metal dendrite surfaces during the liquid-to-solid transformation process. The shrinkage stresses induced by cooling cause cracks to open between the dendrite surfaces.

Center Line. Center line hot cracking follows the longitudinal center line of the deposited weld bead (Fig. 22). During weld deposition, solidification of the progressing weld pool occurs from the outside in, beginning at both toes and meeting at the center. The low melting point impurities are pushed ahead of these two joining solidification fronts where they are concentrated at the center line and open up as a longitudinal hot crack under transverse solidification shrinkage stresses. The likelihood of this occurrence is increased by high travel speed, high depth-to-width ratio of the weld bead and a small weld bead, particularly in the root pass.

Liquation. Liquation cracking, or hot tearing, originates in the heat affected zone of a weld during or just after the hardening of the metal when shrinkage is excessive. When narrow, deep welds are made, the weld metal does not stay molten long enough to fill the normal shrinkage spaces. Constituents (inclusions or segregated alloying elements) with low melting points form a liquid grain boundary film unable to support the shrinkage stresses of the welding process. Such cracks are often microscopic in size but may link up under applied stresses to

FIGURE 22. Section through weld joint exhibiting center line solidification cracking, form of hot cracking.

form a continuous surface or subsurface crack. In general, hot cracking is associated with steels having high sulfur content and the effect is accentuated as carbon content increases. Hot cracks are visible if they break the surface.

Crater Cracking

A variation on the hot crack is the crater crack (Fig. 23). A crater crack occurs if the heat of the electrode is removed too soon at the point where a weld is terminated. Crater cracks are typically star shaped, occurring as sets of radial cracks. The molten metal cools rapidly, producing cracks.

Lack of Fusion

Lack of fusion occurs when some portion of the weld filler metal fails to coalesce with the adjacent base metal or the weld metal from a previous pass. In welding processes that use no filler metal, lack of fusion refers to incomplete coalescence between the two base metal components being joined.

Lack of fusion is caused when the base metal surface or previous weld pass fails to reach melting temperature when the weld metal is deposited. This condition may occur when welding large components that can rapidly transfer heat away from the weld area because of its mass, particularly if the base material is not properly preheated before welding. In such cases, the base metal will not melt and the weld metal will not fuse into the base metal. Lack of fusion is often seen at the beginning of the first weld pass, when the base metal is at its lowest temperature during the welding process. When this condition occurs, it is commonly called a *cold start.*

One welding process particularly susceptible to this discontinuity is gas metal arc welding (GMAW) in the short circuiting arc mode, because of its inherently low heat input. Another frequent cause of lack of fusion is attempting to weld on top of a previously deposited weld pass that has been inadequately cleaned of slag or welding on a dirty base metal surface, so that the heat of the arc is unable to reach the underlying metal.

Lack of fusion may occur at any depth in a weld. Lack of fusion is usually oriented parallel to the direction of welding, and the indication appears on the joint groove surface.

Lack of fusion in autogenous welds (welds without filler metal) may result from large inclusions in the base metal or impurities that become trapped between the surfaces of the joint before welding. Susceptible processes are those that produce a relatively shallow melted zone at the joining surfaces and then expel most of that zone by a subsequent upsetting force (high frequency resistance welding, projection welding, flash welding, friction welding). Other causes of lack of fusion in autogenous welds include inadequate heating and insufficient upsetting force.

FIGURE 23. Location and typical appearance of crater cracks.

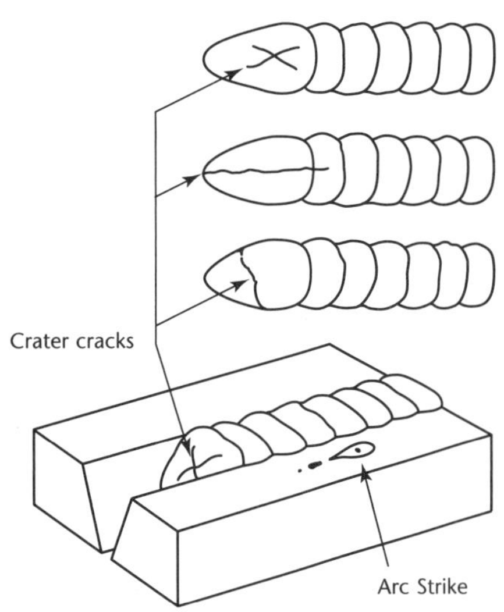

Lack of Penetration

Lack of penetration occurs at the root of a weld when the weld metal does not fully penetrate the root (Fig. 24). This condition can result from any of several incorrect parameters, most related to welding technique. These include low amperage, oversized electrode, excessive travel speed, improper electrode angle, improper arc manipulation and inadequate preweld cleaning.

Often, the joint design does not facilitate good penetration because of too

FIGURE 24. Lack of penetration.

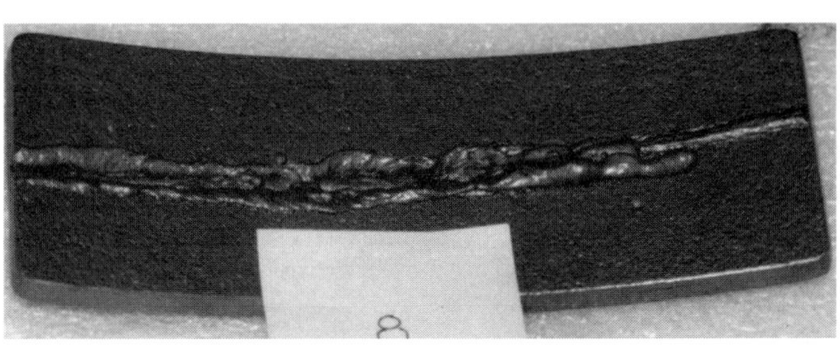

large a root land, too narrow a root gap or too small a bevel angle. Many procedures for double-groove welds specify back gouging of the first pass on the first side before deposition of the first pass on the second side. If back gouging is inadequate during joining, lack of penetration will likely occur.

Lack of penetration has an appearance similar to a longitudinal crack and is usually found at an edge of the original root joint. On open root, single-V welds where the back side (root) of the weld is accessible, lack of penetration may be found visually. However, on double-V or single-V welds with backing bars, this condition cannot be seen.

Slag

Many weld processes use flux shielding, including shielded metal arc welding (SMAW), submerged arc welding (SAW) and flux cored arc welding (FCAW). Welds produced by these methods are particularly susceptible to discontinuities known as *slag inclusions*. Slag can be entrapped in the weld metal during solidification if it is unable to float out while the pool is still liquid. The factors that promote slag entrapment include high solidification rate, high weld pool viscosity, use of an oversized electrode and improper joint geometry.

Slag allowed to remain on the surface of a deposited weld bead is rarely completely dissolved by subsequent passes. Therefore, it is essential to remove all slag from each pass. Joint designs that exhibit a high depth-to-width ratio and weld beads with an excessively convex profile are promoters of slag entrapment.

Tungsten inclusions are found in the weld metal deposited by the gas tungsten arc welding (GTAW) and plasma arc welding (PAW) processes and are usually the result of allowing the molten weld pool or the filler metal to come in contact with the tip of the tungsten electrode.

Oxide inclusions are particles of high melting point oxides present on the base metal surface. During welding, these oxides are then mixed into the weld pool.

Lamellar Tearing

A lamellar tear is a base metal crack that occurs in plates and shapes of rolled steel exhibiting a high nonmetallic inclusion content. These inclusions are rolled flat in the steel plate manufacturing process, severely reducing strength and ductility in the through-thickness direction. When the shrinkage stresses induced by weld solidification are imposed in that direction on the base metal plate, separation of the base metal at the flattened inclusions might occur, as may shearing between those lamellar planes, resulting in a terraced fracture (Fig. 25). Lamellar tearing is detectable on the surface and is most often seen in base metal on the edge of a steel plate or structural shape, adjacent to a deposited weld bead.[6] Lamellar tearing is also found in fillet welds or plate or welded pipe made from plate over laminations.

Undercut

Undercut is a groove melted into the base metal next to a weld bead or root when the base metal thickness is reduced. Essentially, a narrow crevice is formed in the base metal, paralleling the weld and immediately adjacent to it. Undercut lessens joint strength in the static sense by reducing the base metal section thickness. It also creates a stress concentration that reduces the impact, fatigue and low temperature properties of the joint.

Undercut is a processing fault caused by an oversized molten weld pool, which is in turn related to excessive amperage, travel speed and electrode diameter. Undercut may be repaired by adding an extra, narrow weld bead.

Undercut if external is easily detected by visual testing. Figure 26 shows undercut on an outside surface. In pipe welds, however, undercut is found on the inside surface. When undercut occurs in assemblies, the original surface may not be visible.

In some standards, a certain amount of undercut is permitted. As an example, AWS D1.1[8] in some cases permits undercut to a depth of 1.6 mm (0.063 in.)

FIGURE 25. Typical location and appearance of lamellar tearing. This view is parallel to rolling direction of steel plate base metal.

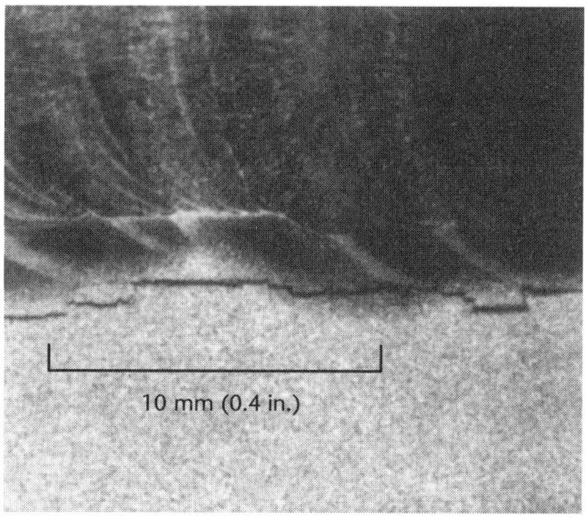

for material thicker than 15.9 mm (0.625 in.). In more critical cases, no undercut is permitted.

Overlap

Overlap is the protrusion of weld metal over the weld toe, producing a form of lack of fusion that creates a sharp mechanical notch or stress concentration. The condition is caused by insufficient amperage or travel speed.

Overlap is illustrated in the chapter on electric power applications and is often detectable by visual testing.

Arc Strikes

Arc strikes are discontinuities that result from establishing the welding arc in zones other than a weld. They consist of remelted metal or portions of electrode metal in unscheduled places. Their potential danger arises from steep changes in metal properties that develop when a material such as steel has been subjected to very rapid heating and cooling. Excessive hardness can result, leading to possible fracture during welding or service.

The condition is identified by its position (away from the weld metal) and by a small patch of extra thickness that is often intermittent but linear.

Arc strikes are usually cause for rejection on critical weldments because of the possible effect on service life. Repairs are possible and would generally involve grinding through the thin affected layer, with fine grinding as a finishing operation and etching to reveal any heat affected zone; no further welding would be scheduled.

FIGURE 26. Undercut.

PART 5. Discontinuities from Processes Other than Welding[2]

Discontinuities that originate from grinding, heat treating, machining, plating and related finishing operations are categorized as secondary processing discontinuities (Table 4). Such discontinuities may be the most costly because all previous processing and material costs are lost when the test object is diverted from service.

Heat Treating

To obtain a specific hardness and microstructure, certain materials are heat treated. During this operation, the metal is heated and cooled under controlled conditions. In some cases, this process produces stresses that exceed the material's tensile strength and cause it to crack. Similarly, when an object is heated to a very high temperature and then rapidly cooled (in air, oil or water), quench cracks may develop. Quench cracks serve as stress concentration sites for fatigue crack initiation and propagation. They may also serve as the initiation site for overload failures. Some quenching operations are so severe that objects break up during the process.

When an object is quenched following heat treating, an initial transformation occurs at the object's surface. When the interior cools and transforms, volumetric expansion takes place but the interior expansion is restrained by the solidified layer. If the solid layer does not expand enough or if the internal expansion is great enough, cracking through the outer layer results.

The amount of volumetric expansion is affected by the chemistry of the metal, particularly carbon. As the carbon content increases, so does the expansion. The severity of the quench can be lessened by using a lower carbon content material or by quenching in a less harsh medium such as oil or an elevated temperature bath.

A tempering process normally follows the quenching operation. Because of this exposure to a high temperature, the surface of quench cracks become oxidized. Visible oxidation is one sign that a crack was caused by quenching.

Heat treating and quench cracks usually emanate from locations of thin cross section, corners, fillets, notches or material thickness changes because these areas cool more quickly and therefore transform first. Restricted movement of the material also influences the location of cracks during the heat treating or quenching operations. Heat treating or quench cracks are typically forked indications in any direction on the test surface.

Straightening

The uneven stresses caused by heat treating frequently result in distortion or warping and the metal forms must be straightened into their intended shape. If the distortion is too great or the objects are very hard, cracking can occur during straightening.

Grinding

Grinding cracks can be attributed to glazed wheels, inadequate coolant, excessive feed rate or attempting to remove too much material in one pass. Grinding cracks develop where there is localized overheating of the base material.

Surface cracks in hardened objects can be caused by improper grinding operations. Thermal cracks are created by

TABLE 4. Secondary processing discontinuities in ferromagnetic materials.

Discontinuity	Location	Cause
Grinding cracks	surface	localized overheating of material because of improper grinding procedure
Heat treating cracks	surface	stresses from uneven heating or cooling and beyond tensile strength of material
Machining tears	surface	improper machining practice or dull tool
Pickling cracks	surface	residual stress being relieved
Plating cracks	surface	residual stress being relieved
Quench cracks	surface	sudden cooling from elevated temperature

stresses from localized overheating of the surface under the grinding wheel. Overheating can be caused by using the wrong grinding wheel, a dull or glazed wheel, insufficient or poor coolant, feeding too rapidly or cutting too heavily. Grinding cracks are especially detrimental because they are perpendicular to the object surface and have sharp crack tips that propagate under repeated or cyclic loading. Grinding cracks are typically at right angles to the grinding direction, are very shallow and are often forked and sharp at the root.

When located in high stress areas, such cracks may result in fatigue failures caused by residual stresses. Materials that have been hardened or heat treated are susceptible to grinding cracks because uncracked they retain high residual stresses from quenching. During grinding, localized heating added to entrapped stresses can cause surface ruptures. The resulting cracks are usually more severe and extensive than typical grinding cracks.

Machining Tears

A dull machining tool shears metal off in a manner that produces rough, torn surfaces. As a result, the surface is work hardened to a degree that depends largely on the depth of cut, the type and shape of the tool and the material properties.

Heavy cuts and residual tool marks from rough machining act as stress risers and can contribute to premature failure in a component. Stress risers may also occur at a change in section, such as in small fillet radii between two shaft sections of different diameters or the poor blending of fillets with shaft surfaces. Although difficult to detect, machining tears must be thoroughly and meticulously located.

Plating, Pickling and Etching

Hardened surfaces are susceptible to cracking from electroplating, acid pickling or etching processes.

Pickling Cracks

A pickling operation is used to remove unwanted scale (Fig. 27) for the purpose of a more thorough test of the base material. It can also be used to prepare the surface for finishing operations such as plating. Pickling cracks are predominantly found in materials that have high residual stresses (hardened or cold worked metals) and in materials with voids or similar discontinuities.

Acid pickling can weaken the surface structure of the metal, allowing internal stresses from the quenching operation to be relieved by crack formation. During pickling, acid etching or electrodeposition, hydrogen is generated at the surface of the material. The absorption, or interstitial diffusion, of hydrogen into the metal adds to the internal stresses of the object and causes a breakdown of its molecular structure. Cracks result. Internal stresses accelerate propagation of preexisting discontinuities. This mechanism, called *hydrogen embrittlement,* can result in cracking during the etching or plating operation or at some later time when additional service stresses are applied.

Plating Cracks

Plating is used for decoration, corrosion protection, wear resistance and to correct undersized dimensions for a wide variety of steel components. However, specific plating materials produce residual stresses that can be either tensile or compressive. Plating materials that develop residual tensile stresses (chromium, copper and nickel) can reduce the fatigue strength of a component.

FIGURE 27. Rolled-in scale consists of scale partially rolled into surface of sheet: (a) surface; (b) close-up.

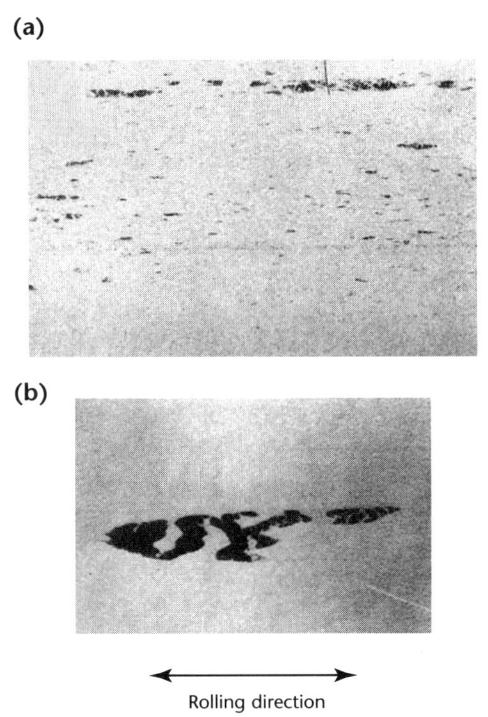

Plating cracks may develop when there is penetration of either hydrogen or hot plating material into the base metal. This action produces crack propagation or initiation. Materials high in hardness or residual stresses are more susceptible to damage from hydrogen absorption during plating or pickling operations. Furthermore, cracks that initiate exclusively in the plating material may act as stress risers and cause cracking in the base material.

Part 6. Service Induced Discontinuities[1,2,9]

The life expectancy of a component depends on its service environment (both mechanical and chemical), the quality of its maintenance and the appropriateness of its design. To perform a visual test accurately, it is helpful for testing personnel to know the service conditions of a component. Although service induced discontinuities appear similar, the mechanisms that cause them are quite different in each case.

The following text briefly describes service induced discontinuities (Table 5) common in metals.

Corrosion[9]

Corrosion is the deterioration of a metal resulting from electrochemical reactions with its environment. Corrosion and corrosion prevention cost hundreds of billions of dollars annually in the United States alone. Although corrosion typically is not catastrophic, it can be dangerous when it leads to fracture.

Corrosion is a natural process that reverses the chemical actions of refining. In their natural stable state, metals are found primarily either as oxides or as sulfides in ore. During refining, the addition of large amounts of energy strips the oxides or sulfides and produces relatively pure metals in a more unstable state. These refined metals are used singly or by alloying with other metals.

To illustrate for a common metal, iron is found as iron oxide in the ore. It is refined to iron or steel for most uses but eventually reverts to iron oxide (rust). The thermal or electrical energy added during refining moves the metal from a low energy level in its natural state, an ore, to a chemically unstable condition at a higher energy level. The refined metal may be remelted and cast, hot formed, cold formed or machined into useful shapes in the manufacture of an end product. However, the refined metal tends to deteriorate and revert to the original, chemically stable condition, releasing energy as heat in the process. (The energy released by corrosion is sometimes converted into electrical energy, as it is in a dry cell battery.)

It is commonly known that slight changes in metals, their design or their environment can make significant differences in their corrosive behavior. For this reason, it is important to obtain direct information about the circumstances of a corrosion problem, particularly corrosive effects complicating a fracture or wear condition.

Galvanic Corrosion

Galvanic corrosion is caused by the physical differences between contacting metals or a metal and its environment. Figure 28 shows the reactions found in a simple battery as an illustration of the

Figure 28. Galvanic cell showing electrochemical nature of corrosion.

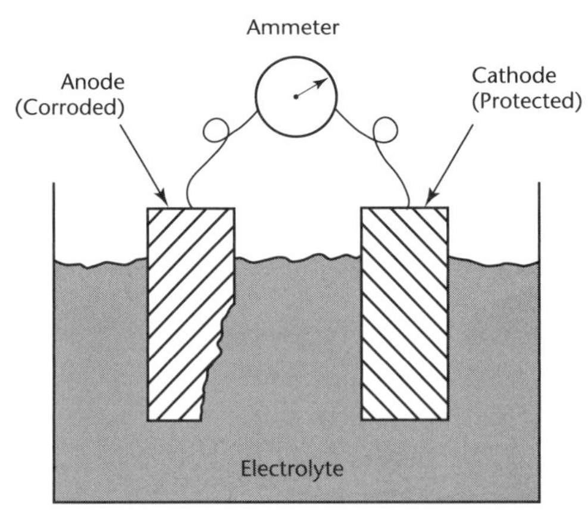

Table 5. Service induced discontinuities in ferromagnetic materials.

Discontinuity	Location	Cause
Creep	surface or subsurface	high temperature and stress below yield strength
Fatigue	surface or subsurface	cyclically applied stress below ultimate tensile strength
Hydrogen cracking	surface or subsurface	applied tensile or residual stress in hydrogen enriched environment
Stress corrosion cracking	surface	static tensile load in corrosive environment

principles of galvanic corrosion. Three components are needed for this electrochemical reaction: two materials (different metals or a metal and graphite) in physical or electrical contact. Electrical connection is achieved with an electrolyte, an electrically conductive liquid or paste. Under these conditions, one of the materials is corroded, hydrogen is released and energy is released in the form of an electric current. The material that corrodes is called the anode. The other material is known as the cathode and does not corrode.

Uniform Corrosion

Many metals corrode uniformly, without obvious galvanic couples. The most common uniform corrosion is rust on iron or steel. Uniform corrosion is a result of microscopic galvanic cells in the surface of the metal. Because of local chemical differences, impurities and alloying intermetallics in the metal, there are microscopic anodes and cathodes ready to corrode if an electrolyte is introduced — the corrosion is uniform only on a macroscopic scale.

Because uniform corrosion is the most common corrosion, it is the most significant economically and damages the greatest tonnage of metal. From a technical standpoint, uniform corrosion is fairly predictable and is relatively easy to control, provided other types of corrosion are not present.

One way to combat uniform corrosion is to use a chemically inert metal or stainless steel. Surface corrosion is also slowed by protective coatings, such as paint. The choice of means to combat corrosion is influenced by considerations such as the physical properties and cost of available materials.

Crevice Corrosion

Crevice corrosion is a kind of galvanic corrosion that is difficult to combat without careful control of design, materials, engineering and quality. *Crevice corrosion* is the commonly used term for differential oxygen concentration cell corrosion, or what is also called *poultice corrosion*.

Because most corrosion is caused by oxidation of reactive metals, areas of high oxygen concentration might be expected to corrode more readily than areas of lower oxygen concentration. However, a crevice between two surfaces, or metal under a poultice of moist debris, is more likely to corrode the more exposed metal. This occurs because there is little oxygen within the crevice or under the poultice. The metal there is anodic and corrodes. Areas exposed to higher oxygen content are cathodic and do not corrode. Concealed metal at the edge of a joint or under debris tends to pit and eventually perforate the metal thickness.

Crevice corrosion is the primary cause of automobile body corrosion, which originates in crevices, joints or under debris in the presence of moisture — frequently laden with electrolytic salts from road deicing. Crevice corrosion also occurs under fasteners, such as bolted or riveted joints, if moisture can penetrate and remain. The condition can occur even if both metals are the same but it is aggravated by dissimilar metals in contact, particularly with a small anode and large cathode. The Statue of Liberty in New York Harbor is subject to crevice corrosion because the copper exterior (cathode) is held in place with a steel framework and bolts (anode) which have corroded at the joints after many years of exposure to sea air.

Personnel should visually inspect exposed surfaces and remove deposits frequently — if materials such as dirt, rust or sand are not present, they cannot contribute to crevice corrosion. Filters, traps or settling tanks can help remove particles from a system but also require periodic maintenance to remove accumulations.

Cracking

Stress Corrosion Cracking[2]

Stress corrosion cracking is a fracture mechanism that results from the combined effects of a static tensile load and a corrosive environment. The stress involved can either be from actual applied loads or from residual stresses. One of the most common causes of this residual stress is the shrinkage that occurs during cooling of weld metal.

What constitutes a corrosive environment varies from material to material (Table 6). Common materials and their corrosive environments include aluminum and austenitic stainless steels exposed to saltwater, copper and its alloys exposed to ammonia (NH_3) and mild steel exposed to sodium hydroxide (NaOH).

The initiation site of a stress corrosion crack may be a preexisting discontinuity or it may be a small pit acting as a stress riser and produced by corrosive attack on the surface. After a crack is formed, the corrosive environment penetrates the surface of the material. The tip of an advancing crack has a small radius and the attendant stress concentration is great. This stress at the crack tip ruptures the normally protective corrosion film and promotes corrosion (Fig. 29).

In addition to this, the formation of corrosion products by local attack in confined areas produces high stress levels in materials if the corrosion products occupy a larger volume than the metal from which they are formed. This wedging action of corrosion products in cracks has been measured to produce stresses over 34 MPa (5000 $lb_f \cdot in.^{-2}$), which aid in the propagation of the crack.

Stress corrosion cracking produces brittle failure, either intergranular or transgranular, depending on the type of alloy or the corrosive environment. In most cases, although fine cracks penetrate into the cross section of a component, the surface shows little evidence of corrosion.

To keep the stress intensity to a minimum, care must be taken to avoid stress concentrations, such as tooling marks, notches, arc strikes and large inclusions near the surface.

In advanced cases of stress corrosion cracking, the metal surface crumbles and the visual inspector can gouge out chunks.

TABLE 6. Some materials that can develop stress corrosion cracks in certain environments.[9]

Material	Environment
Aluminum alloys	$NaCl\text{-}H_2O_2$ solutions
	NaCl solutions
	seawater
	air, water vapor
Copper alloys	ammonia vapors and solutions
	amines
	water, water vapor
Gold alloys	$FeCl_3$ solutions
	acetic acid-salt solutions
Lead	lead acetate solutions
Magnesium alloys	$NaCl\text{-}K_2CrO_4$ solutions
	rural and coastal atmospheres
	distilled water
	hydrofluoric acid
	hydrofluosilic acid
Nickel	fused caustic soda
Nickel chromium steel	caustic soda solutions
Nickel copper alloy	fused caustic soda
Carbon and alloy steels	NaOH solutions
	$NaOH\text{-}Na_2SiO_2$ solutions
	calcium, ammonium and sodium nitride solutions
	mixed acids ($H_2SO_4\text{-}HNO_3$)
	HCN solutions
	acidic H_2S solutions
	moist H_2S gas
	seawater
	molten Na-Pb alloys
Stainless steels	acid chloride solutions such as $MgCl_2$ and $BaCl_2$
	$NaCl\text{-}H_2O2_2$ solutions
	seawater
	H_2S
	$NaOH\text{-}H_2S$ solutions
	condensing steam from chloride waters
Titanium	red fuming nitric acid

Hydrogen Cracking[2]

Hydrogen cracking or hydrogen embrittlement is a fracture mechanism that results from the corrosive environment produced by a hydrogen medium and usually occurs with an applied tensile stress or residual stress. Hydrogen is introduced into a material by processes such as electroplating, pickling, welding in a moist atmosphere or the melting process itself. Hydrogen may also come from corrosion or the presence of hydrogen sulfides, hydrogen gas, water, methane or ammonia.

If no crack or stress riser is present on a material surface, hydrogen can diffuse into the metal and often initiates cracks at subsurface sites, where triaxial stress is maximum. In low strength alloys, this condition can lead to what is known as *hydrogen blistering*.

If a crack is already present, it is quite common to see hydrogen induced cracking initiated at the tips of preexisting cracks.

FIGURE 29. Photomicrograph showing typical stress corrosion crack. Small pit produced by corrosive attack acts as stress riser.

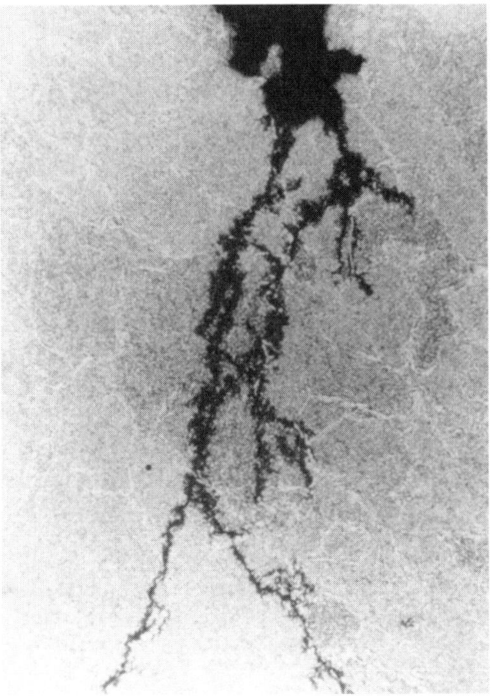

In many instances, hydrogen is already present internally in a metal before it is placed into service. Hydrogen is readily absorbed into molten metal during the initial solidification of the material and during welding processes. The solubility of hydrogen is quite high at elevated temperatures, and in some cases metals can become supersaturated with hydrogen during cooling.

Hydrogen cracking follows grain boundaries and rarely branches (Fig. 30). When such cracking results from blistering or from a static load, it always originates below the test object's surface. Hydrogen cracking from other causes can begin below the surface or at a stress riser.

Wear[9]

Wear is the undesired removal of material from contacting surfaces by mechanical action. Although typically not as serious as fracture, wear is expensive and is often predictable. Contacting surfaces are expected to wear in any machine. In many cases the deterioration can be minimized by lubrication, oil filtering, materials engineering and proper design, among other measures.

In many respects, wear is similar to corrosion. Both have many types, of which two are usually occurring simultaneously. Both are somewhat predictable in stable environments. Both are extremely difficult to evaluate in accelerated laboratory or service tests, with rankings of materials subject to change depending on seemingly minor changes in the test conditions. Finally, wear and corrosion both are of enormous economic importance.

In visual testing, it is necessary to understand the history and operation of the mechanism involved. In many cases, it is not possible to conduct a complete investigation simply by visually testing the worn component. Wear is a surface phenomenon that results from the interaction of other components and materials that also must be studied.

FIGURE 30. Photograph of hydrogen cracking found in heat affected zone next to weld.

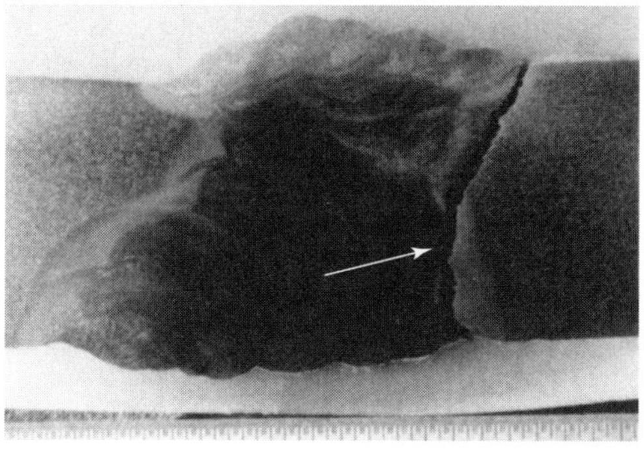

Abrasive Wear

Abrasive wear is characterized by cutting when one surface rolls or slides under pressure against another surface. Machining could be considered abrasive wear except that it is intentional. Another very important characteristic of abrasive wear is the heat generated by friction between the two materials.

Erosive Wear

Erosive wear (or erosion) occurs when particles in a fluid or other carrier slide and roll at relatively high velocity against a surface. Each moving particle contacting the surface cuts a minute particle from the surface. An individual particle is insignificant, but removal of many particles is erosion.

Erosive wear can be marked during visual tests by any of the following conditions, depending on the test object.

1. A general removal of soft surface coatings or material is a common form of wear for fan and propeller blades. In automotive applications, for example, the paint on the trailing end of the concave side of the fan blade is usually removed by the scouring or cutting action of dust and dirt particles in the air.
2. Grooves or channels eroded in the test surface are common in assemblies that move liquids or gases where the design of the component is such that the fluid flows faster or in a different direction at certain locations. Examples include impellers with vanes that push particle laden fluids into various passages. The inside of tubes or pipes is often damaged at curves because the inertia of the particles and the fluid forces them against the outside of the curve. Sudden, sharp curves or bends cause more erosion problems than gentle curves. In textile machinery, high velocity thread or yarn can cause erosion when a sudden change in direction causes grooving in an eyelet. Grooving and channeling are also quite common in various types of nozzles where high speed or high pressure fluids scour through the metal. Drops of liquid can lead to erosive wear, as is frequently seen on the leading edges of high speed aircraft.

3. Erosive wear can make corners rounded. Erosive wear can change the shape of impellers, turbine blades and vanes in ways that substantially impair operating efficiency.

Erosive wear can be expected in metal assemblies such as pumps and impellers, fans, steam lines and nozzles, inside sharp bends in tubes and pipes, sand and shot blasting equipment and similar areas where there is considerable relative motion between the metal and the particles.

Grinding Wear

Grinding wear occurs primarily when particles under high pressure cut small grooves at low speed across a metal surface. High pressure, low speed operation is characteristic of tillage tools (plows, cultivators, rakes) and other ground contact components such as bulldozer track shoes and the cutting edges of blades. In many other industries, similar effects are produced on metals, tending to change their shape and to dull cutting edges, consequently lowering efficiency of operation.

Grinding wear can be recognized in visual tests if the service environment is known and if the wear occurs at high stress locations (particularly points and edges), causing a general change in shape. When two hard metal surfaces slide against each other, frequently with a lubricant, each tends to smooth the other, particularly if fine abrasives are present. When properly controlled, this process may be useful for lapping or polishing. Hard facing (by welding, metal spraying or other means of deposition) is frequently used to improve grinding wear resistance.

Slight, controlled grinding wear can be an advantage for self sharpening of cutting tools. By judicious use of the rat's tooth principle, hardened and soft surfaces may be used together to keep an edge. The front teeth of all rodents have a hard, brittle enamel on the front convex surfaces but relatively soft dentine on the rear concave surfaces (Fig. 31a). When the animal uses its teeth, the dentine on the rear of the teeth wears away, leaving a thin, sharp edge of enamel. The tip of the brittle enamel eventually breaks off, keeping the teeth the proper length.

This same principle can be applied to certain cutting tools. For example, the cutting edges of plowshares can be made self-sharpening if the front surface is soft and the rear surface is faced with a hard material. As the plowshare cuts through the soil, the relatively soft steel on the forward, high stress side is slowly worn away. The hard facing applied to the rear, low stress side is continually exposed at the tip and a sharp edge is retained (Fig. 31b). In the mining industry, digging tools are sometimes hard faced on one side to be self-sharpening.

Gouging Wear

Gouging wear is caused by high pressure impact that lifts large fragments from a metal surface. Gouging is encountered in the fields of earthmoving, mining, quarrying, oil well drilling, steelmaking, cement and clay product manufacture, railroading, dredging and lumbering. When hard, abrasive products are crushed, battered or pounded under high pressure, rapid deterioration of the contact surfaces can be expected unless specific steps are taken to prevent the problem.

Remedies for gouging, as with other types of wear, usually are chosen on the basis of a combination of economics, availability, accessibility and design. Frequently, there are several ways to improve a product's wear resistance; the one is chosen that provides optimum properties at lowest cost.

FIGURE 31. Diagram of self-sharpening components: (a) rodent tooth; (b) self-sharpening plowshare.

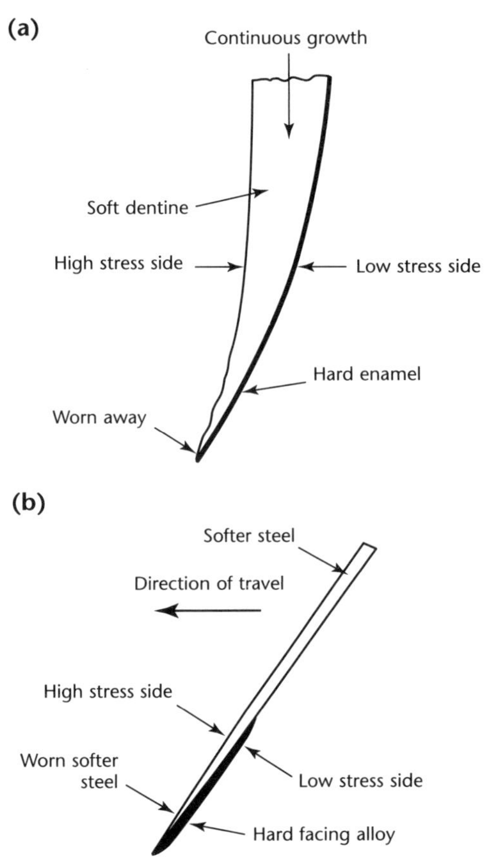

Adhesive Wear

Adhesive wear can be characterized as microwelding. Other terms are sometimes used, including *scoring, scuffing, galling* and *seizing*, but *adhesive wear* is preferred because it accurately characterizes the phenomenon when two surfaces slide against each other, with or without lubricant. When a rough peak (or asperity) from one surface comes in contact with a peak from the other surface, there may be instantaneous microwelding caused by frictional heat.

Figure 32 is an exaggerated view of two surfaces sliding against each other. They may or may not be separated by lubricant. Continued sliding fractures one side of the welded junction, making the asperity on one side higher and the asperity on the other side lower than the original height. The higher peak is now available to contact the peak on the opposite side (Fig. 32).

The peak may either be fractured by the new contact or rewelded on the opposite side and the cycle repeated. In either case, adhesive wear frequently starts out on a small scale but rapidly escalates as the two sides alternately weld and tear metal from each other's surfaces. Also, if a lubricant is present, the debris may be carried to other points on the surfaces. In severe adhesive wear, the debris is composed of free metallic particles. In mild cases, much finer particles may react with the environment to form debris, free particles of metal oxide.

The heat generated by local friction is high enough to cause microwelding, and the temperature is also high enough to cause localized heat treatment of the surface metal. Adhesive wear is similar to grinding burn: both can cause tempering of the subsurface metal and actual rehardening of steel microstructures. This produces white, untempered martensite, extremely susceptible to cracking because of its brittleness. Such cracks can lead to brittle fracture or fatigue fracture, depending on the test object and the application.

A practical way of checking for adhesive wear in hardened steels is with etching, where etching solutions give a high contrast, nondestructive means of checking for adhesive wear and grinding burn. Metallography on a cross section of the surface also can reveal evidence of adhesive wear when studied at moderate to high magnification, depending on the thermal damage.

Fretting Wear

Fretting wear is similar to adhesive wear in that microwelding occurs on mating surfaces. The difference is that adhesive wear is related to moving surfaces and fretting wear is related to stationary surfaces. However, when minute elastic deflections or slight motion does occur, a cyclic motion of extremely small amplitude is enough to cause microwelding on both surfaces. Fretting wear is also known as *fretting corrosion, false brinnelling, friction oxidation, chafing fatigue* and *wear oxidation.*

Fretting frequently occurs in stationary joints that are fixed from shrinking or pressing by interference fits or by bolts, pins, rivets or other mechanisms and also at various contact points in antifriction or rolling elements. This means that nonrotating antifriction bearings are subject to vibration over a period of time and may have fretting wear wherever a ball or roller contacts a raceway under load. If the bearings subsequently rotate in normal service, they may be noisy because of the wear patterns and small indentations present in the raceways and the corresponding flat spots on the rolling elements.

Fretting also is a serious problem on components such as shafts, where it can initiate fatigue cracking on the contacting

FIGURE 32. Illustration of one process by which particle of debris is detached during adhesive wear: (a) bonded junction forms; (b) junction is torn from one peak or asperity; (c) asperity then is sheared off by impact with larger, adjacent peak.

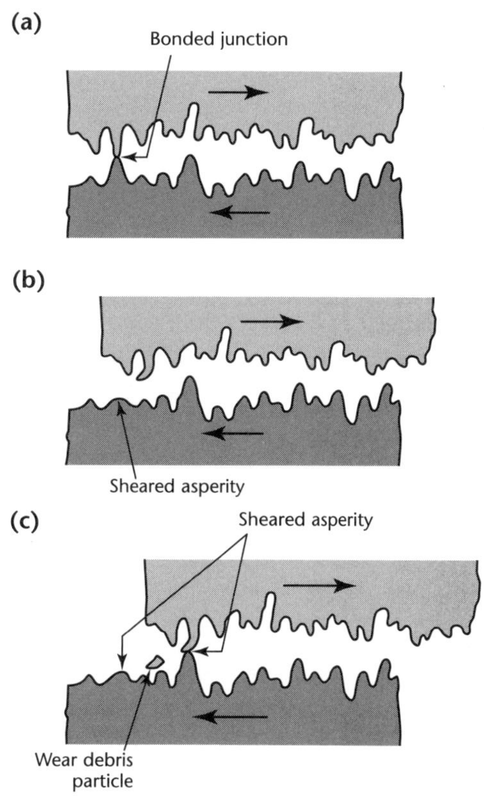

surfaces. In fact, many fatigue fractures of shafts are caused directly by fretting. Because fretting is extremely difficult to prevent, special means must be taken to prevent fracture resulting from fretting, which can occur in unexpected locations.

Because fretting wear is essentially a stationary phenomenon, its debris (usually oxides of the contacting metals) is retained at or near the site of its formation, a condition especially helpful during visual tests. With ferrous metals, the debris is brown, red or black, depending on the type of iron oxide formed. For this reason, ferrous debris is called *cocoa* or, when mixed with oil or grease, *red mud*. Aluminum alloys form a black powder when fretting wear is present.

Fatigue Cracking[2]

Fatigue is a fracture mechanism induced by a cyclically applied stress that is lower in magnitude than the ultimate tensile strength of the material but high enough to initiate a crack or to propagate a preexisting crack. Fatigue cracks can develop from stress risers such as sharp radii, nicks, machining or tooling marks, nonmetallic inclusions present at or near the material surface, pores, holes or notches, keyways and may even develop on a smooth surface (Fig. 33). Surface anomalies such as copper penetration contribute to fatigue cracking. Fatigue cracking typically occurs at the surface.

As a fatigue crack begins to propagate, the stress intensity at the tip of the crack starts to increase. With every incremental growth period of the crack, there is a proportional, incremental increase in the stress intensity. This process continues until the stress intensity K reaches the critical value, where failure occurs. This critical value is known as the *fracture toughness* and is unique for each material. The variance in fracture toughness helps explain the behavior of fatigue cracks: why there is such a rangc of fatigue crack sizes, why some cracks may only propagate a small amount and why others propagate nearly all the way through.

Fatigue Crack Structure

Internally a fatigue crack has unique characteristics but on the test surface resembles any other crack. Fatigue cracks normally originate on the surface but can begin below the surface at discontinuities if the applied and residual stresses exceed the subsurface fatigue strength of the material. Proper care in machining is necessary to ensure that no unanticipated stress risers are introduced. Additional fatigue resistance can be gained by stress relieving a component or by shot peening to introduce a compressive stress on the object's surface.

Elevated Temperature Discontinuities[9]

Elevated temperature discontinuities are the most complex kind of material anomaly because most other discontinuities can occur at elevated temperatures (low temperature brittle fracture is an obvious exception). Elevated temperatures greatly complicate the analysis of the problem and the possible solutions. Figure 34 shows a test object with an extreme example of creep strain, probably thermal deformation, in the alloy's range for plastic deformation.

FIGURE 33. Fatigue cracking in automobile crankshaft, enhanced here with magnetic particles.

FIGURE 34. Plastic deformation.

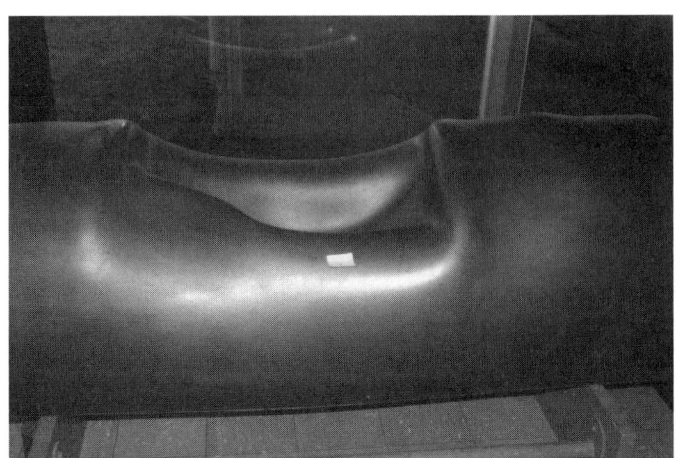

Normally, the useful static strength of a metal is limited by its yield strength. However, as temperature increases, the useful static strength of a metal is limited by the factor of creep, a time dependent strain occurring under stress. Each metal or alloy must be considered individually because of differences in their properties. Approximate thresholds of elevated temperature behavior for several metals and alloy systems are shown in Table 7.

In service at elevated temperature, the life of a metal component is predictably limited, whether subject to static or to dynamic loads. In contrast, at lower temperatures and in the absence of a corrosive environment, the life of a component in static service is unlimited, if the operational loads do not exceed the yield strength of the metal.

The principal types of elevated temperature discontinuity are creep, low cycle fatigue, high cycle fatigue, thermal fatigue, overload failure and combinations of these, as modified by the service environment. Generally, the type of discontinuity is identified by (1) visual testing of fracture surfaces and (2) comparison of operating conditions with data on creep, stress rupture, tension, elevated temperature fatigue and thermal fatigue properties. More thorough analysis may be required when stress, time, temperature and environment change the metallurgical microstructure of the test object.

Creep Cracking[2,9]

At temperatures greater than half the melting point in celsius and at stresses below the yield strength of the material, deformation can occur by the action of grains gradually separating over an extended period of time. This can eventually lead to cracking and finally to failure. This deformation or failure mechanism is called *creep*. By definition, *creep* means the gradual change of shape in a metal under stress. It is a result of tensile stress but creep can and does occur under all types of stress. Gradual change of shape under compressive, torsion bending and internal pressure stresses may or may not lead to fracture.

In addition to alloy selection, heat treatment has an effect on creep properties. Heat treatment generally controls grain size, and a coarser grain at elevated temperatures has higher creep strength than a finer grain. Depending on the alloy, creep fracture may be macroscopically brittle or ductile. In general, lower creep rates, longer rupture times or higher temperatures promote intergranular fracture.

Under certain conditions, some metals may not exhibit all stages of plastic deformation. For example, at high stress temperatures, creep follows loading. At the other extreme, notably in cast alloy, creep may be observed and fracture may occur with minimum extension. Because materials can be subjected to such a variety of loads and temperatures for a particular application, the type of heat treatment should be based on the degree of stability that it imparts to the component initially and throughout its service life.

Creep can be detected and controlled. Periodic tests, particularly those involving field metallography and circumferential measurement can be used to monitor the creep process. By slightly decreasing operating temperature or stress, a substantial decrease in the creep rate yields greater service life.

Thermal Fatigue

Fatigue may be caused either by cyclic mechanical stressing or by cyclic thermal stressing. Thermal fatigue cracks are the result of repeated heating and cooling cycles, producing alternate expansion and contraction. When a metal cools, it contracts, causing residual tensile stresses if restrained from free motion. If this alternate expansion and contraction continues, fatigue cracks form and propagate each time the metal is cooled.

TABLE 7. Approximate thresholds for elevated temperature behavior.[9]

Metal	Temperature (°C)	Temperature (°F)
Aluminum alloys	205	400
Titanium alloys	315	600
Carbon steels	370	700
Low alloy steels	370	700
Austenitic, iron base high temperature alloys	540	1000
Nickel base high temperature alloys	650	1200
Cobalt base high temperature alloys	650	1200
Refractory materials	980 to 1540	1800 to 2800

Thermal cycles may be caused by friction, as in brake drums and clutch plates. Here the surface is frequently heated and expanded by friction but is prevented from expanding freely by the colder, stronger metal below the surface. Compressive yielding occurs in the hot surface layer, causing tensile residual stresses when the metal contracts during cooling. This condition frequently causes thermal fatigue cracks called *heat checking*. This network of cracks on the friction surface may be harmless unless the cracks wear the mating surface or unless the cracks progress to complete fracture.

Engine exhaust manifolds also are subject to thermal fatigue, particularly on heavy duty engines. They may become very hot under certain conditions, then cool when the engine is stopped. If the manifold is not permitted to float or move freely in an axial direction, tensile residual stress may be generated when it cools, eventually causing fatigue fracture.

Thermal fatigue may be prevented in many components by designing curves rather than straight lines into the system. When this is done, heating and cooling cycles simply distort the curves, rather than forming tensile residual stresses on cooling. Expansion loops, bellows in elevated temperature piping and tubing systems operate on this principle.

Metallurgical Instabilities

Stress, time, temperature and environment may act to change metallurgical structures during service, resulting in reduced strength. These microstructural changes are referred to as *metallurgical instabilities*. Sources of instabilities include transgranular or intergranular fracture transition, recrystallization, aging or over aging, intermetallic phase precipitation, delayed transformation to equilibrium phases, order to disorder transitions, general oxidation, intergranular corrosion, stress corrosion cracking, slag enhanced corrosion and contamination by trace elements.

Environmentally Induced Discontinuities

The most important source of elevated temperature discontinuities is environmental degradation. Control of environment or protection of materials is essential to most elevated temperature applications.

General oxidation can lead to premature failure. Grain boundary oxidation may produce a notch effect that also can limit service life. Some environments may be more harmful than others. For example, attack of fireside surfaces of steam boiler tubes by ash from vanadium bearing fuel oils can be quite severe. Vanadium ash attack and hot corrosion in general are equally harmful in gas turbines.

In all elevated temperature discontinuities, the characteristics of the environment must be carefully considered. These include not only the temperature itself but also whether the elevated temperature is steady or fluctuating, the rate of temperature change (which affects differential expansion and contraction), the thermal conductivity of the metals involved, the characteristics of the fluids (both liquid and gases) in contact with the surfaces and the way in which the fluids contact the metal surfaces. Fluid contact is most important in components that have high gas or liquid flow rates at elevated temperatures, causing erosion problems.

Corrosion and Corrosion Erosion

Certain components function in environments where high rates of fluid flow at high temperature are normal. Typical of those in gaseous environments include engine exhaust valves, blades and vanes in the hot sections of gas and steam turbine engines or generators, certain locations (particularly inlets and outlets) in various furnaces and ducts or pipes that conduct hot gases. Typical components in high temperature liquid environments are piping systems, pumps, rotors, propellers and nozzles.

The problem with these components is that the combination of high temperature and high velocity fluid flow often results in erosive wear at critical locations. Such wear frequently destroys the efficiency of the components and their assemblies. Erosive wear is caused by high speed, low stress particles that tend to cut or erode materials in their path. In general, elevated temperatures reduce metal strength and hardness. Any component that changes the direction of high temperature, high velocity fluids is subject both to increased erosion from the fluid's mechanical action and to increased corrosion from the fluid's chemical action.

General Oxidation

In certain applications, the primary elevated temperature problem is general oxidation or scaling (formation of metal oxide layers). This is particularly true when the metal is subjected to repetitive heating and cooling cycles in an oxidizing atmosphere. The oxide scale flakes off when the metal cools because of differences in the thermal expansion characteristics of the scale and the base metal.

As a group, ferritic stainless steels are usually superior in oxidation resistance when compared to iron base alloys. In fact, the main advantage of ferritic stainless steels for high temperature use is their good oxidation resistance, comparable to austenitic grades. In view of their lower alloy content and lower cost, ferritic steels should be used in preference to austenitic steels when stress conditions permit. Oxidation resistance of stainless steel is affected by many factors, including temperature, time, service (cyclic or continuous) and atmosphere. For this reason, selection of a material for a specific application should be based on tests that duplicate anticipated conditions as closely as possible.

Because of the need for good oxidation resistance in automotive exhaust systems and catalytic converters, ferritic stainless steels are widely used. Under favorable conditions, these steels form a tightly adhering oxide scale that expands and contracts with the base metal and are suitable when there is no need for high strength and elevated temperatures.

Carburization

The problem of steel carburization is common to many industrial applications — especially stainless steels in furnace environments. Simultaneous carburization and oxidation of stainless steel heating elements results in a form of attack sometimes referred to as *green rot*. This discontinuity is common to nickel chromium and nickel chromium iron alloys.

Liquid Metal Contact

Liquid metal contact is another problem encountered in high and low temperature service environments. High temperature alloys often cannot tolerate contact with liquid metals because high temperatures cause the precipitation of chromium carbides in the grain boundaries. This condition, called *sensitization*, is the depletion of the surrounding areas of chromium, permitting grain boundary corrosion, cracking and fracture.

Liquid mercury can cause severe stress corrosion cracking by contact with high strength steels and with cobalt, aluminum, nickel, titanium and their alloys.

Many high temperature alloys frequently cannot be used with liquid or molten metals. Molten lead, for example, is highly corrosive to most high temperature alloys. Molten zinc, used in hot dip galvanizing of fabricated components, is commonly contained in tanks or vats made from plain carbon plate steel. Aside from strength, the principal requirement of galvanizing tank material is the ability to resist the corrosive attack of molten zinc. (Some alloying elements dissolve in the liquid metal, changing the base metal alloy.)

A common problem with molds used to die cast zinc, aluminum, magnesium and copper is heat checking or thermal fatigue cracking of surfaces in contact with the cast, liquid metal. This condition produces ridges in the casting when molten metal flows into the mold cracks. It is necessary to keep the die at a high temperature so that there is little differential expansion and contraction that can cause tensile residual stresses and cracking on its surfaces.

Cooling Methods

In gaseous flow mechanisms, it is possible to use air or other gases to cool components. This is commonly done in the hot sections of gas turbines where tremendous air flow is available. Some of the incoming air is routed through holes in the blades and vanes.

Internal combustion engines may be cooled by liquid or air. However, no cooling system can function effectively if its heat transfer properties are impaired. An effective cooling system is critical to engine operation.

References

1. Section 8, "Applications of Visual and Optical Testing in the Metals Industries": Part 1, "Physical Properties of Metals." *Nondestructive Testing Handbook,* second edition: Vol. 8, *Visual and Optical Testing.* Columbus, OH: American Society for Nondestructive Testing (1993): p 193-227, 242.
2. Atkins, D.R., M.A. Urzendowski and R.W. Warke. Chapter 12, "Magnetic Testing of Metals." *Nondestructive Testing Handbook,* third edition: Vol. 8, *Magnetic Testing.* Columbus, OH: American Society for Nondestructive Testing (2008): p 297-321.
3. CT-6-3, *Nondestructive Testing: Magnetic Particle,* second edition. Classroom Training Handbook. Fort Worth, TX: General Dynamics, Convair Division (1977).
4. AISI SPM 3, *Plates. Rolled Floor Plates: Carbon, High Strength Low Alloy, and Alloy Steel.* Washington, DC: American Iron and Steel Institute (1997).
5. *Sheet Steel — Carbon, High Strength Low Alloy, and Alloy Steels.* Steel Products Manual. Iron and Steel Institute (1991). Superseded by *Sheet Steel: A Steel Products Manual* (2004). Warrendale, PA: The Minerals, Metals and Materials Society, Association for Iron and Steel Technology (2004).
6. *Plates: Rolled Floor Plates — Carbon, High Strength Low Alloy, and Alloy Coils and Cut Lengths.* Steel Products Manual. Iron and Steel Society (1991). Superseded by AISI SPM 3, *Plates; Rolled Floor Plates: Carbon, High Strength Low Alloy, and Alloy Steel.* Washington, DC: American Iron and Steel Institute (1997).
7. Rokhlin, S.I. Chapter 18, "Radiographic Testing of Metals": Part 3, "Discontinuities in Welds." *Nondestructive Testing Handbook,* third edition: Vol. 4, *Radiographic Testing.* Columbus, OH: American Society for Nondestructive Testing (2003): p 478-481.
8. AWS D1.1/D1.1M, *Structural Welding Code: Steel.* Miami, FL: American Welding Society (2008).
9. Wulpi, D.J. *Understanding How Components Fail,* first edition. Materials Park, OH: ASM International (1985). Superseded by second edition (1999).

Bibliography

Aizawa, H. and K. Miyagawa. Section 8, "Applications of Visual and Optical Testing in the Metals Industries": Part 2, "Visual and Optical Testing in the Steel Industry." *Nondestructive Testing Handbook,* second edition: Vol. 8, *Visual and Optical Testing.* Columbus, OH: American Society for Nondestructive Testing (1993): p 228-243.

ASM Handbook, ninth edition: Vol. 17, *Nondestructive Inspection and Quality Control.* Metals Park, OH: ASM International (1989).

AWS Welding Handbook, ninth edition: Vol. 1, *Welding Science and Technology.* Miami, FL: American Welding Society (2001).

The Making, Shaping and Treating of Steel, eleventh edition. Warrendale, PA: AISE Steel Technology (1998).

Struk, D. *NDT in the Foundry.* Columbus, OH: American Society for Nondestructive Testing (1995).

Tool and Manufacturing Engineers Handbook: Vol. 4, *Quality Control and Assembly.* Dearborn, MI: Society of Manufacturing Engineers (1986).

9

CHAPTER

Chemical and Petroleum Applications of Visual Testing

Nat Y. Faransso, KBR, Houston, Texas

David R. Bajula, Acuren Inspection, La Porte, Texas (Part 1)

Joseph L. Mackin, Team Industrial Services, Houston, Texas (Part 4)

Roderic K. Stanley, NDE Inspection Consultants, Houston, Texas (Part 2)

Portions of Part 2 are reprinted with permission from *Welding Inspection* (1968), © American Welding Society. ASNT has revised the text in 1993 and 2010, and deficiencies are not the responsibility of the American Welding Society.

Portions of Part 2 are reprinted with permission from the *ASME Boiler and Pressure Vessel Code* (1989) and ANSI/ASME B31, *ASME Code for Pressure Piping* (1990), © ASME International. ASNT has revised the text in 1993 and 2010, and deficiencies are not the responsibility of ASME International.

Portions of Part 2 are reprinted with permission from API 620, *Design and Construction of Large, Welded, Low-Pressure Storage Tanks* (1990), and API 650, *Welded Steel Tanks for Oil Storage* (1988), © American Petroleum Institute. ASNT has revised the text in 1993 and 2010, and deficiencies are not the responsibility of the American Petroleum Institute.

PART 1. Chemical and Petroleum Industry

Petroleum Industry[1]

Oil refining began in the nineteenth century with the production of kerosene and lamp oil. In the twentieth century, gasoline was needed to fuel the combustion engines in automobiles. Refining and chemical plants are vital to the oil and gas industries, producing much needed resources for the world. Modern refineries and petrochemical plants are highly dependent on crude oil and other fossil fuels to produce a wide range of chemicals and products.

Petrochemical plants and refineries are comprised of processing units ranging from simple distillation towers to complex fluid catalytic cracking, hydrodesulfurization units, cokers and other processing units. Process components such as vessels and piping are subject to many forms of service related material degradation. Corrosion and cracking if left unchecked could cause catastrophic failures and loss of life. The history of catastrophic failures in the refining, chemical and other industrial complexes has led to federal regulation of process safety.[2] Part of the regulatory document covers mechanical integrity.

Nondestructive testing is vital in ensuring the mechanical integrity and serviceability of this equipment. The chemical industries use many nondestructive test methods to maintain and ensure safe operation of their production facilities. Nondestructive testing is important early, during fabrication; later, during maintenance and servicing of vessels and piping; and finally, in onstream inspection. The goal of such testing programs is to achieve and maintain capacity production. Owner/user teams and engineering staff use nondestructive test results to plan their maintenance activities, to assess their risks and to implement risk based inspection. Risk based inspection philosophies prioritize inspections, minimize failures and maximize the performance of equipment.

Test Procedures[1]

Visual testing is an important nondestructive testing method for the chemical and petroleum industry. When using nondestructive methods, chemical and petroleum industry inspectors will give consideration to the following inservice concerns.

1. What is the fabrication and service history of the tested parts? Has the equipment, component or material had any previous fabrication requirements or nondestructive test requirements?
2. What requirements does the customer's inspection contract specify? Are there preexisting manufacturer's or purchaser's specifications to a national or local code?
3. What visible indications correspond to material conditions of interest?
4. Is the surface visible? Does corrosion, scale, flaking or coating need to be removed? What is the surface condition before and after preparation for testing?
5. How will the test object be used?

When establishing a procedure, it can be difficult to correlate the materials with original procurement specifications. In many cases, the acceptance criteria exceed those originally specified. In the resulting confusion, high quality materials may be rejected and low quality materials could be accepted. Accurate procedures are established through reference standards with artificial discontinuities.

Tube testing falls into two categories: ferrous and nonferrous. Ferrous metals include carbon steel, stainless steel and metals with similar magnetic properties; nonferrous metals are nonmagnetic and include copper, brass, nickel and austenitic stainless steel. The choice of test method is mainly influenced by the type of service damage that needs to be detected, but often the technique is dictated by tube cleanliness. Several damage mechanisms and discontinuities can occur, on either the inside or outside diameter surface; other discontinuities are volumetric and not connected with either surface. However, the primary discontinuities typically break either the outside or inside surface.

Pressure vessels are continually subject to testing and are considered one of the most critical pieces of equipment in a petrochemical plant or refinery. Traditional inservice tests include visual testing. Industry practices for inservice tests of pressure vessels call for visual, ultrasonic, electromagnetic and wet fluorescent magnetic particle testing.

Pipelines connect field production (gas and oil extraction) with refineries and petrochemical plants where gas and crude petroleum are processed into usable products (Fig. 1). Because pipelines cross state lines in the United States, they are governed by the Department of Transportation. The construction, maintenance and testing of these pipelines are critical to the safety of the environment and the general public. Buried pipelines not only have the potential for catastrophic failure but could contaminate lakes, rivers and underground water sources if leakage occurs. Damage to environment and health can result also from leakage or failure of aboveground storage tanks (Fig. 2).

Traditional preservice tests include radiographic and ultrasonic testing during fabrication to ensure the quality of the welding. Once a pipeline is in service, the pipeline companies depend largely on inservice testing to assess corrosion.

Tasks of Visual Testing

Virtually all written inspection procedures call for visual testing and often call it by terms other than *visual testing.* The standard instead may say that a particular component should be inspected or that the inspector should verify certain details of fabrication. Such directions are calling for visual testing. The written procedure may assign the following sorts of tasks to visual testing.

1. Before joining or painting, a surface is visually inspected for cleanliness. In weld end preparation, bevels, alignment and faying surfaces must meet specification requirements.
2. Dimensional gaging is not considered nondestructive testing but is often assigned to the visual inspector, who uses an array of checklists, calipers, gages and other devices to ensure that weldments and other components including fasteners are fabricated and installed according to specification.
3. The visual inspector is expected to detect surface conditions, such as corrosion or deformation, that may lead to component failure. Components of concern can be identified and logged in an inspection schedule so that processes such as strain, erosion and corrosion can be tracked.
4. The visual inspector is expected to detect apparent surface indications, such as cracks and oil stains, that may be caused by subsurface discontinuities. The inspector then recommends further inspection by one or more other test methods to locate and size discontinuities and to evaluate the component's material condition (Fig. 3).

FIGURE 1. Carbon steel, 0.75 m (30 in.) outside diameter, gas transmission pipeline.

FIGURE 2. Aboveground storage tank for petroleum products.

FIGURE 3. Compared to visual tests, ultrasonic tests have advantages for detection of subsurface fatigue cracking at base of slip marks in walls of drill pipe.

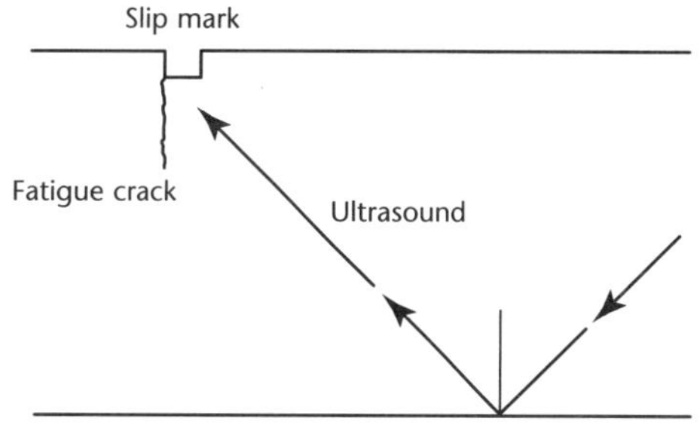

5. Because of shared concerns about vision acuity and the appearance of surface indications, the visual inspector may work closely with inspectors who evaluate surface breaking discontinuities by other nondestructive test methods, such as liquid penetrant and magnetic particle testing. The visual inspector may use a photometer to verify adequate illumination, for example, or may administer vision examinations.
6. The visual inspector may work with an array of devices and machines for indirect or remote viewing of interior surfaces. These techniques could be as simple as rigid borescopy to look inside a valve or as sophisticated as guiding a robotic crawler and receiving images from its video camera, many kilometers distant.
7. The visual inspector may need to maintain an archive of inspection images (photographs and video files) for training or quality assurance.

Like other surface methods, visual testing is subjective, or qualitative: the diagnostics rely on the training and experience of the inspector. Qualitative tests are followed by objective methods that provide quantitative information about discontinuity dimensions, such as crack depth or wall thickness. Accept/reject decisions are reached by applying specific criteria.

Part 2. Visual Acceptance Criteria for Welds[3]

The following information is provided for instructional purposes only and includes information from various standards. A source document for each section is indicated by the reference appended to its heading.

Those who perform work according to code or contract need to refer to the applicable published standard or specification and identify the edition in their written procedure.

Other anomalies in metals are discussed in the chapter on visual testing of metals.

Weld Discontinuities[4]

A discontinuity is an interruption of the continuity of a typical structure, such as heterogeneity in the mechanical, metallurgical or physical characteristics of a material. A discontinuity is not necessarily a defect, but all defects are discontinuities. Discontinuities associated with welds may be divided as follows: (1) dimensional, (2) process, (3 mechanical or chemical and (4) base metal properties. Dimensional discontinuities include distortion; incorrect weld size, profile or proportions; and excess weld reinforcement. Process discontinuities include porosity, inclusions, incomplete fusion, inadequate joint penetration, undercut, cracks and surface irregularities.

Dimensional Anomalies

Although dimensional checks are usually considered to be distinct from nondestructive tests, in practice there is often a need to consider them together. Dimensional characteristics are directly related to the material and mechanical quality of a test object and to the very sorts of performance, such as service life, that nondestructive testing is used to evaluate. For these reasons, some welding situations call for dimensional gaging, subjective dimensional visual tests and conventional nondestructive tests. Welding typically involves the application of heat and the melting of metal. Stresses of high magnitude result from thermal expansion and contraction during welding. Some stresses may remain in the weldment after cooling and can cause discontinuities such as distortion (Fig. 4). The size of a normal equal leg fillet weld is expressed as the leg length of the largest isosceles right triangle that can be inscribed within the fillet weld cross section. The size of a groove weld is the joint penetration (depth of the groove face plus the root penetration when specified). Welds that are not adequate in size may be detected visually using a weld gage or by comparison with approved reference standards. The profile of a finished weld may have considerable effect on its performance under load. In addition, the profile of one pass in a multipass weld may increase the tendency toward certain discontinuities (incomplete fusion or slag inclusions) when subsequent layers are deposited. Specific requirements concerning the acceptability of discontinuity types are usually included in the specifications. Overlap is another dimensional anomaly — weld metal is deposited beyond the weld toe (Fig. 5) in butt joints or is not fused into the base

FIGURE 4. Weld alignment: (a) incorrect angular alignment; (b) correct alignment using proper control methods.

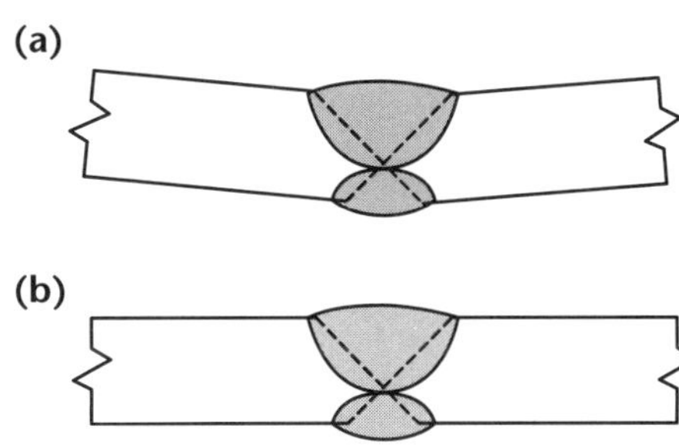

FIGURE 5. Weld cold lap.

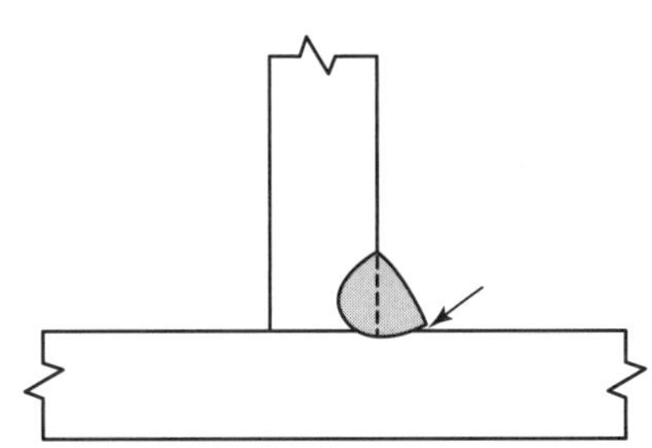

metal in fillet welds. Overlap tends to produce notches that serve as stress concentrators under load. Overlap can also occur at the toe of a completed weld's reinforcement.

Visual Testing Acceptance Criteria

Power Boilers[4]

Surfaces to be welded should be free of paint, oil, rust, scale or other foreign materials before welding. Cleanliness is verified by visual inspection before welding. Offset tolerances of butt welded edges are standardized.[5] Any offset within the allowable tolerance is faired at a three-to-one taper over the width of the finished weld or, if necessary, by adding additional weld metal beyond the edge of the weld.

The thickness of reinforcement on each face of a weld should not exceed specified values. The root opening of joints should be made within weld procedure tolerances.

The discontinuities found by visual testing of power boiler welds include those shown in Table 1.[6]

Pressure Vessels and Storage Tanks[7]

The edges of butt joints are restrained during welding so that the maximum offset[8] is not exceeded in the completed joint.

When fitted girth joints have deviations exceeding permitted tolerances, the head or shell ring, whichever is out of true alignment, is refit, reworked or reformed until the alignment is within the specified limits. When fillet welds are used, the lapped plates must fit closely and are kept in contact during welding. For all fillet welds, leg dimensions and theoretical throat dimension must lie within the cross section of the deposited weld metal (Fig. 6).

Attachments such as lugs, brackets, saddle nozzles, manhole frames and reinforcements around openings must fit reasonably well to the curvature of the shell or surface to which they are attached. When pressure components such as saddle nozzles, manhole frames and reinforcement around openings, extend over pressure retaining welds, the portion of the weld to be covered is ground flush with the base metal and should be visually tested before being covered. When nonpressure components such as lugs, brackets and supports, extend over pressure retaining welds, the welds are ground flush or the components may be notched or coped to clear the welds. For circumferential head-to-shell butt welds, the length of the taper does not extend beyond the tangent line of the head and the misalignment of the centerlines of the shell and head should be no greater than half the difference in the thicknesses of the two.

The thickness of the weld reinforcement on each face of the weld should not exceed the specified values.

Typical discontinuities found by visual testing of welds in pressure vessels and storage tanks are shown in Table 2.[6]

FIGURE 6. Fillet welds with unequal legs: (a) convex; (b) concave.

(a)

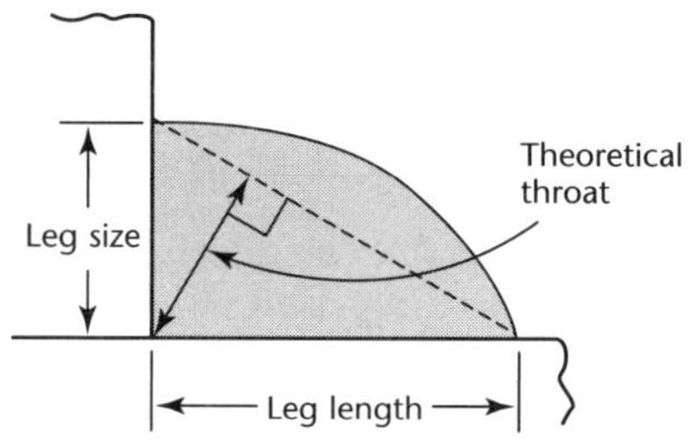

(b)

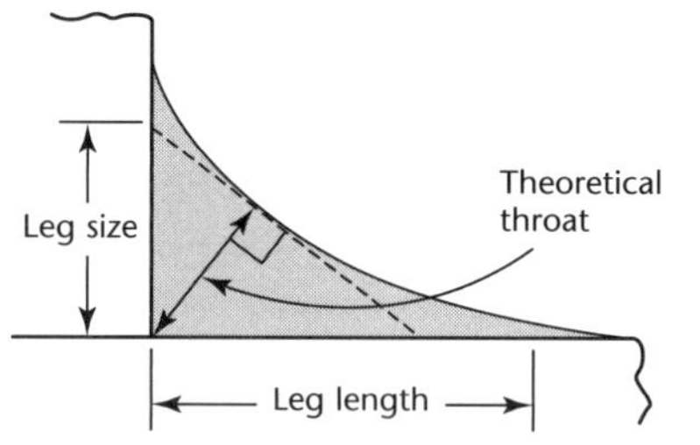

TABLE 1. Typical disposition of weld discontinuities detected by visual testing of power boiler welds.[6]

Discontinuity	Limitations
Incomplete penetration	none permitted
Lack of fusion	per specification
External porosity	none permitted
External inclusions	none permitted

TABLE 2. Typical weld discontinuities located by visual testing of pressure vessels and storage tanks.

Discontinuity	Limitations
Cracking	none permitted
Incomplete penetration	none permitted
Lack of fusion	none permitted
Undercut in vertical joints	0.4 mm (0.016 in.)
Undercut in horizontal joints	0.75 mm (0.033 in.)
External porosity	none permitted
External slag inclusions	none permitted
Concave root surface	weld as thick as base metal

Pipe and Socket Welds[7]

For assembly of a joint before welding, a pipe or tube may be inserted into a socket to the maximum then withdrawn about 1.5 mm (0.06 in.) away from contact between the end of the pipe and the shoulder of the socket. In sleeve joints without internal shoulders, there is typically a distance of about 1.5 mm (0.06 in.) between the butting ends of the pipe or tube.

The inside diameters of piping components should be aligned as accurately as practical within existing commercial tolerances on diameters, wall thicknesses and out-of-roundness. Where ends are to be joined and the internal misalignment exceeds 1.5 mm (0.06 in.), the component with the wall extending internally may be internally trimmed (Fig. 7) so that adjoining internal surfaces are nearly flush. However, such trimming should not result in a piping component wall thickness less than the minimum design thickness, and the change in contour should not exceed 30 degrees.

Fillet pipe welds may vary from convex to concave (Fig. 6).

Figure 8 shows a typical branch connection with and without added reinforcement. Various types of weld attachments are used in the fabrication of branch connections, and branch connections may be made by welding half couplings or adapters directly to the run pipe. The cover fillet welds should have minimum throat dimensions not less than that indicated in the applicable standard. The typical reinforcements of girth and longitudinal butt welds are standardized, relative to the temperatures for which the component was designed. Discontinuities detected with visual tests may include cracking, undercut, lack of fusion, external slag, porosity, incomplete penetration and concave root surface.

Girth and Miter Joints[9]

If component ends are trimmed for fitting backing rings or consumable inserts or as shown in Figs. 7 and 9 for correcting internal misalignment, such trimming may not result before welding in a wall thickness less than minimum wall thickness t_m. Where necessary, weld metal may be deposited on the inside or outside of the component to provide alignment or sufficient material for machining to ensure satisfactory seating of rings or inserts. It is also permissible to size pipe

FIGURE 7. Joint trimming and permitted misalignment in butt welds. Thicker pipe is bored for alignment.

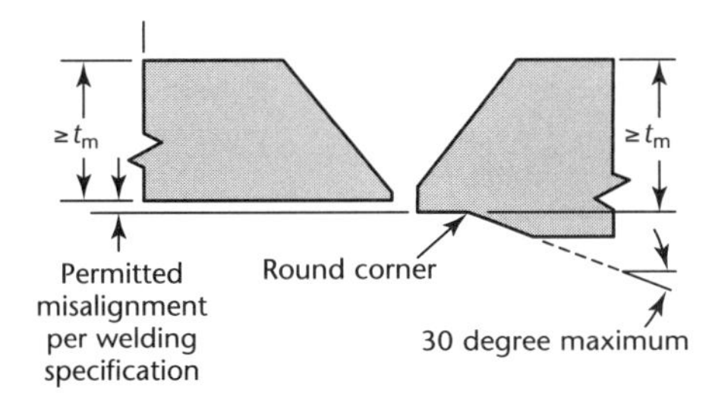

FIGURE 8. Typical welded branch connection: (a) without additional reinforcement and (b) with additional reinforcement.

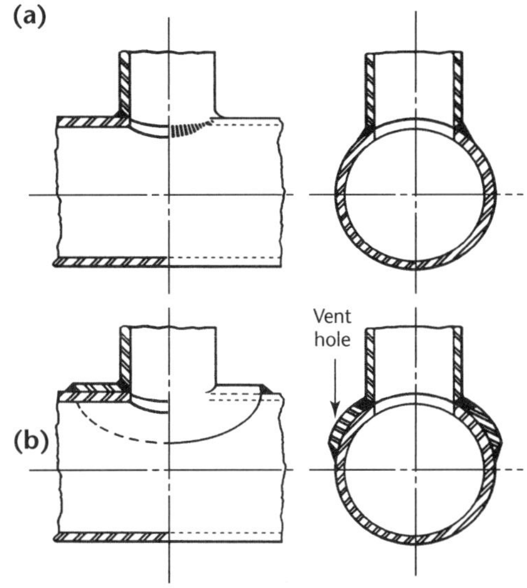

FIGURE 9. Typical backing rings and inserts: (a) butt joint with bored pipe ends and solid or split backing ring; (b) butt joint with taper bored ends and solid backing ring.

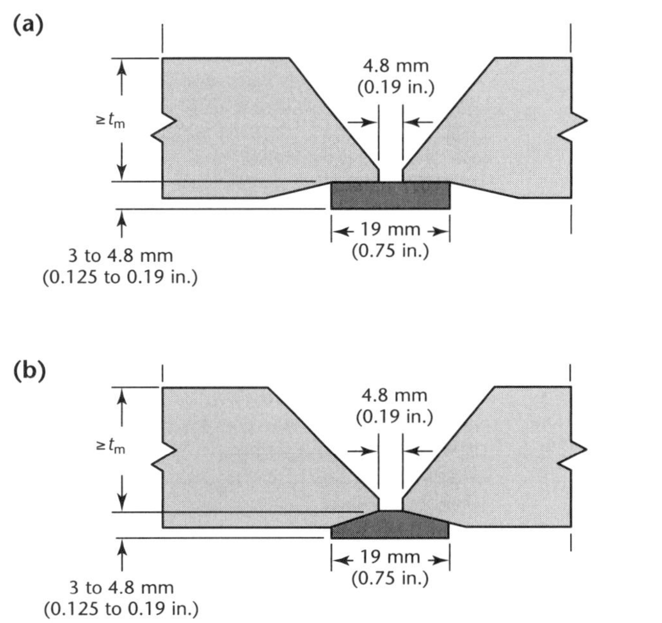

ends of the same nominal size to improve alignment, if the above wall thickness requirements are maintained.

For alignment of girth and miter joint butt welds, inside diameters of components at the ends to be joined are aligned within the dimensional limits in the welding procedure and the engineering design. If the external surfaces of the two components are not aligned, the weld is tapered between the two surfaces. Preparation for longitudinal butt welds shall conform to the requirements of the welding specification. Branch connection welds that abut the outside surface of the run wall are contoured to meet the welding specification requirements.

Fillet welds (including socket welds) and slip-on flanges should be welded according to the applicable standard.

Transmission Pipe Welds[10]

End preparations and flange welds for butt welding of pieces having unequal thickness or unequal yield strength are detailed in applicable standards. ASME B 31.8 is given here as an example.[10] Branch connections with fillet welds must meet or exceed minimum dimensions.

Branch connections are attached by a weld for the full thickness of the branch or header wall plus a fillet weld. Concave fillet welds are preferred to further minimize stress concentration at the toe. Ring or saddle reinforcements are attached as shown in ASME B 31.8.[10] When a full fillet is not used, it is recommended that the edge of the reinforcement be relieved or chamfered at about 45 degrees to merge with the edge of the fillet. Discontinuities found by visual testing of pipe welds are listed in Table 3.[6]

Storage Tank Welds[11]

The thickness of the weld reinforcement on each face of the weld should not exceed the values specified. Nor should the thickness of the reinforcement on each side of a welded storage tank plate exceed the values specified. Discontinuities found by visual testing of storage tank welds are shown in Table 2.[6]

The maximum out of plumbness of the top of the shell relative to the bottom of the shell must not exceed 0.5 percent of the total tank height. For roundness, radii measured at 0.3 m (1.0 ft) above the bottom corner weld do not exceed specified tolerances. With a horizontal sweep board 0.9 m (3.0 ft) long, peaking does not exceed 13 mm (0.5 in.). With a vertical sweep board 0.9 m (3.0 ft) long, banding does not exceed 13 mm (0.5 in.).

Storage tanks are addressed also in other standards issued by the American Petroleum Institute.[12,13]

Table 3. Typical pipe weld discontinuities located by visual testing.[6]

Discontinuity	Limitations
Cracking	none permitted
Incomplete penetration	as specified for component
Lack of fusion	none permitted
Undercut	minimum dimensions as specified
External porosity	none permitted
External slag inclusions	none permitted
Concave root surface	as specified for component
Weld reinforcement	as specified for component

Pipelines[14]

Typically a weld must be free of cracks. However, shallow crater or star cracks located at the stopping point of weld beads resulting from weld metal contraction during solidification are not considered injurious discontinuities unless their length exceeds 4 mm (0.16 in.). At no point should the crown surface be below the outside surface of the pipe, nor should it be raised above the parent metal by more than 1.5 mm (0.06 in.). Two beads are not started at the same location. The face of the completed weld is about 3 mm (0.125 in.) greater than the width of the original groove. When visual and mechanical means are used to determine depth, undercutting adjacent to the cover or root bead cannot exceed the values specified.

Note that when both mechanical and radiographic measurements are available, the mechanical measurements are used to determine if the undercut is acceptable or rejectable. Inadequate penetration of the weld root should not exceed 25 mm (1 in.). The total length of all occurrences of such a condition in any continuous 300 mm (12 in.) length of weld cannot exceed 25 mm (1 in.). If the weld is less than 300 mm (12 in.) long, then the total length of such a condition cannot exceed 8 percent of the weld length. Inadequate penetration from high-low, when one edge of the root is exposed (or unbonded), cannot exceed 50 mm (2 in.) at individual locations or 75 mm (3 in.) in any continuous 300 mm (12 in.) length of weld.

Incomplete fusion cannot exceed 25 mm (1 in.) in length at individual locations. The total length of such a condition in any 300 mm (12 in.) length of weld metal cannot exceed 25 mm (1 in.). If the weld is less than 300 mm (12 in.) long, then the total length of such a condition cannot exceed 8 percent of the weld length. Internal concavity, incomplete fusion from cold lap, slag inclusion, internal porosity or gas pockets and burn-through are inspected by radiography. When automatic or semiautomatic welding is used, clusters of surface porosity, bead starts and high points are removed by grinding before depositing weld metal over them.

PART 3. Petroleum Tubular Specifications[15]

Oilfield pipes and tubes are visually tested according to specifications and recommended practices of the American Petroleum Institute (API) and according to test specifications written by oil and gas companies. Visual testing organizations are expected to have written practices for the tests they perform. The tests of tubular materials are covered by documents published by the American Petroleum Institute.[16-22]

Pipe grades and their minimum yield strengths are shown in Table 4.[16] Generally, the lower the grade, the less critical is the pipe's service. For example, shallow wells might be drilled with grade E-75 drill pipe and be completed with H-40, J-55 or K-55 casing and tubing. Deep wells, 4.5 to 6 km (15 000 to 20 000 ft), may be drilled with S-135 drill pipe and completed with P-110 or Q-125 casing and tubing. Many proprietary grades of pipe are also available.

The level of testing in a manufacturing plant must meet or exceed relevant specifications. Lower grade pipe or tube may receive only hydrostatic, dimensional, drift and visual testing for relatively gross discontinuities. The end user may, however, require other nondestructive tests, such as magnetic particle, electromagnetic, ultrasonic and, in the case of line pipe (pipe for pipelines), radiographic testing. Higher grade tubulars receive different levels of automated nondestructive tests and it is common for them to be inspected as many as four times before use.

Field tests for casing, tubing, plain end drill pipe and line pipe include visual testing.

Full Length Visual Testing of Line Pipe

Full length visual testing of line pipe is an inspection of the total length, including bevel and root face, to detect gouges, cuts, flats, dents, ground areas, mechanical damage, lack of straightness or other visually detectable discontinuities. Special attention is given to the weld line for undercut and off-joint weld. Rolling each length and viewing the entire external surface is required. The entire inside surface is inspected using a high intensity light source or borescope on small diameter pipe.

Typical Visual Testing for New Pipe

Specifications typically require that all new pipe be visually tested. In so doing, the following conditions might be

TABLE 4. Pipe grades and their minimum yield strengths.

Grade	Pressure	
	MPa	(10^3 $lb_f{\cdot}in.^{-2}$)
Casing and Tubing		
H-40	276	(40)
J-55	379	(55)
K-55	379	(55)
M-65	517	(75)
C-75	517	(75)
L-80	552	(80)
N-80	552	(80)
C-90	621	(90)
C-95	655	(95)
T-95	655	(95)
P-110	758	(110)
Q-125	862	(125)
Drill Pipe		
E-75	517	(75)
X-95	655	(95)
G-105	724	(105)
S-135	931	(135)
Line Pipe		
A, A-25	172	(25)
B	241	(35)
X-42	290	(42)
X-46	317	(46)
X-52	359	(52)
X-56	386	(56)
X-60	414	(60)
X-65	448	(65)
X-70	483	(70)
X-80	552	(80)
Coil Line Pipe		
X-52C	359	(52)
X-56C	386	(56)
X-60C	414	(60)
X-65C	448	(65)
X-70C	483	(70)
X-80C	552	(80)
X-90C	620	(90)

detected: out-of-roundness; excessive height of an inside flash weld or excessive trim of the inside surface weld on electric resistance welded pipe; out-of-straightness; three-dimensional discontinuities such as slugs, gouges and pits; excessive mill ripple or roll marks, which can lead to out-of-tolerance pipe walls; stretch indentations in tubing, which may lead to out-of-tolerance pipe walls; unacceptable threads; seams, cracks, porosity, pits, gouges and grip marks on couplings; workmanship problems in belled ends of line pipe; dents; offset plate edges in line pipe; out-of-line weld beads; arc burns; undercuts of welded pipe; blisters; scabs and slivers; upset underfill or materials.

Detection of such conditions can initiate further testing or evaluation and may lead to rejection of the tube, pipe or its coupling. Outside surface eddy current testing may be substituted for coiled tubing.

Straightness

In the middle of the twentieth century, oil field pipe was only required to be reasonably straight. However, later specifications define the degree of permissible bend in pipe with an outer diameter of 114 mm (4.5 in.) and larger. Comparison with a taut string or a 1.5 m (5 ft) straight edge is required such that deviation from chord height or straight does not exceed either 0.2 percent of the total pipe length or 3 mm (0.125 in.) in the 1.5 m (5 ft) length at each end. (Note: chord height applies to string and straight applies to straight edge.)

Couplings

Couplings require a visual test for the detection of discontinuities such as visible seams and cracks. Certain depths of pits and round bottom gouges are permitted, requiring a mechanical pit depth gage. Sharp bottom imperfections have separate criteria for acceptance and also require depth measurement.

Borescopy

Where borescope testing is required, lamp wattage rather than surface brightness is recommended for the pipe inside diameter (Table 5). A working requirement may be to require reading, through the borescope, a vision acuity test card or the date on a penny or dime.

Typical Visual Testing of Used Drill Pipe

API RP 7G outlines the testing of drill pipe, much of which is performed with rulers and mechanical gages.[18] Wall gaging, either by ultrasonic or radiation absorption, and magnetic flux leakage are also commonly performed. Because many discontinuities found in used drill pipe are service induced, those on the outside surface can usually be visually detected. Exceptions are wear and fatigue cracking. (Wear is hard to detect without a diameter gage.) The inspection of threads is discussed in detail below.

External Surface Visual Testing

Visual testing of the sealing shoulders (Fig. 10) often leads to dimensional measurement and flatness gaging. Visual testing of threads may detect indications of insufficient torque or overtorque, lapped threads, galling and stretching. A profile gage is often used to help detect thread discontinuities. Pin stretch is

FIGURE 10. The sealing shoulders on drill pipe occur at left ends of sections 2. Flat faces at these points are essential. Note that threads on drill pipe and drill collars do not seal.

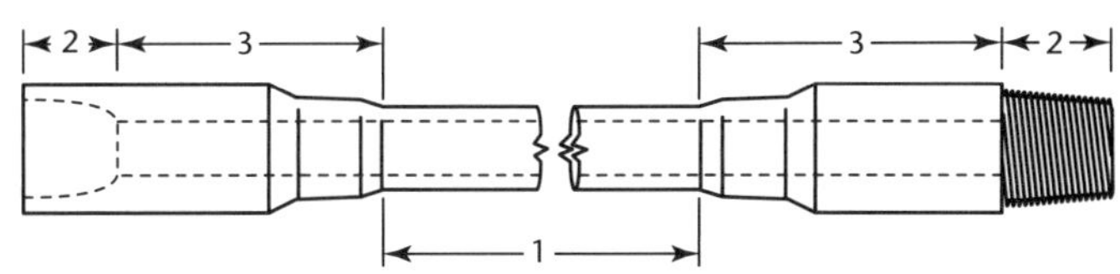

Legend

1. Length covered under drill pipe classification system.
2. Length covered under tool joint testing standard.
3. Caution: length not covered by testing standard.

TABLE 5. Incandescent lamp power for borescopic visual tests of various pipe diameters.

Inside Diameter		Minimum Power
mm	(in.)	(W)
0 to 25	(0 to 1)	10
25 to 75	(1 to 3)	30
75 to 125	(3 to 5)	100
>125	(>5)	250

shown in Fig. 11. Box swell is also detected during dimensional inspection of the connections.

Visual testing of the pipe body may reveal such conditions as washes, mashes, string shot, necking, crushing, gouges, slip marks and outside diameter pitting. Washes or washouts occur when a fatigue crack travels through the pipe wall during drilling and high pressure drilling mud is forced through the opening to rapidly abrade the crack walls into a three-dimensional hole. Washouts are often detected by loss of pressure while drilling. Necking arises from excessive overpull beyond the material's local elastic limit. Crushing arises from excessive inward force on the pipe while in the slips of the drilling rig.

String shot is an expansion of the pipe wall during a downhole controlled explosion performed to break out a pin box connection. The pipe material is taken past its elastic limit. Stretching indicates excessive pulling on a stuck pipe. Slip marks are circumferential grooves cut into the pipe's outside diameter when the pipe is rotated in the slips of the rig. Slip marks act as stress risers and encourage formation of fatigue cracks. Such cracks in the bases of slip marks (Fig. 3) are very difficult to detect with visual or conventional wet fluorescent magnetic particle testing but are often detected by transverse wave ultrasonic testing.

FIGURE 11. Stretched pin end threads on drill pipe (sealing face is just to left of gage).

Internal Surface Visual Testing

One of the oldest forms of inservice drill pipe testing is the use of an internal borescope to study magnetic particle indications. In this test, the inspector is searching for transverse fatigue cracks on the inside diameter wall, often in the presence of pitting.

Visual Testing of Oil Field Hoisting Equipment

API RP 8B[19] covers testing of crown and traveling blocks, block-to-hook adapters, connectors, link adapters, drilling hooks, tubing and sucker rod hooks, elevator links, casing, tubing and drill pipe elevators, sucker rod elevators, swivel bail adapters, rotary swivels, spiders, deadline tie downs, kelly spinners, rotary tables and slips, heave compensators and tension members of underwater handling equipment. The document recommends routine field testing, periodic field tests, critical load testing and disassembly tests for cracks, loose fits, elongation of parts, wear, corrosion and overloading. Much of the recommended testing is visual.

Part 4. Visual Testing of Pipe Threads[23]

Before the assembly of the downhole tubular structure of an oil well, casing and tubing threads must undergo visual testing. These types of tubular products in the oil field are referred to as oil country tubular goods. The term casing applies to the many strings of pipe that are used to line the hole during and after drilling. This pipe protects the hole from formation collapse, keeps the formation fluids out of the hole and — perhaps most importantly — keeps the oil well fluids out of the water tables. The casing strings are a permanent part of the well and many are cemented into the formation.

The tubing string is the production string, the pipe through which the oil or gas is brought to the surface. To do this, it is important that the connections not allow the fluids to leak out.

Requirements

Threads on both tubing and casing are required to perform two functions: seal the connection and provide the strength to support the weight of the string as it is lowered into the well.

Extensive service data for the older connections and extensive design testing for new connections together help predict the lifetime of properly manufactured and undamaged threaded connections. Inspectors visually testing threads are looking for manufacturing errors or damage caused by handling or for corrosion that would affect the ability of the connection to seal.[24] The second function of a visual test is to detect any discontinuity that would interfere with the ability of the connection to be properly "made up," that is, screwed together.

Figure 12. Casing and tubing round thread — nominal tolerances.

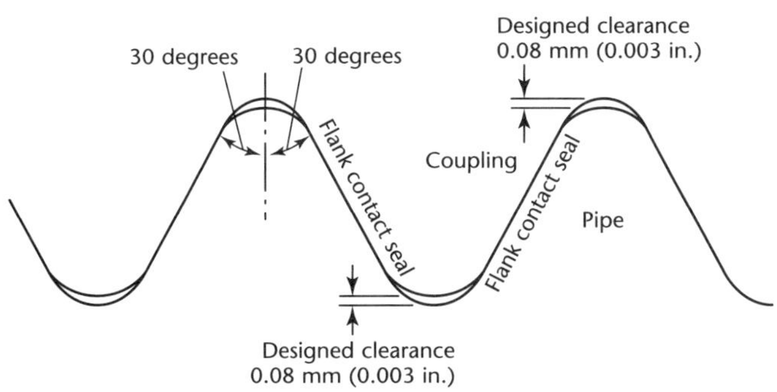

Types of Seal

There are three types of seals used on oil field tubing and casing: interference sealing threads, gasket seals and metal to metal seals. The interference sealing threads, or interference fitted threads, use a tapered connection made up under great pressure, forcing the mating surfaces together more tightly than is possible by hand alone. Figure 12 shows the flank engagement for American Petroleum Institute (API) round threads, and Fig. 13 shows the root and crest engagement for API buttress threads.

Interference Seal

Because two mass produced machined parts cannot be made to fit perfectly, there are designed clearances in the mating pieces. If these clearances are not plugged they would provide a helical leak path through the connection. Thread lubricant is used to close the gap. If properly placed in the gaps the thread lubricant, a heavy grease, will complete the seal by plugging up these gaps.[25] The smallness of the gap and the long helical distances make an effective seal.

Gasket Seal

The gasket seal uses a ring of resilient material somewhere in the connection (Fig. 14). The ring is ductile enough to form itself to the shape of the mating piece. This type of seal is always used with at least one other seal.

Figure 13. Buttress casing thread — nominal tolerances.

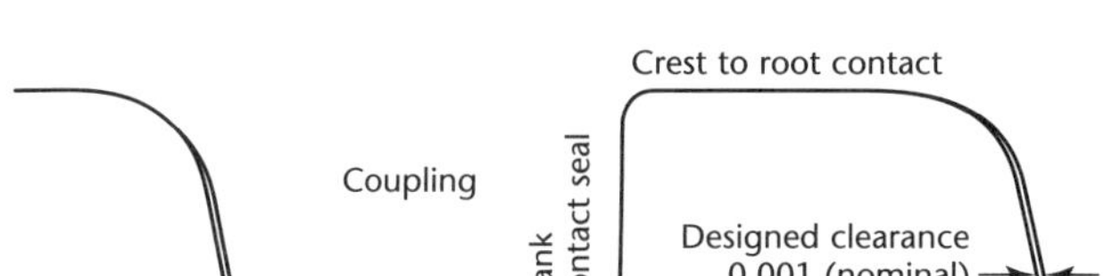

Metal-to-Metal Seal

Metal to metal seals are considered the premium seals in the oil field. The mating surfaces on the external connection (the pin) and internal connection (the box) are machined to provide a pressured interference fit 360 degrees around the connection (Fig. 15).

These three types of seals are used either alone or with others in the various connections used in oil country tubular goods.

Specifications

American Petroleum Institute (API) threads are public property governed by API SPEC 5B, where the inspection guidelines are very well defined.[26] Also, over a hundred thread designs used on oil country tubular goods are proprietary, that is, the design is owned by someone and in many cases patented. For these non-API threads, the inspection criteria are confidential, usually closely guarded by the manufacturer.

The third party inspector of these connections can only examine the threads and set aside any imperfect thread for the manufacturer's evaluation. To recognize any deviation from normal in proprietary connections, the third party inspector must be familiar with the published literature on the connection.

Visual Testing Procedures

Threads of oil country tubular goods are visually tested as a separate service or with other services such as mechanical gaging of the threads for API threads or magnetic particle or liquid penetrant testing of the ends, depending on whether the host material is ferromagnetic.

Before inspection, the threads must be cleaned with solvents and brushes. The waste materials must be captured for proper disposal. During cleaning, the inspector should begin his inspection of the threads. Any obvious imperfections should be marked as soon as they are found.

FIGURE 14. Gasket seal using resilient ring.

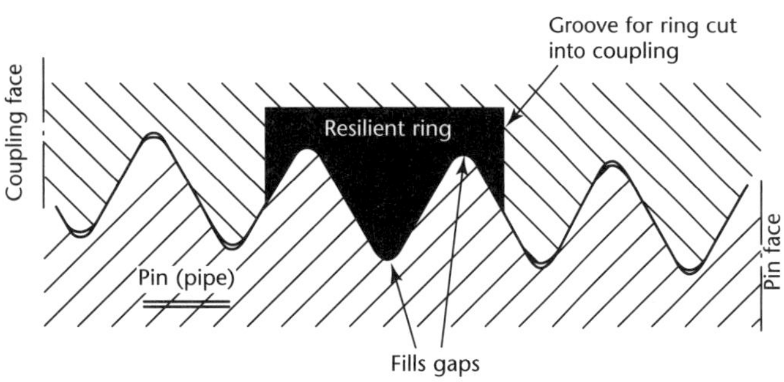

Pin Threads

The pin thread of API round and buttress threads have four distinct areas with different criteria for each (Figs. 16 and 17). The threads toward the end of the pipe are the sealing threads, having the minimum length L_c of full crested threads.

Pin Sealing Area Criteria

The L_c length is a measured distance that is closely tested visually for anything that might cause a leak path through the connection. The L_c thread must be free of visible tears, cuts, grinds, shoulders or any other anomaly that breaks the continuity of the threads.[26] All threads in the L_c must have full crests on round threads — threads with less than full crests are called black crested threads, so called because threading has not removed the dark, carburized original mill surface (Fig. 18).

FIGURE 15. Metal to metal seals are machined to provide pressured interference fit 360 degrees around connection.

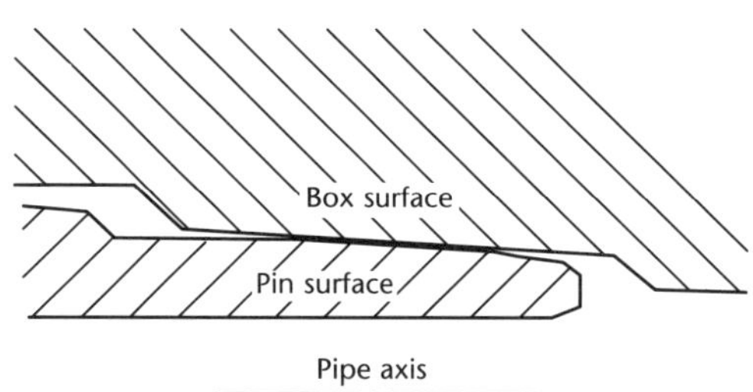

FIGURE 16. Round threads: (a) field end; (b) mill end. L_c indicates minimum length of full crested threads.

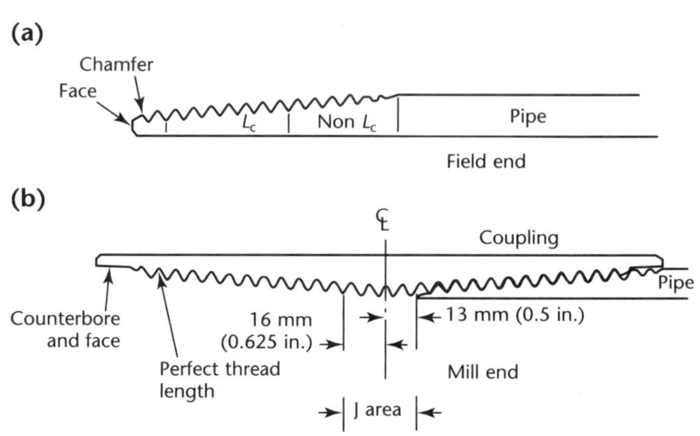

On buttress threads, there may be two black crested threads in the L_c area as long as neither is longer than 25 percent of the pipe circumference. The tables in API SPEC 5B give the distance from the end of the pipe to the end of the L_c area.

While there may be superficial discoloration in this area, such discoloration is the limit of discontinuities allowed in the L_c area. The most critical consideration throughout the threaded area, including the L_c area, is that there be no protrusions on the thread surfaces that could score the mating surface. Because the thread flanks slide past each other during makeup, the surfaces must be smooth. If the surface has a protrusion, the pressure, instead of being distributed across the flank, will be concentrated in the high spot causing friction and galling. Minor repair of high spots with a hand file may be permitted with the pipe owner's permission.

Pin Nonsealing Threads Criteria

The threads between the end of the L_c area and the vanish point of the threads are not considered as sealing threads so they are allowed to have imperfections that would be considered to be leak paths. In fact, the manufacturer may repair threads in this area by grinding as long as the grind does not go either below the root cone of the thread or 12.5 percent of the specified pipe wall thickness measured from the projected pipe surface, whichever is greater. (After imperfection removal, the threaded area must be at least 87.5 percent of wall thickness.) The most critical factor in this area of the threads is that there be no protrusions on the thread flanks that will remove the protective coating or score the mating surfaces.

Chamfer Criteria

The chamfer area on the end of the pipe is beveled to provide a place for the thread to start. This 65 to 70 degree bevel must be present for 360 degrees around the pipe face and the starting thread must run out on the chamfer.[26] This requirement is to prevent a feather edge, which could be folded over during stabbing at the rig floor. If a ridge were present and folded over, the coupling thread would not have a place to go since the fold would occupy the thread groove designed for the coupling. The required length of the chamfer, to be sufficiently long to allow the thread to run out on the chamfer, must be tempered with the industry standard that the chamfer not come to a sharp edge.[27]

If the starting thread is not continuous, that is, if a portion of the groove is missing, this condition in itself is acceptable but may be a sign that the pipe was misaligned during the finishing of the thread. There are tolerances for angular alignment and this condition must be evaluated. A false starting thread is not acceptable if it runs into the true starting thread.[27] The chamfer smoothness is not critical for it does not contribute to the thread after it provides a place to start makeup.

Pin Face Criteria

The fourth and final area where threads are critical is the pipe end. The ends must be free of burrs on the inside and outside.[28] Freedom from burrs is actually important to the entire threaded area because burrs might be dislodged during makeup. If they become dislodged they could interfere with makeup and promote galling (Fig. 19).

FIGURE 17. API buttress threads: (a) field end; (b) mill end.

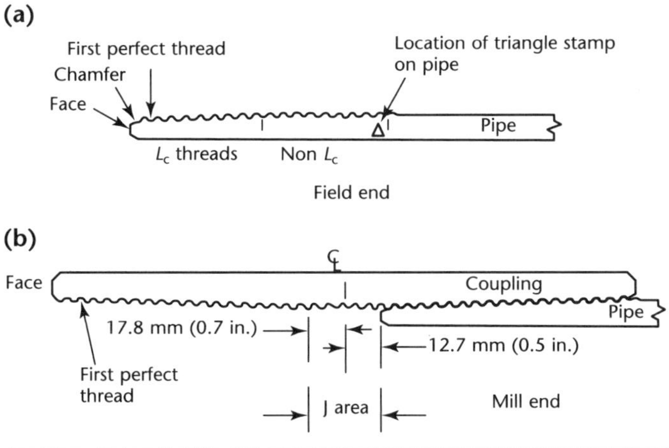

FIGURE 18. Black crested threads.

Coupling Threads

There are three areas on the coupling threads (Figs. 16 and 17). The area on the coupling referred to as the *perfect thread length* is longer than the L_c area on the pin. It starts with the first threads in the coupling and continues to the plane located near the made up position of the first full thread of the pin threads. The length of this area provides for a good thread throughout the travel area during makeup of the pin thread. This area has the same criteria as the L_c on the pin threads. These threads must be nearly perfect.

On smaller diameter pipe, a mirror is required to view the load flanks, that is, the flanks facing the center of the coupling. The repair of minor anomalies in the coupling threads is normally not as practical as repairs on the pin threads. To

FIGURE 19. Galling on pipe thread.

FIGURE 20. Burr on pipe end.

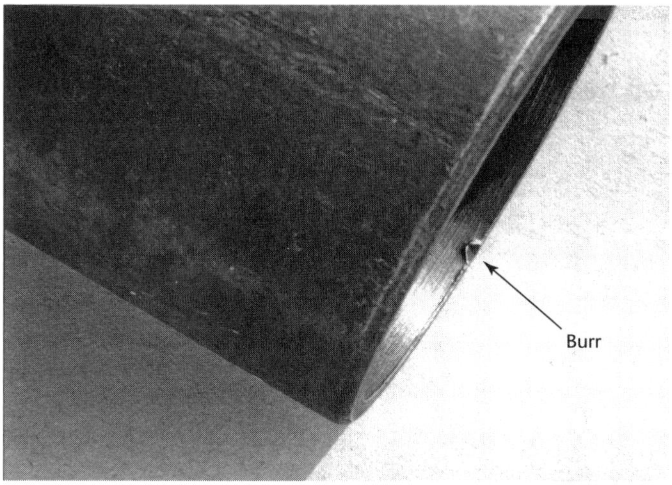

TABLE 6. Criteria for visual testing of round threads.[26] Anomalies described below are considered defects.

Area	Anomalies
Pin face	knife edge
	any burrs
Chamfer	feather edge
	any burrs
	chamfer not present for full 360 degrees around pipe
	starting thread running out to end of pipe
Pin, L_c area	any imperfection or distortion of thread form which will produce longitudinal or helical leak path
	black crested threads
	any imperfection that visibly bulges flanks or results in protrusion of metal from one or more threads
Pin, not L_c area	any imperfection through root of thread or 12.5 percent of specified wall thickness measured from projected pipe surface, whichever is greater. Remaining wall not less than 87.5 percent of specified wall thickness.
	any metal protrusion that may prevent proper makeup
Coupling	threads not extending to center of coupling
	any metal protrusion that may prevent proper makeup
	any imperfection or distortion of thread form which will produce longitudinal or helical leak path
	black crested threads
	any imperfection that visibly bulges flanks or results in protrusion of metal from one or more threads
Coupling face and counterbore	any metal protrusion that would prevent proper makeup

improve corrosion resistance, antigalling and sealing ability, the internal threads are coated with zinc, tin or metallic phosphate.[26]

The threads in the center of the coupling are required only to be present. Sometimes these are not full threads but are considered acceptable if the thread root is present. Seams, laps or cracks in the coupling threads are always considered rejectable but are not normally found by visual testing alone because of the coating applied to the internal threads. Double cut threads in this area are acceptable. While cutting the second side, the thread cutter may clip the crests of the threads from the first side. This condition is acceptable unless the cutting extends into the perfect thread area.[27]

The counter bore and face are the other areas of the coupling. The diameter of the recess shall be sufficient to prevent cutting ghost thread roots on the surface of the recess.[25] Also, there should be no burrs (Fig. 20) or protrusions in the counter bore area that could damage the pipe threads during stabbing at the rig site.

Presence of Makeup Triangle

The visual thread inspector checks for the presence of the makeup triangle (a manufacturer's stamp indicating where makeup should stop) on buttress threads and round threads larger than 400 mm (16 in.). The lack of a triangle is not normally cause for rejection, but the customer should be notified since the triangle is used to aid proper makeup on the rig floor. The thread area stops at the apex of the makeup triangle. These criteria are summarized in Table 6 for round threads and Table 7 for buttress threads.[26]

Makeup Connections

The completely tightened makeup is also checked visually during a visual thread test. The pin thread face should be made up to 13 mm (0.5 in.) from the center of the coupling (Fig. 21) for most pipe. The center of the coupling can usually be visually located. By counting threads

TABLE 7. Criteria for visual testing of buttress threads.[26] Anomalies described below except triangle stamps are considered defects. Triangle stamp errors need to be reported.

Area	Anomalies
Pin face	knife edge
	any burrs
Chamfer	a feather edge
	any burrs
	chamfer is not present for full 360 degrees around pipe
	starting thread running out to end of pipe
Pin, L_c area	any imperfection or distortion of thread form which will produce longitudinal or any imperfection or distortion of thread form that will produce longitudinal or helical leak path
	black crested threads, no more than two with each no longer than 25 percent of circumference
	any imperfection that visibly bulges three flanks, crests, or results in protrusion of metal from one or more threads
	superficial corrosion covering more than half circumference of pin area
Pin, not L_c area	any imperfection or distortion of thread form which will produce longitudinal or any imperfection traveling through root of thread or 12.5 percent of specified wall thickness measured from projected pipe surface, whichever is greater. Remaining wall not less than 87.5 percent of specified wall thickness.
	any metal protrusion that may prevent proper makeup
	makeup triangle present on pin end
Coupling	threads not extending to center of coupling
	any imperfection or distortion of thread form that will produce longitudinal or helical leak path
	black crested threads
	any imperfection that visibly bulges three flanks, crests or results in protrusion of metal from one or more threads
	superficial corrosion covering more than half circumference of pin area
Coupling face	any metal protrusion that would prevent proper makeup
	any feather edges
Coupled end	proper makeup to triangle

between the center of the coupling and the face of the pin on the opposite side, the distance can be quite accurately estimated. Further evaluation by measurement may be required for classification if visual evaluation shows a significant error.

Thread Profile

As an aid in detecting thread form problems, a profile gage (Fig. 22) must be used on each thread tested.

Profile gages may also help determine whether an anomaly is causing a protrusion or whether all protruding metal has been removed.

FIGURE 21. Power tight makeup is visually inspected by counting threads between the center of the coupling and the face of the pin on the opposite side (not to scale).

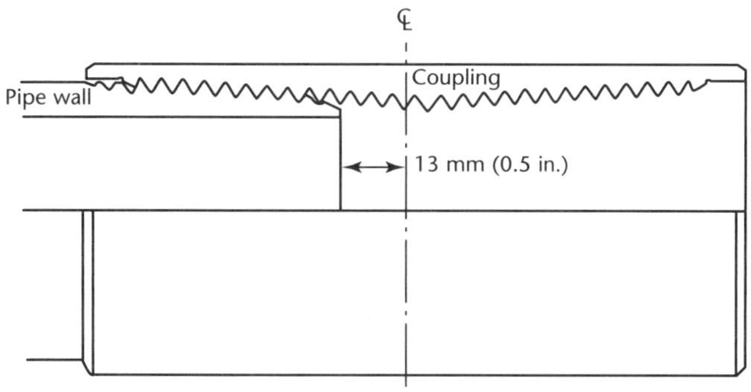

FIGURE 22. Handheld pipe thread gage.

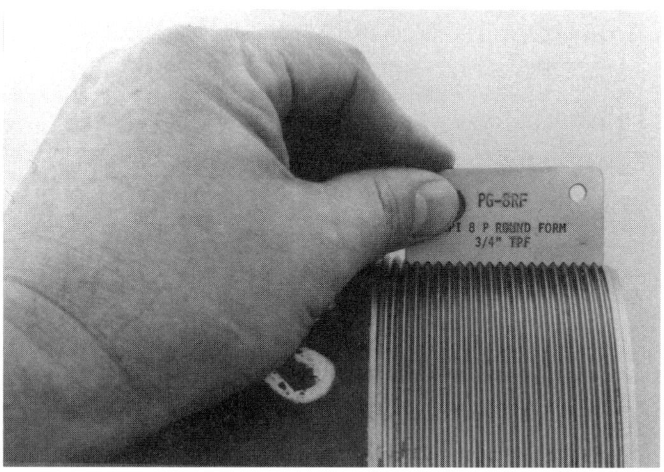

Arc Burns

Arc burns are localized points of surface melting caused by arcing between electrode and ground and the pipe surface.[29] (The condition can be caused during magnetic particle testing by passing current into the pipe through steel thread protectors. Magnetic particle testing is no longer required for new tubing.[27]) Because arc burns cause localized changes in the metallurgical structure and frequently cause protrusions, they always cause the thread to be rejected.

Shoulders

API round threads are designed to run out at the pipe surface. Excessive metal, machined for threading but in fact not threaded, where the thread stops on the outside surface of the pipe is referred to as a shoulder.[29] If the shoulder goes all the way around the connection, it indicates that either the pipe is too big or the thread is too small.

Further investigation is required to determine which of these conditions exists. A small thread size is a serious condition because makeup and sealing depend on both thread members' being the proper size. Because of the thread design, shoulders do not occur on buttress threads but threads that are too small may be recognized by the lack of black crested threads (described above) near the L_c area. If there are not black crested threads in this area, either the pipe is too big or the threads are too small.

An alignment problem may be indicated by shoulders on only one side for round threads or by lack of black crested threads on one side for buttress. The threads may be angularly or axially misaligned. There are API limits for both conditions because the angular misalignment (hooked threads) would cause makeup problems in the field and the axial misalignment would cause joint strength problems.

References

1. Bajula, D.R. Chapter 13, "Chemical and Petroleum Applications of Magnetic Testing": Part 1, "Chemical and Petroleum Industry." *Nondestructive Testing Handbook,* third edition: Vol. 8, *Magnetic Testing.* Columbus, OH: American Society for Nondestructive Testing (2008): p 324-326.
2. CFR 1910.119, *Process Safety Management Regulations.* Washington, DC: Government Printing Office.
3. Section 6, "Visual and Optical Testing Procedures": Part 3, "Visual Testing Acceptance Criteria for Welds." *Nondestructive Testing Handbook,* second edition: Vol. 8, *Visual and Optical Testing.* Columbus, OH: American Society for Nondestructive Testing (1993): p 163-175.
4. *Welding Inspection.* Miami, FL: American Welding Society (1968). Superseded by *Welding Inspection Handbook.* Miami, FL: American Welding Society (2000).
5. *ASME Boiler and Pressure Vessel Code:* Section I, *Power Boilers.* New York, NY: ASME International (1989, 2007).
6. *ASME Boiler and Pressure Vessel Code:* Section VIII, *Rules for Construction of Pressure Vessels.* New York, NY: ASME International (1989, 2007).
7. ANSI/ASME B 31.1, *ASME Code for Pressure Piping: Power Piping* (1990, 2007).
8. *ASME Boiler and Pressure Vessel Code:* Section X, *Fiber Reinforced Plastic Pressure Vessels.* New York, NY: ASME International (1989, 2007).
9. ASME B 31.3, *ASME Code for Pressure Piping: Process Piping.* New York, NY: ASME International (1990, 2008).
10. ASME B 31.8, *ASME Code for Pressure Piping: Gas Transmission and Distribution Piping Systems.* New York, NY: ASME International (1990, 2007).
11. API STD 650, *Welded Steel Tanks for Oil Storage.* Dallas, TX: American Petroleum Institute (1988). Superseded by API 650, *Welded Tanks for Oil Storage.* Dallas, TX: American Petroleum Institute (2007).
12. API STD 620, *Design and Construction of Large, Welded, Low-Pressure Storage Tanks.* Dallas, TX: American Petroleum Institute (2008).
13. API STD 653, *Tank Inspection, Repair, Alteration, and Reconstruction,* third edition. Washington, DC: American Petroleum Institute (with errata, 2008).
14. API STD 1104, *Welding of Pipelines and Related Facilities.* Dallas, TX: American Petroleum Institute (2008).
15. Stanley, R.K. Section 7, "Codes, Standards and Specifications for Visual and Optical Testing": Part 3, "Oil Field Tubular Specifications." *Nondestructive Testing Handbook,* second edition: Vol. 8, *Visual and Optical Testing.* Columbus, OH: American Society for Nondestructive Testing (1993): p 188-191.
16. ISO 3183/API SPEC 5L, *Specification for Line Pipe.* Dallas, TX: American Petroleum Institute (2009).
17. API RP 5L8, *Recommended Practice for Field Inspection of New Line Pipe.* Dallas, TX: American Petroleum Institute (1996).
18. API RP 7G, *Recommended Practice for Drill Stem Design and Operating Limits.* Dallas, TX: American Petroleum Institute (2009).
19. ISO 13534/API RP 8B, *Recommended Practice for Hoisting Tool Inspection and Maintenance Procedures.* Dallas, TX: American Petroleum Institute (2005).
20. API SPEC 5D, *Specification for Drill Pipe.* Dallas, TX: American Petroleum Institute (2001).
21. API SPEC 7, *Specification for Rotary Drill Stem Elements.* Dallas, TX: American Petroleum Institute (2009).
22. API 570, *Piping Inspection Code: In-Service Inspection, Rating, Repair, and Alteration of Piping Systems.* Dallas, TX: American Petroleum Institute (2009).
23. Mackin, J.L. Section 11, "Other Applications of Visual and Optical Tests": Part 4, "Visual Testing of Threads in Oil Country Tubular Goods." *Nondestructive Testing Handbook,* second edition: Vol. 8, *Visual and Optical Testing.* Columbus, OH: American Society for Nondestructive Testing (1993): p 322-327.
24. Raulins, Max. "How Loading Affects Tubular Thread Shoulder Seals." *Petroleum Engineering International.* Dallas, TX: Otis Engineering Corporation, Premium Threading Services Division (March 1984).

25. API RP 5B1, *Recommended Practice for Gaging and Inspection of Casing, Tubing, and Line Pipe Threads,* fifth edition. Dallas, TX: American Petroleum Institute (2004).
26. API SPEC 5B, *API Specification for Threading, Gaging, and Thread Inspection of Casing, Tubing, and Line Pipe Threads,* thirteenth edition. Dallas, TX: American Petroleum Institute (2008).
27. API RP 5A5 [ISO 15463-2003], *Recommended Practice for Field Inspection of New Casing, Tubing and Plain End Drill Pipe.* Dallas, TX: American Petroleum Institute (2005).
28. API SPEC 5CT, *Specification for Casing and Tubing,* fourth edition. Dallas, TX: American Petroleum Institute (1992). Superseded by ISO 11960 [API SPEC 5CT], *Petroleum and Natural Gas Industries — Steel Pipes for Use as Casing or Tubing for Wells.* Geneva, Switzerland: International Organization for Standardization (2006).
29. API STD 5T1, *Standard for Imperfection Terminology.* Dallas, TX: American Petroleum Institute (2003).

Additional Standards

AWS B 1.11, *Guide for the Visual Examination of Welds.* Miami, FL: American Welding Society (2000).

ASME B 31.4, *Pipeline Transportation Systems for Liquid Hydrocarbons and Other Liquids.* New York, NY: ASME International (2006).

ASME B 31.5, *Refrigeration Piping and Heat Transfer Components.* New York, NY: ASME International (2006).

ASME Boiler and Pressure Vessel Code: Section V, *Nondestructive Examination.* Article 9, *Visual Examination.* New York, NY: ASME International (2009).

ASME Boiler and Pressure Vessel Code: Section V, *Nondestructive Examination.* New York, NY: ASME International (2007).

ASME Boiler and Pressure Vessel Code: Section XII, *Rules for Construction and Continued Service of Transport Tanks.* New York, NY: ASME International (2007).

AWS D 1.1M, *Structural Welding Code — Steel.* Miami, FL: American Welding Society (2008).

AWS D 18.2, *Guide to Weld Discoloration Levels on Inside of Austenitic Stainless Steel Tube.* Miami, FL: American Welding Society (1999).

ISO 3058, *Non-Destructive Testing — Aids to Visual Inspection — Selection of Low-Power Magnifiers.* Geneva, Switzerland: International Organization for Standardization (1998).

ISO 10407, *Petroleum and Natural Gas Industries — Rotary Drilling Equipment — Inspection and Classification of Used Drill Stem Elements.* Geneva, Switzerland: International Organization for Standardization (2010).

ISO 17637, *Non-Destructive Testing of Welds — Visual Testing of Fusion-Welded Joints.* Geneva, Switzerland: International Organization for Standardization (2003).

Compressed Gas

CGA C-6, *Standards for Visual Inspection of Steel Compressed Gas Cylinders.* Chantilly, VA: Compressed Gas Association (2007).

CGA C-6.1, *Standards for Visual Inspection of High Pressure Aluminum Compressed Gas Cylinders.* Chantilly, VA: Compressed Gas Association (2006).

CGA C-6.2, *Guidelines for Visual Inspection and Requalification of Fiber Reinforced High Pressure Cylinders.* Chantilly, VA: Compressed Gas Association (2005).

CGA C-6.3, *Guidelines for Visual Inspection and Requalification of Low Pressure Aluminum Compressed Gas Cylinders.* Chantilly, VA: Compressed Gas Association (1999).

CGA C-6.4, *Methods for External Visual Inspection of Natural Gas Vehicle (NGV) Vehicle Fuel Containers and Their Installations.* Chantilly, VA: Compressed Gas Association (2007).

CGA C-13, *Guidelines for Periodic Visual Inspection and Requalification of Acetylene Cylinders.* Chantilly, VA: Compressed Gas Association (2006).

Personnel Qualification

ASNT Central Certification Program (ACCP), Revision 4 (March 2005). Columbus, OH: American Society for Nondestructive Testing (2005).

ANSI/ASNT CP-189, *Standard for Qualification and Certification of Nondestructive Testing Personnel.* Columbus, OH: American Society for Nondestructive Testing (2006).

ASNT Standard CP-105, *Topical Outlines for Qualification of Nondestructive Testing Personnel.* Columbus, OH: American Society for Nondestructive Testing (2006).

ASNT Recommended Practice No. SNT-TC-1A, *Personnel Qualification and Certification in Nondestructive Testing*. Columbus, OH: American Society for Nondestructive Testing (2006).

Chemical and Petroleum Industry Inspection Technology Topical Conferences

Disser, J. "Fully Connected Remote Visual Inspection (RVI)." *ASNT's International Chemical and Petroleum Industry Inspection Technology (ICPIIT) X Conference* [Houston, TX, June 2007]. Columbus, OH: American Society for Nondestructive Testing (2007): p 128-132.

Hirtle, B. "Automated Computerized Thread Inspection." *Petroleum Industry Inspection Technology* [Houston, TX, June 1989]. Columbus, OH: American Society for Nondestructive Testing (1989): p 115-119.

Jones, D.N. and J.A. Jones. "Optical and Other Electronic Systems for Oilpipe Thread Sizing." *Petroleum Industry Inspection Technology* [Houston, TX, June 1989]. Columbus, OH: American Society for Nondestructive Testing (1989): p 136-140.

Karlin, R.A and A. Kindrick. "Thread Gauging." *Petroleum Industry Inspection Technology* [Houston, TX, June 1989]. Columbus, OH: American Society for Nondestructive Testing (1989): p 124-129.

Morris, C., J.[L.] Rose and H.J. Sin. "Boreside Ultrasonic Guided Wave and Rotating Mirror Techniques for Rapid Screening and Detailed Characterization of Heat Exchanger Tubing and Piping." *ASNT's International Chemical and Petroleum Industry Inspection Technology (ICPIIT) V Topical Conference* [Houston, TX, June 1997]. Columbus, OH: American Society for Nondestructive Testing (1997): p 53-68.

O'Brien, D., M. Gupta, R.M. Gray and J. Hagene. "Automatic Classification of Images from Internal Optical Inspection of Gas Pipelines." *ASNT's International Chemical and Petroleum Industry Inspection Technology (ICPIIT) VIII Topical Conference* [Houston, TX, June 2003]. Columbus, OH: American Society for Nondestructive Testing (2003): p 163-167.

Somers, T. and K.F. Schmidt. "Robotic System for Tank Inspection." *ASNT's International Chemical and Petroleum Industry Inspection Technology (ICPIIT) IV Topical Conference* [Houston, TX, June 1995]. Columbus, OH: American Society for Nondestructive Testing (1995): p 81-85.

Woodward, J. "Enhanced Visual Inspections of Railcar Equipment." *ASNT's International Chemical and Petroleum Industry Inspection Technology (ICPIIT) IV Topical Conference* [Houston, TX, June 1995]. Columbus, OH: American Society for Nondestructive Testing (1995): p 150-151.

10

CHAPTER

Electric Power Applications of Visual Testing

Richard T. Nademus, Exelon Corporation, Oyster Creek Generating Station, New Jersey (Part 2)

Part 1 and portions of Part 2 are reprinted with permission from *Visual Examination Technologies* (EPRI learning modules), © [ca. 1982 and] 1996, the Electric Power Research Institute (EPRI), Charlotte, NC. ASNT has revised the text in 1993 and 2010, and deficiencies are not the responsibility of the Electric Power Research Institute.

PART 1. Visual Testing of Welds[1,2]

To perform accurate visual tests of metallurgical joints, the following general information should be known: (1) joint configurations, (2) welding processes, (3) weld joint terminology and (4) the fabrication process. Although the visual inspector may be examining only the final welds, a more thorough test can be done by understanding how a weld is made. This understanding is especially valuable if the visual inspector is involved with weld repair or replacement.

Metal Joint Configurations

The geometry of metal joints is determined by the requirements of their host structure. The five basic joint types are the butt, corner, edge, tee and lap joints (Fig. 1).

FIGURE 1. Five basic joint types: (a) butt; (b) corner; (c) edge; (d) T; (e) lap.

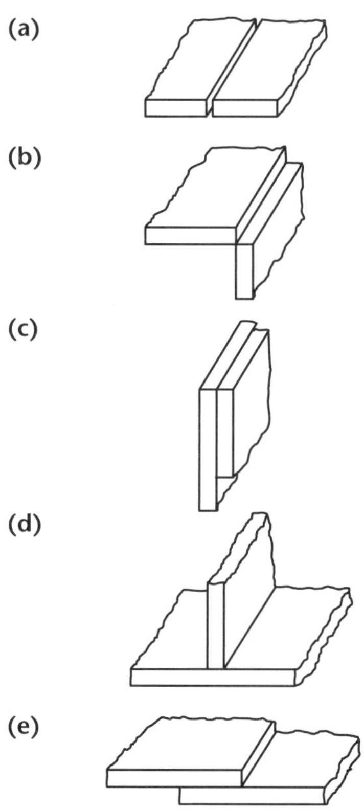

The purpose of a weld joint is to transfer forces from one member to the other through the joint. The forces may be introduced at various points and transmitted to different areas throughout the weld joint (Fig. 2). The amount of stress transferred across the joint depends on the type of loading and on the service of the weldment. These factors affect joint preparation designs, which are typically classified as (1) complete joint penetration and (2) partial joint penetration.

Complete Penetration

Complete joint penetration (Fig. 3) is joint penetration in which the weld metal completely fills the groove and is fused to the base metal throughout its thickness.

Assuming that the weld metal strength is equal to or greater than that of the base metal, which is nearly always the case, the weld joint in Fig. 3 could be considered

FIGURE 2. Forces may be transmitted through various weld preparations: (a) square; (b) single J; (c) single bevel; (d) single V cap; (e) double bevel; (f) single U.

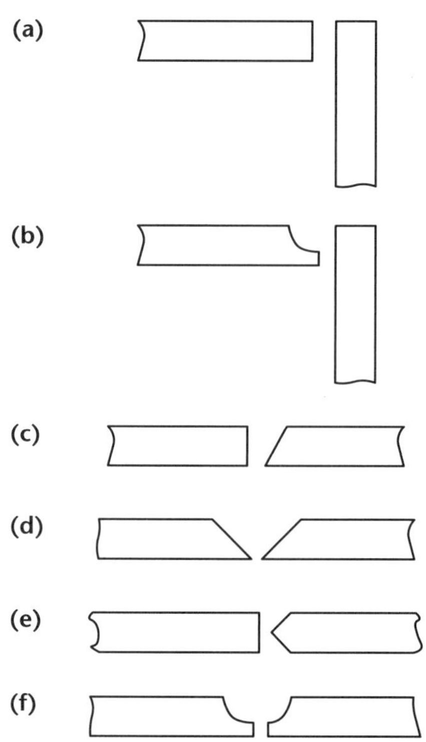

100 percent efficient. That is, for purposes of loading, the member could be considered a uniform solid structure. Complete joint penetration is usually required if weld joints are subject to static loads as well as dynamic, reversing or impact loads at varying temperatures.

Partial Penetration

Partial joint penetration (Fig. 4) leaves an unfused area — the weld does not completely penetrate the joint. The load rating or joint efficiency is based on the percentage of the weld metal depth to the total joint depth. Partial penetration joints are reliable for specific loads and particular service environments.

Basic Welding Processes

Welding is a materials joining process that produces coalescence of materials by heating them to suitable temperatures — with or without the application of pressure, by the application of pressure alone and with or without filler metal. The processes shown in Fig. 5 are grouped according to the means of energy transfer. A secondary consideration is the influence of capillary attraction in distribution of filler metal in the joint.

Metallurgical joints are typically formed by: (1) soldering, (2) brazing, (3) oxyacetylene gas welding, (4) resistance welding and (5) arc welding. Some other processes, including diffusion bonding and electron beam welding, are widely used in certain industries. There are also metallurgical joints that are mechanical in character (bolted or riveted connections, for instance).

FIGURE 3. In full penetration joints, weld metal completely fills groove and is fused to base metal throughout its thickness: (a) double V; (b) single V.

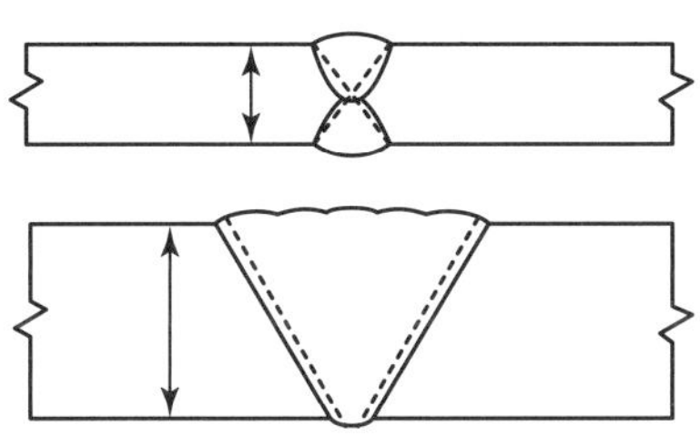

FIGURE 4. Partial joint penetration: (a) single V butt joint; (b) double V butt joint.

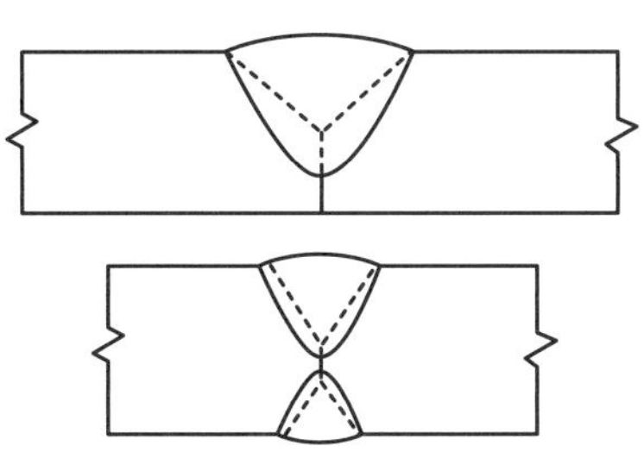

Soldering

Soldering joins materials by heating them to a suitable temperature and by using filler metal with a liquid state below 450 °C (840 °F) and below the solid state threshold, or solidus, of the base metals. The filler metal is distributed between the closely fitted faying surfaces of the joint by capillary action. Solder is normally a nonferrous alloy used to accelerate wetting and to remove oxides. Many metals can be soldered, including aluminum, copper base alloys, nickel base alloys, steel and stainless steel. The mechanism for soldering involves three closely related phenomena: (1) wetting, (2) alloying and (3) capillary action.

Wetting is the bonding or spreading of a liquid filler metal or flux on a solid base metal. When molten solder leaves a continuous permanent film on the surface of a base metal, it is said to *wet* the surface. Wetting occurs when there is a stronger attraction between certain atoms of the solder and the base metal than between the atoms of the solder. Wetting is essentially a chemical process.

The ability of a solder to alloy with the base metal is related to its ability to wet the surface. Alloying also is dependent on the cleanliness of the base metal — there must be intimate contact between the solder and the base metal for alloying to occur at the interface.

The fluidity of the molten solder must be such that it can flow into narrow spaces by capillary action. Fluidity is the property that influences the spreading of the solder over the base metal surface.

The strength of a soldered joint depends on the joint design and its clearance. The joint and filler metal can be heated by a number of means, including dipping, electric induction, infrared irradiation, or application of a torch or soldering iron.

Brazing

Brazing joins materials by heating them to a suitable temperature and by using a filler metal with a liquidus above 450 °C (840 °F) and below the solidus of the base metal. The filler metal is distributed between the closely fitted faying surfaces of the joint by capillary action. This is considered a metallurgical joint because the members are held together by the adhesion of the filler metal to the joint surfaces. Most metals can be joined by brazing. Filler metals commonly used for low carbon and low alloy steels are silver alloys and copper zinc alloys.

The selection of the filler metal, the nominal alloy composition and the melting and brazing temperature depend mainly on the joint design and the means of assembly. Heating of the joint and filler metal can be accomplished by a gas furnace, induction heating, resistance heating, infrared heating or immersion into molten salt. Fluxes and inert atmospheres must be used to prevent surface oxidation and to ensure wetting action. Surface preparation and precleaning are also important in this process.

Oxyacetylene Gas Welding

Oxyacetylene gas welding joins materials by heating with an oxyacetylene gas flame with or without the application of pressure and with or without filler metal. The process involves melting of the base metal and a filler metal, if used, by means of a welding torch. This process normally uses acetylene as the fuel gas and uses pure oxygen instead of air. Molten metal

FIGURE 5. Welding and allied processes.

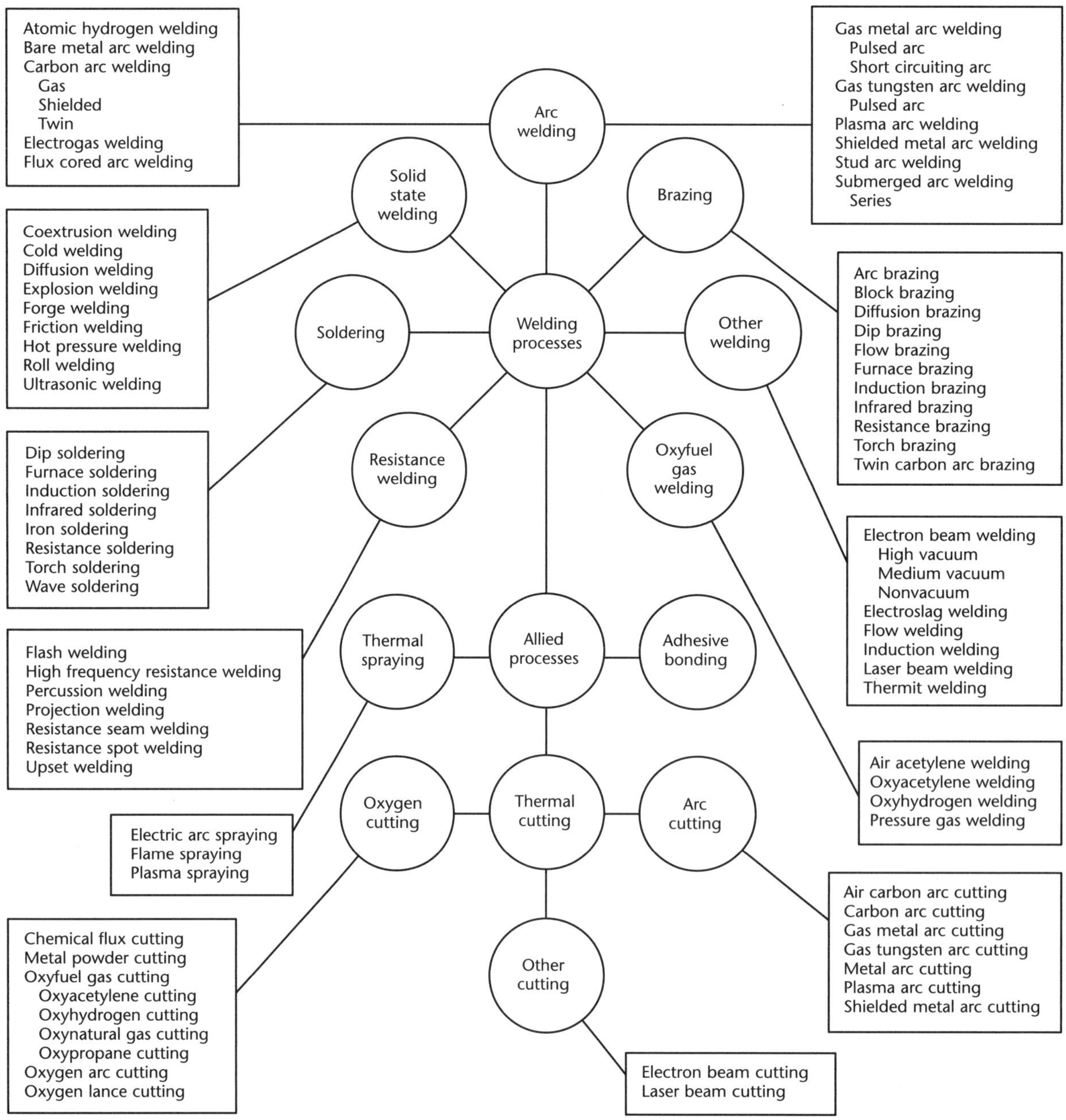

from the base material edges and filler metal intermix in a common pool and coalesce to form a continuous material.

An advantage of this welding process is that the rate of heat input can be controlled, as is the temperature of the weld zone. Because the filler metal is added independently of the heat source, weld bead size and shape are also controlled. Oxyacetylene gas welding is ideally suited for thin sheet, tubes and small diameter pipes, as well as for repair welding.

Resistance Welding

Resistance welding joins metals with the heat obtained from resistance of the metal to electric current and by the application of pressure.

In both spot and seam welds, a nugget (the weld metal) is produced at the electrode site. Seam welding is a variation of spot welding in which a series of overlapping nuggets is produced to obtain a continuous seam. Projection welding is similar except that the nugget location is determined by a projection on one closely fit (faying) surface or by the intersection of parts. These resistance processes can be used to make corner, tee, edge and the common lap joints.

Flash, upset and percussion welding are also resistance welding techniques. These techniques can be used to produce a butt joint between components with similar cross sections by making a weld simultaneously across the entire joint without the addition of filler metal. A force is applied before, during or after the heat energy is applied to bring the parts into intimate contact. These three processes are distinguished by the means of heating and by the time of the force application.

Arc Welding

Arc welding joins metals by heating them with an electric arc, with or without pressure and with or without filler metal. All forms of arc welding shield the arc to block out harmful elements in the atmosphere.

The work piece serves as one electrode of a circuit. The other electrode can be a consumable or a nonconsumable material. The electric arc is generated between these two electrodes. Nonconsumable electrodes do not melt in the arc and filler metal is not transferred across the arc. Welding processes that can use a nonconsumable electrode are carbon arc welding (CAW), plasma arc welding (PAW) and gas tungsten arc welding (GTAW). Consumable electrodes melt in the arc and are transferred across the arc to become deposited filler metal. Welding processes that use consumable electrodes are shielded metal arc welding (SMAW), gas metal arc welding (GMAW), flux cored arc welding (FCAW) and submerged arc welding (SAW).

Common Types of Arc Welding

Shielded Metal Arc Welding

Shielded metal arc welding joins metals by heating them with an arc between the work piece and a covered metal electrode. Shielding is provided by decomposition of the electrode covering. Pressure is not used, and filler metal is obtained from the electrode.

This process, sometimes called *stick welding,* is manual, which accounts for its high versatility. Shielded metal arc welding uses the heat of the arc to melt the base metal and the tip of a consumable flux covered electrode. The electrode and the work piece are part of the electrical circuit (Fig. 6a).

This circuit includes a power supply with controls, welding cables, an electrode holder and an arc welding electrode. One of the cables from the power supply is attached to the work piece and the other is connected to the electrode holder.

When an arc is struck between the electrode tip and the work piece, the intense heat of the arc melts the electrode

Figure 6. Shielded metal arc welding: (a) setup, where electrode and work piece are part of electrical circuit; (b) interface with base metal.

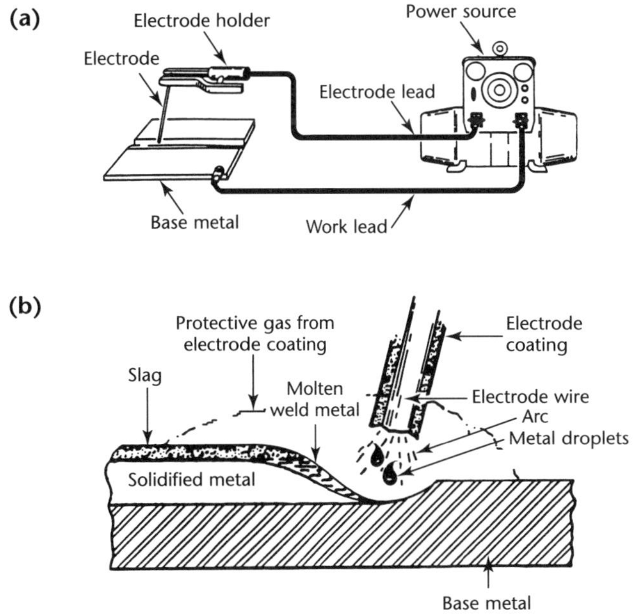

tip as well as the work piece beneath the arc. Tiny globules of molten metal rapidly form on the tip of the electrode and are then transferred across the arc into the molten weld pool. In this manner, filler metal is deposited and the electrode is consumed. The arc is one of the hottest commercial heat sources, with temperatures above 5000 °C (9000 °F) at its center. Melting is instantaneous when the arc contacts the metal. If the weldment is flat and horizontal, the metal transfer is induced by gravity, gas expansion, electric and electromagnetic forces and surface tension. For welds in other positions, gravity works against the other forces.

Shielded metal arc welding (Fig. 6b) is one of the most widely used welding processes for the following reasons:

1. The equipment is simple, inexpensive and portable.
2. The shielding of the electrodes make the process less sensitive to drafts than some other processes.
3. The position of welding is limited only by the size and type of electrode.
4. The process is suited for most of the commonly used metals and alloys.
5. It is the most tolerant arc welding process in regard to undesirable fit up conditions, such as wide root openings.

One disadvantage is that slag removal is required before performing shielded metal arc welding.

Gas Tungsten Arc Welding

Gas tungsten arc welding joins metals by heating them with an arc between a tungsten, nonconsumable electrode and the work piece. Shielding is obtained from gas; pressure may or may not be used and filler metal may or may not be used. This process, sometimes called *tungsten inert gas* (TIG) welding, can be manual, semiautomatic machine or automatic (Fig. 7a).

The equipment needed for gas tungsten arc welding are the welding machine, the welding electrode holder, the tungsten electrode, the shielding gas supply and controls. Several accessories are optional: arc pulsers, wire feed systems, a foot rheostat (to let the welder initiate or extinguish the arc and control the welding current) and water circulating systems (to cool the electrode holder).

Pure (99.5 percent) tungsten electrodes are less expensive and are generally used on less critical operations than tungsten alloy electrodes containing thorium or zirconium. A pure tungsten electrode has a relatively low current carrying capacity with alternating current power and a low resistance to contamination. Tungsten electrodes with 1 or 2 percent thorium have some advantages over pure tungsten electrodes, including higher electron emissivity, better current carrying capacity, longer life, greater resistance to contamination, easier arc initiation and a more stable arc.

The electric arc is produced by the passage of current through the ionized inert shielding gas. To prevent oxidization, the heated weld zone, the molten metal and the tungsten electrode are shielded from the atmosphere by a blanket of inert gas fed through the electrode holder. The inert shielding gas is usually helium, argon or a mixture of the two. Figure 7b shows the positions of the gas tungsten arc welding torch, the arc, the tungsten electrode, the shielding gas and the filler metal.

Because of its excellent control of heat input, gas tungsten arc welding is good for joining thin base materials, small diameter pipe, tubing and root passes of piping for critical applications. Because the electrode is nonconsumable, the process can be used to weld by fusion alone or with the addition of filler metal in the form of inserts or wire. It can be used on almost all metals and is especially

FIGURE 7. Gas tungsten arc welding (tungsten inert gas welding): (a) setup; (b) interface with base metal.

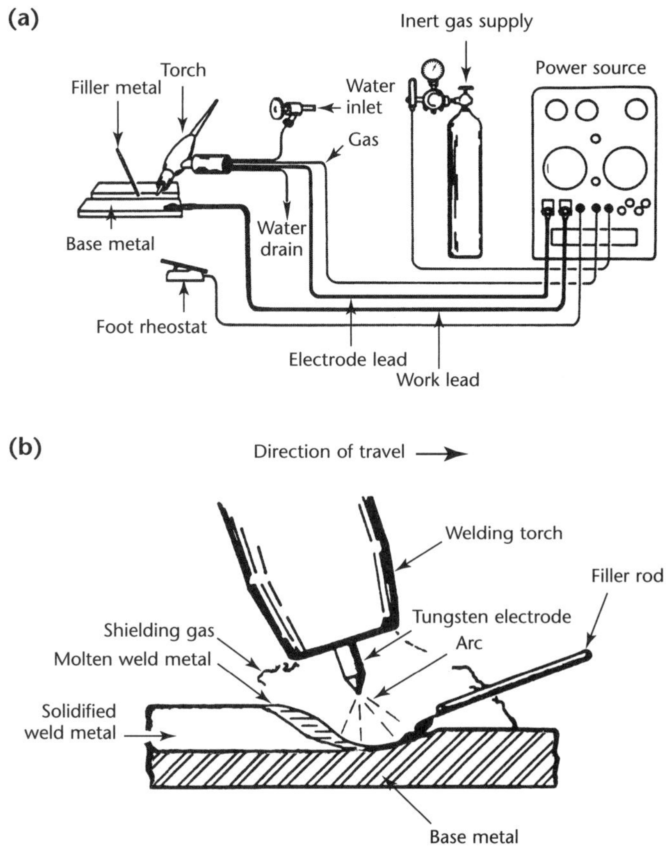

useful for joining aluminum and magnesium which form refractory oxides. The process lends itself to high quality welding but removal of surface contaminants before welding is very important. Some limitations of gas tungsten arc welding include (1) speed slower than consumable electrode arc welding processes, (2) possible transfer of tungsten from the electrode to the weld, causing contamination (the resulting tungsten inclusion is hard and brittle), (3) exposure of the hot filler rod to air causing weld metal contamination, (4) additional costs of inert shielding gases and tungsten electrodes and (5) higher equipment costs.

Gas Metal Arc Welding

Gas metal arc welding joins metals by heating them with an arc between a consumable electrode and the work piece. Shielding is achieved using an externally supplied gas. Gas metal arc welding can be semiautomatic or automatic. The process is used particularly where high production quantities are needed. Important metals such as carbon steel, aluminum and copper can be welded with this process in all positions by proper choice of shielding gas, electrode and welding parameters.

The filler metal can be transferred from the electrode to the work piece in two ways: (1) short circuiting transfer, in which the electrode contacts the molten weld pool and establishes a short circuit; or (2) drop transfer, in which discrete drops are moved across the arc gap under the influence of gravity or electromagnetic forces. Drop transfer can be globular or a spray. The type of transfer is determined by the type and magnitude of the welding current, the current density, the electrode composition and the shielding gas.

Equipment used for gas metal arc welding includes a welding gun, a power supply, a shielding gas supply and a wire drive system that pulls the wire electrode from a spool and pushes it through a welding gun. After passing through the gun, the wire becomes energized by contact with a copper contact tube, which transfers current from a power source to the arc (Fig. 8).

High quality welds are produced by this process when proper welding procedures are used. The absence of flux or electrode covering eliminates slag inclusions in the weld. Some dross formation may occur when highly deoxidized steel electrodes are used and it should be removed before the next weld bead or pass is made. The inert gas shielding provides excellent protection of the weld area from oxygen and nitrogen contamination. Hydrogen is virtually eliminated as a concern in the weld and heat affected zones of low alloy steels. On the other hand, the process permits low cost welding of carbon steels with inexpensive carbon dioxide gas shielding.

One of the chief advantages of gas metal arc welding is that, in general, it does not require the degree of operator skill essential to shielded metal arc welding or gas tungsten arc welding. Other advantages include (1) high deposition rates, (2) good use of filler metal, (3) no slag and flux removal, (4) reduction of smoke and fumes and (5) versatility.

Fabrication Process

For a visual inspector to fully understand a metallurgical joint and to effectively communicate test results, welding terminology and symbols must be used. Figures 9 to 12 show terms commonly used for describing the features of welds.

In making a welded joint, the various configurations must first be fabricated and suitably prepared. In most instances, it is also necessary to fit up or hold one or more joint members in place with fixturing to prevent movement during welding. The base metal and filler metal are usually melted and re-fused so that a metal stringer bead develops along the length of the weld (Fig. 13).

FIGURE 8. Gas metal arc welding (metal inert gas welding): (a) setup; (b) interface with base metal.

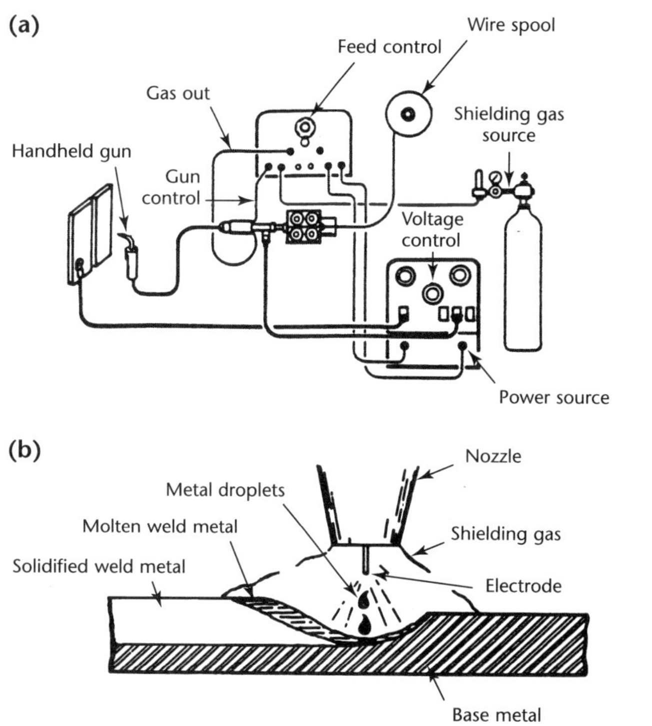

Laying down a longitudinal bead along the entire weld length is called a *pass*. The weld depicted in Fig. 13 is called a *multipass weld*. In those welding techniques using flux, slag removal is required after each pass or after each stop, if welding is interrupted during a pass.

Because there is usually a depression (crater) at the end of a weld bead, the welder must take special measures to ensure filling this crater when restarting the bead after a stop. The welder must

FIGURE 9. Features of fillet welds.

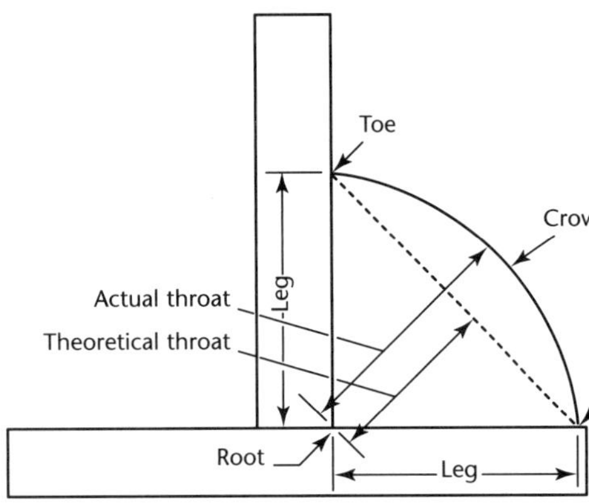

FIGURE 10. Components of fillet weld: (a) concave face; (b) convex face.

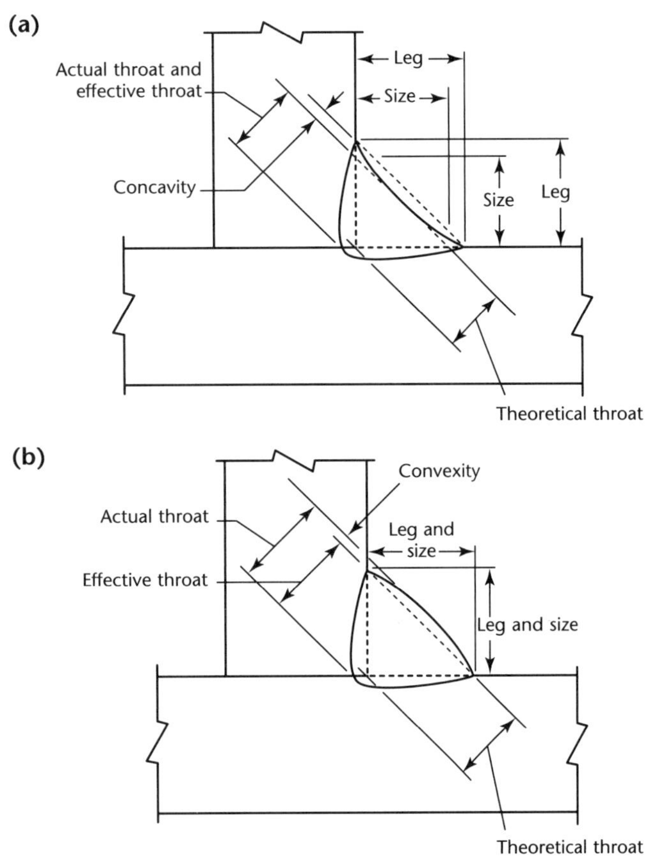

FIGURE 11. Components of typical weld.

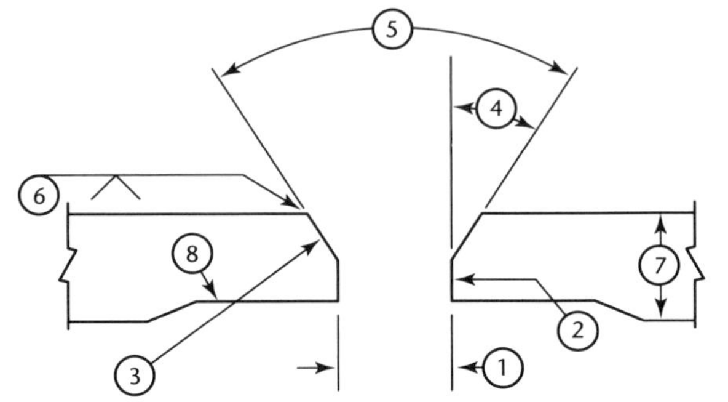

Legend

1. Root opening: separation between members to be joined at root of joint (sometimes called *gap*).
2. Root face: groove face adjacent to root of joint (also called *land*).
3. Groove face: surface of member included in groove (also called *bevel face*).
4. Bevel angle: angle formed between prepared edge of plane perpendicular to surface of member.
5. Groove angle: total included angle of groove between parts to be joined by groove weld (also called *included angle*).
6. Size of weld: joint penetration (depth of chamfering plus root penetration when specified).
7. Plate thickness: thickness of welded plate.
8. Counterbore: boring of pipe inside diameter to correct for out of roundness caused during manufacture.

FIGURE 12. Common weld terminology.

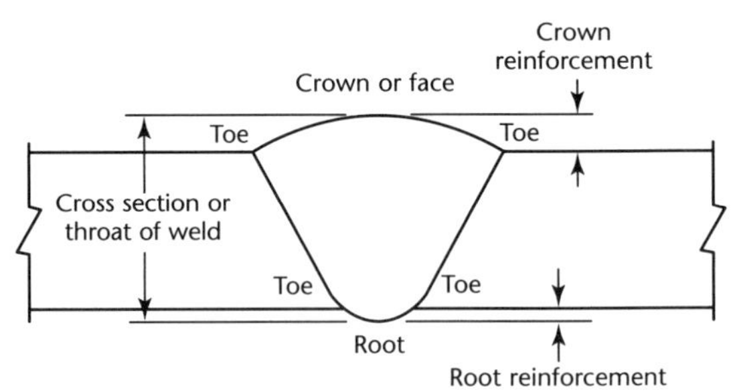

FIGURE 13. Stringer bead weld pattern.

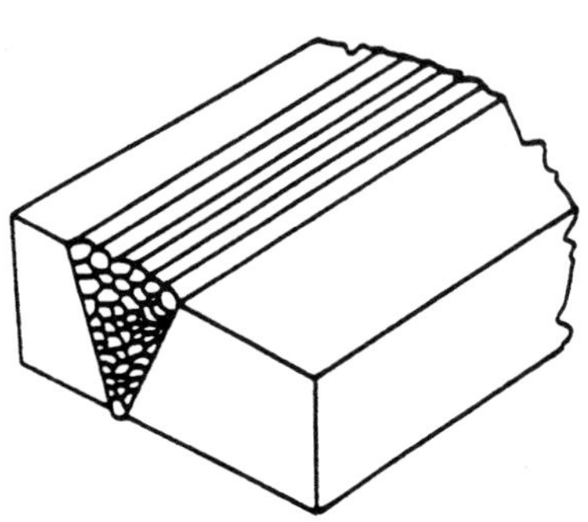

also avoid melting a groove in the base metal near the toe of the weld. This undesirable groove is called *undercut* (see discussion below of weld joint discontinuities).

Visual Tests of Metal Joints

The text below is provided for educational purposes and does not present actual visual test requirements. All nondestructive tests must conform to standards (with supplements) specified by law and by contract.

Visual tests of welds are normally performed before the start of fabrication, in process, after completion of fabrication, periodically in service and after repair of discontinuities found during the course of inservice tests. The durations of intervals between inservice tests are governed by the applicable code. Although inservice inspection may not specifically call for visual testing, a preliminary visual test under suitable conditions before required surface or volumetric tests can detect serious material discontinuities.

The text below discusses typical requirements for visual tests, the inspector's role in the examination process and some techniques for determining the quality of welded joints. Before performing an inservice visual test, the inspector needs to verify that the following requirements have been met: (1) personnel qualifications, (2) testing procedures, (3) drawings, (4) equipment, (5) surface preparation and (6) test conditions.

Personnel Qualification

Typically, the applicable code is consulted to determine requirements for personnel qualification and vision acuity. The *ASME Boiler and Pressure Vessel Code,*[3] in Section XI, IWA-2300, requires training and qualification to levels of competency given in ANSI/ASNT CP-189 or central certification.[4,5] Personnel qualification entails education, training, experience, testing, periodic recertification and vision examinations.

Procedures

Procedures in visual testing of welds are often governed by codes. The *ASME Boiler and Pressure Vessel Code*[3] and the American Welding Society's *Structural Welding Code*[6] are widely used. A written procedure based on the applicable code should be prepared by the organization responsible for performing the visual test. The procedure should include (1) visual test performance guidelines, (2) surface condition, (3) technique of surface preparation and tools used, (4) type of viewing (direct or remote), (5) special illumination, instruments or equipment, (6) sequence of test steps, (7) data to be tabulated and (8) report forms or statements needed.

The inspector should refer to a copy of the visual test procedure at the test site. A reference standard can be a valuable tool for judging preweld fitup, welding technique and completed weld quality (Fig. 14).

Drawings

A thorough understanding of all drawing requirements is necessary before a visual test. The inspector not only is responsible for detecting surface discontinuities in the completed weld but also must verify that all weld size and contour requirements are met. Included is the responsibility for ensuring that the full extent of welding specified in the drawing has been performed. These tasks require a thorough knowledge of welding terminology and symbols.

FIGURE 14. Typical welding reference standard.

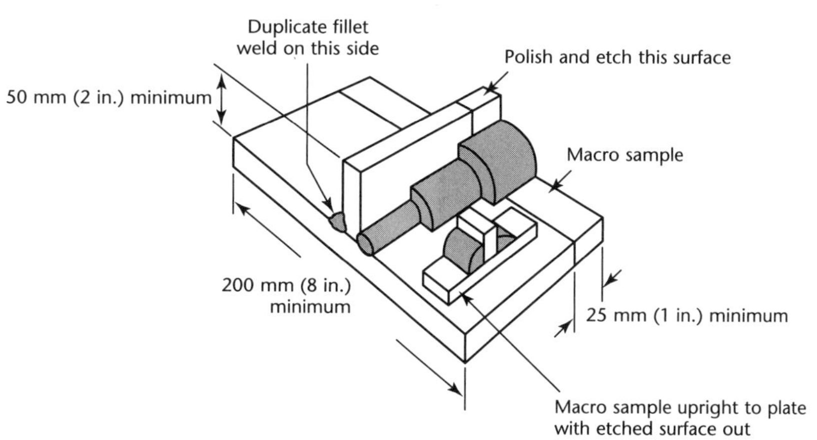

Equipment

Basic tools for direct visual testing include (1) an artificial light source, (2) mirrors, (3) magnifiers, (4) straight edges or rules and (5) weld gages (Fig. 15).

FIGURE 15. Typical weld gages: (a) gage for concave fillet weld; (b) high low gage for inside diameter mismatch of small diameter pipe; (c) cambridge gage for checking alignment; (d) weld throat gage; (e) high low welding gage for misalignment after welding.

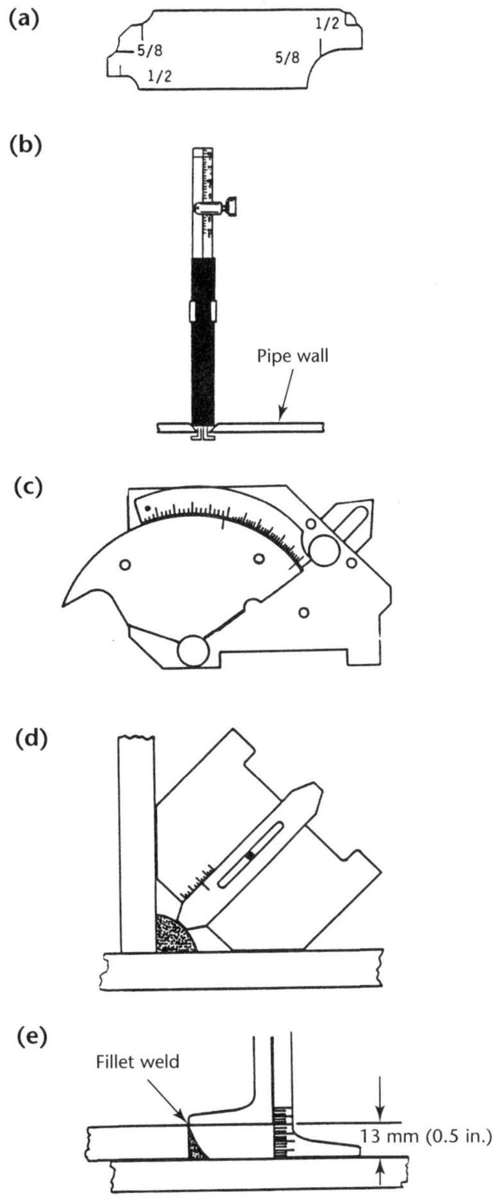

Surface Preparation

The test surface should be free of slag, dirt, grease, weld spatter or other contaminants that might make it obscure to the unaided eye. Surface preparation may also include steps needed for valid interpretation of subsequent nondestructive tests.

The test area normally consists of 100 percent of the readily accessible exposed surfaces of the test object, including the entire weld crown surface at a specified distance such as 25 mm (1 in.) from the adjacent base metal (that is, from the test surface of the weldment).

Test Conditions

Direct visual tests may be used if there is access to the area of interest without personal injury and if the unaided eye may be placed within 600 mm (24 in.) of the test surface at an angle no less than 30 degrees. Mirrors can be used to improve the angle of vision.

Natural or artificial lighting of sufficient intensity and placement is needed to illuminate the test areas and to allow proper reading of weld gages and other equipment. Criteria for vision acuity are specified in various procedures and in the *ASME Boiler and Pressure Vessel Code*[3] — for example, in Section XI, IWA-2211.

Remote visual test systems may be needed if access is impaired or if personal injury could result from direct visual testing.

Testing Procedure

The following are some issues that a visual inspector should address before welding.

1. Verify the procedure's qualification.
2. Verify welder or welding operator qualification status and limits of qualification, including limits of base and welding material type, limits of base material thickness and diameter limits, limits of welding position and currency of qualification date.
3. Verify that lighting on the test surface is sufficient.
4. Verify identification of base material and visually examine joint preparation. This part of examination should verify that base material (including backing ring, if used) are compatible with the detailed weld procedure; that the welding follows the drawing; that the weld preparation is free from base material discontinuities such as laminations, laps, nonmetallic inclusions or pinhole porosity.

5. Verify that the weld preparation geometry is to the dimensions required on the joint design drawings. This should include: alignment of parts to be welded; size of root face (land) and root gap; groove angle; identification mismatch of butt joints; clearance of backing rings, strips or consumable inserts.
6. Verify that the conditions of the general welding procedure and the conditions of the detailed welding procedure are followed.
7. Verify that tack welds are completely removed or prepared for incorporation into the final weld.
8. Visually examine prepared tack welds for discontinuities.
9. Visually examine inside of piping (when applicable) for cleanliness.

During welding, the visual inspector should do the following.

1. Verify preheat temperature as required by the detailed welding procedure.
2. Visually inspect for cleanliness, weld spatter, slag and oxide removal between passes.
3. Visually inspect for discontinuities in surfaces of weld beads and side walls of preparation.
4. Verify interpass temperatures as required by the detailed weld procedure.
5. Verify that amperage and voltage specifications where appropriate are being met.

After welding, the inspector should do the following things.

1. Verify postheat temperature as required by the detailed welding procedure.
2. Visually inspect the finished weld for adequacy of the dimensions required in the test procedure, including the following: leg size, throat and profile of fillet welds; root concavity and convexity where possible; weld reinforcement; transition of weld metal for thick to thin sections; and acceptable weld slope for weld joint offsets.
3. Visually inspect the weld surface for discontinuities and workmanship.
4. Verify the removal of fitup lugs and any other temporary attachments and the proper preparation of the affected base metal.

Visual Testing for Weld Discontinuities

The purpose of visual weld testing is to identify critical surface discontinuities at a point in the fabrication process when repair is still possible. The principal tools needed by the visual inspector are training, good vision acuity and the ability to distinguish relevant discontinuities.

Weld Joint Configuration Discontinuities

Configuration discontinuities occur when the weld is not within the specified size or the shape required by the governing documents. This condition includes a butt joint reinforcement (Fig. 16) where weld metal extends above the base metal.

Most codes closely control butt joint reinforcement height. Many weld gages can be used for measurement of reinforcement height. One leg of the gage is placed on the base metal and the other is moved to touch the tip of the reinforcement. The reinforcement height is then read off a sliding scale. To comply with most acceptance standards, the gage should be read to 7.5 mm (0.3 in.). Most inspectors are interested in adequacy, not actual dimensions. In practice, it is convenient to set the gage to the allowable height and then use it as a go/no-go gage. If the gage clears the reinforcement, fine. If it hits, then the reinforcement is excessive.

For an equal leg fillet weld, size is the leg length of the largest isosceles right triangle that can be inscribed within the weld cross section (Fig. 17). The size of an unequal leg fillet weld is the leg lengths of the largest right triangle that can be inscribed within the fillet weld cross section (Fig. 18). In each case, the weld size is based on the length of the fillet weld leg.

Fillet weld throat dimension requirements are commonly given for three different throats, two of which are very important to the visual inspector.

FIGURE 16. Butt joint reinforcement.

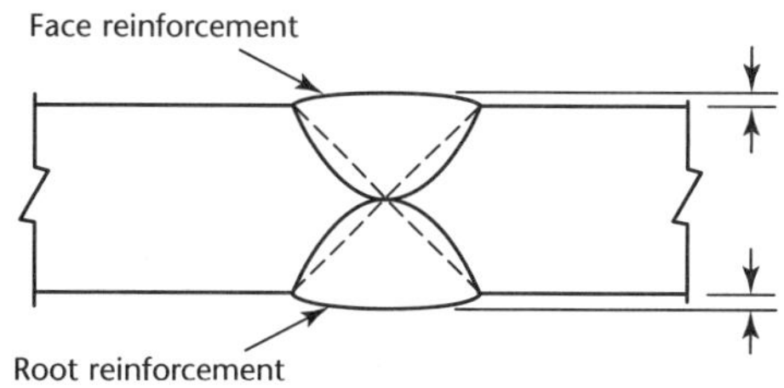

1. Theoretical throat is 0.707 times the fillet weld size for an equal leg weld.
2. The actual throat is the shortest distance from the root of the weld to the face of the weld. It must be equal to or greater than 0.707 times the fillet weld size. The visual inspector does not have access to the root area of the fillet weld and cannot consider base metal penetration.
3. The effective throat includes the amount of weld penetration but ignores excess metal between the theoretical face and the actual face (the visual inspector does not consider this throat). The actual fillet weld throat of a T joint can be measured by placing the perpendicular surfaces of a weld gage in contact with the base metal of the joint. The effective throat is then read off the sliding pointer of the gage.

Fillet weld length and spacing are critical design parameters when intermittent fillet welding is specified in a construction drawing. The welding symbols contain all of the information necessary to determine the extent of welding required. Fillet weld length and spacing is easily measured with a ruler.

FIGURE 17. Determination of equal leg fillet weld of size a: (a) convex face; (b) concave face.

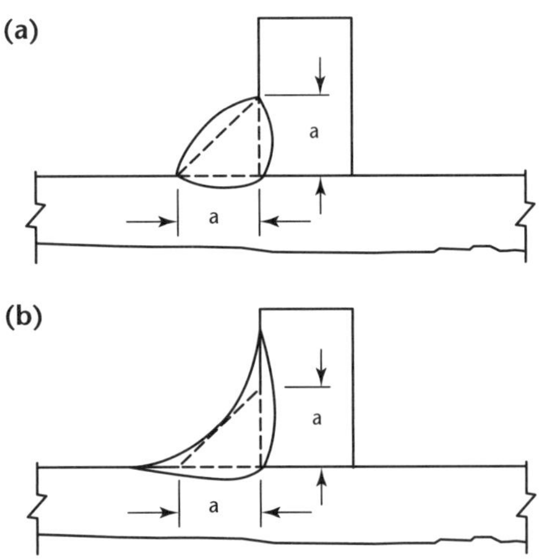

FIGURE 18. Determination of unequal leg fillet weld size, an a × b weld.

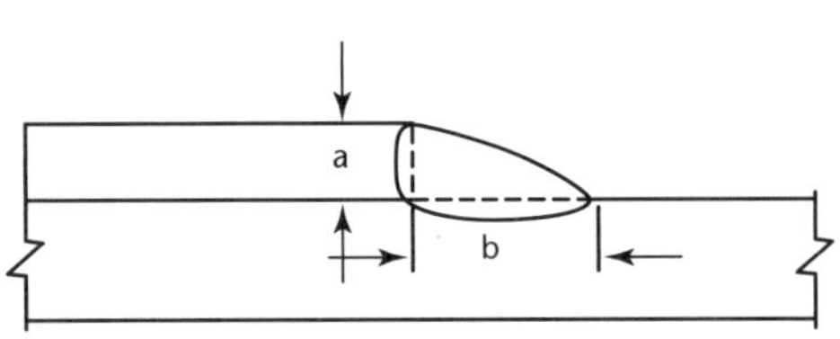

A concave fillet or butt joint groove weld surface is one that curves inward (Fig. 10a). Concavity is a smooth transition in thickness with complete fusion at both sides of the joint. This should not be confused with inadequate fusion or overlap, which is an abrupt change in thickness where weld metal is not fused to the base metal.

Concavity in a butt joint is unacceptable when the weld thickness is less than the thinnest member being joined. As shown in Fig. 19, this situation is equivalent to underfill.

The acceptability of concavity in a fillet weld is based on the actual throat being equal to or greater than the theoretical throat calculated from the specified weld size. One way to verify this is to use a standard fillet weld gage. The appropriate gage (Fig. 20) for the specified weld size may be used to square the edges of the gage with the welded parts. If all three points of the gage's double arc make contact with the weld and base metal, the

FIGURE 19. Concavity in butt joint: (a) unacceptable concavity with underfill; (b) thinnest section of base metal with acceptable concavity and no underfill.

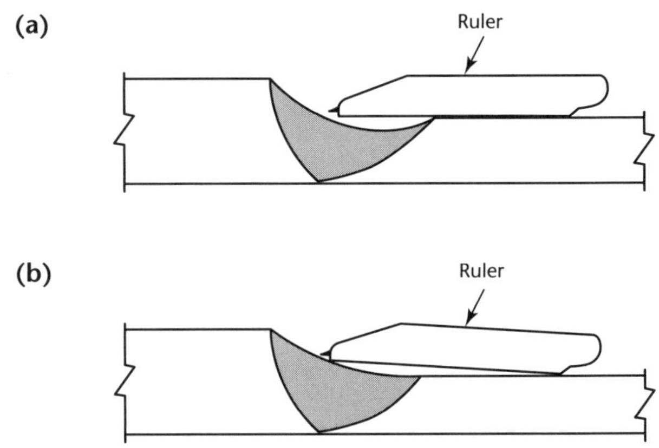

FIGURE 20. Gage to determine fillet weld concavity.

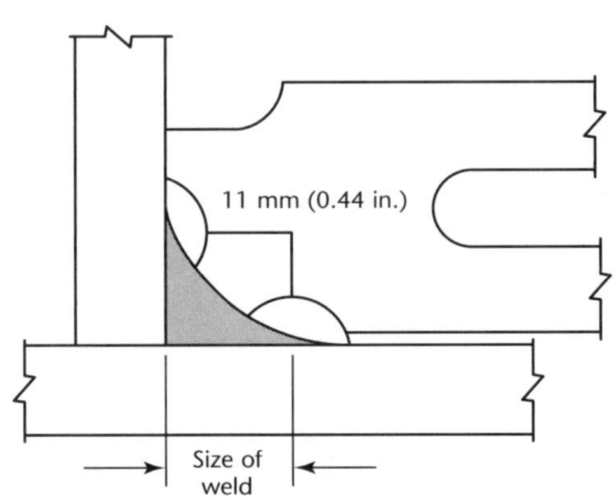

weld concavity is acceptable (Fig. 21). If the center point does not make contact with the weld metal, the concavity is excessive and the weld is unacceptable.

A convex weld surface bulges or curves outward (Fig. 10b). Excessive convexity in a butt joint is the same as excessive reinforcement. A few codes have accepted convexity not greater than 0.1 times the measured fillet weld leg length (or the longer leg in the case of an unequal leg fillet weld) plus 0.75 or 1.5 mm (0.03 or 0.06 in.), depending on the standard being used. The actual weld throat minus the allowable convexity must be less than or equal to the theoretical throat (Fig. 22).

In axially aligned members of different materials, thicknesses are required so that the slope in the transition zone does not exceed some specified amount. The transition is accomplished by chamfering the thicker part, tapering the wider part, sloping the weld metal or a combination of these. Figures 23 to 25 show examples of transition weld requirements adapted from the American Welding Society's *Structural Welding Code*,[6] which requires that the transition zone not exceed 25 mm (1 in.) in 64 mm (2.5 in.).

FIGURE 21. Desirable fillet weld profiles: (a) diagram of weld components; (b) weld cross section.

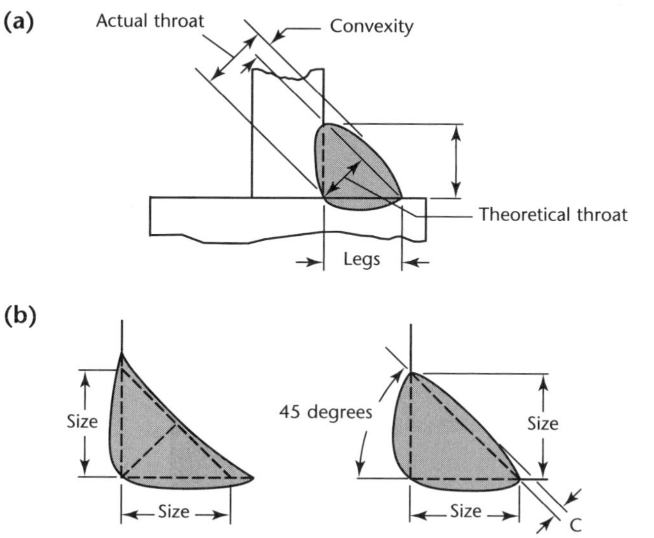

FIGURE 22. Acceptable fillet weld profiles: (a) diagram of weld components; (b) weld cross section.

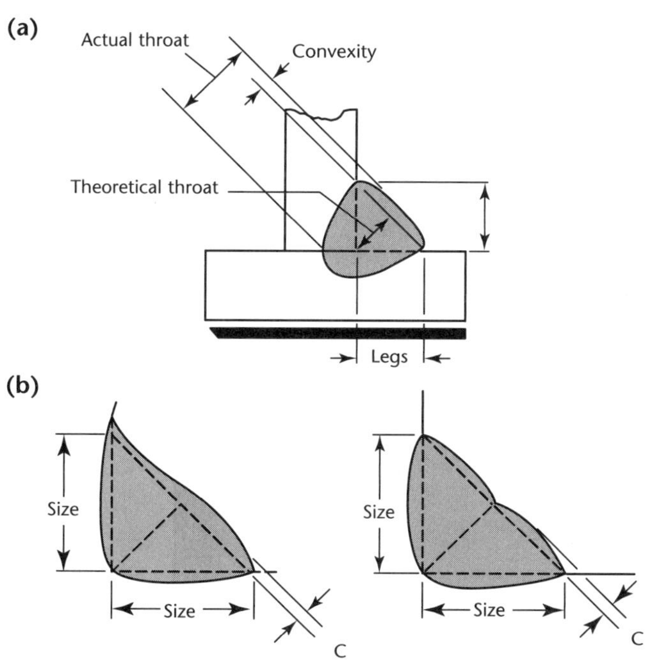

FIGURE 23. Transition formed by sloping weld surface: (a) center line alignment; (b) offset alignment.

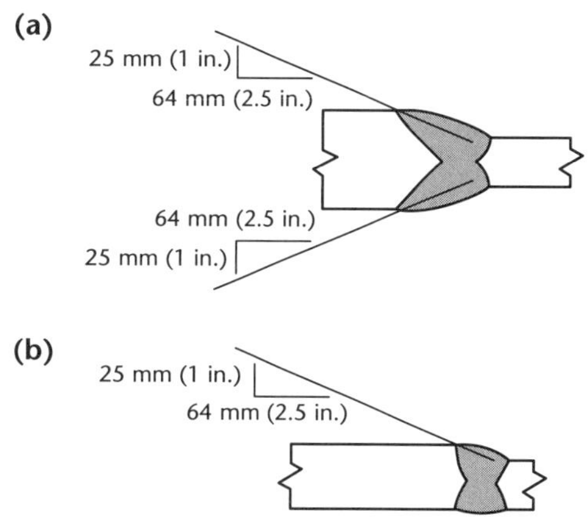

FIGURE 24. Transition formed by sloping weld surface and chamfering: (a) center line alignment; (b) offset alignment.

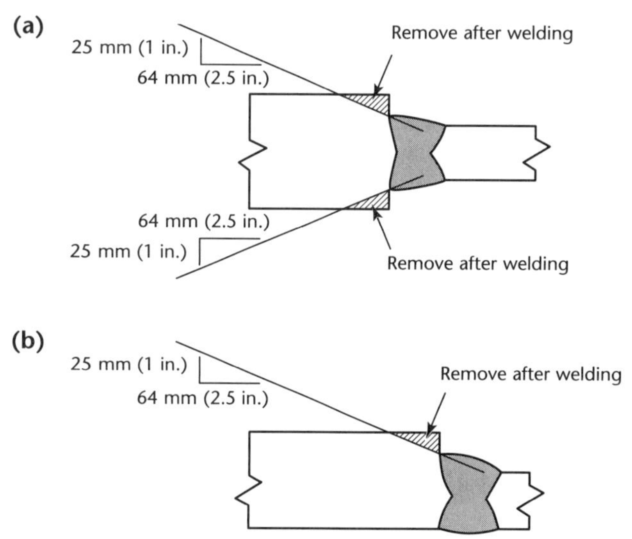

A simple gage (Fig. 26) may be used for determining acceptable transitions. The gage is positioned with the flat edge on the thicker base metal and the sloped edge over the transition zone. To be acceptable, the slope of the transition zone must follow the slope of the gage.

Weld Joint Discontinuities

Porosity

Porosity is a group of gas pockets or voids (Fig. 27), inside or on the surface of weld metal. Porosity is commonly observed as spherically shaped voids that are uniformly scattered, clustered or linear. Uniformly scattered porosity consists of individual pores which can vary in diameter from microscopic to easily visible. Cluster porosity is a grouping of small pores. Linear porosity typically occurs in the root pass. When porosity has tails some standards, such as Section VIII of the *ASME Boiler and Pressure Vessel Code,* base the size of the bore to include the tail.

Surface porosity is usually visible with the naked eye if the test area is properly illuminated. Surface porosity is generally undesirable, but the fabrication document should be consulted for acceptance standards. In cases where standards are given for porosity (Sections III and VIII of the *ASME Boiler and Pressure Vessel Code,*[3] ANSI/ASME B31.1,[7] AWS D1.1M[6]), acceptance limits are based on either maximum single pore diameter, number of pores below a given size per unit area, or aggregate pore diameter per unit area. A rule with increments of 0.8 mm (0.03 in.) or less is needed to make this determination. A low power magnifying lens may also be useful for measuring pores.

FIGURE 25. Transition by chamfering thicker part: (a) center line alignment (particularly applicable to web plates); (b) offset alignment (particularly applicable to flange plates).

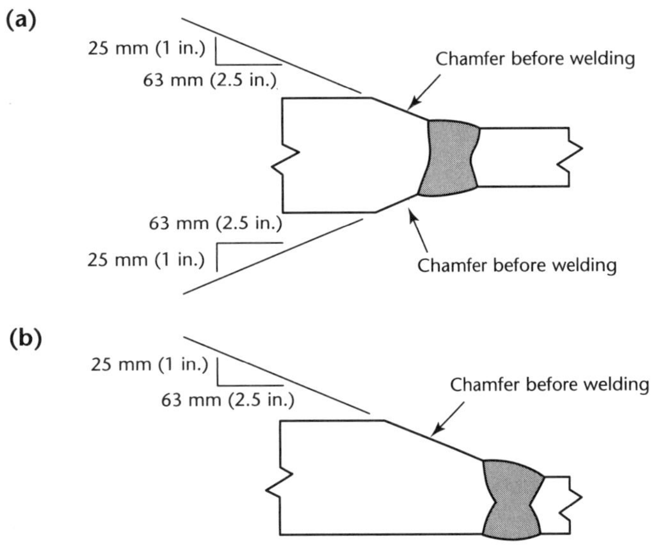

FIGURE 26. Example of acceptable slope of taper where 4:1 slope is specified.

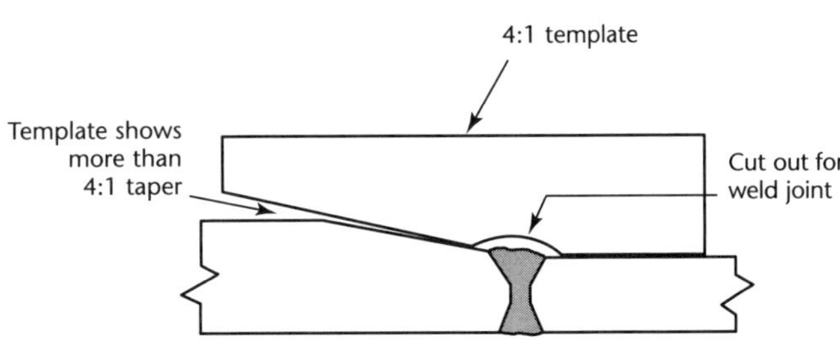

Overlap

Overlap or cold lap is the protrusion of weld metal beyond the toe or root of the weld (Figs. 28 and 29). Overlap is generally not allowed in a completed weld because the notch formed between the crown and base metal concentrates stresses.

The presence of overlap is determined by visually examining the weld metal-to-base metal transition at the toe of the weld. The transition should be smooth, without rollover of weld metal.

Consider a tangent to the weld metal at the intersection between the weld metal and base metal. If no overlap is present, the angle between the tangent to the weld metal and the base metal is greater than or equal to 90 degrees. If overlap is present, the angle between the tangent to the weld metal and the base metal is less than 90 degrees.

FIGURE 27. Porosity beneath weld surface: (a) longitudinal section view; (b) cross section view.

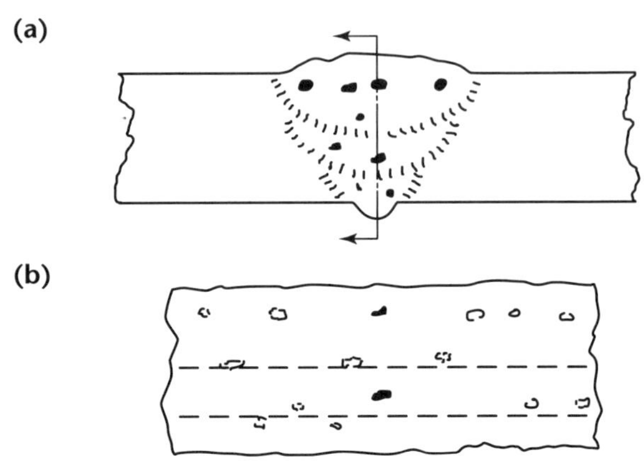

Undercut

Undercut is a groove formed at the toe or root of a weld when base metal is melted away and left unfilled by weld metal (Fig. 30). Undercut can be difficult to measure accurately. When a gage is used to measure the depth of undercut in butt joints, the body of the gage is held to straddle the weld crown and a pointed tip on a semicircular dial is placed in the undercut.

Undercut results in a reduction of base metal thickness and may materially reduce the strength of a joint, particularly the fatigue strength. For this reason, excessive undercut is undesirable in a completed weld and most codes specify acceptance criteria for undercut. For example, Section III of the *ASME Boiler and Pressure Vessel Code*[3] has required that undercut be less than 0.8 mm (0.03 in.) deep; welders and inspectors must consult the edition specified by contract. AWS D1 gives acceptable limits for undercut.[6]

Inadequate Joint Penetration

Inadequate penetration is a failure of base material to fuse with filler metal in the root of the weld joint (Fig. 31). Where full penetration is specified by the construction drawing, the inspector must visually determine that the root has been completely filled with weld metal.

Cracks

Surface cracks detectable by visual testing may occur longitudinally or transversely along the weld root, face, toe and heat affected zone (Fig. 32) or in arc strikes outside the weld. A low power (5×) magnifying lens is the best tool for visually detecting surface cracks. Care must be exercised when using a magnifier — minor surface irregularities can be accentuated, affecting the accuracy of interpretation.

With few exceptions, no visible surface cracks are allowed in completed welds.

FIGURE 28. If overlap is present, the angle between the tangent to the weld metal and the base metal should be less than 90 degrees. Acceptable profiles without overlapping: (a) in V joint; (b) in butt joint.

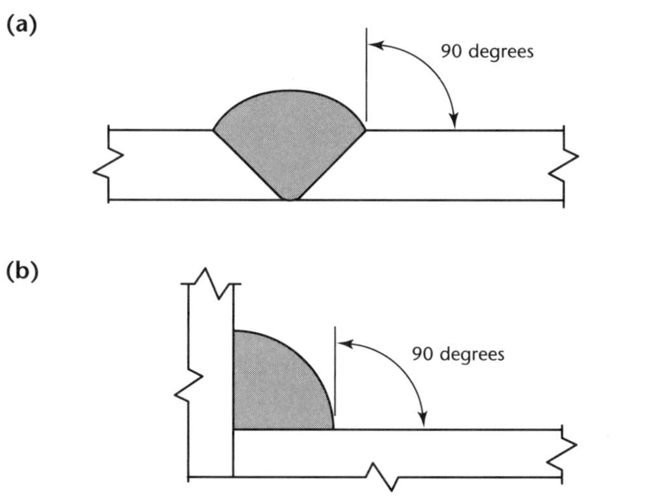

FIGURE 29. Presence of unacceptable overlap: (a) in V joint; (b) in fillet joint.

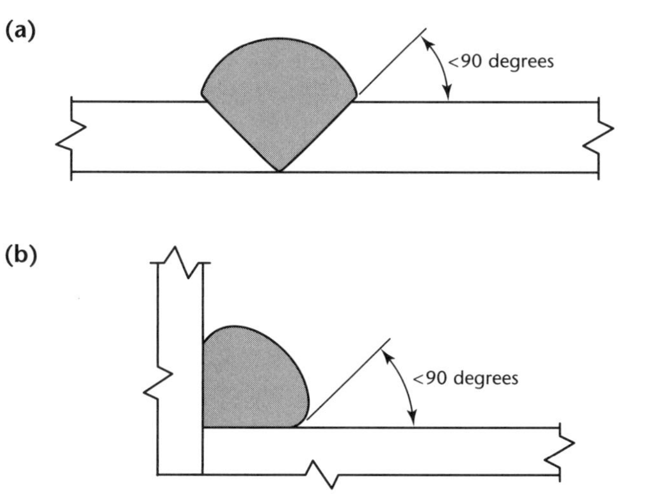

FIGURE 30. Presence of undercut: (a) in fillet weld joint; (b) in butt weld joint.

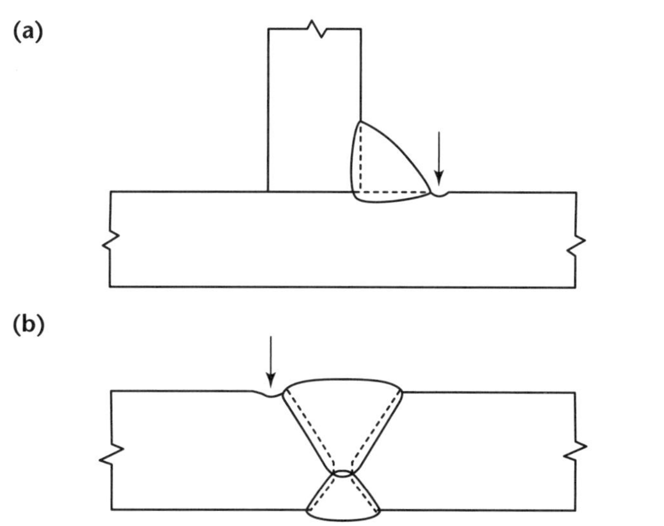

FIGURE 31. Inadequate joint penetration.

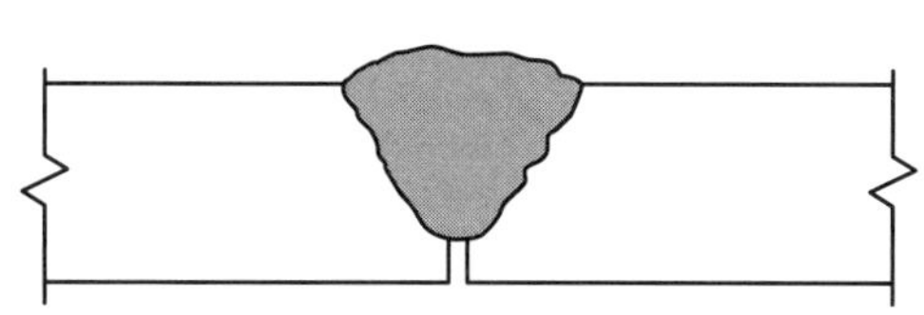

Arc Strikes

Arc strikes are caused by unintentional rapid heating of the base metal or weld metal and subsequent rapid cooling of the molten material (Fig. 33). Melting of the base metal and deposition of filler metal are often associated with strikes caused by a welding arc.

Arc strikes may also be caused by a poorly connected welding ground clamp or by instruments used for magnetic particle testing. The extremely high heat input that occurs during an arc strike can cause localized hardness and cracking.

Visual Testing of Brazed and Soldered Joints

Brazed Joints

In power plants, brazed joints are commonly used on instrumentation lines in low criticality systems, called Class III systems in the *ASME Boiler and Pressure Vessel Code.*[3] Lap joints are the most common configurations for these applications. Before brazing, the soundness of these joints is ensured by control of cleanliness, of joint overlap and of the joint gap.

The objective in visual testing of brazed lap joints is to determine that a continuous flow of braze filler metal has been achieved. If filler metal is applied from the same side of the joint as is visually examined, there is no assurance of penetration within the joint, even if the appearance of the fillet is good. Ultrasonic testing may be used to determine filler metal penetration in such cases.

To simplify verification of a sound braze, good joint design requires that filler metal is placed inside the joint before brazing. Visual testing is then conducted on the outside fillet to determine if it is continuous. If the outside fillet is continuous, there is reasonable assurance that a sound joint has been made.

Soldered Joints

Inadequate wetting is the most common problem encountered with soldering, resulting in incomplete coverage of the surface with solder. Nonwetting is apparent during visual testing by the appearance of the original surface finish. Dewetting is the flow and retraction of solder, caused by contaminated surfaces, dissolved surface coatings or overheating before soldering.

Dewetting may look like nonwetting but it is identified by a colored residual film with beads of solder where the solder receded. Excessive movement of the soldered joint during solidification may materially weaken the joint and is visually determined by a frosty appearance.

Visual testing is the most widely used method for nondestructive inspection of soldered joints. Reference standards are used to aid the inspector in checking for general design conformance, wetting, finished cleanliness of the product and the quality and quantity of solder.

Acceptance Standards

Every feature of the joint configurations and dimensions subjected to visual testing has an associated quantitative tolerance limit or qualitative criteria of acceptability. These are established with the drawing or incorporated into the requirements of the governing document. Discontinuities also have quantitative and qualitative criteria.

The inspector should record all test results. Acceptability may be determined at the site or later, depending on the specified requirements for the application.

Figure 32. Typical weld cracks.

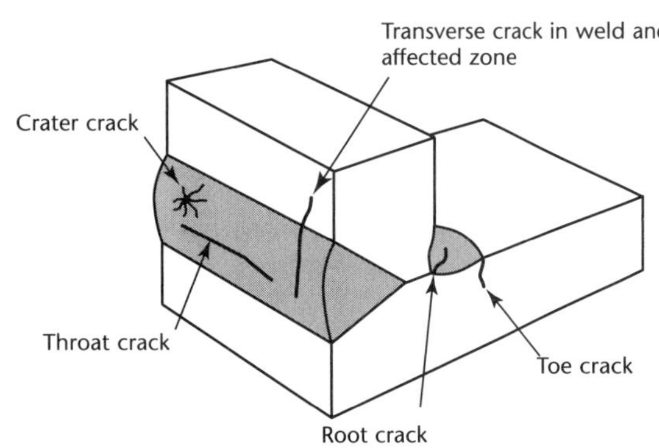

Figure 33. Arc strikes on work pieces and portions of multipass weld.

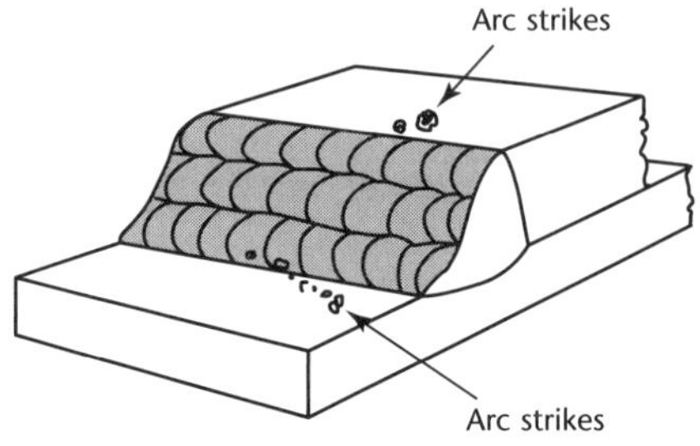

Recording and Reporting Visual Test Results

Visual test results are recorded by using forms developed for specific applications. If no form is provided by the customer or regulation, the inspector should develop one that provides the following data: (1) joint identification and description, (2) joint inspector's identification, (3) testing dates, (4) instrument and equipment identification (may refer to procedure), (5) procedure identification (and checklist) and (6) record of the test results.

In the record of results, identify all measurements to be recorded, provide space for all recorded measurements with identification, and provide space for recording indications.

Through standard recording formats and prepared forms, the inspector and supervisor can determine if all required observations have been made properly.

Part 2. Visual Testing of Various Components[8,9]

Visual Testing of Reactor Pressure Vessels

Visual testing of a reactor vessel and its internal components is one of the most critical operations in any inspection program. Not only is the test closely audited by the Nuclear Regulatory Commission and the American Nuclear Institute (ANI), it is usually done on a critical outage schedule, so there is no opportunity to substitute operations if there are problems with the inspection.

A Level III inspector is typically involved in all phases of the test, beginning with the selection of equipment, the training of test personnel, supervision of the test itself and the final interpretation and review of the results. The text below presents situations typical of such a visual inspection and suggests techniques that might help resolve anticipated problems.

Applicable Code

Visual tests of reactor vessel internals are performed according to the rules of the plant's inservice test program, which is in turn based on visual tests required in appropriate standards. It is the responsibility of the inspector to perform the tests so that the requirements of the standard and Nuclear Regulatory Commission bulletins are satisfied.

The primary requirement is that the visual test be done according to a written procedure and that a checklist be used to plan and perform the test. Possible checklist items are in Table 1.

Visual tests of the vessel are performed to determine the condition of critical inner surfaces. Such critical surfaces include high stress points at the junctions of nozzles and the vessel or nozzles and the cladding. Cladding is critical because it protects the vessel from corrosion.

The visual inspection of a reactor vessel may involve many techniques. Vessel tests may be made with the vessel empty — or filled with water with the internal components in place — by remote viewing or direct viewing (in any such test, it is impossible for the inspector to place his unprotected eye close to the test site).

The purpose of the test is to determine how the pressure vessel system has been operating. Many small component failures have been detected by such tests, preventing serious failures. It is almost impossible to make a complete visual test of anything as complex as pressure vessel internals. The visual inspector has limited time available because time in the vessel may be expensive during a refueling outage. Some limitations are set by equipment, and each test technique has drawbacks.

Boiler Code Visual Test Categories

Nuclear plants are examined on a routine basis according to the requirements of their license, the laws of the state in which they are operated and special requirements of the Nuclear Regulatory Commission. Most of these tests must meet *ASME Boiler and Pressure Vessel Code* requirements in the inservice test program. In certain instances, visual tests are requested by the Nuclear Regulatory Commission when component failures are suspected.

Before the 1977 edition of the *ASME Boiler and Pressure Vessel Code*, visual tests were conducted to determine the general condition of a part, component or surface. These tests were made according to Section V, Article 9, with an additional requirement for lighting sufficient to resolve a 0.75 mm (0.03 in.) wide black line on an 18 percent neutral gray card.

After the 1980s, Section XI of the *ASME Boiler and Pressure Vessel Code* divided visual tests into three categories: VT-1, to determine the condition of a part, component or surface; VT-2, to detect leakage from pressure retaining components or abnormal leakage during system pressure or functional inspection; VT-3, to determine the general mechanical and structural condition of components and supporting structures and to determine conditions relating to component or device operability.

VT-1. The VT-1 test (for condition of part, component or surface) is required on the following components, according to Section XI of the *ASME Boiler and Pressure Vessel Code*.

1. The reactor vessel bushings and closure washers are available for close, direct visual observation when the head is removed from the reactor. The bushings, part of the head, may require that the inspector wear protective clothing to avoid inhaling loose contaminant. The washers should present no contamination problem because they are usually kept separate from the head in special bins. Volumetric examination of the vessel bushing area and ultrasonic testing of vessel bushing areas may be required.
2. All bolts, studs and nuts of the reactor vessel are kept clear of the head and are straightforward to inspect. The root of the threads in the studs may be fouled with a threading compound nearly impossible to remove, making a close visual test impractical. The eddy current technique for surface testing of these components may be used. Volumetric and surface examinations of the studs and surface examination of the nuts are needed, in addition to ultrasonic testing of reactor vessel head bolts.
3. The reactor vessel internal attachments within the beltline region can be visually tested by using a video monitor.

The visual inspector is expected to be (1) trained and qualified and (2) able to demonstrate vision acuity and color discrimination. Details of qualification and certification should be specified in the contract, written practice and regulations.

VT-2. The VT-2 test (for evidence of leaks) can be performed every time an inspector works around the reactor vessel, particularly in the area of penetrations or underneath the vessel (in boiling water reactors). The VT-2 test is recorded during a pressure test of the system.

The VT-2 inspector looks for stains, dislodged insulation and water on the floor or dripping from overhead insulation. A leak in a vessel nozzle or safe end can result in water traveling along the insulation until it has a path for escape, so the vessel inspector should be alert to conditions in piping around the vessel.

Vertical insulated components may be examined at the lowest point where leakage may be detected. Horizontal insulated components are examined at each insulation joint. If accessible, the VT-2 test is performed directly on the vessel. If this is not possible, the test may be made on the surrounding areas and floor, looking for evidence of leakage.

Much of the VT-2 leak test is performed by the visual inspectors when the plant is cold. These individuals have close access to the nozzles and safe ends when they are uncovered. The inspector should form the habit of making a very close inspection for leaks during other visual tests.

There is a tendency for some inspectors to limit themselves to the specialized test underway, neglecting other visual evidence. Inspectors should remember that the reactor is in theory a watertight and gastight system. Any water where none is expected, no matter how small the amount, is cause for concern.

VT-3. The VT-3 test (for mechanical and structural conditions) is used to determine the general condition of the vessel outside of the beltline region and the core supports. This test can be video recorded. The object of the test is to disclose corrosion or broken, worn or missing components and may require physical measurements. Test personnel are qualified by the same program as VT-1 and VT-2.

Table 1. Typical visual inspection checklist for pressurized water nuclear reactor's vessel internals. Each component listed gets complete visual inspection.

Core barrel assembly

1. Core barrel alignment key
2. Core barrel thermal pad mounting
3. Core barrel irradiation specimen basket
4. Core barrel lower internals head alignment pin
5. Core barrel outlet nozzle/vessel interface surface
6. Core barrel midplane girth weld
7. Instrumentation mounted in core barrel
8. Upper internals alignment key in core barrel
9. Baffles in core barrel
10. Guide tube attachment to upper core plate
11. Upper core plate
12. Upper core plate alignment keyway
13. Alignment of keyway upper support plate
14. Core barrel

Supporting components

15. Upper support plate instrument conduits and supports
16. Lower core plate showing fuel assembly alignment pins, flow holes, in core instrument guides, supports and attachments
17. Lower core support columns and instrument guides
18. Support column instrument guide tube attachments to lower core support plate
19. Upper internal keyway insert attachments
20. Fuel pin and support column attachment upper internals
21. Upper internals guide types and support columns
22. Guide tube note: split pin attachments
23. Baffle bolting
24. Upper tube assemblies on upper support plate with instrument port guide post
25. Secondary core support assembly and instrument guide tubes
26. Reactor vessel internal surfaces
27. Interior attachments and core support structures
28. Core support structures

Proper performance of a remote visual test depends on appropriate fixturing and lighting. A video camera on a long pole is like a pendulum — unless care is taken to stabilize it, the inspector can spend as much time waiting for the camera to stop swinging as for completing the test itself.

Test Procedures

A direct VT-1 test should be performed with the eyes no more than 600 mm (24 in.) away from the test object, at an angle no greater than 30 degrees to the surface. The illumination on the object is such that a 0.75 mm (0.03 in.) wide line inscribed on an 18 percent neutral gray card can be seen by the inspector. Remote viewing aids such as video cameras work if conditions equivalent to direct inspection can be met.

It is very unlikely that a gray card test could be used with an underwater closed circuit television system. In some cases, the closed circuit television must resolve a stretched wire of a known diameter placed against the test material (Fig. 34).

Pressurized Water Reactor Internals

All of the internals of the pressurized water reactor (Fig. 35) may be removed from the vessel for testing. To do this, the fuel must also be removed. Consequently, the inservice inspection program recommended for this type of reactor calls for visual testing of internals at ten year intervals, as defined in the *ASME Boiler and Pressure Vessel Code.*

During a normal refueling operation on a pressurized water reactor, the head and control rod drive assembly are

FIGURE 34. Calibration standard for remote visual test.

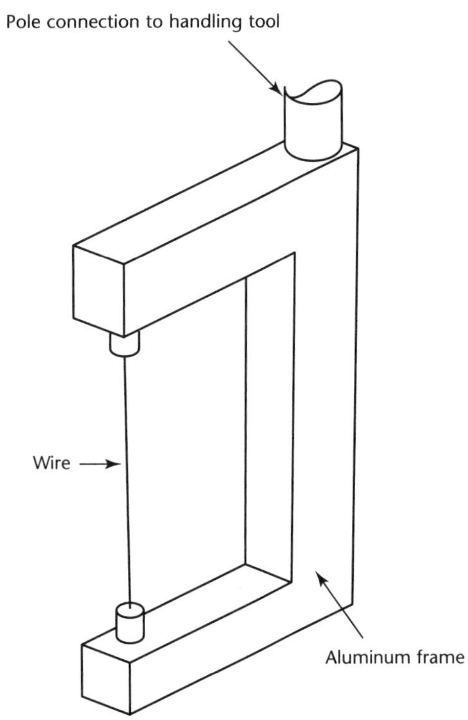

FIGURE 35. Pressurized water reactor vessel.

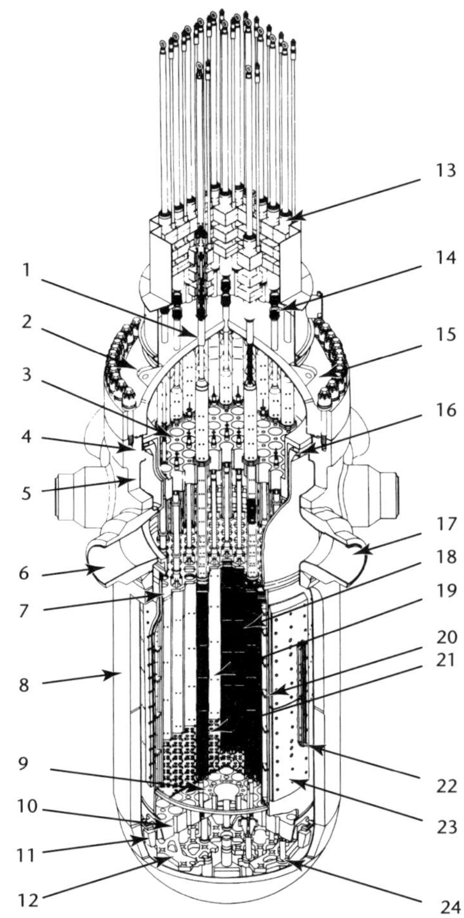

Legend

1. Control rod drive shaft.
2. Lifting lug.
3. Upper support plate.
4. Internals support ledge.
5. Core barrel.
6. Outlet nozzle.
7. Upper core plate.
8. Reactor vessel.
9. Lower instrumentation guide tube.
10. Bottom support forging.
11. Radial support.
12. Tie plates.
13. Control rod drive mechanism.
14. Thermal sleeve.
15. Closure head assembly.
16. Hold-down sharing.
17. Inlet nozzle.
18. Fuel assemblies.
19. Baffle.
20. Former.
21. Lower core plate.
22. Irradiation specimen guide.
23. Neutron shield pad.
24. Core support columns.

removed and stored. Access to the lower portions of the vessel must, in many cases, be through openings left by fuel that has been removed. In some cases, access openings for a television camera may be provided so that a small video unit may be inserted between the thermal shield wall and the vessel.

Before beginning the test, the inspector should have a written procedure and a checklist. The checklist helps to ensure that nothing is overlooked (Table 2).

At the end of the ten year interval, the pressurized water reactor internals are completely removed from the vessel and stored separately. At this time, they are examined in detail. Because this test is repeated over such long time intervals, it is important that complete records be maintained. Because an index and inspection logs may become separated from a video tape, it is a good idea to make the tape self identifying. If the recorder has a voice channel, it can be used. A title block, made by writing the essential data on a card before the taping, should be required.

Boiling Water Reactor Internals

When the boiling water reactor is opened for refueling, the head, the steam separator and the steam dryer are removed and placed on stands. The head is stored out of the water on the refueling building floor and the other components are on stands in a holding pool, which may be gated off from the reactor vessel. Fuel may or may not be completely removed. A cross section of a typical boiling water reactor is shown in Fig. 36. It is over 24 m (80 ft) from the refueling bridge to the bottom of the vessel and it is 6 m (20 ft) from the vessel flange to the refueling bridge.

With the components removed from the vessel, a television camera can view the upper nozzles and interior piping. Visual inspection of these items is very important and must be carefully performed. A positioning fixture is required for close examination of the piping.

The remaining fuel in the vessel can be a cable trap. The tops of the fuel bundles have many closely grouped springs, nuts, antirotation devices and other hardware. It is very easy to snag a small cable if it is allowed to loop into the fuel. This can be prevented by attaching floats to the cable.

The inspector working over the open reactor vessel must take extreme care that equipment does not become dislodged and fall into the vessel. Operating personnel make sure that the inspector has all pockets taped, eyeglasses taped on and all small tools moved away from the vessel. It may be worthwhile to situate the inspector and the data recorder away from the vessel and to have an operator position the camera on the bridge, using a small monitor and intercom.

Two types of remote camera positioning equipment are a camera positioner mounted to the trolley of the refueling bridge and a submersible remotely operated vehicle.

1. A video camera positioner (Fig. 37) supports and positions an underwater camera and lighting system in X, Y and Z coordinates. The unit's design allows stable positioning of the camera. The horizontal inspection boom (Fig. 38) attaches to the bottom-most extension tube and is designed for scanning the reactor pressure vessel circumference from the center of the vessel. This position obviates moving the refueling bridge or trolley for circumferential examinations.

Table 2. Reactor vessel and internals visual testing.

Test Object	Test Category	Extent Every 10 Years
Partial penetration weld	VT-2	25 percent of nozzles
Pressurizer heater penetration	VT-2	all nozzles
Vessel nozzles	VT-2	25 percent of nozzles
Control rod nozzles	VT-2	25 percent of nozzles
Instrumentation nozzles	VT-2	25 percent of nozzles
Flange surfaces	VT-1	all when disassembled
Nuts, bushings and washers	VT-1	all
Bolts, studs, nuts	VT-1	all
Vessel interior	VT-3	accessible areas
Interior attachment within beltline	VT-1	accessible welds
Interior attachment beyond beltline	VT-3	accessible welds
Core support structure	VT-3	accessible surfaces
Pressure retaining boundary	VT-2	coolant system leak test
Pressure retaining boundary	VT-2	coolant system hydrotest

FIGURE 36. Boiling water reactor vessel.

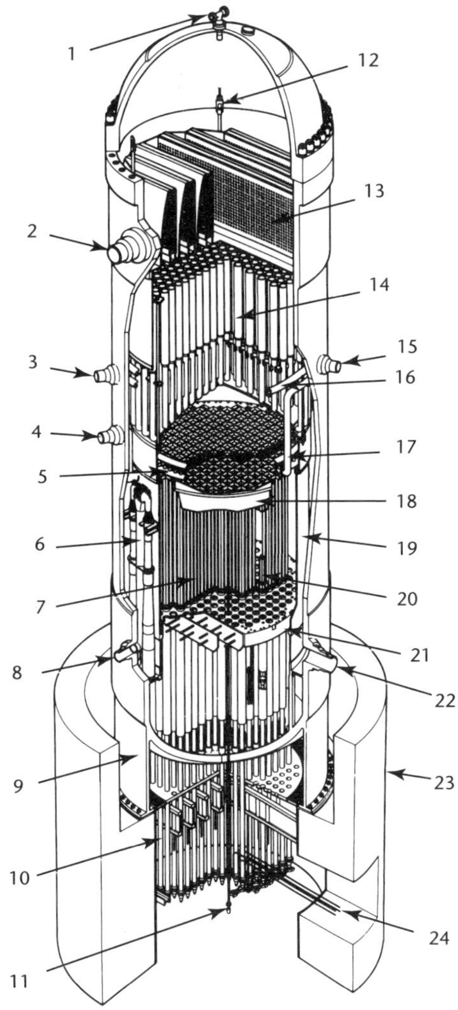

Legend

1. Vent and head spray.
2. Steam outlet.
3. Core spray inlet.
4. Low pressure coolant injection inlet.
5. Core spray sparger.
6. Jet pump assembly.
7. Fuel assemblies.
8. Jet pump/recirculation water inlet.
9. Vessel support skirt.
10. Control rod drives.
11. In-core flux monitor.
12. Steam dryer lifting lug.
13. Steam dryer assembly.
14. Steam separator assembly.
15. Feedwater inlet.
16. Feedwater sparger.
17. Core spray line.
18. Top guide.
19. Core shroud.
20. Control blade.
21. Core plate.
22. Recirculation water outlet.
23. Shield wall.
24. Control rod drive hydraulic lines.

2. The remotely operated vehicle is controlled manually, with the controller connected to the vehicle by a coaxial cable. A power supply is connected to the vehicle by an umbilical cable. The vehicle has a pan and tilt color camera, twin lights, two horizontal thrusters and one vertical thruster.

Inservice inspection components include core spray sparger and core spray piping, feedwater sparger, top guide, core plate, steam dryer, moisture separator, shroud head bolts, fuel support pieces,

FIGURE 37. Video camera positioner.

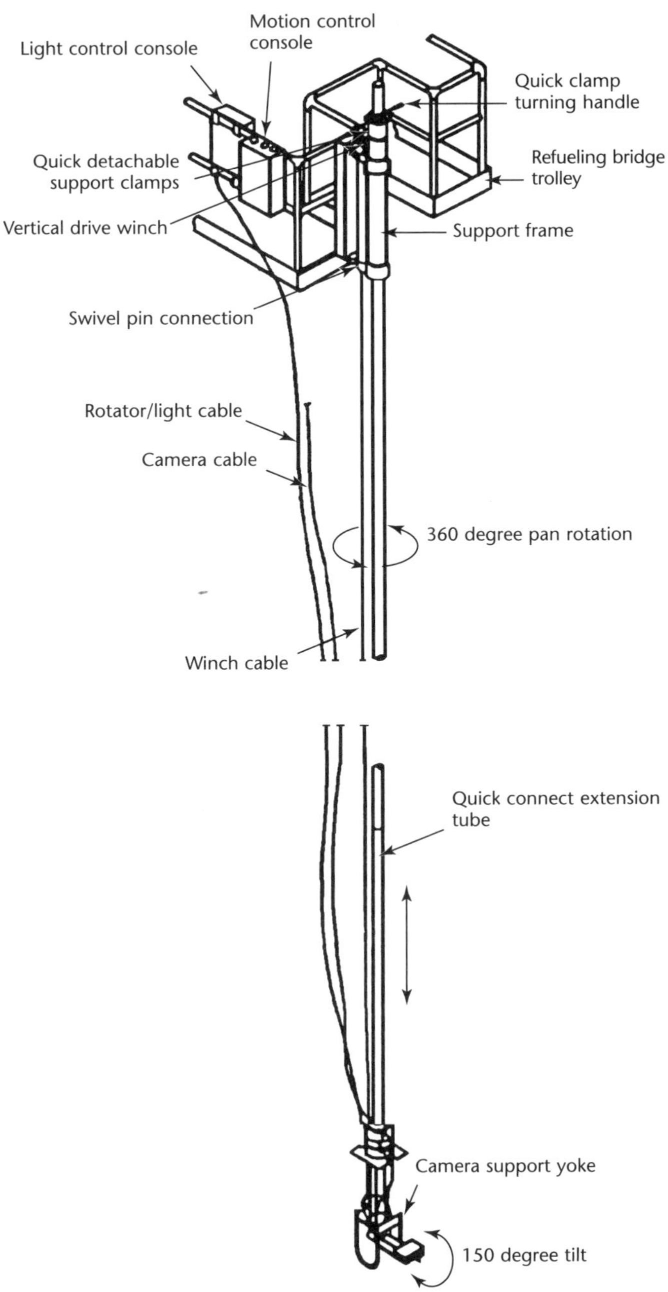

stub tubes and clad patches. Components requiring visual testing because of license agreements or regulatory requirements vary from plant to plant. Typical augmented inspection components are the intermediate range monitors, source range monitors, local power range monitors, control rod drive return nozzle shroud and shroud separator mating surface.

The expense of remote camera handling equipment is justified when inexperienced personnel are visually testing the internals of a reactor vessel. Often the remote handling equipment can provide steadier pictures and can speed up the visual tests. Some remotely operated vehicles are designed specifically for the nuclear industry. These vehicles are small enough to get to areas difficult to access with handheld cameras and produce pictures superior in steadiness and clarity.

The environment of the visual test mandates camera selection. Charge coupled device cameras start to lose the image as the radiation field increases and would not be a good choice for visual testing of reactor vessel internals. Metal oxide semiconductor chip cameras provide color images and have been used in remotely operated vehicles. For high radiation fields, however, metal oxide semiconductor chip cameras can be replaced with tube cameras.

FIGURE 38. Horizontal inspection boom for video camera.

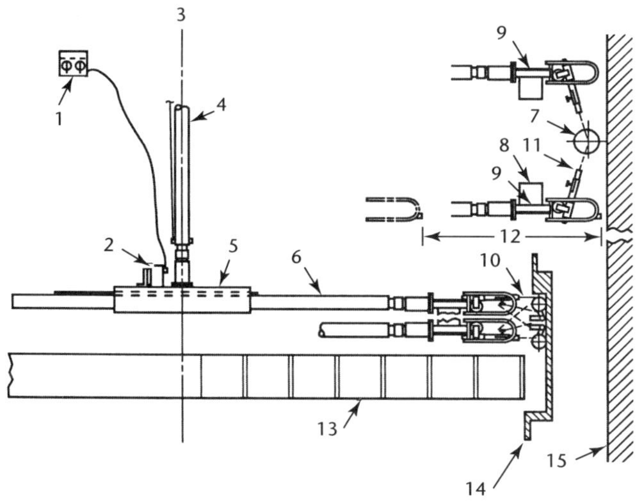

Legend

1. Boom control console.
2. Boom drive motor.
3. Reactor pressure vessel and vertical extension tubes.
4. Bottom extension tube.
5. Boom adapter.
6. Horizontal boom.
7. Top view of feedwater sparger or core spray header.
8. Buoyancy chamber.
9. Camera support yoke.
10. Core spray spargers.
11. Camera.
12. Travel.
13. Top guide.
14. Core shroud.
15. Pressure vessel wall.

Handheld cameras will probably always be needed. The technicians performing the visual testing of reactor vessel internals should get hands-on training with the camera equipment as well as training on what to look for during a visual test. Skill is needed to position a camera and record a steady picture in a timely manner.

Familiarization with Visual Recording Means

Visual test data are recorded as video. The inspector should remember that light is slower in water than in air, so the focal settings on the camera must be modified. It also means that a test cannot be qualified on the bench: it must be qualified underwater.

Where possible, a zoom lens on the camera simplifies the test because it is very difficult to increment a camera on a long pole. A camera fixture lets the camera be tilted manually.

A video test should be planned and performed systematically. If the test is recorded, the counter readings can be marked on an appropriate drawing to index each view.

The video record has the advantage of recording motion. This is important when attempting to analyze certain visual discontinuity indications. For example, a crack may cause a surface disruption that produces shadows whereas a stain or small scratch does not.

It must be remembered, however, that the detail in video images may be set to a low quality. Older analog video screens had fewer than 300 horizontal lines, much less than digital images. Even with digital images, an image which is a single frame captured from a video file may have a lower resolution than expected in a still digital photograph.

For some records, it is possible to make usable photographs from a video monitor if the camera exposure suits the monitor screen and the speed of the camera is kept slower than half the frame frequency. Monitor brightness usually does not match the surroundings, however, and a proper exposure of the entire scene is unfavorable for the monitor. Even if the picture is not as detailed as desired, it may still be used as a guide when the video is replayed later.

Visual Test Reporting

The test report should be a record of all pertinent data resulting from the test, as well as the procedures, personnel and equipment used. The report may be a specific form for each test object or a general form made specific by the entries of the inspector. The test report and photographic or video files may be filed as hard copy and archived electronically. If possible, the filed copy of the report should indicate the file names and locations of the remaining records of the test for future reference.

Drawings included as part of the test report should indicate the file names and location of photos taken or recorded scans of the video camera (Fig. 39). Video counter readings can be entered on the drawing to facilitate review of the test.

Visual Testing of Pumps

Condensate and Boiler Feed Pumps

Always refer to the dismantling procedure outlined in the manufacturer's instructions. During disassembly, check the impellers for signs of erosion and cavitation damage. Typically, the areas most affected are near the inlet of the impeller vane. The vane inlet should be smooth and radiused. Vanes exhibiting abnormal wear may be blunt edged, jagged and, if cavitation was present, porous. Each impeller is also inspected using liquid penetrant techniques over 100 percent of its surface. All cracks are repaired or the impeller is replaced.

FIGURE 39. Drawing with video and photographic record locations in horizontal cross section of reactor vessel.

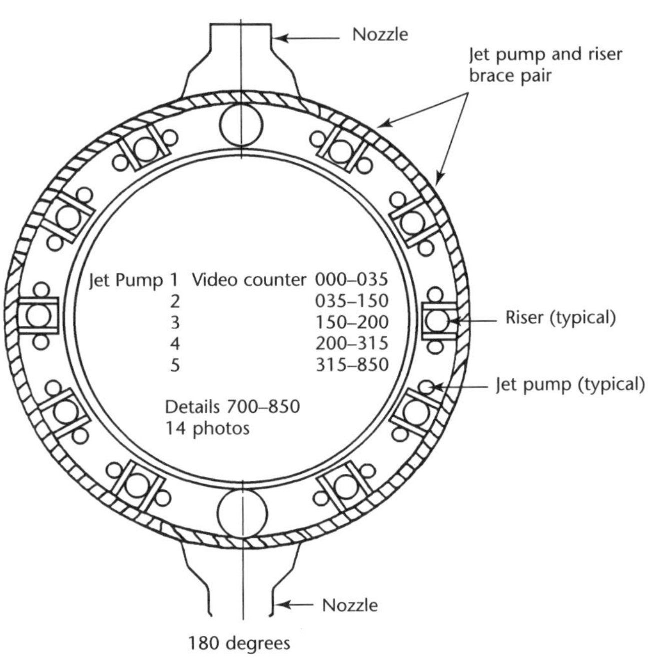

Diffuser elements for pumps are visually inspected for erosion and cracks in the same way as the impeller. Discontinuity indications are referred to the pump manufacturer for disposition if the component is to be reused.

The dimensions of sleeves and rings should be verified against the manufacturer's tolerances and replaced if worn beyond tolerance. Excessive clearances reduce the pump's efficiency and can cause hydraulic imbalance.

A total indicator reading should be taken on the pump shaft to ensure its straightness. Even slightly warped or bent shafting may precipitate excessive vibration during use. Shaft bearing journals should be checked for proper finishes. Total indicator reading or runout is the deviation from a perfect form and is normally detected by full rotation. A runout tolerance applied to a surface means that the considered surface must lie within a tolerance zone normal to the true profile of the component. The zone limits are separated by a distance equal to the specified total tolerance.

Because runout is applied as a composite form of related features having a common axis, measurements should be taken under a single setup. Runout is a geometric form control that can include such qualities as roundness, straightness, flatness and parallelism of surfaces. When reviewing the condition of a shaft or element, never try to mount the shaft in the original machining centers. It is likely that the centers could be damaged through erosion or corrosion, making true readings difficult.

Always try to set journal surfaces of the shaft on a set of V blocks. The readings should be taken on all bearing and ring surfaces with a dial indicator. In any case, the pump manufacturer should be consulted for acceptable runout tolerances.

Babbitted surfaces must be inspected for smoothness and wear. Babbitt, also called *white metal*, is a soft alloy of lead or tin and can be scored by tiny particles. Liquid penetrant tests are also used to check babbitt for lamination or separation from its backing.

The pump casing should be visually inspected for erosion or washout. Casing contours should be smooth and continuous. Pits or ridges can reduce efficiency and accelerate casing wear. Casing joints should be checked for erosion and effective flange sealing.

All sleeve bearings should be visually inspected for pitting, finish, scoring and size. The design clearance should be checked with feeler gages. Frictionless ball or roller bearings must be inspected for surface finishes on the rotating assemblies.

Circulating Water Pumps

Always refer to the dismantling procedure outlined in the manufacturer's instruction manual. If the pump has rubber bearings, use only fresh water for cleaning; solvents may damage the rubber. After all components have been cleaned, check them and the pump casing for corrosion. Deposits or scaling must be removed before reassembly.

The components are checked for cracks and signs of erosion or cavitation damage. Particular attention is given to the inlet and discharge areas of each impeller vane. The vane tips and edges are inspected on vertical column pumps with an impeller-to-impeller cone design.

The diffuser vanes are inspected in the same manner as the impeller vanes. Areas of wear or erosion should be repaired and ground smooth.

Check the shaft sleeves for scoring. A scored sleeve accelerates bearing wear and should be replaced. Polishing is acceptable if the scoring is minor.

Sleeve, rubber or babbitted bearings should be checked for wear or damage. Worn or damaged bearings are replaced. Visual inspection is also used to check rubber and babbitt for adherence to the shell. The preferred method of checking babbitt adherence is liquid penetrant testing.

Remove the packing and check the shaft sleeve for scoring. Make sure that the stuffing box and the seal water piping lines are flushed.

Check all running clearances with a feeler gage. Consult manufacturer's instructions for allowable wear limits.

Inspect the water pump for erosion or loss of protective coatings (if used). Pits and ridges can reduce efficiency and can accelerate wear. Check flanges and welded joints for signs of erosion. Shafts should be checked for runout. Visually inspect shafts for corrosion or signs of washout.

Visual Testing of Valves

A valve is a mechanical device that controls flow into, inside of or out of enclosed conduits such as piping and tubing. When fully open, the perfect valve offers no more flow resistance than an equal length of pipe. Closed, the perfect valve permits no fluid to pass. In addition, it completely resists distortion from internal fluid gas pressure. Also necessary is resistance to fluid dynamic effects, temperature, pressure drop, vibration, corrosion, wear, erosion by small particles and damage from large objects in the fluid stream. This performance has to be constant over the life of the valve.

Because a valve rarely approaches perfect performance, and because valve failure can be costly from an operational and safety standpoint, periodic tests are performed during the life of the valve to minimize failures.

Before visual testing of any valve, the inspector must first know the type of valve, its function, temperatures and pressures, how long it has been in service and its maintenance history. With this information, the inspector can more accurately evaluate questionable conditions.

Valves are dismantled always according to a detailed procedure or in accordance with the manufacturer's technical instruction manual. Documentation of the as-found condition is very important step for determining continuance of service. The as-found condition is also compared to the valve's maintenance record.

Valves in Service

Inservice testing allows the performance of a complete visual test of the valve, plus any other nondestructive test that may be needed. It is best to remove the valve from the line. This permits a more detailed inspection by exposing more of the internal surfaces. More reliable seat or shell leak tests are also possible with the valve offline.

The first step in performing an inservice test is to review the preservice, maintenance and past inspection records.

Gate Valves

Sealing surfaces of the wedge and body are inspected for evidence of physical damage (cracks, scratches, galling, wire drawing, pits, indentations). If facilities are available, liquid penetrant testing is also performed. Guide surfaces (stuffing box and yoke area) are inspected for evidence of wear or galling. The wear pattern on guide surfaces often reveals misalignment of working components. Excessive clearance in guide surfaces can lead to excessive rubbing between seat faces when the valve is closing, causing premature leakage failure (Fig. 40).

In gate valves, the upstream seat of the wedge and downstream body seat are the most likely places for erosion and wear. The sealing surface behind removable seat rings are good candidates for leakage from erosion through wire drawing or steam cutting. The stuffing box is the second most vulnerable spot in a valve. Remove the packing and check for corrosion in the walls of the box. Check the stem in the area passing through the bonnet guide surface and the packing gland for evidence of rubbing or corrosion. The stem in this area should have a surface finish no greater than 0.8 µm (3.2×10^{-5} in.) to ensure reasonable packing life. Another area of concern is the stem-to-wedge connection.

In an outside stem and yoke gate valve, the stem-to-wedge connection must, in addition to opening the valve, prevent the stem's turning for the stem nut to operate the valve. Check this connection for wear or erosion. Excessive wear, allowing the stem to turn or to become disengaged, renders the valve inoperative. Check the stem-to-stem nut threads for excessive wear. Excessive wear can be corrected only by replacing the worn components.

Because stem threads can be observed during an inservice test, a judgment can be made regarding the rate of wear and the need for replacement. Stem nuts can usually be replaced while the valve remains in service but stem replacement requires removal of the valve from service. Wear and refitting the wedge of a gate valve make the wedge fit lower in the body. Thus, when inspecting a valve, it is important to be sure that the reconditioned wedge fits properly. There must be adequate contact between the wedge seats and body seats to prevent leakage.

Globe Valves and Stop Check Valves

Sealing surfaces of the disk and body are inspected for evidence of physical damage: cracks, scratches, galling, wire drawing, pits and indentations. Liquid penetrants are used, if possible. Inspect the guide surfaces of the disk and seat ring or body for evidence of galling and wear. In a globe valve, excessive clearance in the guides increases the possibility of the bottom edge of the disk catching on the top edge of the seat in the body and damaging one or both.

This is especially significant if the stem is not vertical during service. Check the body seat for leakage behind the ring when performing an offline pressure test. If it is necessary to resurface seat faces, verify that the clearance between the hand wheel and yoke permits full closure of the reconditioned valve. Remove the packing and check the walls of the stuffing box for corrosion. Check the stem in the stuffing box for evidence of rubbing or corrosion. A 0.8 µm (3.2×10^{-5} in.) finish is the maximum permitted in the area through the packing. In a globe valve, the stem-to-disk connection is especially important for proper operation. The fit should be tight but not rigid. Excessive clearance at this point leads to erratic operation, excessive noise and accelerated wear and this connection should be carefully inspected for excessive clearance. The stem-to-nut threads are an important test site. Excessive wear is corrected by replacement of the component (Fig. 41).

FIGURE 40. Bolted bonnet gate valve.

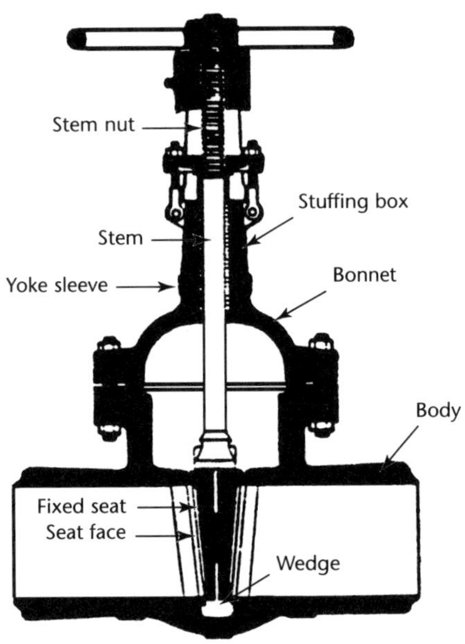

FIGURE 41. Bolted bonnet globe nonreturn valve.

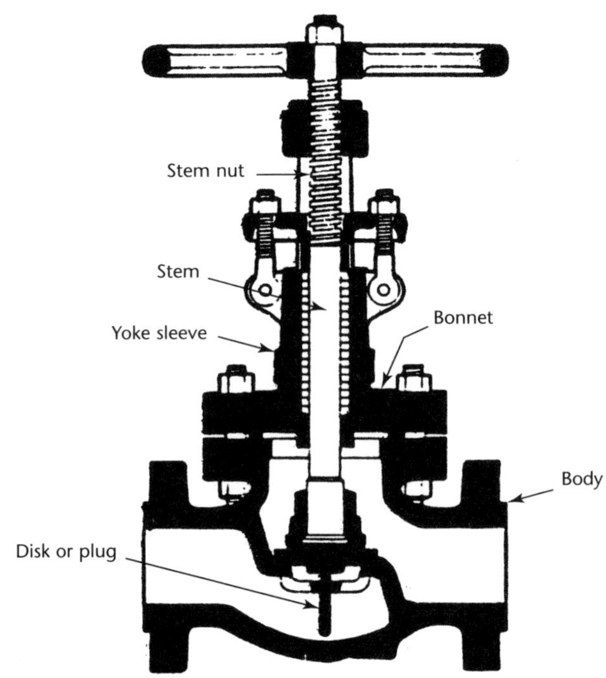

Lift Check Valves and Swing Check Valves

Lift check valves are inspected the same way as globe valves, without the stem or stuffing box (Fig. 42).

In a swing check valve, inspect sealing surfaces of the disk and body for evidence of physical damage and perform a liquid penetrant test if possible. Check for leakage behind the seat ring and verify that the disk is not installed backward. Inspect the hinge pin, hinges and disk for evidence of wear. Excessive clearance at these points can lead to misalignment of the seat surfaces and leakage (Fig. 43).

Ball Valves

Sealing surfaces are visually inspected for evidence of physical damage, wear or corrosion that causes leakage. The horizontal plane at the flow centerline is the first location to show wear because it always seals against the full differential pressure. Check the stuffing box and stem thrust bearing surfaces for evidence of corrosion or wear (Fig. 44).

FIGURE 42. Bolted bonnet lift check valve.

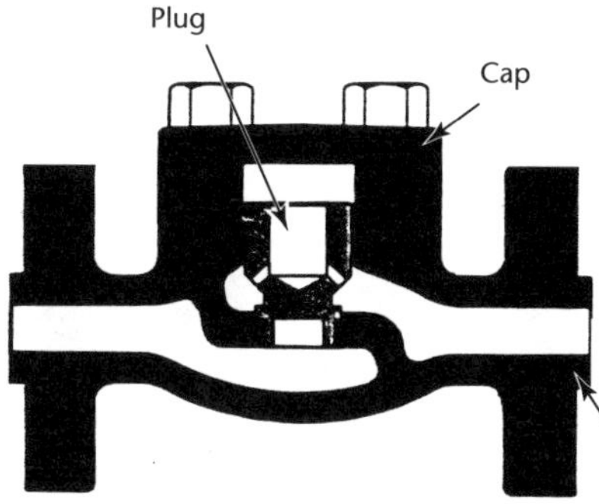

FIGURE 43. Bolted bonnet swing check valve.

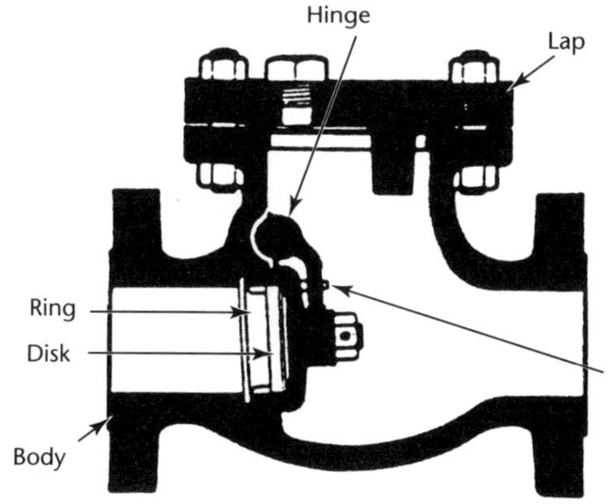

Plug Valves and Butterfly Valves

Plug valves are tested the same way as ball valves. In a tapered plug valve, wear and refitting of the plug in the body causes the plug to fit lower. When inspecting a valve of this type, it is important to be sure that the reconditioned component fits properly. Proper alignment between the components in the plug and body prevent excessive turbulence and pressure drop that may cause accelerated erosion.

In butterfly valves, inspect the seating surface of the disk for evidence of physical damage, wear or corrosion. Again, the horizontal plane at the centerline is the most vulnerable point. Inspect the plastic seal or body liner for evidence of damage or cold flow. Alignment of the seal and disk when closed is vital to the satisfactory operation of a butterfly valve. Inspect the bearing surfaces and position stops for excessive wear (Fig. 45).

Diaphragm Valves

Inspect the sealing surface of the body partition for evidence of corrosion, erosion or damage. Most importantly, visually inspect the diaphragm for evidence of aging and cracking, especially where it is retained by the body and bonnet or on the surface in contact with the body partition. The stuffing box is inspected the same way as it is in a globe valve.

Visual Testing of Bolting

Bolts (Fig. 46) are visually tested to detect conditions such as cracks, wear, corrosion, erosion or physical damage on the surfaces of the components.

The test setup and environment are specified in detail. For example, tests of pressure vessel bolts may be done with direct techniques when access is sufficient to place the eye within 600 mm (24 in.) of the test surface at an angle not less than 30 degrees to the surface. Mirrors may be used to improve the viewing angle. Lighting, natural or artificial, must

FIGURE 44. Floating ball valve.

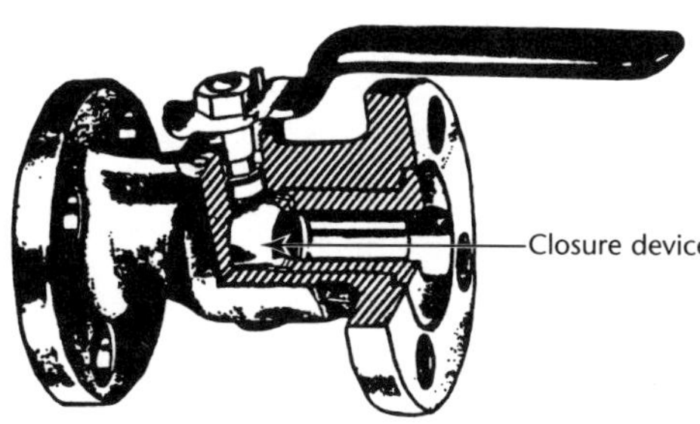

be sufficient to resolve a 0.75 mm (0.03 in.) black line on an 18 percent neutral gray card.

Remote visual tests may be substituted for direct visual testing, using devices such as telescopes, borescopes, cameras or other suitable instruments. Such devices must have resolution at least equivalent to that of direct visual testing.

Equipment and Preparation

Steel rules, micrometers, vernier calipers, depth micrometers, thread gages and magnifying glasses are necessary for direct visual testing of bolts.

Visual tests often require clean surfaces or decontamination for valid interpretation of results. Precautions should be taken before any cleaning process.

Discontinuity Classification

Discontinuities in bolts, studs, washers and nuts fall into four classes.

1. Inherent discontinuities originate from solidification of metal in the ingot. Pipe and nonmetallic inclusions are common and can lead to other types of discontinuities in service.
2. Primary processing discontinuities are produced from the hot or cold working of the ingot into rod and bar.
3. Secondary processing discontinuities are produced during manufacture of studs, washers, bolts and nuts. Secondary processing includes machining, grinding, heat treating, plating and other finishing operations.
4. Service induced discontinuities may be caused by vibration, tensioning and corrosion.

The presence of inherent, primary processing and secondary processing discontinuities is sometimes revealed in service.

The most common locations for fastener failures are in the head-to-shank fillet, through the first thread inside the nut on threaded fasteners or at the transition from the thread to the shank. Sources of failure are discontinuities in the metal caused either by segregation in the form of inclusions in the ingot or by folds, laps or seams that have formed because of faulty working in the semifinishing or finishing mills (Fig. 47).

FIGURE 45. Butterfly valve.

FIGURE 46. Diagram of nondestructive test for bolting.

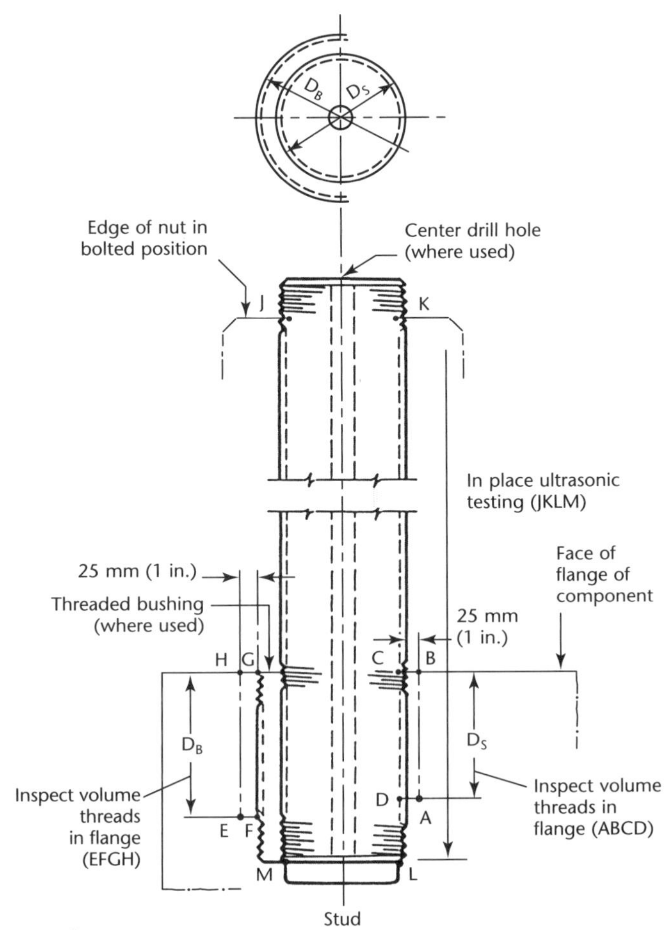

Legend

ABCD, EFGH = threaded area requiring visual testing
D_B = diameter including threads
D_S = diameter not including threads
LM = unthreaded distal end

Discontinuities Visible in Bolting

Bursts in bolting materials may be internal or external. Found in bars and forgings, internal bursts are caused by rupturing of metal extruded or forged at temperatures too low or too high. External bursts often occur where forming is severe or where sections are thin (Fig. 48).

Seams are generally inherent in the raw material from which a fastener is made. Seams are usually straight or smoothly curved discontinuities running parallel to the longitudinal axis.

Laps are surface discontinuities caused by folding of the metal. If laps occur in threads, they generally show a pattern of consistency — they are located in the same place on all threads of the nut or bolt.

Surface tears occur along the length of a bar or threaded fastener and are caused by faulty extrusion dies or inadequate lubrication during extrusion. Surface tears can resemble seams.

FIGURE 47. Common failures in threaded fasteners: (a) tension failures; (b) shear failures; (c) cracked nut.

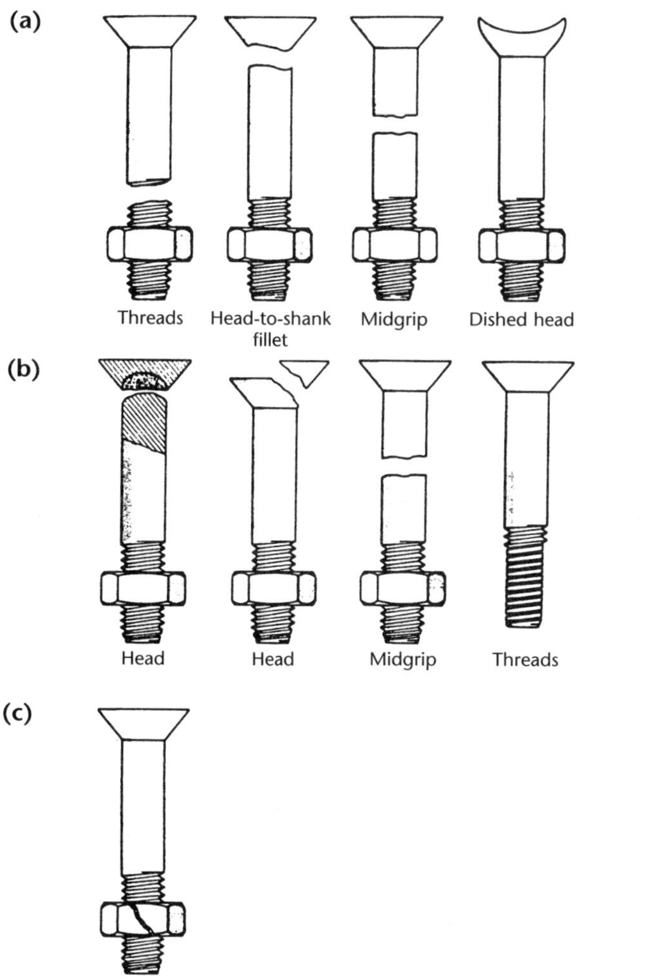

Folds may occur during forging, at or near the intersection of diameter changes.

Tool marks are longitudinal or circumferential shallow grooves produced by the movement of manufacturing tools over the fastener surface. A nick or gouge is an indentation on the surface of a fastener produced by forceful abrasion or by impact of the fastener against other components.

Shear breaks or shear cracks are open breaks in the metal at the periphery of a bolt or nut, at about a 45 degree angle to the long axis. Shear breaks occur most often with flanged products. They can be caused by overstressing the metal during forging, insufficient ductility and high strain rates.

Necking down is a localized reduction in area of a component in overload conditions.

Erosion is destruction of metals or other materials by the abrasive action of moving fluids, usually accelerated by the presence of solid particles in suspension.

Crevice corrosion cracks are a type of concentration cell corrosion — corrosion of a metal caused by the concentration of dissolved salts, metal ions, oxygen or other gases. It occurs in crevices or pockets remote from the principal fluid stream, with a resulting build up of differential cells that ultimately cause deep pitting.

Recording

Any area where a visual test reveals surface discontinuities (physical damage, wear, cracks, gouges, corrosion, erosion, misalignment, nicks, oxidation, scratches) on studs, washers, nuts or bolts is recorded on a data sheet regardless of discontinuity size. Unspecified movement and the looseness of bolts is also recorded.

If there are areas or indications that cannot be easily recorded on a data form, a sketch or photograph is included with the report, to clarify the results.

FIGURE 48. Visible bolting discontinuities.

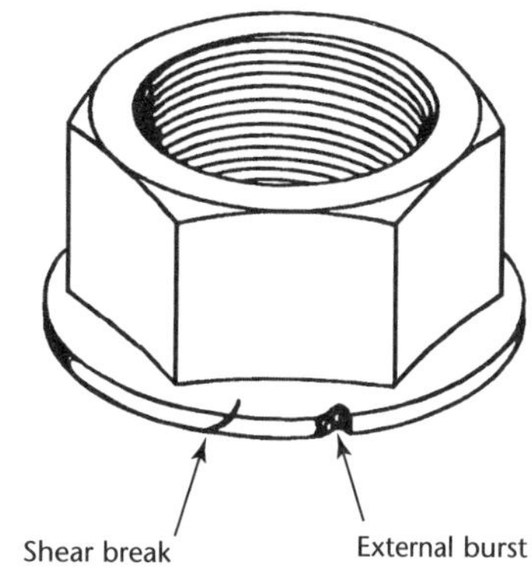

Visual Tests for Casting Discontinuities

Because casting is a primary process, the discontinuities associated with casting are considered to be inherent. These discontinuities include (1) hot tears, (2) inclusions, (3) porosity, (4) unfused chills and unfused chaplets and (5) cold shuts.

1. Hot tears appear as ragged cracks or, in severe cases, as a group of cracks. Tears are always open to the surface and can be detected visually.
2. Inclusions in a casting may be detectable during a visual test — a visual inspector must draw conclusions based on a small portion of the discontinuity being visible at the surface. Inclusions are usually sand or refractory material and appear as irregularly shaped cavities containing a nonmetallic material.
3. Porosity appears as a series of hemispherical depressions in the surface of a casting. This pockmarked appearance is easily recognized during visual examination.
4. Unfused chills and chaplets appear as irregularly shaped cavities on the surface of the casting. The cavity varies from the entire shape of the chill or chaplet, such as a square, to a portion of the shape, depending on the amount of fusion. Chills and chaplets differ in shape, so their cavities also differ. Each foundry selects the chills and chaplets it prefers, so the shapes of these discontinuities cannot be defined here.
5. Cold shuts appear as folds or smooth cracklike discontinuities, depending on the location and the severity. They are usually visible on the surface of a casting.

Visual Testing of Forgings

Forgings are often large but simple in shape. They can usually be visually inspected without complex viewing equipment. The forging process usually occurs at an elevated temperature, so scaling or oxidation can sometimes be found inside the discontinuity.

The processing discontinuities that a visual inspector can detect on a forging are bursts, laps and cracks.

These primary processing discontinuities are a result of the forging process. Inherent discontinuities caused by pipe, porosity or inclusions could also be detected if the forging process moves them to an exposed surface. In visual testing, a discontinuity must appear on an accessible surface to be detectable. It is important to plan and perform the visual test during the manufacturing cycle to provide opportunities for viewing all finished surfaces.

Bursts are internal forging discontinuities. They appear as scaly, ragged cavities inside the forging. Large forgings generally receive secondary processing during the manufacturing cycle. Trimming, descaling and machining are all processes that could expose a hidden discontinuity such as a burst. A trim cut on the end of a forged shaft is a good example. By scheduling visual testing during the manufacturing cycle after each individual operation, discontinuities may be detected that would otherwise remain hidden.

Laps are folds of metal forced into the surface of the component during forging. A lap can be shallow or very deep and its appearance is that of an oddly shaped crack on the surface. A lap indication can vary from a tight, straight, linear discontinuity to a wide U shaped indication. When a lap is viewed through a low magnification lens, the inside surfaces often contain oxidized scale — a gray, porous material.

Cracks are different from laps in that cracks follow the stress distribution within the forging whereas laps do not. Both laps and cracks can appear on the surface of a forging as thin, jagged, linear indications.

It is often helpful for the inspector to use a 5× to 10× magnifier. Lenses higher than 10× are typically large and difficult to keep steady.

Rolling

Rolled products are probably the most common visual test objects. A visual inspector should become familiar with the rolling processes to identify discontinuities by their location on the component. Visual tests of rolled products are complicated by the fact that many complex structures are fabricated from plate and rolled shapes.

Inherent discontinuities such as pipe, inclusions and gas holes are affected by rolling, forming laminar discontinuities parallel to the rolling direction. When any of these discontinuities are moved to the surface of a rolled product, a seam (Fig. 49), stringer (Fig. 50) or crack can form.

The location of the discontinuity on the component helps to classify it. Seams, cracks and stringers can appear anywhere on a rolled product. Seams and stringers follow the direction of rolling. Laminations detected by visual inspection appear on the edges of plate or the ends of pipe (Fig. 51). They are linear and parallel with the top and bottom surfaces of the plate. On rolled shapes, laminations are parallel to the rolling direction and appear on the edges of the shape. Pipe causes laminations oxidized inside. Gas holes also cause laminations that may be oxidized if oxygen was in the hole. Inclusions cause laminations that contain a layer of the included material.

Processing discontinuities encountered in rolled product include tears and cracks. Such discontinuities exhibit characteristics similar to some forging discontinuities, including an oxidized, scaly interior.

Visual testing must be performed before fabrication hides portions of the material. In addition to magnifiers (5× to 10×), a mirror is useful for visual testing of rolled shapes with obstructed areas.

FIGURE 49. Seam in billet: (a) typical location; (b) magnified view of rough, oxidized surface seam.

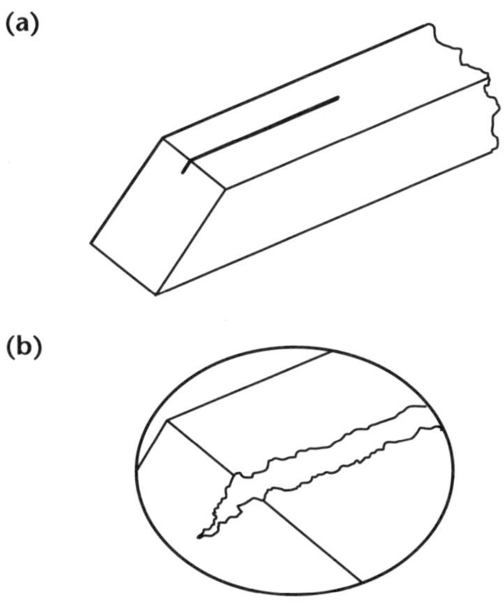

Drawing, Extruding and Piercing

The discontinuities associated with drawing, extruding and piercing are all on the surfaces of the component and therefore detectable during a visual test.

Drawn products usually exhibit gross failure if any discontinuity is present. Because most drawn products have thin walls, failure usually appears as a through-wall break.

Extrusions can have surface discontinuities that appear as scrapes and tears.

Pierced pipe can contain slugs of metal that are easily identified. Severe score marks usually lead to the slug.

FIGURE 50. Rolled bar containing stringers or inclusions elongated during the rolling process.

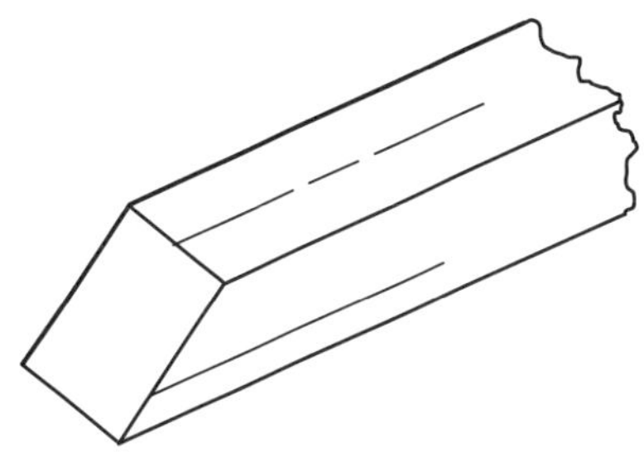

FIGURE 51. Typical I-beam lamination and seam locations.

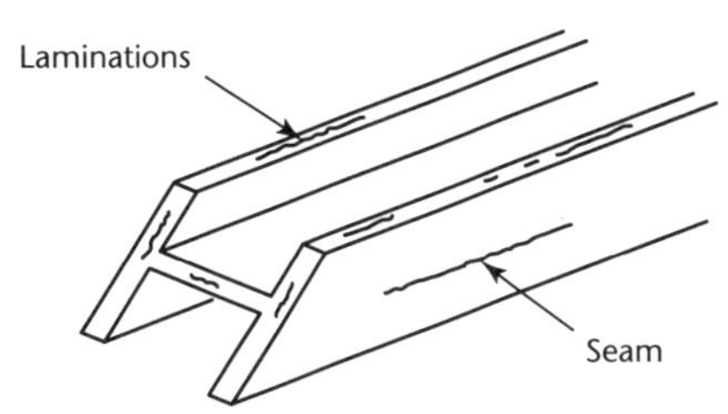

References

1. Section 9, "Applications of Visual and Optical Testing in the Electric Power Industries": Part 1, "Joining Processes." *Nondestructive Testing Handbook,* second edition: Vol. 8, *Visual and Optical Testing.* Columbus, OH: American Society for Nondestructive Testing (1993): p 247-262.
2. EPRI TM101PS, *Visual Examination Technologies:* Level 1, *Practical/Specific.* Charlotte, NC: Electric Power Research Institute (1996).
3. *ASME Boiler and Pressure Vessel Code.* New York, NY: ASME International (2008).
4. ANSI/ASNT CP-189, *Standard for Qualification and Certification of Nondestructive Testing Personnel.* Columbus, OH: American Society for Nondestructive Testing (2006).
5. *ASNT Central Certification Program* (ACCP), Revision 4 (March 2005). Columbus, OH: American Society for Nondestructive Testing (2005).
6. AWS D1.1M, *Structural Welding Code — Steel.* Miami, FL: American Welding Society (2008).
7. ANSI/ASME B31.1, *Power Piping Systems.* New York, NY: ASME International (2007).
8. Hedtke, J. and R.[T.] Nademus. Section 9, "Applications of Visual and Optical Testing in the Electric Power Industries": Part 2, "Specific Visual Inspection Applications." *Nondestructive Testing Handbook,* second edition: Vol. 8, *Visual and Optical Testing.* Columbus, OH: American Society for Nondestructive Testing (1993): p 263-276.
9. EPRI TM102CG, *Visual Examination Technologies:* Level 2, *General.* Charlotte, NC: Electric Power Research Institute (1996).

Bibliography

Abbott, P.J., K. von Felten and B.A. Pellegrino. EPRI 2603, "Development of Next-Generation Inner-Bundle Imaging Device for Triangular-Pitch, Secondary-Side Steam Generator Visual Inspection." *21st Annual Steam Generator NDE Workshop* [Berkeley, CA, July 2002]. Charlotte, NC: Electric Power Research Institute.

Anderson, M.T., S.E. Cumblidge and S.R. Doctor. "An Assessment of Remote Visual Testing System Capabilities for the Detection of Service Induced Cracking." *Materials Evaluation.* Vol. 63, No. 9. Columbus, OH: American Society for Nondestructive Testing (September 2005): p 883-891.

ASME Boiler and Pressure Vessel Code: Section V, *Nondestructive Examination.* Article 9, *Visual Examination.* New York, NY: ASME International (2007).

ASME NQA 1, *Quality Assurance Requirements for Nuclear Facilities Applications.* New York, NY: ASME International (2008).

AWS B1.11, *Guide for the Visual Examination of Welds.* Miami, FL: American Welding Society (2000).

Chang, Pan-Tze, I. Kaufman and D.-Y. Shyong. "Detection and Imaging of Surface Cracks by Optical Scanning." *Materials Evaluation.* Vol. 45, No. 8. Columbus, OH: American Society for Nondestructive Testing (August 1987): p 943-950.

Chuang, K.C. "Application of the Optical Correlation Measurement to Detection of Fatigue Damage." *Materials Evaluation.* Vol. 26, No. 6. Columbus, OH: American Society for Nondestructive Testing (June 1968): p 116-119.

Cumblidge, S.E., M.T. Anderson and S.R. Doctor. NUREG/CR 6860, *An Assessment of Visual Testing.* Washington, DC: Nuclear Regulatory Commission (2004).

Davis, Renée S. "Remote Visual Inspection in the Nuclear, Pipeline and Underwater Industries." *Materials Evaluation.* Vol. 48, No. 6. Columbus, OH: American Society for Nondestructive Testing (June 1990): p 797-798, 800-803.

Pellegrino, Bruce A. "Remote Visual Testing for Internal Pressure Vessel Inspection." *Materials Evaluation.* Vol. 56, No. 5. Columbus, OH: American Society for Nondestructive Testing (May 1998): p 606-609.

Samsonov, Peter. "Remote Visual Inspection for NDE in Power Plants." *Materials Evaluation.* Vol. 51, No. 6. Columbus, OH: American Society for Nondestructive Testing (June 1993): p 662-663.

11 CHAPTER

Aerospace Applications of Visual Testing

Donald R. Christina, Boeing Company, Charleston Air Force Base, South Carolina

Bruce L. Bates, Yucaipa, California

Part 2 is a work of the United States government and not subject to copyright.

Part 1. Visual Testing of Aircraft Structure[1,2]

Visual testing is the oldest and most economical form of nondestructive testing. Visual testing is the primary method used in aircraft maintenance, and such tests can reveal a variety of discontinuities. Generally, these tests cover broad surfaces of the aircraft.

Detailed tests of small areas are conducted with optical instruments. These instruments have two functions: (1) to magnify discontinuities that cannot be detected by the unaided eye and (2) to permit visual checks of areas not accessible to the unaided eye. Such tests include the use of magnifiers and borescopes. The text below details optically aided tests of structures where access is poor for other nondestructive tests. A short description of optical instruments is given below, with examples where such devices are used to inspect aircraft structures and operating mechanisms.

It is important to know the type of discontinuities that may develop and to recognize the areas where such problems may occur. Many discontinuities are revealed during visual tests of aircraft. Rust stains reveal corroded steel. Paint blisters around fasteners in wing skins reveal underlying intergranular corrosion of the skin. And fuel leaks in the lower wing skin can indicate a cracked spar cap or skin. Even the pilot's walk-around visual inspection can reveal serious faults. For example, an improper attitude of the aircraft can indicate a failed main landing gear attachment fitting, and a floating spoiler can reveal a failed torsion bar. The following text discusses and illustrates corrosion pitting, stress corrosion cracks and fatigue cracks.

Optical Aids

Magnifying devices and lighting aids are used and the general area is checked for cleanliness, presence of foreign objects, security of the component, corrosion and cracks or other damage. In many cases, the area to be inspected is cleaned before visual inspection.

Magnifying Lenses

An optical microscope is a combination of lenses used to magnify an image. The object is placed close to the lens in order to obtain as great a magnification as desired. The distance from lens to object is adjusted until the object is in the lens's depth of field and is in focus. The simplest form of a microscope is a single converging lens, often referred to as a simple magnifier.

Magnifying lens are discussed in this volume's chapter on direct visual testing.

Borescopes

A borescope is a precise optical instrument with built-in illumination. It can be used to visually check internal areas and deep holes and bores. Borescopes are available in rigid and flexible models from 2.5 mm (0.1 in.) in diameter to 19 mm (0.75 in.) in diameter and meters (several feet) in length. These devices are generally provided with fixed diameters and fixed working lengths, with optical systems designed to provide direct, right angle, retrospective and oblique vision. These orientations are described in this volume's chapter on indirect techniques of visual testing. Generally, the diameter of the borescope is contingent on the diameter of the hole or bore being inspected. The borescope length is governed by the distance between the available access and the distance to the test area. The choice of viewing angle is determined by discontinuity type and location.

Microborescopes

A microborescope is an optical device, of a kind resembling a medical endoscope. It is much like a borescope but with superior optical systems and high intensity cold light piped to the working tip through fiber optic bundles. Other useful features include constant focus from about 4 mm (0.15 in.) to infinity. Actually, when the tip is about 4 mm (0.15 in.) from the test surface, a magnification factor of about 10× is achieved. Microborescopes are available in diameters down to 1.7 mm (0.07 in.) and in lengths from 100 to 150 mm (4 to 6 in.).

Flexible Fiber Optic Borescopes

Flexible fiber optic borescopes permit manipulation of the test instrument around corners and through passages with several directional changes. Woven stainless steel sheathing protects the image relay bundle during repeated flexing and maneuvering. These devices are designed to provide sharp, clear images of components and interior surfaces that are normally impossible to inspect. Remote end tip deflection allows the viewer to thread the fiber optic borescope through a complex series of bends. The end tip is deflected with a rotating control mounted on the handle. Most of these devices have a wide angle objective lens that provides a 100 degree field of view and tip deflection of ±90 degrees. They all have a fiber optic image bundle and are equipped with a focus control to bring the subject into sharp focus over a wide range of viewing distances. The working lengths are normally from 0.6 to 3.7 m (2 to 12 ft) with diameters from 3 to 13 mm (0.12 to 0.5 in.).

Microelectronic Video Borescopes

An electronic sensor embedded in the movable tip of the probe transmits signals to a video processor, where the image is sent to a monitor. The video borescope has a bright, high resolution color image with no distortions or spots. The device does not have an eyepiece like other borescopes. It has a freeze frame feature that allows closer viewing of the image. The image can be electronically transferred for permanent documentation. Grids or measurement references may be entered into the margin of the image and can become part of the permanent record. The image may be magnified for precise viewing. The field of view is up to 90 degrees and the probe tip has four-way articulation. Presently, the smallest probe diameter is 10 mm (0.37 in.) with working lengths up to 30 m (100 ft).

Diffracted Light

A technique using diffracted light has been developed for visualizing surface distortions, depressions or protrusions as small as 10 µm (0.0004 in.). A real time technique particularly applicable to rapid inspection of large surfaces, the diffracted light technique has been used to inspect automobile body panels and metal working dies. Commercial versions of diffracted light equipment have ranged from a manual handheld system, in which the operator directly views the inspected part, to systems with video cameras and computer based image processors. This technique has been used to inspect composite structures for barely visible impact damage.[3,4]

Computer based image processing has been applied to diffracted light technique images. An image from a previous inspection can be directly compared to current results for quick identification of areas where surface features have changed. The optical setup for the diffracted light technique consists of a light source, a retroreflective screen and the object being inspected (Fig. 1). The surface being inspected must be reflective. Both flat and moderately curved surfaces can be inspected using this technique. The diffracted light effect can be explained with geometric optics. If a flat surface with an indentation is inspected, the light striking the indentation is deflected. It then strikes the retroreflective screen at a point removed from the light rays reflected from the area surrounding the indentation. The retroreflective screen attempts to return all these rays to the points on the inspected surface from which they were first reflected. However, the screen, consisting of numerous glass beads, returns a cone of light to the surface. This imperfection of the retroreflective screen creates the diffracted light effect. By backlighting the discontinuity, the technique increases the light intensity on one side of the indentation and reduces it on the opposite side.

The edge of light technique is a related test. Raking light from an oblique angle creates shadows from irregularities on the test surface. Image processing helps identify surface anomalies such as pitting, impact damage, fatigue deformation and pillowing of laminates.[4,5]

FIGURE 1. Optical setup for diffracted light technique.

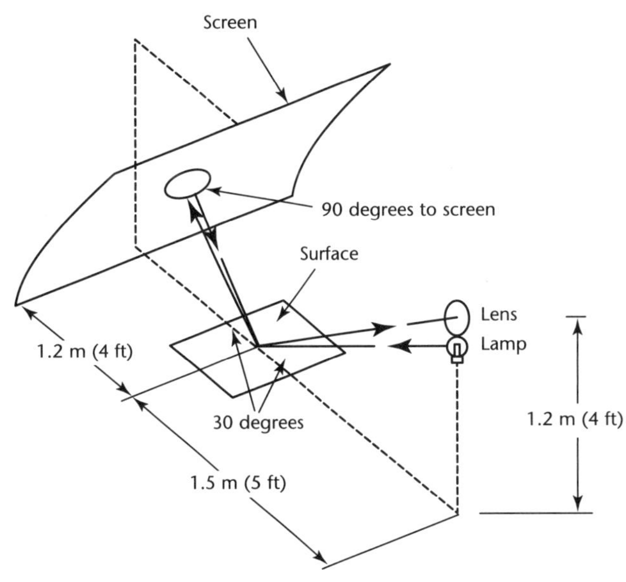

Typical Applications

Following are examples of visual tests used by airline maintenance personnel to ensure the structural integrity of transport aircraft.

Torsion Bar Core Corrosion Pitting

Stress corrosion cracks can cause failure in high strength steel spoiler torsion bars. In one investigation, corrosion pitting on the inside surface of the torsion bar cavity (bore) was found to have led to stress corrosion cracks and subsequent failure (Fig. 2). It was determined that torsion bar failure can be greatly minimized by removing corrosion from the bore within repairable limits and resealing the cavity with a polysulfide sealant (to prevent moisture ingress). For this particular case, a service bulletin was issued requesting operators to remove the torsion bars, clean the exterior surface and perform a magnetic particle test for cracks. If the test shows the bar is cracked, it is scrapped. If no cracks are detected, sealant (if present) and primer are removed from the bore and it is visually inspected for corrosion pits using a 70 or 90 degree borescope. If corrosion pits are detected, the bore is reamed to a maximum oversize and reinspected. If the second borescope test reveals pits, the component is scrapped. If no pits are detected, the bore is cleaned with solvent, given two coats of corrosion inhibiting primer and then filled with polysulfide sealant. The torsion bar is then radiographed to verify sealant integrity.

Slat Drive Crank

Instances of slat bell crank failures were reported by operators. Investigation revealed that failures were caused by fatigue cracks that initiated at the bell crank-to-collar attachment holes (Fig. 3). With the bell crank installed on the aircraft, access to the crack location is extremely poor, preventing the use of ultrasonic, eddy current or radiographic nondestructive testing. This leaves the fiber optic borescope as the only option for inspecting this component. A right angle flexible borescope with deflecting tip, 5 mm (0.2 in.) in diameter and about 1 m (40 in.) in length, is recommended for this test. Before the inspection, the bore of the bell crank is cleaned with solvent to remove grease. The tip of the flexible borescope is placed in the forward end of the bell crank and fed aft until it is aligned with the two attachment holes. To control the position of the flexible borescope, it can be placed within a stiff but flexible plastic or thin metal tube. When the operator has the tip aligned with the attachment bolts, it can be rotated to inspect for cracks originating at the attachment holes. If cracks are detected, the bell crank is removed from the aircraft. If no cracks are detected, repetitive tests can be conducted until the bell crank is replaced with a better component.

Spoiler Lubrication Hole Cracks

Failures of the link or fitting assemblies of the slat drive mechanism may be reported by operators. In one investigation, the failures were found to be caused by fatigue cracks generated at the inner surface of the lubrication holes in the link

FIGURE 2. Borescopic visual test of spoiler torsion bars.

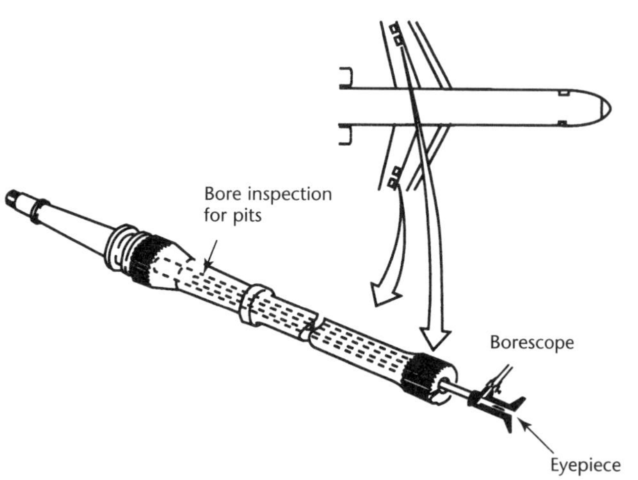

FIGURE 3. Flexible borescopic visual test of slat drive bell crank.

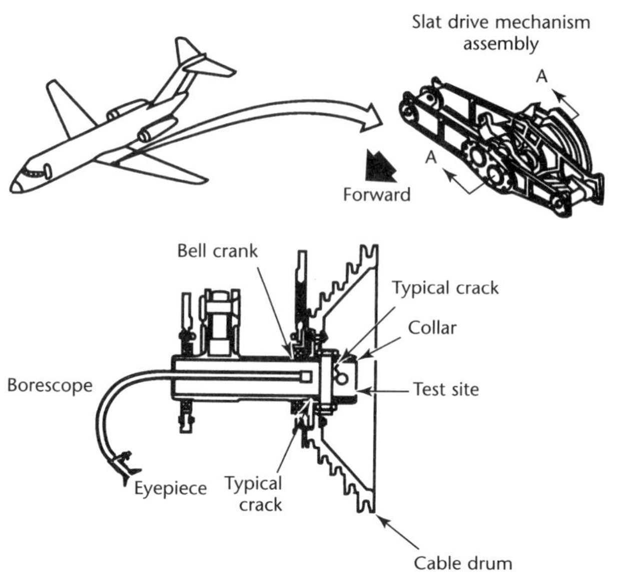

and fitting assemblies (Fig. 4). The initial service bulletin called for an ultrasonic transverse wave test to detect subsurface cracks adjacent to the lubrication holes. This test proved to be unreliable (cracks were missed) and was replaced by an eddy current test of the bores of three holes.

FIGURE 4. Borescopic visual test of spoiler actuating mechanism: (a) link assembly; (b) fitting asembly.

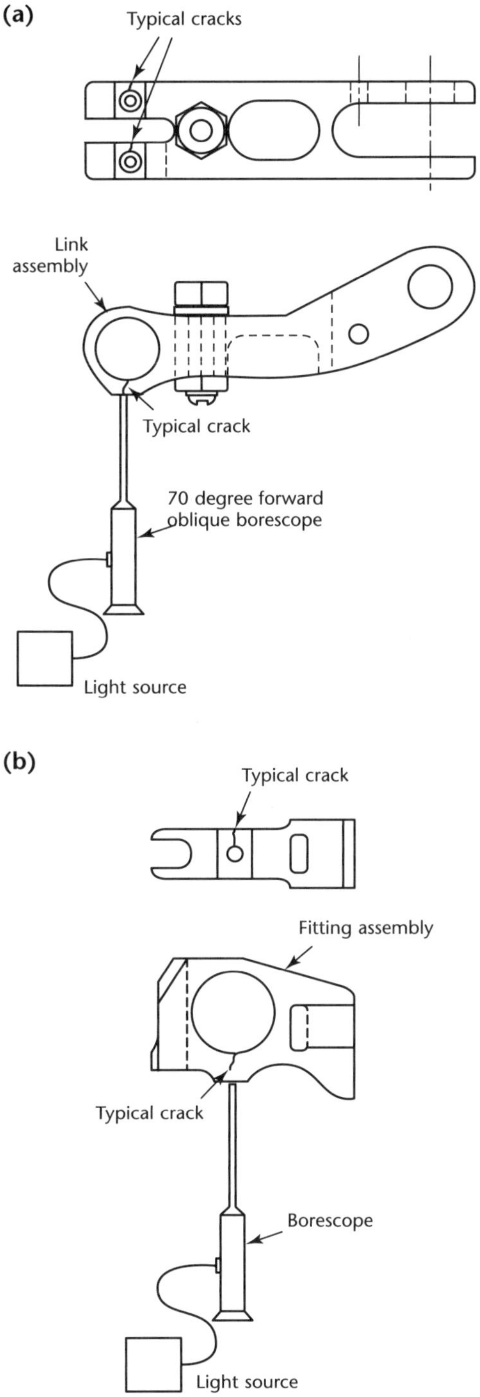

This test required removal of the lubrication fittings and grease before inserting the 3 mm (0.125 in.) eddy current probe into the holes. Borescope tests were considered when the microborescopes discussed above became available commercially. If the visual test is performed, the lubrication fittings and grease are removed and the bores are cleaned with solvent. The visual test is performed using a 70 degree forward oblique borescope, 2.7 mm (0.11 in.) in diameter by 180 mm (7 in.) long. Crack indications may appear at the inboard and outboard sides of the lubrication bores. Cracks, if detected, are not acceptable and the component must be replaced.

Wing Cracks under Panel

Fatigue cracks may occur in the wing rear spar cap web and doubler under the trapezoidal panel attachment fitting. These fatigue cracks may occur in both the web and doubler or in each member separately. The cracks originate at the lower edge of both members (Fig. 5). Direct access to the cracked areas of the web and doubler require removal of the trapezoidal fitting. If both members are

FIGURE 5. Wing rear spar doubler and web crack location: (a) from above; (b) from side.

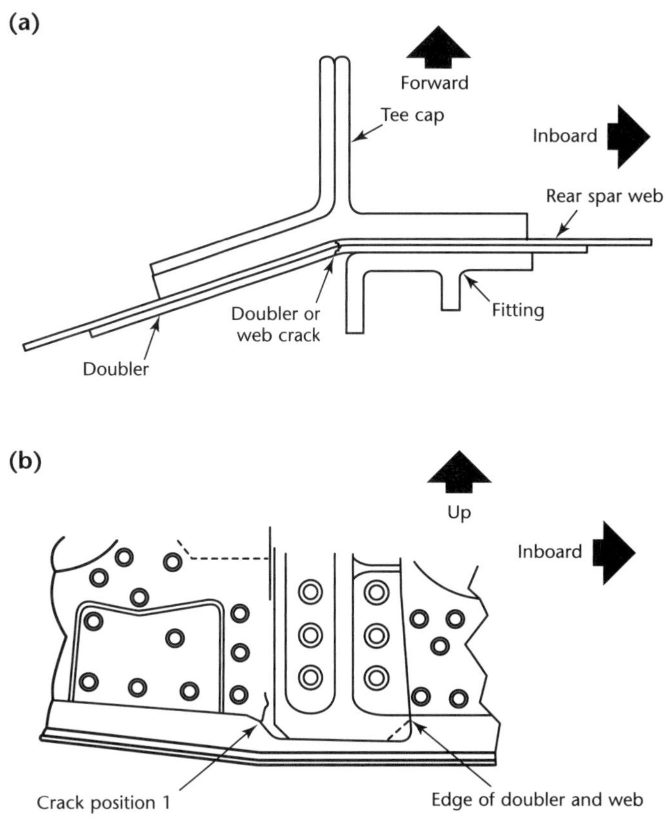

cracked, fuel may leak from under the fitting, indicating the existence of through-thickness cracks. Other methods of nondestructive testing cannot be used because of the poor access caused by the fitting. The doubler and web are made from clad, wrought aluminum sheet about 4 mm (0.16 in.) thick. Access to the area is aft of the rear spar plane, inboard and outboard of the trapezoidal panel and fitting. The visual test can be performed with a 0 degree borescope along with a 70 to 90 degree borescope, 300 to 480 mm (12 to 19 in.) in length and 4 to 5 mm (0.16 to 0.20 in.) in diameter. The test area is cleaned using a cotton swab wetted with solvent. The area is first viewed using the 0 degree borescope to check for cracks in the radius areas of the doubler and web (position 1, Fig. 5b). If a crack exists and it does not run under the fitting, its length in the doubler may be determined by a liquid penetrant or high frequency eddy current test using a shielded surface probe. To measure the length of a crack in the web (foremost member) under an uncracked doubler requires removal of the fitting and application of a low frequency eddy current test. Detecting a crack in the doubler or web hidden by the fitting (Fig. 6b, positions 2 and 3) requires the use of the 70 or 90 degree borescope. The area is cleaned with a cotton swab wetted with solvent and the borescope is inserted through the small opening between the forward side of the fitting and the aft side of the doubler (Fig. 6). To measure the length of doubler cracks under the fitting at position 2 or 3 requires ultrasonic transverse wave (angle beam) techniques.

Rudder Flange Cracks

Radiographic tests can reveal cracks developing in the rib flanges of the rudder. In one case, analysis determined that cracking of the rib flanges resulted from acoustically induced vibration. It was also determined that installation of stiffeners on the rudder ribs strengthens the rudder and minimizes the possibility of further crack development. A service bulletin was issued giving criteria for flyable crack lengths based on the number of cracked ribs and the length of the cracks. Unrepaired rib flange cracks may cause cracks to occur in the rudder skins, thereby requiring more extensive repairs. A radiographic test is first conducted and if cracks are detected in the rib flanges at or adjacent to the skin attachment fastener holes, their lengths must be determined. The cracks may run upward into the flange radius and progress into the rib area, where their lengths may be difficult to determine from radiographs alone (Fig. 7). If cracks occur within or progress into the flange upper radius, their lengths must be determined by use of a 3.2 mm (0.125 in.) maximum diameter rigid borescope or flexible borescope (Fig. 7).

To perform this test, the fastener common to the rudder skin and the rib flange (opposite the cracked flange) must be removed. A 0 degree borescope is inserted through the open hole and the opposite flange radius and web are inspected to determine crack position and length. An alternative approach requires removal of a fastener common to the rudder skin and the cracked flange that is judged to be about 50 mm (2 in.) beyond the crack image on the radiograph. A flexible borescope or rigid retrograde borescope is inserted through the open

FIGURE 6. Borescopic visual test of wing rear spar doubler and web for cracks under fitting: (a) view looking inboard; (b) view looking forward at lower portion of rear spar left side.

(a)

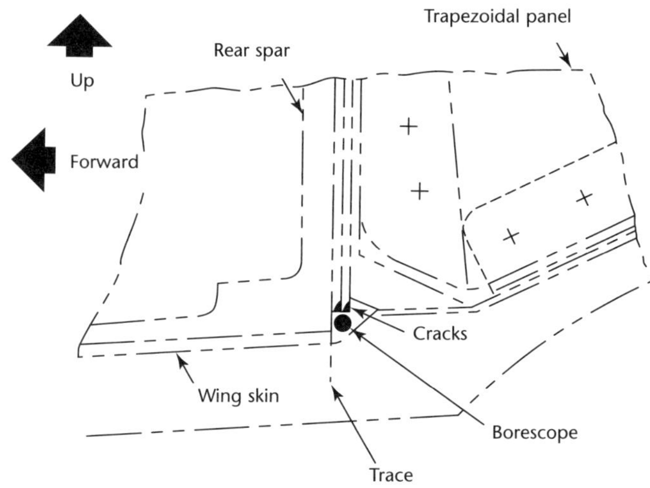

(b)

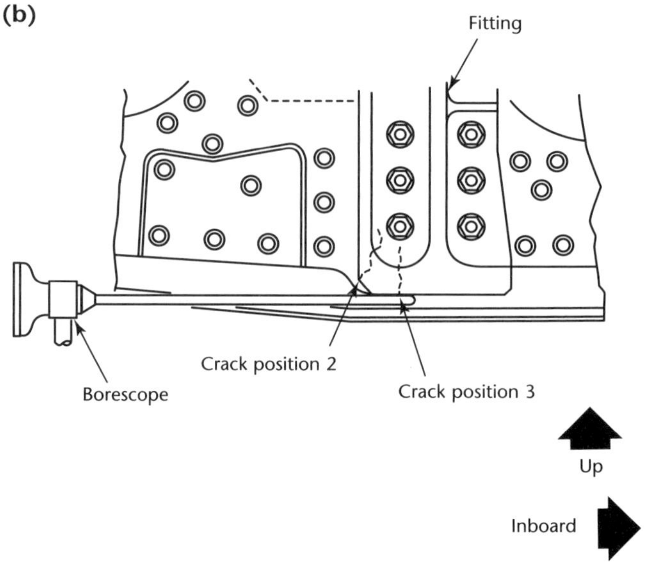

hole and articulated to allow viewing of the rib flange, radius and web to determine crack position and length. If cracks exceed flyable length, the ribs are repaired. If cracks are of tolerable length, an easily replaced fastener is installed in the open holes to allow for repetitive evaluation.

Landing Gear Pitting Corrosion

Three instances of main landing gear truck beam assembly failures were reported as resulting in major secondary damage to the aircraft. Investigation revealed that failure was a result of stress corrosion fracture that initiated at or immediately adjacent to the intersection of the lubrication hole and the pivot bore (Fig. 8). The stress corrosion fracture in the lubrication hole results from severe pitting caused by inadequate lubrication. If this condition is not corrected, the truck beam assemblies are vulnerable to failure. Removing corrosion or pitting from the surface of the four pivot bore lubrication holes and increasing the frequency of lubrication minimizes the possibility of failure and extends the service life of the beam assemblies. Inservice testing for corrosion pitting requires removal of the lubrication fitting and grease from each of four holes in the subject beam. The internal surface of each bore is checked using a 0 degree (forward looking) 2.8 mm (0.11 in.) diameter borescope. If corrosion or pitting is revealed, the hole is checked a second time with a 70 or 90 degree (lateral) borescope. When heavy pitting is detected, the beam is removed from the aircraft and the pits are removed by reaming the affected holes to a maximum oversize. Beams showing slight corrosion may continue in service for a limited time provided that periodic borescope and ultrasonic transverse wave tests are made to detect possible stress corrosion cracks that may originate at a pit. Figure 9a shows an acceptable condition at the inner bore chamfer. Figure 9b shows a typical heat treat pit — an acceptable condition away from the inner bore chamfer. Figure 9c shows unacceptable corrosion pitting at the inner pivot bore intersection. Figure 9d shows acceptable machining marks (rifling) in the bore outer edge, close to the lubrication fitting threaded area.

Wing Spar Cap Cracks

The purpose of this test is to check the lower spar cap forward tang for fatigue cracks at the fastener locations. The area of interest is located at the four wing pylons that support the jet engines. The total inspection for fatigue cracks at the pylons includes X-ray, ultrasonic transverse wave and borescope tests. In order to accomplish the ultrasonic and borescope tests inboard of the pylons, an access hole is made in the wing leading edge, as shown in Fig. 10. In addition, a smaller access hole is made in the footstool fitting, so that the fasteners located under the fitting can be visually inspected. Inspection of the six fasteners at the inboard side of the fitting is

FIGURE 7. Determining crack length in rudder ribs by visual test.

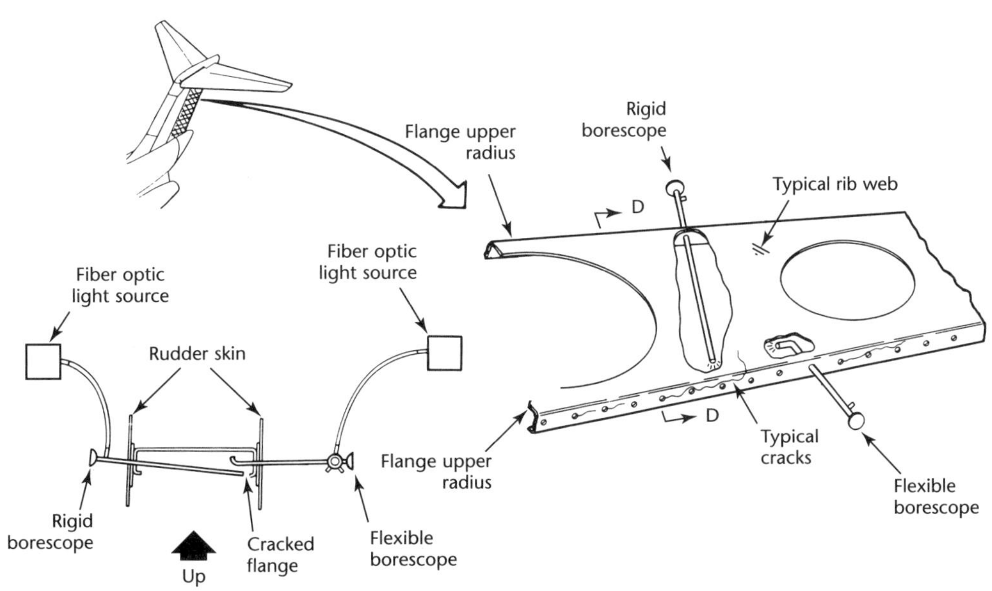

accomplished using a flexible borescope with a minimum length of 0.6 m (2 ft).

The inspector places his hand through the access hole and positions the tip of the 90 degree flexible borescope forward and aft of each fastener to detect fatigue cracks in the spar cap forward tang. The area under the footstool fitting is inspected using two techniques: (1) the seven fastener locations are inspected using a 0.6 m (2 ft) long, 90 degree, rigid borescope or a flexible borescope supported by semirigid plastic tubing and (2) the forward edge of the spar cap is inspected with a flexible borescope in a semirigid tube bent into a J shape (Fig. 10). These tests have been successful in detecting small cracks at the forward and aft side of the fasteners and larger cracks that propagate to the leading edge (forward) or vertical leg (aft) of the cap.

FIGURE 8. Case history of corrosion leading to component failure: (a) main landing gear truck beam; (b) fractured surface of failed component caused by failure to apply lubricant. The lubrication hole is visible opening on the inside surface.

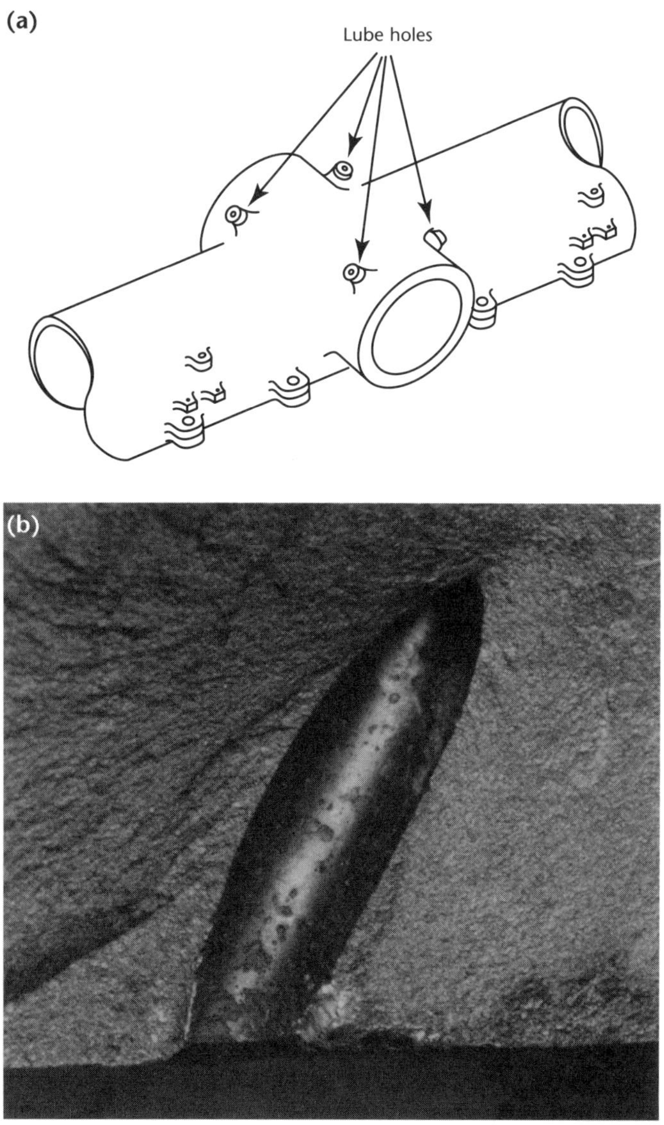

FIGURE 9. Surface conditions possible in typical main landing gear truck beam: (a) acceptable bore chamfer; (b) typical heat treat pit; (c) severe bore chamfer corrosion; (d) machining marks (rifling).

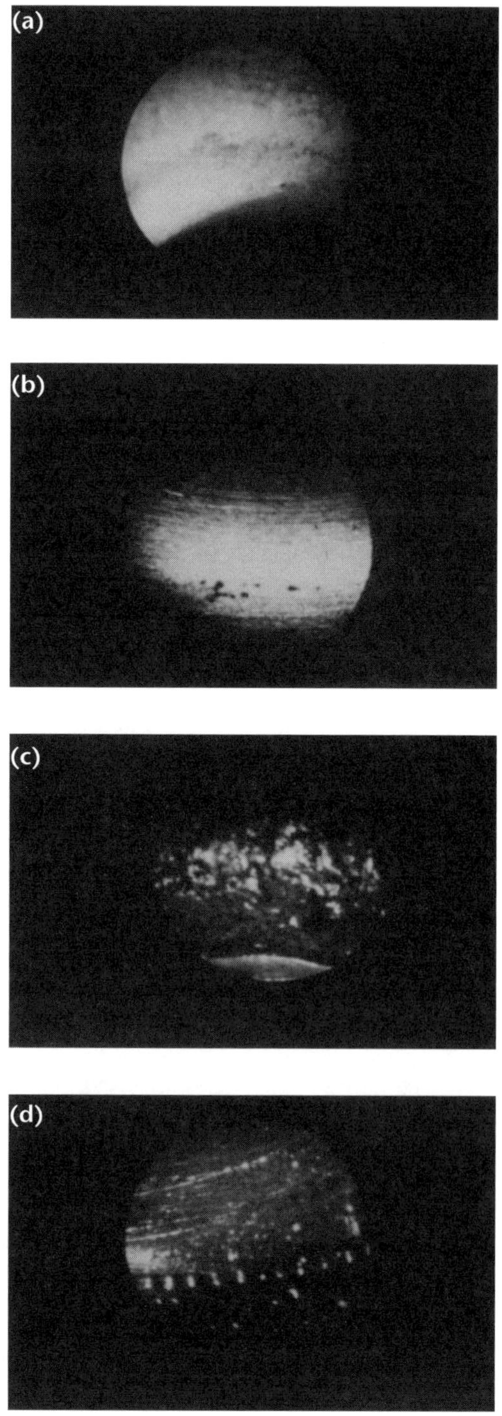

Closing

As indicated above, visual tests use magnifiers and borescopes. These devices are typically used on aircraft to detect corrosion and cracks where access is limited. The examples given are typical of other such applications for maintenance testing of aircraft structures and operating mechanisms. Visual testing is a viable and economical method that helps monitor the structural integrity of inservice aircraft.

Figure 10. Borescopic visual test of wing lower forward spar cap tang for fatigue cracks.

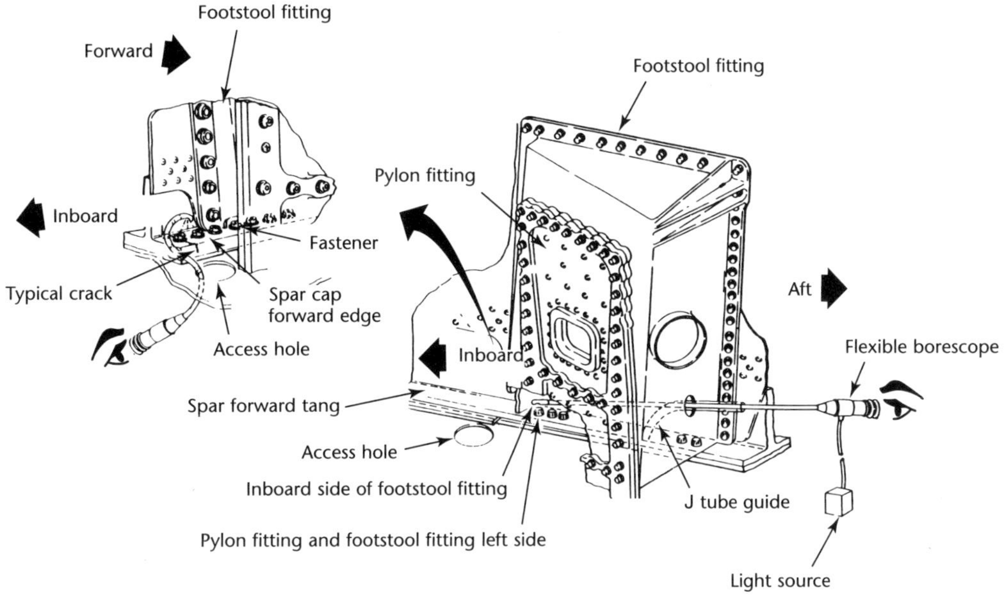

Part 2. Visual Testing of Jet Engines[6]

Distinctive Features of Engine Inspection

Details observed in engine inspection differ from airframe inspection and generally relate to surface finish and the means for inferring underlying conditions. Inspections within protective hoods and housings often require complex visual aids such as light guides, borescopes and special inspection fixturing.

The engine components have surface textures different from that of the rolled and formed surfaces in airframes and so are inspected with different criteria. Many engine anomalies are related to surface features of the material, and engine manuals describe visible surface characteristics in terms of the process that caused them and its effect on satisfactory performance of the part. As a result, much guidance relating to engine visual inspection is related to manufacturing rather than operational and maintenance inspection. Some anomalies, such as forging cracks or rolling marks, are found first in the manufacturing process; other anomalies, such as corrosion and fatigue damage, may be found by maintenance inspectors after some flights.

Another difference related to engine inspection is associated with the environment in which engines are expected to operate. Many engine parts are expected to survive intense heat. Some superficial clues to engine condition are leaks and evidence of smoke. Peening tends to remove evidence of heat discoloration and to smooth damage, making it less apparent.

Turbine Blades

Blades are evaluated from a standpoint of structural integrity. The blade is engineered to transfer combustion energy to the engine compressor (providing mostly air as driving force) and, in the process, experiences severe thermal and mechanical loading. In an aerospace engine, the material integrity of turbine blades must be high to sustain the service life. All blade surfaces must be smooth, and all evidence of previous leading and trailing edge repairs should be smoothed, or blended, with a minimum edge specified for that component.

Blades may be damaged by impact with foreign objects. The use of a substantial number of blades repaired to or near maximum limits, or use of blades having many repaired areas, may adversely affect compressor efficiency and therefore engine performance. If two or more blended blades are found grouped, an attempt should be made to redistribute these blades evenly throughout the engine.

When checking blades for foreign object damage, they should be visually tested for arc burn. Arc burn is evidenced by a small circular or semicircular heat affected area on the blade surface that may contain shallow pitting, remelting or cracking. Blades with arc burn are not acceptable for further service.

Blending is intended to ensure that all subsurface as well as surface damage is removed. It is critically important that the blending procedures be followed carefully whether in the shop or on the wing. Proper blending must be performed in successive stages: removal of the visible damage, confirmation by eddy current or fluorescent penetrant testing that all surface damage has been removed and, finally, removal of additional material to ensure the removal of subsurface deformed material.

After refurbishment or repair, the blade should be visually tested with bright light, mirrors, a 3× magnifier, fingertips and fingernails, looking specifically for small depressions in surface contour. Observed conditions may be compared with standard documentation and reference photographs.

Electrical Discharge Damage

Electrical discharge damage occurs in two forms.

1. Strong shapes such as ovals, crescents or rectangles depressed into the blade surface might be caused by contact with a rod or wire carrying an electrical charge.
2. Small overlapping craters or pits are coarse linear indications.

Regardless of form, electrical discharge damage is local, unlike erosion pitting.

Damage boundaries do not have raised material as does damage from foreign objects.

Blades suspected of electrical discharge damage must be removed from service immediately and blue etch anodize inspected to confirm damage. Electrical discharge damage is not repairable by blending. Electrical discharge damage produces a heat affected area that may extend 1.5 mm (0.06) in.) deeper than visible damage. Blades exhibiting electrical discharge damage should be removed from service.

Low Pressure Blade Assembly

Consult applicable procedures whenever foreign object damage is suspected. Consult applicable specifications for functional arrangement and part number applicability, and see general inspection data.

Inspect low pressure compressor blade assembly for damage.

Inspect airfoil for damage: (1) sustained indentations with compressed material and raised edges, (2) deformation with small radii or ragged edges and (3) tip curl damage.

Damage should be blended within limits specified for the part. Limits shown in Fig. 11 refer to maximum material removal allowed during blending and not to actual depth of damage before blending. Damage at or near maximum limits shown are not repairable if blending will exceed these limits.

Inspect leading and trailing edges for foreign object damage such as nicks and dents. These discontinuities must be blended.

1. Damage outside critical area should be blended if it exceeds 0.51 mm (0.020 in.) in depth.
2. Damage within critical area should be blended if it exceeds 0.127 mm (0.005 in.) in depth.
3. Damage not exceeding 0.127 mm (0.005 in.) in depth need not be blended if material is not torn.

No visible cracks or tears are permitted.

Injection Molded Blades[7]

Ceramic materials have been developed for many uses. Visual techniques are used for ceramic blades, along with liquid penetrant, radiographic and ultrasonic techniques. For a ceramic turbine blade of silicon nitride and 6 percent glass, liquid penetrant tests are used to detect and locate cracking and porosity.[8] Radiography is used to detect several types of inclusions as well as subsurface cracking. Visual techniques are also commonly used for inspection of these injection molded blades.

FIGURE 11. Airfoil inspection limits.

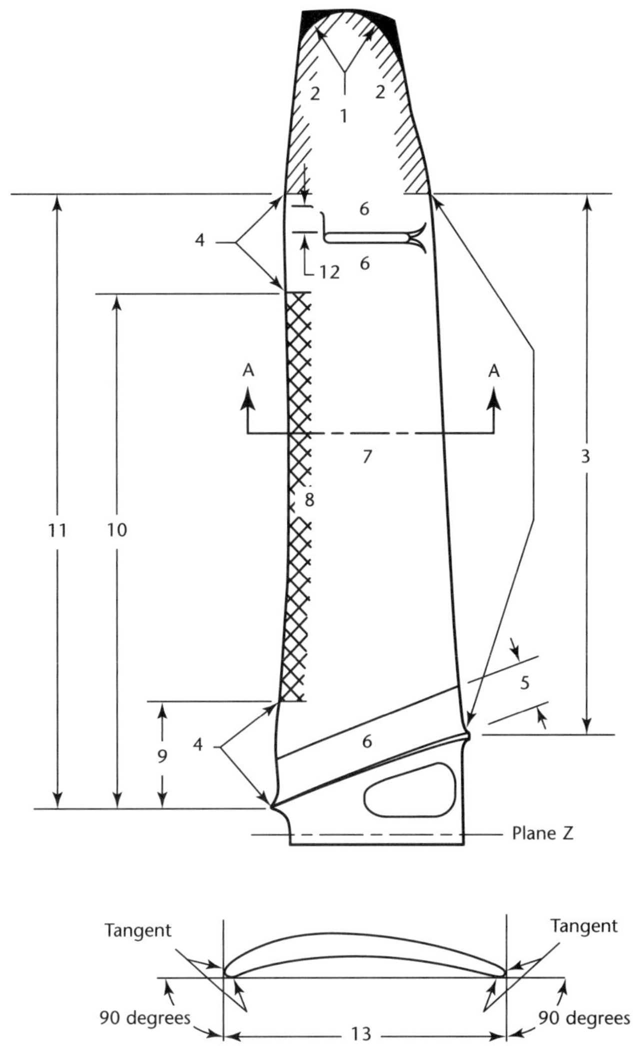

Section A-A
chord length measurement

Legend

1. Blade tip blend radius — 38.100 mm (1.500 in.) maximum.
2. Leading or trailing edge and tip blend area — blend to be 19.050 mm (0.750 in.) deep maximum.
3. 495.681 mm (19.515 in.).
4. Critical area — depth of blend to be 6.350 mm (0.250 in.) maximum except for airfoil root section. See index 5 dimension.
5. 25.400 mm (1.000 in.) — damage in airfoil-to-platform radius area cannot be blended except per airfoil root section blend limits.
6. No damage or blending permitted in shroud-to-airfoil radius or platform-to-airfoil radius.
7. Concave and convex blade surface blend area — depth of blend in this area to be 0.762 mm (0.030 in.) maximum (round bottom).
8. Leading edge blend area — depth of blend to be 12.77 mm (0.500 in.) maximum.
9. 160.020 mm (6.300 in.).
10. 469.900 mm (18.500 in.).
11. 571.800 mm (22.500 in.).
12. 12.700 mm (0.500 in.), pour places.
13. Minimum chord length to be 176.530 mm (6.950 in.) within 160.020 mm (6.390 in.) of leading edge platform. Chord length must be measured tangent to blade leading and trailing edges and parallel to phase Z.

Note: Damage which affects both index 1 and index 2 must be blended by combining both limits to produce a smooth blend.

There are several ways visual and optical tests may be used with advanced ceramic components. In one application, optical magnifications from 5× to 40× have been used for crack detection.[8] Cracking previously located with penetrants is typically not detected visually but open surface cracks along a turbine blade's split lines are visible (Fig. 12). It is possible that such cracks occur during removal from the mold and are subsequently sintered smooth so that they do not hold liquid penetrant.

Visual techniques may also be used in this application to locate pores in the turbine blade's thick sections.

Borescopy

Borescopic testing of the engine represents a significant aid to effective maintenance. Borescope access ports are incorporated at many locations along the engine gas path, allowing detailed examination of critical internal engine areas (Fig. 13).

Indirect visual testing with video borescopes or other electric viewing devices must be performed in a protected area. In wet weather, precautions must be used to prevent damage to equipment or electric shock to the operator. Also, operation of visual test equipment beyond its recommended limits may not be economically justifiable because of impaired performance or reduced service life.

The regular inspection interval is part of the written maintenance procedure. A reduced inspection interval may be half that value. Visual test intervals may be adjusted on the basis of operator experience and consultation with the local regulatory agency.

Engines get very hot. Temperatures of parts to be borescopically tested should be permitted to cool after engine operation. Damage may result to borescopic equipment if equipment is exposed to gas path or metal temperatures above 66 °C (150 °F). A rule of thumb is that, if adjacent cases are too hot to touch, then visual testing should be delayed.

Borescopic equipment for visual testing should satisfy the standards, procedures and specifications in force for the engine and workplace. Equipment and associated hardware includes power source, light cables and adapter required during borescopy. The specification sets the quality and functional standards for this equipment. For example, the following borescopic equipment has been recommended: (1) low magnification rigid borescope with 6.8 mm (0.270 in.) barrel diameter, (2) high magnification rigid borescope with 11.3 mm (0.444 in.) barrel maximum diameter and (3) flexible borescope 6.8 mm (0.270 in.) cable maximum diameter. Optional borescope equipment enhances the test capability and eases the task for the inspector. These items range from clamping equipment to hold the borescope in position to optical video equipment to record the visual test for later review.

For some engines, full 360 degree borescopic testing of the high pressure rotor can be accomplished by rotor

FIGURE 12. Visually detected open crack on aerospace engine ceramic turbine blade.

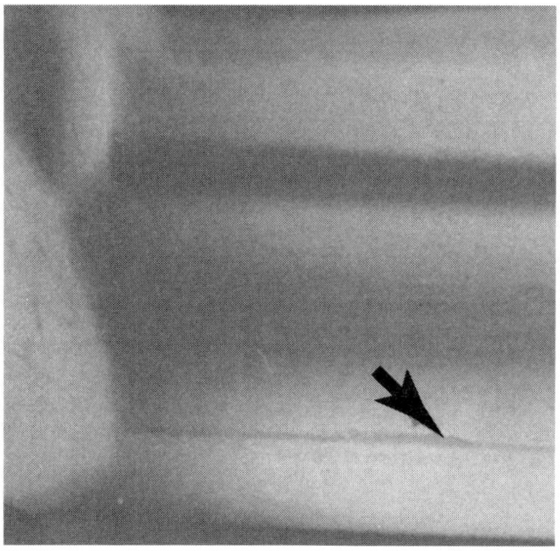

FIGURE 13. Borescope access ports (right side).

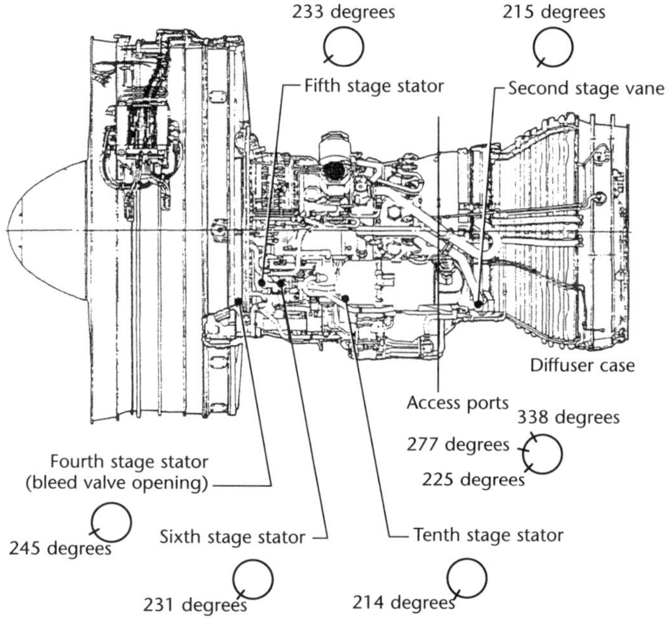

cranking of the gearbox. A pneumatic motor actuated by foot or hand will free the operator to use both hands to position and adjust the borescope. It is recommended that rotation be stopped as each rotor blade is positioned properly for inspection. Tilting will allow thorough appraisal of each blade's condition.

PART 3. Visual Testing of Composite Materials[9]

Visual and optical testing are important methods to the composite materials industry as primary inspection methods, as quick checks and as backups to other forms of nondestructive testing. The need for visual testing often results from several associated needs, such as the desire to visualize an anomalous situation, the desire to perform a low cost evaluation and the desire to visualize or document the findings of another nondestructive evaluation method.

Visual testing is often used with other forms of nondestructive evaluation — looking at a composite structure's surface for pits or voids, for example, before performing radiography. Despite its low technology and low cost, a visual evaluation may be all that is required by engineering personnel in assessing quality or service life conditions. Several topics will be covered in the following text: problems in the applications of visual testing, what visual testing can detect in composite materials, anomalous situations calling for visual testing, specific applications and recent image processing techniques for visual testing.

Problem Areas

Visual testing of composite materials is more difficult than that of most other engineering materials, due to the heterogeneous nature of composites. A perfectly good structure may have variations in resin, causing color variations on the exterior surface, surface roughness or waviness due to woven materials used to apply pressure during the layup, or benign surface features such as small voids or wrinkles.

Preparation of the part to be viewed is important in visual testing of composites. Often, structures cannot be viewed immediately after manufacture and edges and bag side surfaces must be trimmed extensively to remove excess resin. After these trimming operations there may be dust and residue on the surfaces. If the edge of the structure must be viewed, sanding or polishing of the edge may be necessary. After the cleaning, good lighting is essential. Straight-on lighting may be needed to see the features at the bottom of a void; or oblique lighting, to distinguish between a protruding or concave surface feature.

Because of inhomogeneity, it is important to choose an appropriate scale factor when viewing composite materials with magnification. Many visual tests are performed with 5× to 10× power. This relatively low power evaluation is a balance of close viewing of the small surface imperfections and retention of the ability to view a significant surrounding area for perspective and comparison. Figure 14 shows a questionable area on a cross section of an aramid polyester composite. Without proper lighting and scale factor, it appears that the area in the center of the figure has two matrix cracks. Figure 15 shows the same area, with higher magnification and with better lighting. The two lines are not matrix cracks. Figure 16 shows actual matrix cracks (notice the prominent crevice).

Because a visible light image can be readily viewed, documentation of an evaluation is important. Often the visual test will be requested and documentation required because of an anomaly detected with another form of nondestructive evaluation. Both film based and video photography are used extensively. Low power magnification film photography is useful to document surface discontinuities (Fig. 17); microscope photography is useful in imaging discontinuities that have been cross sectioned and polished (Fig. 18). Video photography is useful in documenting inaccessible areas, particularly when moving the imaging device from the external structure into the inaccessible area while the video recorder is turned on.

FIGURE 14. Possible matrix cracks.

1 mm (0.04 in.)

Conditions of Interest in Composite Materials

Visual testing is the first evaluation a composite structure receives after manufacture. This may be simply a quick evaluation of the condition of the structure after removal from the autoclave. After removal of the tooling and bagging materials, the surfaces of the component may be viewed for excessive wrinkles, holes, gaps between adjacent plies, surface voids, cracks, buckling and other conditions. This visual check is normally done without magnification. Just as the actual part itself is viewed, the tooling and bond forms may be viewed for discontinuities and alignment problems. These evaluations are normally performed by manufacturing personnel.

A more thorough evaluation may be required by engineering personnel. This may entail looking for resin rich or resin starved areas, blisters or delaminations in the exterior plies, surface voids and edge separations. If the matrix material is semitransparent, this evaluation may also include looking for internal voids by shining a high intensity light through the structure and viewing in a darkened area.

Visual evaluations often are required at fixed inservice intervals. Structures that experience stresses from weather, temperature fluctuations, handling or impact damage are routinely checked for damage initiation or growth. Delaminations, cracks and buckling can occur. This visual evaluation may reveal suspicions about an area and may necessitate followup nondestructive testing. Often the visual indication is the

FIGURE 15. Coloration change only, no cracks.

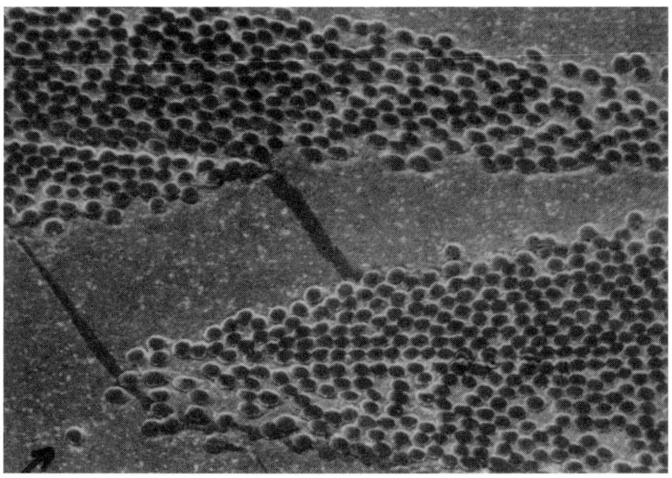

FIGURE 16. Actual matrix cracks.

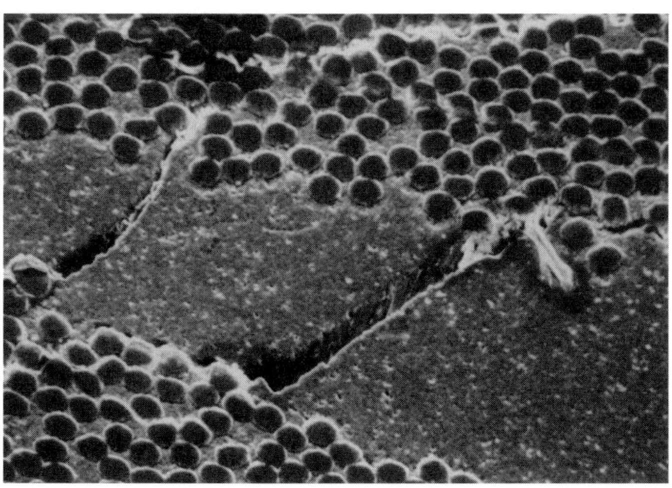

FIGURE 17. Voids exposed on composite surface.

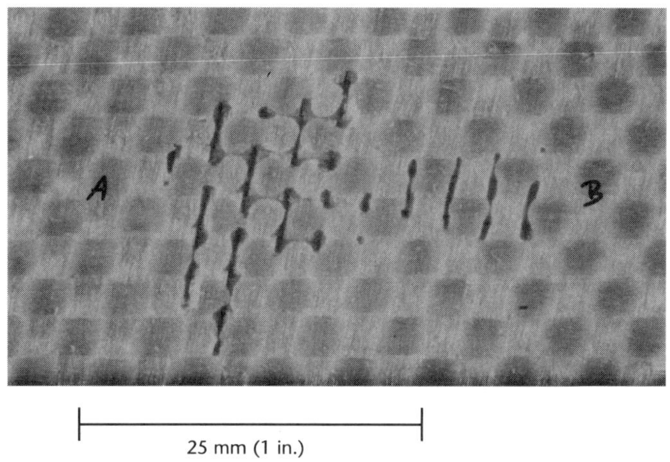

FIGURE 18. Voids in cross section of composite.

first sign of something wrong, and then the followup nondestructive evaluation reveals the details.

Applications

Visual testing may be all that is required to evaluate a structure in low stress applications. For example, thin sheets of composite material bonded to a honeycomb core may show visibly the large voids and delaminations that need to be detected. Even though the structure may be used in an aerospace application, it may not be necessary to spend large dollar amounts performing sophisticated nondestructive evaluation procedures. An example of this type of structure may be the fold down door on the overhead baggage compartment in an airliner. As an added test, a sonic tap test may be performed (also a low cost evaluation).

Visual testing may be done on large, costly aerospace structures as a first look to confirm that extensive discontinuities do exist, making it uneconomical to continue with more nondestructive testing. If no surface voids, excessive wrinkles or buckles are noted, ultrasonic testing may be used to look for delaminations and radiographic techniques may be used to look for voids/porosity. The cost of the ultrasound and radiography may be avoided if the first visual evaluation discovers rejectable discontinuities.

Borescope assisted visual testing has been used to augment other methods of nondestructive evaluation, where probes or film placements are in areas difficult to access. Some ultrasound composite testing applications have mounted probes on borescope devices that are inserted into accessible apertures of closed structures. Viewing the borescope image allows placement of the probe in an otherwise inaccessible area.

Visual testing is often the first opportunity to detect damage to aircraft from hail impact, runway debris or lightning strikes. Looking at the structure for dents, buckled areas or discoloration is part of preflight inspection. Some aircraft radomes (antenna housings) have had excessive static discharges that have resulted in burns to the composite. These burns are visually detected; then the composite structure is measured by ultrasound and radiography, and the results are provided to engineering personnel.

Magnifiers, light sources, inking dyes and wetting agents such as alcohol are all necessary tools for performing visual testing on composite materials. As noted above, 5× to 10× seems to be an appropriate viewing magnification. A common light source is a small, bright flashlight. It is portable and its beam is easily directed to the viewing area. The alcohol is used as a visible penetrant to enhance the contrast on edge discontinuities and voids. After wiping the alcohol onto the composite surface, the surface will dry very quickly, whereas the alcohol trapped in cracks or voids will take longer to dry. The dark marks go away as the alcohol evaporates. The inking dyes are useful in imaging small surface breaking cracks or voids, particularly when trying to document the cracks with photography. A small amount of dye is poured onto a cloth and wiped over the area. After the residual surface dye is wiped away, some dye stays trapped in the cracks, contrasting them against the background color. Figure 19 shows surface cracks in polyester resin after an impact damage event.

Of course, contacting materials must be compatible, not triggering corrosion for example.

Impact Damage and Exfoliation in Composite Panels

Impact damage and exfoliation are conditions that cause anomalies in the surface contours of composite panels like those used in aviation. The diffracted light technique described above has been investigated for possible use in evaluating the condition of such panels.

FIGURE 19. Dye enhancement of surface cracks.

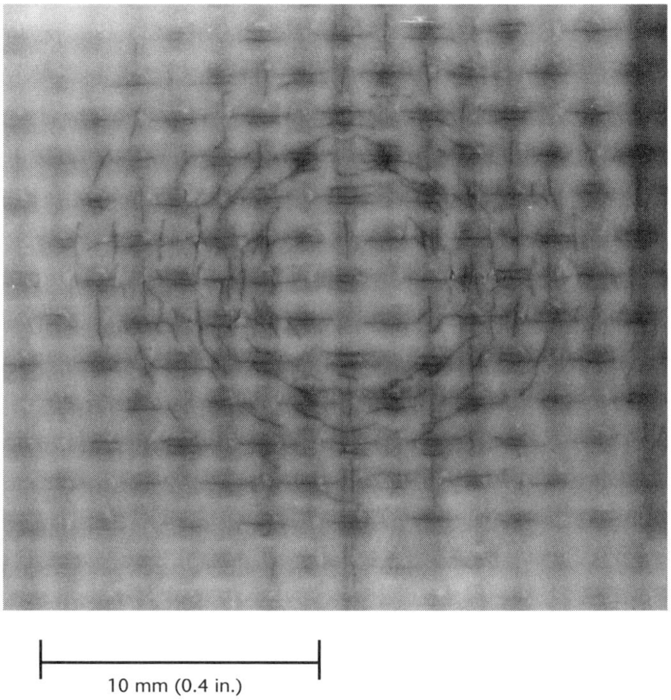

Graphite epoxy panels of three thicknesses (8, 24 and 48 plies) using a [0/45/90/–45]s layup were numbered and each was cut into three specimens (A, B and C). The specimens were then painted and ultrasonically C-scanned. No discontinuities were found. The specimens were inspected with the diffracted light technique before being subjected to impact energy ranging from 0.7 to 21.7 J. The resulting indentations ranged from nonvisible to barely visible. The depths of the indentations were measured and the specimens were ultrasonically C-scanned to detect and measure delaminations. The diffracted light images of the specimens are shown in Fig. 20 along with the C-scans.[10] The test results clearly show that the diffracted light technique can detect barely visible impact damage in composite laminates and can also detect cold worked holes in wrought aluminum panels and fatigue cracks extending from holes in wrought aluminum panels.[3,4,10]

Figure 21 shows the result of exfoliation corrosion detection under ambient lighting, ultrasonic C-scan and diffracted light technique imaging. The test object is not a composite but illustrates the techniques. The diffracted light technique shows corrosion around six of the twelve fastener holes. C-scan definitely shows corrosion in four of the six. The remaining two fastener holes probably would have indicated corrosion if a more sensitive ultrasonic technique had been used.

FIGURE 20. Damage 6.1 µm deep and 18.6 mm wide caused by 4.1 J impact on carbon epoxy panel: (a) ultrasonic C-scan; (b) diffracted light image.

(a)

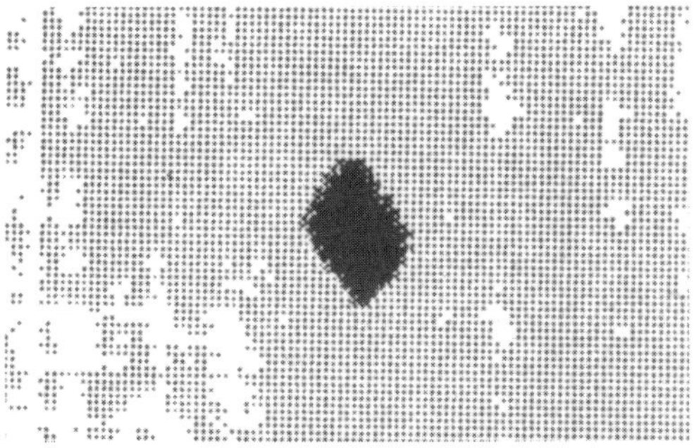

(b)

FIGURE 21. Ultrasonic and visual test results for exfoliation corrosion around steel fasteners in wrought aluminum plate: (a) ultrasonic C-scan; (b) ambient light; (c) diffracted light.

(a)

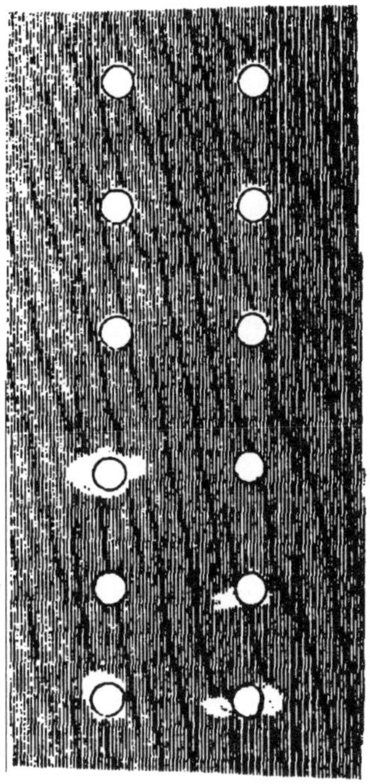

(b)

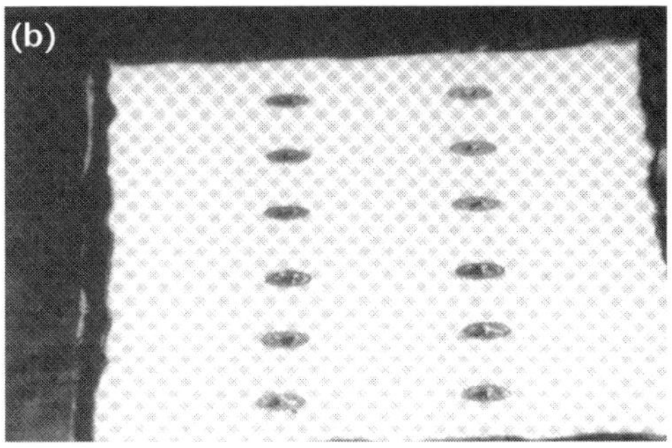

(c)

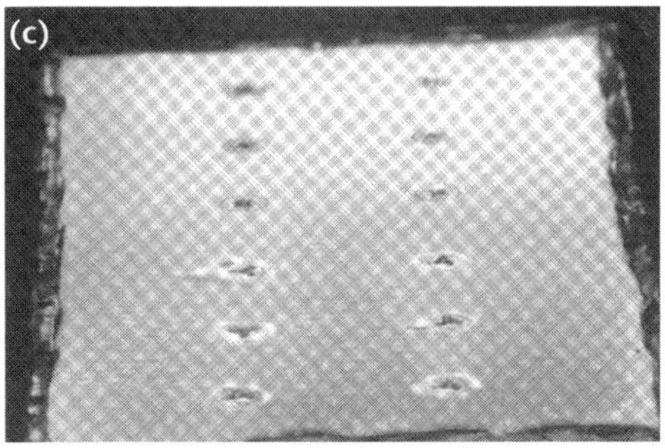

Image Processing

Recent advances and lowering costs of image processing tools have made their use in visual testing more popular. Some commercially available video borescopes now can be attached to computer systems with the ability to capture, store and manipulate the images. This ability to manipulate the images makes visual testing very powerful in visualizing faint images, in tracking features over time by storing and comparing images and by using measurement capabilities.

Captured images of a composite structure can be filtered, denoised and enhanced. Faint matrix cracks can be scaled up, contrast enhanced, pseudocolored, have an edge detecting filter applied to them—all to make them more visible. The captured image can be measured to determine the length of a void or crack.

Summary

Visual testing as applied to composite structures has been discussed, showing the importance of the method as a primary nondestructive evaluation method and as a supplement to one or more other nondestructive evaluation methods. Composite structures have some unique problem areas that require attention during visual testing. Several applications were briefly presented, as was the importance of visual testing to the verification of discontinuities noted by other methods. Looking to the future, it can be said that image processing techniques will play an increasing role in the choice of visual testing as applied to composite materials.

References

1. Bates, B.[L.], D.[R.] Christina and D.[J.] Hagemaier. "Optically Aided Visual Testing of Aircraft Structure." *Nondestructive Testing Handbook,* second edition: Vol. 8, *Visual and Optical Testing.* Columbus, OH: American Society for Nondestructive Testing (1993): p 292-301.
2. Hagemaier, D.J., B.[L.] Bates and D.R. Christina. "Optically Aided Visual Inspection of Aircraft Structure." *Materials Evaluation.* Vol. 46, No. 13. Columbus, OH: American Society for Nondestructive Testing (December 1988): p 1696-1701, 1707.
3. Komorowski, J.P., D.L. Simpson and R.W. Gould. "Enhanced Visual Technique for Rapid Inspection of Aircraft Structures." *Materials Evaluation.* Vol. 49, No. 12. Columbus, OH: American Society for Nondestructive Testing (1991): p 1486-1490.
4. Komorowski, J.P. and D.S. Forsyth. "The Role of Enhanced Visual Inspections in the New Strategy for Corrosion Management." *Aircraft Engineering and Aerospace Technology.* Vol. 72, No. 1. Bradford, West Yorkshire, United Kingdom: Emerald (2000): p 5-13.
5. Liu, Z., D.S. Forsyth and T. Marincak. "Preprocessing of Edge of Light Images: Towards a Quantitative Evaluation." *Nondestructive Evaluation and Health Monitoring of Aerospace Materials and Composites II* (San Diego, CA, March 2003). SPIE Proceedings Vol. 5046. Bellingham, WA: International Society for Optical Engineering (Society of Photo-Optical Instrumentation Engineers) (2003): p 30-38.
6. FAA 43-204, *Visual Inspection for Aircraft.* Advisory circular. Washington, DC: United States Government Printing Office, for Federal Aviation Administration (1997).
7. "Visual Tests of Ceramics." *Nondestructive Testing Handbook,* second edition: Vol. 8, *Visual and Optical Testing.* Columbus, OH: American Society for Nondestructive Testing (1993): p 328-330.
8. Dunhill, T. "Some Experiences of Applying NDE Techniques to Ceramic Materials and Components." *Nondestructive Testing of High-Performance Ceramics.* Westerville, OH: American Ceramic Society (1987): p 268.
9. Bailey, D. "Visual Testing of Composite Materials." *Nondestructive Testing Handbook,* second edition: Vol. 8, *Visual and Optical Testing.* Columbus, OH: American Society for Nondestructive Testing (1993): p 328-330.
10. Komorowski, J. and R. Gould. "A Technique for Rapid Inspection of Composite Aircraft Structures for Impact Damage." *AGARD Conference Proceedings.* No. 462. Neuilly sur Seine, France: North Atlantic Treaty Organization, Advisory Group for Aerospace Research and Development (1990).

Bibliography

Bailey, W.H. "Requirements for Visual Acuity in the Aerospace Industry." *Materials Evaluation.* Vol. 46, No. 13. Columbus, OH: American Society for Nondestructive Testing (December 1988): p 1625-1630.

Boyd, Cliff and M. Lyman. "Video Borescope Technology Helps Aircraft Engine Maintenance." *Materials Evaluation.* Vol. 51, No. 5. Columbus, OH: American Society for Nondestructive Testing (May 1993): p 543.

Drury, C.G. "Human Reliability in Civil Aircraft Inspection." Research and Technology Organization Meeting Proceedings 32, *The Human Factor in System Reliability — Is Human Performance Predictable?* [Human Factors and Medicine Panel Workshop, Siena, Italy, December 1999]. Hanover, MD: National Aeronautics and Space Administration, Center for AeroSpace Information, for North Atlantic Treaty Organization (2000).

Good, G.W. and V.B. Nakagawara. Final Report [FAA Grant 02-G-031], *Vision Standards and Testing Requirements for Nondestructive Inspection (NDI) and Testing (NDT) Personnel and Visual Inspectors*. Washington, DC: Federal Aviation Administration, Aviation Maintenance and Human Factors (2005).

Katzoff, S. "The Surface-Tension Method of Visually Inspecting Honeycomb-Core Sandwich Plates." *Nondestructive Testing [Materials Evaluation]*. Vol. 18, No. 2. Columbus, OH: American Society for Nondestructive Testing (March-April 1960): p 114-118.

McGarry, G. "Penetrants and Fiber Optic Borescope Help Ensure Safety of Shuttle Training Aircraft." *Materials Evaluation*. Vol. 52, No. 10. Columbus, OH: American Society for Nondestructive Testing (October 1994): p 1180.

12 CHAPTER

Techniques Allied to Visual Testing

Stephen L. Meiley, Champion International, West Nyack, New York (Part 2)

Donald Parrish, Southern Company Services, Birmingham, Alabama (Part 3)

Donald J. Roth, National Aeronautics and Space Administration, Glenn Research Center, Cleveland, Ohio (Part 1)

Michael A. Urzendowski, Valero Energy, San Antonio, Texas (Part 2)

PART 1. Indications Not from Visual Testing

As a nondestructive test method, visual testing is defined by its interrogating and indicating energy: light, in the visible part of the electromagnetic spectrum. Light and images are used to display, observe, analyze, communicate and record the results of all nondestructive tests.

Visibility criteria are specified for magnetic particle tests, liquid penetrant tests and some leak tests. Vision acuity is verified for some inspectors who use magnetic particle and ultrasonic testing. Light levels, indication sizes, viewing angles, color sensitivity and many other phenomena pertaining to human vision are strictly controlled to achieve reliable accuracy in visual tests as well as other nondestructive testing techniques.

Visual testing is linked to these other methods by shared hardware.

1. Borescopes, the basic tools of visual testing, are often used to view obstructed magnetic particle or liquid penetrant test indications.
2. Magnifiers are used to visually test material surfaces, as well as to study the details in radiographs and to measure indications in magnetic particle and liquid penetrant testing.
3. Still and video cameras are used in cineradiography, in the photographic recording of various test results, in machine vision, in remote television pickups and in virtually all the automated nondestructive test methods.
4. Image processing software enhances the display and interpretation of, for example, radiographic, ultrasonic or microwave test results.

Visual testing may be used with another method, either before or after it.

1. Visual testing may be used qualitatively to identify problem areas that need follow-up with a quantitative method. A corroded spot identified visually may be flagged for ultrasonic thickness testing, for example.
2. Visual tests are often used to inspect or verify the data of the other tests. It might be said, for example, that radiography is ultimately a visual test of the radiograph — to determine that the radiographic images are properly exposed, that the areas of interest are free of artifacts and that the images can show the characteristics of interest. As another example, wet magnetic particle tests are not reliable unless the bath has passed a preliminary visual test so that the inspector can see discontinuity indications.

Visual Aspects of Leak Testing

Leak testing is done by detecting a tracer medium, a gas or liquid, that has escaped from confinement. The tracer can be an added fluid or in some cases the fluid that the vessel is designed to hold. Testing is done visually, aurally or electronically. Occasionally, tracers are designed to interact with materials applied or naturally present outside the vessel, to produce highly visible evidence of leakage. The visual portion of a typical leak test is that which determines the presence and location of leakage. The rate of leakage and its effect on fluid flow may be determined by visual observation of meters and gages.

Visual tests are conducted to locate leakage from pressure retaining components. In nuclear power plants, visual tests are required for locating abnormal leakage from components with or without leakage collection systems. For certain nuclear power components, visual testing is performed using the reactor coolant (water) as the tracer medium. The visual testing of noninsulated pressure retaining components is performed by inspecting external, exposed surfaces for visible evidence of leakage.

Components whose external surfaces are inaccessible for direct viewing are examined by visually checking the surrounding area, including drip pans or surfaces beneath the components of interest. Color detection and color differentiation are essential in certain leak testing procedures. For example, bromocresol purple is a dye used in chemical reaction leak testing with ammonia gas tracers. The dye is sprayed or brushed onto the outside surfaces of pressure vessel welds and allowed to dry. After drying, the dye turns into a yellow powder. The vessel is then pressurized with ammonia gas. If a leak exists, it is indicated by change in the color of the powdered dye — from a light yellow to a vivid purple.

Various visible color and fluorescent dyes are also used in leak testing. These materials are sensitive to the concentration of active ions, either acid or alkaline. Such ions determine the hydrogen potential (pH) of a solution, and the pH value can be shifted by addition of an acid or an alkali. A number of useful indicator dyes are sensitive to small changes in pH and the most direct way to use these effects is to use the dyes with a tracer gas that produces a change in pH. This type of gas-phase leak indicator typically employs a liquid applied onto areas suspected of leakage. One dye suitable for producing visible color leak indications is phenolphthalein.

Color differentiation is also needed in high voltage discharge leak testing. In this technique, a spark coil is used to excite a visible glow discharge in systems where the pressure is between 1 and 1000 Pa (10^{-2} to 10 torr). The tracer can be a gas such as carbon dioxide or a volatile liquid such as benzene, acetone or methyl alcohol. When the tracer gas enters the system through a leak, the color of the discharge changes from blue or purple (the color of air) to the color characteristic of the tracer (Table 1). Visual detection of this color change is required to interpret the leak test.

Visual Aspects of Liquid Penetrant Testing

In liquid penetrant tests, visual techniques are used to compare sensitivities of penetrant systems, to detect discontinuity indications and to verify the cleanliness of the testing materials. When a hydrophilic emulsifier bath is used, visual monitoring of the bath can provide clues to its condition. A fresh solution of emulsifier in water exhibits a typical color in visible light and in ultraviolet light. Traces of fluorescent penetrant contamination in the bath will darken its color and cause the fluorescence to shift. High levels of penetrant contamination cause a hydrophilic emulsifier solution to become cloudy and, at higher levels, free penetrant can be seen to float on the bath surface.

Penetrant developers may also suffer degradation that is visually detectable. The best test of a developer is a visual comparison with new material under visible and ultraviolet light. Penetrant contamination in dry developer causes bright color spots or bright fluorescent spots.

During liquid penetrant tests, color detection and color differentiation are often critical to completion of the test. Certain visual data produced by penetrant testing procedures may correlate with specific aspects of material discontinuities. In many codes, it is specified that penetrant indications may be visually detected in natural or artificial light.

Because penetrant tests normally rely on an inspector's visual detection, the lighting for this procedure is very important. It not only affects the sensitivity of the test method but it is also an important factor in inspector fatigue. Visible light sources for penetrant tests are identical to those specified for other visual test applications. Spectral characteristics are usually not critical for visible light sources but it may be better to use a light source deficient in the light reflected by the penetrant and rich in the other components of the visible spectrum. When such a light is used on a test object having good white developer background, the penetrant indication appears darker and maximum contrast is obtained. Floodlights are advantageous for large and relatively flat test surfaces. On intricate or small test objects, manually directed spotlights may be the most effective visible light source.

TABLE 1. Discharge colors in gases and vapors at low pressures.

Gas	Negative Glow	Positive Indication
Air	blue	red
Argon	blue	deep red (violet)
Bromine	yellow/green	red
Carbon dioxide	blue	white
Carbon monoxide	green/white	white
Chlorine	green	light green
Helium	pale green	violet/red
Hydrogen	pink/bright blue	pink (rose)
Iodine	orange/yellow	peach
Krypton	colorless	green
Lithium	white	bright red
Mercury	green (gold/white)	green/blue (green)
Methane	colorless	red/violet
Neon	red/orange	red/orange (blood red)
Nitrogen	blue	yellow (red/gold)
Oxygen	yellow/white	lemon
Potassium	green	green
Sodium	yellow/green (white)	yellow
Xenon	colorless	blue/white

The proper intensity of visible light is determined by the requirements of the penetrant test. For gross discontinuities, a brightness level of 300 to 550 lx at the surface is typically sufficient. Intensity levels of 1000 lx are necessary for small but critical discontinuities.

Because liquid penetrant tests use the human eye as a detection device, the condition of the inspector's eyes is an important test specification. For fluorescent penetrant testing, the inspector must be dark adapted and the intensity of the ultraviolet source must reach specified minima. Photosensitive eye glasses must not be worn. The eye itself will fluoresce under certain conditions, causing temporarily clouded vision.

Optical technologies are used in penetrant tests in a variety of ways. For example, a flying spot laser can be used for detection of penetrant test indications. When the laser beam strikes fluorescent penetrant materials, a pulse of different wavelength light is generated. A photodetector converts this pulse into an electrical signal that is analyzed, using pattern recognition techniques, to determine the discontinuity's shape and size.

Visual Aspects of Radiography

Vision acuity is vital to the radiographic interpretation process. Individual vision can and does vary from test to test depending on physiological and psychological factors. Annual vision acuity examinations cannot detect daily fluctuation or its influence on interpretation and the frequency of a vision acuity examination may be specified.

In radiographic testing, the physical measure of interest for vision acuity is the discontinuity as displayed on the film, regardless of how much it may differ from the actual discontinuity in the test object. Vision acuity tests can be based on microdensitometric scans of discontinuities taken directly from actual radiographs. Vision acuity exams for radiographers may include factors such as the figure-to-ground relationship, background luminance, contrast, line width, line length, viewing distance, blur, line orientation and characteristics of the light source.

During radiographic tests, viewing conditions are very important. A finished radiograph should be inspected under conditions that afford maximum visibility of detail together with maximum comfort and minimum fatigue for the interpreter. Subdued lighting in the viewing area is preferable to total darkness. The room lighting must be arranged so that there are no reflections from the surface of the film.

Magnifiers and densitometers may help to view industrial radiographs. Scanning microdensitometry equipment can be useful for certain industrial applications, including focal spot measurements and determination of total radiographic unsharpness. The relationship of density differences to material thickness differences can be graphed and studied with an item of known thickness (a shim, for example) on the radiograph.

For an image to be digitally enhanced, it must be in digital form. Digital radioscopy and computed tomography are inherently digital. Radioscopic images from luminescent screens (fluoroscopy) or cathode ray tubes may be scanned and presented in a two-dimensional digital array of picture elements (pixels). Conventional film radiographs also may be digitized. Once the radiographic image is digitized, a variety of enhancement methods can be used, including brightness transfer functions, gradient removal, digital filtering, field flattening, smoothing and a number of other transforms.

The coding of several scale intervals with a different color results in an enhancement technique known as *pseudocolor*. Though these color images are striking to the viewer, the technique has the disadvantage of presenting color changes that do not necessarily correspond to abrupt changes in optical density on the image.

Visual Aspects of Magnetic Particle Testing

In magnetic particle tests, the vision acuity of the inspector and the visibility of the test results are as important as they are for liquid penetrant tests.

Dry magnetic particles are commercially available in visible colors, fluorescent colors and daylight fluorescent colors. Visible particles are typically available in gray, red, black, yellow, blue and metallic pigments. Colors of visible particles are chosen for individual testing applications based on the highest contrast with the test object surface.

Daylight fluorescent colors were developed to increase visibility in those applications where somewhat lessened sensitivity is acceptable. Daylight fluorescent colors have enhanced visibility in light from the sun, blue mercury vapor lamps or white fluorescent tubes. The yellow light from sodium vapor sources does not excite fluorescence in these magnetic particles. The dyes absorb photons of one energy and emit photons at a lower energy.

Much of the specified technique for magnetic particle and other tests is established to enhance visibility of the discontinuity indications. Choice of color, magnetization levels and application of the particles themselves are all done in ways that maximize visibility. In certain magnetic particle applications, a thin white lacquer is applied to the test object surface to make dark colored particles more visible (smaller particles are more responsive to magnetism and the addition of a layer of pigment to the particles decreases their sensitivity).

Wet method particles are available in fluorescent and nonfluorescent forms. Particle color is again chosen to increase contrast with the test object surface.

Not only are particle characteristics chosen to maximize visibility but light levels are also carefully monitored. A light intensity of 1000 lx (100 ftc) can be used for viewing nonfluorescent magnetic particle indications. The optical light level, however, is sometimes a compromise between operator fatigue and visibility. On bright or reflective surfaces, high light intensities can cause glare that interferes with vision and subsequent interpretation. On darker surfaces or those covered with thin scale, the 1000 lx (100 ftc) level may be barely adequate for visibility.

In wet method fluorescent tests, standards require a minimum ultraviolet light intensity of 1000 $mW \cdot mm^{-2}$, with a maximum allowable visible light intensity at the surface of 20 lx (2 ftc). Even small amounts of visible light can lower the contrast of a fluorescent indication.

Appropriate light levels are so critical to the completion of magnetic particle tests that visible light intensity measurements are routinely performed and logged. Typically, such entries are made from a light meter that is itself calibrated every six months or at intervals prescribed by the relevant specification.

Visual capabilities are critical to magnetic particle indication detection but they are also used to verify the quality of the testing materials. In an agitation system, where the particle bath constantly passes through a centrifugal pump, particles are subject to constant high speed impact and shearing from the pump's impeller. As breakdown of the particles increases, test indications become dimmer and background fluorescence is seen to increase (indication-to-background contrast diminishes).

Optical equipment used for magnetic particle tests is the same as for liquid penetrant tests: illuminated rigid borescopes, fiber optic borescopes and video borescopes. These devices are used in the same way, with the same advantages and precautions, as in a purely visual test. The difference is that the object of interest is a magnetic particle indication of a material discontinuity. Light intensities for borescopes and remote viewing instruments must be set to include losses between the eyepiece and the distal tip of the device. Quartz fiber borescopes typically need less light intensity than conventional borescopes used for viewing visible particle indications. In a rigid borescope, light transmission is governed by an optical system. For video borescopes, the sensitivity of the camera chip is the critical factor. Reference standards and comparative tests with known discontinuities are used to verify remote viewing of magnetic particle indications.

Visual Aspects of Ultrasonic Testing

Ultrasonic imaging can be performed by several techniques. The most common is that used by commercially available ultrasonic immersion C-scanning systems. An ultrasonic wave of known amplitude is transmitted through a test material. The final C-scan image is a representation of amplitude attenuation or energy lost by the ultrasonic wave during its trip from the transmitter to the receiver.

C-scan images represent the energy lost by the ultrasonic wave but, unfortunately, not solely because of the internal structure of the test object. The ultrasonic reflection coefficients at the water-to-object interfaces are generally assumed to be constant but actually can vary widely. These reflection coefficients must be included in the analysis to determine the actual energy loss or attenuation caused by the inner structure of the test object. They must also be used in determining the accuracy of the acoustic image as well as the accuracy of the imaging technique.

An ultrasonic transducer is scanned over the test object surface while a load cell gages coupling pressure. After collection and subsequent fourier analysis of the appropriate waveforms, velocity and attenuation can be determined accurately at different positions on the test object, in an organized array. The resulting data yield velocity and attenuation maps that are displayed on a video system.

A typical data set, from which about 25 images can be obtained, contains about 2000 waveforms, 2000 fourier spectra and 40 000 attenuation and velocity values.

The original acquired data use about 20 megabytes of computer memory. The reduced data set containing the attenuation, velocity and reflection coefficient[2] images occupies about 5 megabytes.

Evaluating an image often requires tests of both the raw and reduced data for a particular point in the image. A data retrieval program provides the necessary interface to allow immediate retrieval of data displayed in a video image overlay. For example, the data for one point (determined by the cursor location) in an ultrasonic image of velocity are displayed in an overlay. This display contains waveforms and their fourier spectra, reflection and attenuation coefficients (may be used to determine accuracy), phase velocity as a function of frequency and group velocity (determined by cross correlation).

All these data are inspected for irregularities when evaluating a specific point on the ultrasonic image. For example, if a quantitative measure of the density difference between two points is needed, it can be determined from the group velocity at these points on the appropriate overlays. Also, the frequency dependence of attenuation has been related to the subsurface pore or discontinuity sizes,[3] the state of the recrystallization in metals[4] and the mean grain size.[5] Easy access to this type of information is made possible by the use of a video system.

Role in Data Interpretation

Video and computer systems also play a crucial role in the interpretation of ultrasonic data. In the past, only a few measurements were taken at a small number of positions to evaluate a material ultrasonically. The data from these measurements were averaged and the standard deviation determined. These two values (the average and the standard deviation) were used to classify the test object but the two quantities may not contain sufficient information for providing a full understanding of the internal structure of the test material.

For example, from nine measurements, the average ultrasonic attenuation and its standard deviation at 100 MHz in a ceramic test object can be calculated as 0.11 and 0.03 $Np \cdot mm^{-1}$. But this information does not describe all cloudlike attenuation structures. Ultrasonic attenuation in ceramics can be caused by diffractive scattering from individual grain and pore boundaries. An increase in the number or size of pores produces an increase in attenuation. High ultrasonic attenuation may also be caused by diffractive scattering at regions exhibiting large velocity (density) gradients, not increased porosity.

Closing

The importance of optical technologies to the broad field of nondestructive testing is typified in the interplay of visual and optical techniques with ultrasonic technology, as detailed above. Visual testing has a long and vital history. Despite their origins and their value in the past, the visual and optical techniques are also directly linked to the most complex nondestructive testing technologies of the present.

PART 2. Replication[6]

Replication is a valuable tool for the analysis of fracture surfaces and microstructures and for documentation of corrosion damage and wear. There is also potential for uses of replication in other forms of surface testing.

Replication is a method used for copying the topography of a surface that cannot be moved or one that would be damaged if transferred. A police officer making a plaster cast of a tire print at an accident scene or a scientist making a cast of a fossilized footprint are common examples of replication. These replicas produce a negative topographic image of the subject known as a *single stage replica.* A positive replica made from the first cast to produce a duplicate of the original surface is called a *second stage replica.*

The technology of replication is described in the literature.[7-9]

Many replicating mediums are commercially available. Most replica types in nondestructive testing are cellulose acetate or silicone rubber. Both have advantages and limitations, and either can provide valuable information without altering the test object.

Cellulose Acetate Replication

Acetate replicating material is used for surface cleaning, removal and evaluation of surface debris, for fracture surface microanalysis and for microstructural evaluation. Single-stage replicas are typically made, creating a negative image of the test surface. A schematic diagram of microstructural replication is shown in Fig. 1.

Cleaning and Debris Analysis

Fracture surfaces should be cleaned only when necessary. Cleaning is required when the test surface holds loose debris that could hinder analysis and that cannot be removed with a dry air blast.

Cleaning debris from fracture surfaces is useful when the test object is the debris itself or the fracture surface. Debris removed from a fracture can be coated with carbon and analyzed using energy dispersive spectroscopy. This provides a semiquantitative analysis when a particular element is suspected of contributing to the fracture.

Removal of loose surface particles is usually done by wetting a piece of acetate tape on one side with acetone, allowing a short period for softening and applying the wet side of the tape to the area of interest. Thicker tapes of 0.013 mm (0.005 in.) work best for such cleaning applications (thin tapes tend to tear). Following a short period, the tape hardens and is removed. This procedure is normally repeated several times until a final tape removes no debris from the surface.

Fracture Surface Analysis

The topography of fracture surfaces can be replicated and analyzed using an optical microscope, scanning electron microscope or transmission electron microscope. The maximum useful magnification obtained using optical microscopes depends on the

FIGURE 1. Acetate tape replication producing negative image of surface: (a) microstructure cross section; (b) softened acetate tape applied; (c) replica curing; (d) replica removal.

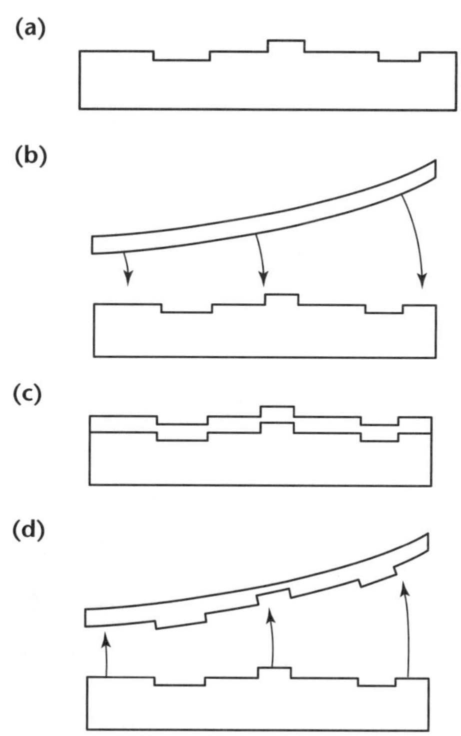

roughness of the fracture but seldom exceeds 100×. The scanning electron microscope has good depth of field at high magnifications and is typically used for magnification of 10 000× or less. The transmission electron microscope has been used to document microstructural details up to 50 000×.

In general, scanning electron microscope analysis of a replica provides information regarding mode of failure and, in most instances, is sufficient for completion of this kind of analysis. An example of a replicated fracture surface is shown in Fig. 2. The transmission electron microscope is used in instances where information regarding dislocations and crystallographic planes is needed. Both single stage (negative) and second stage (positive) replicas can be used for failure analysis. Some scanning electron microscope manufacturers offer a reverse imaging module that provides positive images from a negative replica. This eliminates the need to think and interpret in reverse. This feature has also proven valuable for evaluating microstructures through replication.

FIGURE 2. Fracture surface replication shows fatigue striations on surface: (a) 2000×; (b) 10 000×. Scale has been modified during publication.

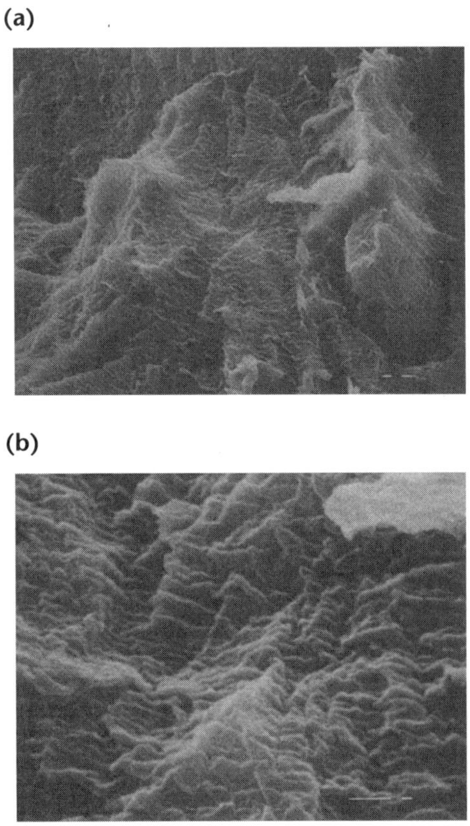

As with the removal of surface debris, it has been found that the thicker replicas provide better results, for the same reasons. The procedure for replication of fracture surfaces is identical to that for debris removal. On rough surfaces, however, difficulty may be encountered when trying to remove the replica. This can cause replication material to remain on the fracture surface but this can easily be removed with acetone.

Replicas, in the as-stripped condition, typically do not exhibit the contrast needed for resolution of fine microscopic features such as fatigue striations. To improve contrast, shadowing or vapor deposition of a metal is performed. The metal is deposited at an acute angle to the replica surface and collects at different thicknesses at different areas depending on the surface topography. This produces a shadowing that allows greater resolution at higher magnifications.

Shadowing with gold or other high atomic number metals enhances the electron beam interaction with the sample and greatly improves the image in the scanning electron microscope by reducing the signal-to-noise ratio.

Microstructural Interpretation

To date, the greatest advances in the use of acetate replicas for nondestructive testing have come from their use in microstructural testing and interpretation. Replication is an integral part of visual tests in the power generation industries as well as in refining, chemical processing and pulp and paper plants. Replication, in conjunction with microstructural analysis, is used to quantify microstrain over time and to predict the remaining useful life of a component. Future applications are not limited by material type.

In industry, tests are carried out at preselected intervals to assess the structural integrity of components in their systems. These components can be pressure vessels, piping systems or rotating equipment. Typically these components are exposed to stresses or an environment that limits their service life. Replication is used to evaluate such systems and to provide data regarding their metallurgical condition.

Microstructural replication is done in two steps: surface preparation followed by the replication procedure. Surface preparation involves progressive grinding and polishing until the test surface is relatively free of scratches (metallurgical quality). Depending on the material type and hardness, this can be obtained by using a 1 to 0.05 mm (4×10^{-5} to 2×10^{-6} in.) polishing compound as the final step. Electrolytic polishing can increase efficiency if many areas are being tested. Surfaces can be electropolished with a 320 to 400 grit finish.

The disadvantages of electropolishing are that (1) the equipment is costly, (2) with most systems only a small area can be polished at one time and (3) pitting has been known to occur with some alloy systems containing large amounts of carbides.

Next, the polished surface is etched to provide microstructural topographic contrast which may be necessary for evaluation. Etchants vary with material type and can be applied electrolytically, by swabbing or spraying the etchant onto the surface. With some materials, a combination of etch-polish-etch intervals yields the most favorable results.

To replicate the surface microstructure, an area is wetted with acetone and a piece of acetate tape is laid on the surface. The tape is drawn by capillary action to the metal surface, producing an accurate negative image of the surface microstructure. Thin acetate tape at 0.025 mm (0.001 in.) provides excellent results and gives the best resolution at high magnifications. Thicker tapes must be pressed onto the test surface and, depending on the expertise of the inspector, smearing can result. Thicker tapes are more costly and the resolution of microscopic detail does not match thinner tapes. Studies of carbide morphology and creep damage mechanisms have been performed at magnifications as high as 10 000× with thin tape replicas.

Before removal of the tape from the test object, the back is coated with paint to provide a reflective surface that enhances microscopic viewing. The replica is removed and can be stored for future analysis.

If analysis with the scanning electron microscope is needed, replicas should be coated to prevent electron charging. This is accomplished by evaporating or sputter coating a thin conductive film onto the replica surface. Carbon, gold, gold-palladium and other metals are used for coating. There are differences in the sputtering yield from different elements and this should be remembered when choosing an element or when attempting to calculate the thickness of the coating.

The main advantage of sputter coating over evaporation techniques is that it provides a continuous coating layer. Complete coating is accomplished without rotating or tilting the replica. With evaporation, only line of sight areas are coated and certain areas typically are coated more than others.

Some examples of replicated microstructures, documented with both a scanning electron microscope and with conventional optical microscopy, are shown in Figs. 3 to 6. Replication is used for detection of high temperature creep damage, stress corrosion cracking, hydrogen cracking mechanisms, as well as the precipitation of carbides, nitrides and second phase precipitates such as sigma or gamma prime. Replication is also used for distinguishing fabrication discontinuities from operational discontinuities.

Strain Replication

Attempts have been made to replicate strain to evaluate localized strain in materials exposed to elevated temperatures and stresses over time (materials susceptible to high temperature creep). The purpose is to monitor accumulated strain before detectable microstructural changes occur. Strain replication entails inscribing a grid pattern onto a previously polished surface. A reference grid pattern is replicated by using material with a shrinkage factor quantified through analysis. This known shrinkage factor is

FIGURE 3. Comparison of optical microscopy to scanning electron microscopy in documentation of replicated microstructure. Evidence of creep damage is visible in grain boundaries. Etchant is aqua regia. Magnification was 100× before publication; (a) optical microscope image; (b) scanning electron microscope image.

(a)

(b)

included in future numerical analysis of strain. The grid is then coated to prevent surface oxidation during use.

FIGURE 4. Documentation of creep damage: (a) linked creep voids can be observed in microstructure weld consisting of austenitic matrix, precipitated nitrides and carbides, as viewed originally at 500× with optical microscope; (b) grain boundary carbides, creep voids and particles believed to be nitrides can be observed in matrix of alloy in Fig. 3, as viewed originally at 1000× with scanning electron microscope. Scale has been modified during publication.

(a)

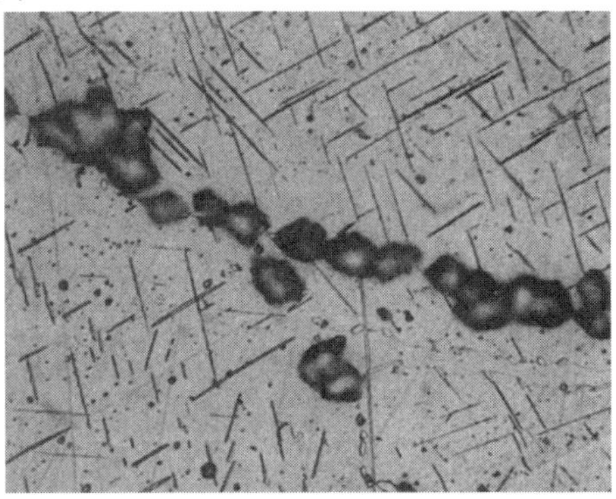

(b)

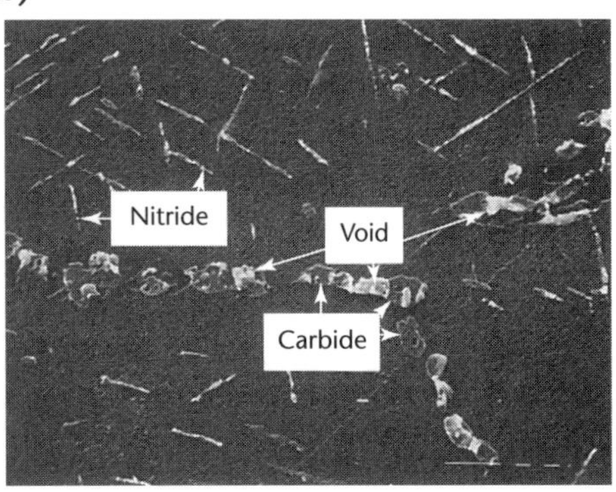

FIGURE 5. Documentation of stress corrosion cracking found in welds of anhydrous ammonia sphere. Three percent nital etch at 200× before modification for publication.

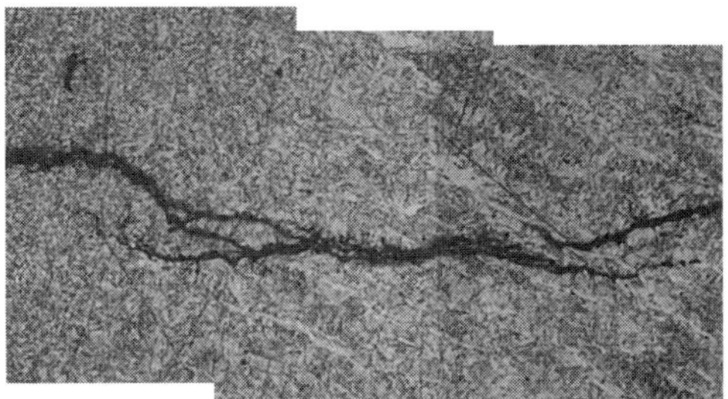

After a predetermined period of operation, the coating is removed and the area is again replicated. The grid intersection points on the two replicas are compared for dimensional changes, and the changes are then correlated to units of strain. This technique does not yield absolute values of strain but gives information to help project the component's service life.

Strain replication may be attempted for materials that do not exhibit creep void formation until late in their service life (Fig. 7).[10] When the relationship deviates from linearity, the strain rate can become unstable.

In addition to qualitatively assessing by replication the degree of creep damage, whenever creep damage is suspected, an engineering assessment consistent with API 579 and ASME FFS[11] should be conducted to determine if the component is fit for continued operation and for how long it will be safe to operate.

Silicone Rubber Replicas

Silicone impression materials have been used extensively in medicine, dentistry and in the science of anthropology. In nondestructive testing, silicone materials are used as tools for documenting macroscopic and microscopic material detail. Quantitative measurements can be obtained for depth of pitting, wear, surface finish and fracture surface evaluation.

Silicone material is made with varying viscosities, setting times and resolution capabilities. Compared to an acetate replica, the resolution characteristics of a silicone replica is limited. With a medium viscosity compound, fine features visible

FIGURE 6. Documentation of heat-affected zone cracking in boiler grade steel. Cracking associated with nonrelieved stress. Repair weld was unknown until in-field metallography and replication were performed. Three percent nital etch at 100× before magnification was altered for publication.

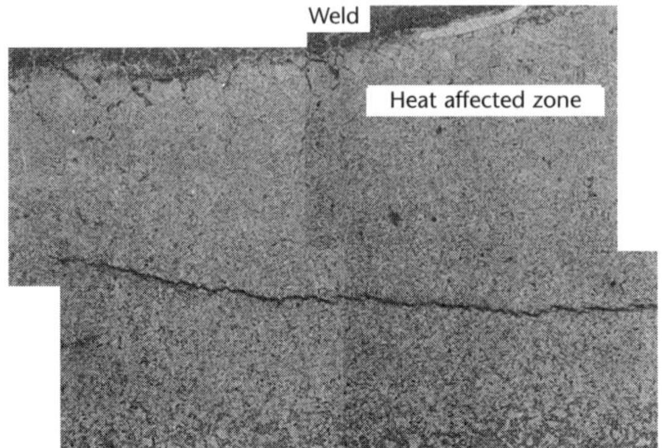

at 50× can be resolved but difficulties are encountered at higher magnifications. With a low viscosity compound, slightly better resolution is obtained but curing times are long and not suited to field applications. The lower viscosity medium is also known to creep with time and is not recommended for applications where very accurate dimensional studies are needed.

Silicone Replicating Materials

Silicone replicating materials are supplied in two parts: a base material and an accelerator. Although it is best to follow the recommended mixing ratios, these can be altered slightly to change the working time of the material. The two parts are mixed thoroughly and spread over the subject area. Additional material can be added to thicken the replica. Molding clay can also be used to build a dam around a replicated area. The dam supports the replica as its sets and allows thicker replicas to be made.

Measurements of pit depth and surface finish can be obtained easily because of the silicone's ability to flow into crevices on the test object. To evaluate pit depth and surface finish, the replica is cut and the cross section is examined with a microscope or a macroscopic measuring device (a micrometer or an optical comparator).

FIGURE 7. Creep damage curve showing typical relationship of strain to time for material under stress in high temperature atmosphere. Development of creep related voids in this alloy occurs early in service life — their eventual linkage is shown schematically on curve.[10]

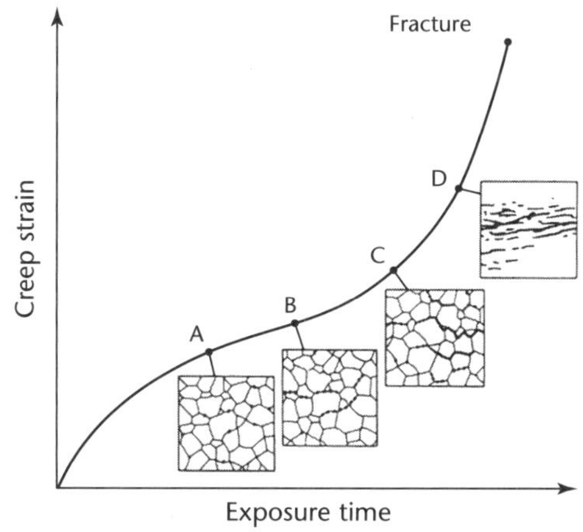

Legend

A. Isolated cavities.
B. Oriented cavities.
C. Microcracks.
D. Macrocracks.

Wear can be determined in a similar manner by replicating and comparing a worn surface to an unworn surface (Fig. 8). Fracture surfaces with rough contours can be easily replicated with silicone (taking an acetate replica of such surfaces is difficult). However, the resolution characteristics of a silicone replica are not as good as acetate replicas and this limits the amount of interpretation that can be performed. Macroscopic details such as chevron markings can be easily located with the silicone technique to determine crack propagation direction or to trace a fracture path visually to its origin.

Conclusion

Cellulose acetate tape and silicone impression materials are commonly used for nondestructive visual tests of surface phenomena such as corrosion, wear, cracking and microstructures.[8] Both types of replicating material have advantages and limitations but when used in the correct application, can provide valuable information.

In terms of resolution, the silicone replica typically does not have the capability to copy fine detail above 50×.

FIGURE 8. Silicon replicas used to determine wear variance on failed pinion gear.

The acetate replica can reveal detail up to 50 000× on a transmission electron microscope. The acetate replica is limited, however, by the roughness of the topography it can copy. On rough fracture surfaces, difficulty is encountered in both applying and removing an acetate replica. The silicone material is not as restrictive in terms of the surface features it can copy. The need for fine, resolvable detail versus macroscopic features normally indicates whether acetate or silicone replicas are best for the application.

PART 3. Etching[12]

The information contained in this text is simplified for general instruction. Local authorities for health and environment (including the Occupational Safety and Health Administration and Environmental Protection Agency) should be consulted about the proper use and disposal of chemical agents. For reasons of safety, all chemicals must be handled with care, particularly the concentrated chemicals that aid visual testing.

In visual nondestructive testing, chemical techniques are used to clean and enhance test object surfaces. Cleaning processes remove dirt, grease, oil, rust and mill scale. Contrast is enhanced by chemical etching before visual testing. Etching is described by standards[13-15] and is well documented in the literature.[16,17]

Macroetching is the use of chemical solutions to attack material surfaces to improve the visibility of discontinuities for visual inspection at normal and low power magnifications. Caution is required in the use of these chemicals — the use of protective clothing and safety devices is imperative. Test object preparation and the choice of etchant must be appropriate for the inspection objectives. Once the desired etch is achieved, the metal surface must be flushed with water to avoid excessive etching.

Test Object Selection

Figure 9 shows typical test objects removed from their service environment. Governing codes, standards or specifications may determine the number and location of visual tests. Specific areas may contain discontinuities from forming operations such as casting, rolling, forging or extruding. Weld tests may be full length or random spots and typically cover the weld metal, fusion line and heat-affected zone. The service of a component may also indicate problem areas requiring inspection.

Location of the test site directly affects surface preparation. The test site may be prepared and nondestructively inspected *in situ*. Removal of a sample for laboratory examination is a destructive alternative test method that typically requires a repair weld.

FIGURE 9. Components removed from service for visual testing.

Surface Preparation

Preparation of the test object before etching may require only cleaning or a process including cleaning, grinding and fine polishing (improper grinding is shown in Fig. 10). The extent of these operations depends on the etchant, the material and the type of discontinuity being sought.

Solvent Cleaning

Solvent cleaning can be useful at two stages in test object preparation. An initial cleaning with a suitable solvent removes dirt, grease and oil and may make rust and mill scale easier to remove.

One of the most effective cleaning solvents is a solution of detergent and water. However, if water is detrimental to the test object, organic solvents such as ethyl alcohol, acetone or naphthas have been used. These materials generally have low flash points and their use may be

FIGURE 10. Improper surface preparation. Grind marks mask indications, and even severe etchant does not give good test results.

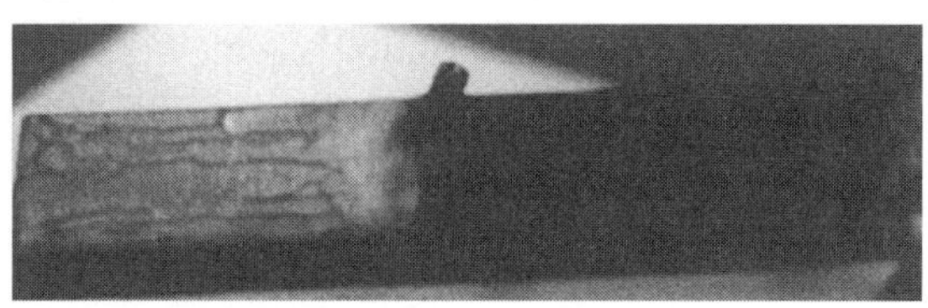

prohibited by safety regulations. Safety solvents such as the chlorinated hydrocarbons and high flash point naphthas may be required to meet safety standards.

Removing Rust and Scale

Rust and mill scale are normally removed by mechanical methods such as wire brushing or grinding. If appropriate for a particular test, the use of a severe etchant requires only the removal of loose rust and mill scale. Rust may also be removed chemically. Commercially available rust removers are generally inhibited mineral acid solutions and are not often used for test object preparation.

Most surface tests require complete removal of rust and mill scale but a coarsely ground surface is often adequate preparation before etching.

Grinding may be done manually or by belt, disk or surface grinding tools. Surface grinders are usually found only in machine shops. Hand grinding requires a hard flat surface to support the abrasive sheet. Coolant is needed during grinding and water is the preferred coolant but kerosene may be used if the test material is not compatible with water.

Grinding and Polishing

Fine grinding and polishing are needed for visual tests of small structural details, welds and the effects of heat treatment. Finer grinding usually is done with 80 to 150 abrasive grit followed by 150 to 180 grit and finally 400 grit (in a gaging of grit size in the United States, 400 being the finest). At each stage, marks from previous grinding must be completely removed. Changing the grinding direction between successive stages of the process aids the visibility of previous coarser grinding marks. Coolant is required for grinding and typical abrasives include emery, silicon carbide, aluminum oxide and diamond.

If the required finish cannot be achieved by fine grinding with 400 grit abrasive, the test surface must be polished. Polishing is generally done with a cloth-covered disk and abrasive particles suspended in paste or water. Common polishing media include aluminum oxide, magnesium oxide, chromium oxide, iron oxide and diamond with particle sizes ranging from 0.5 to 15 mm (0.02 to 0.6 in.).

During polishing, it is critical that all marks from the previous step be completely removed. If coarser marks do not clear, it may be necessary to repeat a previous step using lighter pressure before continuing. Failure to do so can yield false indications.

Etching

Choice of Etchant

The etchant, its strength, the material and the discontinuity all combine to determine surface finish requirements (Table 2). Properly selected etchants chemically attack the test material and reveal welds (Fig. 11), pitting (Fig. 12), grain boundaries, segregation, laps, seams, cracks and heat-affected zones. The indications are highlighted or contrasted with the surrounding base material.

Safety Precautions

Etchants are solutions of acids, bases or salts in water or alcohol. Etchants for macroetching are water based. Etching solutions need to be fresh and the primary concerns during mixing are purity, concentration and safety.

TABLE 2. Chemicals for etchants.

Formula	Name	Concentration
Acid		
HCl	hydrochloric (muriatic)	37 percent
HF	hydrofluoric	48 percent
HNO_3	nitric	70 percent
H_3PO_4	phosphoric	
H_2SO_4	sulfuric	
Alcohol		
C_2H_5OH	ethanol	
CH_3OH	methanol	
Aqueous		
H_2O_2	hydrogen peroxide	30 percent
Base		
NH_4OH	ammonium hydroxide	
NaOH	sodium hydroxide	
Element		
I_2	iodine	
Salt		
$(NH_4)2S2O_8$	ammonium persulfate	
$(NH_4)2SO_4$	ammonium sulfate	
$2NH_4Cl \cdot CuCl_2 \cdot 2H_2O$	copper ammonium chloride	
$CuCl_2$	cupric chloride	
$FeCl_3$	ferric chloride	
KI	potassium iodide	

FIGURE 11. Example of contrast revealing weld in stainless steel.

Safety precautions are necessary during the mixing and use of chemical etchants.[18] Chemical fumes are potentially toxic and corrosive. Mixing, handling or using etchants should be done only in well ventilated areas, preferably in an exhaust or fume hood. Use of an exhaust hood is mandatory when mixing large quantities of etchants. Etching large areas requires the use of ventilation fans in an open area or use of an exhaust hood.

Contact of etchants with skin, eyes or clothes should be avoided. When pouring, mixing or handling such chemicals, protective equipment and clothing should be used, including but not limited to glasses, face shields, gloves, apron or laboratory jacket. A face-and-eye wash fountain is recommended where chemicals and etchants are sorted and handled. A safety shower is recommended when large quantities of chemicals or etchants are in use.

Should contact occur, certain safety steps must be followed, depending on the kind of contact and the chemicals involved. Skin should be washed with soap and water. Chemical burns should have immediate medical attention. Eyes should be flushed at once with large amounts of water and immediate medical attention is mandatory. Hydrofluoric and fluorosilic acids cause painful burns and serious ulcers that are slow to heal. Immediately after exposure, the affected area must be flooded with water and emergency medical attention sought.

FIGURE 12. Effect of etching: (a) unetched component with shiny appearance at rolled area; (b) pits visible in dulled area after etching with ammonium persulfate.

(a)

(b)

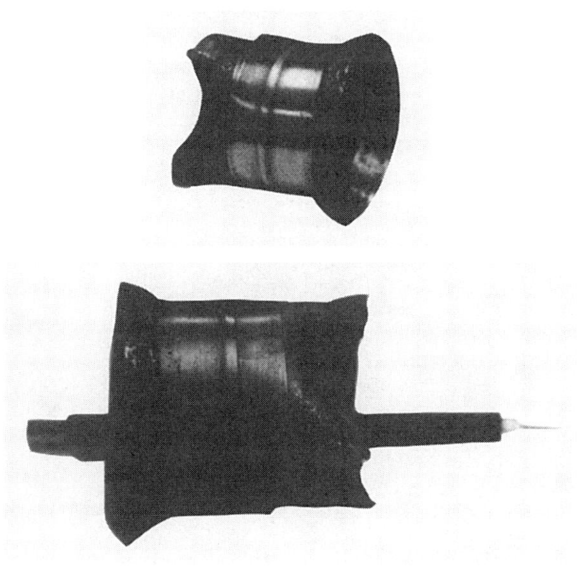

Other materials that are especially harmful in contact with skin are concentrated nitric acid, sulfuric acid, chromic acid, 30 and 50 percent hydrogen peroxide, sodium hydroxide, potassium hydroxide, bromine and anhydrous aluminum chloride. These materials also produce vapors that cause respiratory irritation and damage.

Containers

Containers used with etchants must be rated for mixing, storing and handling of chemicals. Glass is resistant to most chemicals and is most often used for containment and stirring rods. Hydrofluoric acid, other fluorine based materials, strong alkali and strong phosphoric acids can attack glass, requiring the use of inert plastics.

Generation of Heat

Heat may be generated when chemicals are mixed together or added to water. Mixing chemicals must be done using accepted laboratory procedures and caution. Strong acids, alkalis or their concentrated solutions incorrectly added to water, alcohols or other solutions, cause violent chemical reactions. To be safe, never add water to concentrated acids or alkalis.

In general, the addition of acidic materials to alkaline materials will generate heat. Sulfuric acid, sodium hydroxide or potassium hydroxide in any concentration generate large amounts of heat when mixed or diluted and an ice bath may be necessary to provide cooling. Three precautions in mixing can reduce or prevent a violent reaction.

1. Add the acid or alkali to the water or to a weaker solution.
2. Slowly introduce acids, alkali or salts to the water or the solution.
3. Stir the solution continuously to prevent layering and a delayed, violent reaction.

Concentration

The strength of an acid, base or salt in solution is expressed by its concentration or composition. Etchants are typically mixtures of liquids or solids in liquids. The concentration of liquid mixtures is expressed as parts or percent by volume.

For solids in liquids, units of concentration are parts or percent by weight. Generally, etchants are mixed in small quantities. Table 2 lists a variety of etchants.

Chemical Purity

Chemicals are available in various grades of purity ranging from technical to very pure reagent grades. For etchants, the technical grade is used unless a purer grade is specified. For macroetchants, the technical grade is generally adequate.

Water is the solvent used for most macroetching solutions and water purity can affect the etchant. Potable tap water may contain some impurities that could affect the etchant. Distilled water has a significantly higher purity than tap water. For macroetchants using technical grade chemicals, potable tap water is usually acceptable. For etchants in which high purity is required, distilled water is recommended.

Disposal

Before disposing of chemical solutions, check environmental regulations (federal, state and local) and safety department procedures. The steps listed here are used only if there are no other regulations for disposal. Spent etchants are discarded and must be discarded separately — mixing of etchant materials can produce violent chemical reactions.

Using a chemical resistant drain under an exhaust hood, slowly pour the spent etchant while running a heavy flow of tap water down the drain. The drain is flushed with a large volume of water.

Using Etchants

After proper surface preparation and safe mixing of etchants, the application of etchants to the test object may be done with immersion or swabbing. The technique is determined by the characteristics of the etchant being used.

Immersion

During immersion, a test object is completely covered by an etchant contained in a safe and suitable material — glass can be used for most etchants except hydrofluoric acid, fluorine materials, strong alkali and strong phosphoric acid.

A glass heat resistant dish on a hot plate may be used for heated solutions. The solution should be brought to temperature before the test object is immersed. Tongs or other handling tools are used and the test object is positioned so that the test surface is face up or vertical to allow gas to escape. The solution is gently agitated to keep fresh etchant in contact with the test object.

Swabbing

Etching may also be done by swabbing the test surface with a cotton ball, cotton tipped wooden swab, bristled acid brush, medicine dropper or a glass rod. The cotton ball and the cotton tipped wooden swab generally are saturated with etchant and then rubbed over the test surface.

Tongs and gloves should be used for protection and the etchant applicator must be inert to the etchant. For example, strong nitric acid and alkali solutions attack cotton and these etchants must be applied using a fine bristle acid brush. A glass or plastic medicine dropper may be used to place etchants on the test object surface and a suitable stirring rod can be used to rub the surface. The test object may be immersed in etchant and swabbed while in the solution.

Etching Time

Etching time is determined by (1) the concentration of the etchant, (2) the surface condition and temperature of the test object and (3) the type of test material. Details about the chemicals and their concentrations, handling and applications are available in the literature.[12-17] During etching, the material surface loses its bright appearance and the degree of dullness is used to determine when to stop etching. Approximate dwell times are given in the procedures but experience is important as well.

Test Object Preservation

Cleaning and coating may be required for preservation of the test object. Rinsing removes the etchant by flushing the surface thoroughly under running water. Cold water rinsing usually produces better surface appearance than hot water rinsing. Hot water rinsing does aid in drying.

If adherents are a problem, the test object can be scrubbed with a stiff bristled brush or dipped in a suitable desmutting solution. The test object should be dried with warm dry air. Shop air may be used if it is filtered and dried. After visual inspection, the test surface may be coated with a clear acrylic or lacquer but such coatings must be removed before subsequent tests. If the component is returned to service, a photographic record of the macroetched area should be made.

Closing

Visual testing is performed in accordance with applicable codes, standards, specifications and procedures. Chemical aids enhance the contrast of discontinuities making them easier to interpret and evaluate. This enhancement is attained by macroetching — a controlled chemical processing of the surface. Macroetching gives the optimum results on a properly cleaned and prepared surface. Chemicals for etching must be mixed, stored, handled and applied in strict accordance with safety regulations.[18]

References

1. Generazio, E.R. and D.J. Roth. Section 1, "Other Applications of Visual and Optical Tests": Part 1, "Interface of Visual Testing with Other Nondestructive Testing Methods." *Nondestructive Testing Handbook,* second edition: Vol. 8, *Visual and Optical Testing.* Columbus, OH: American Society for Nondestructive Testing (1993): p 304-312.
2. Generazio, E.R. "The Role of the Reflection Coefficient in Precision Measurement of Ultrasonic Attenuation." *Materials Evaluation.* Vol. 43, No. 8. Columbus, OH: American Society for Nondestructive Testing (July 1985): p 995-1004.
3. Truell, R., C. Elbaum and B.B. Chick. *Ultrasonic Methods in Solid State Physics.* New York, NY: Academic Press (1969).
4. Generazio, E.R. "Ultrasonic Attenuation Measurements Determine Onset, Degree and Completion of Recrystallization." *Materials Evaluation.* Vol. 46, No. 9. Columbus, OH: American Society for Nondestructive Testing (August 1988): p 1198-1203.
5. Generazio, E.R. "Scaling Attenuation Data Characterizes Changes in Material Microstructure." *Materials Evaluation.* Vol. 44, No. 2. Columbus, OH: American Society for Nondestructive Testing (February 1986): p 198-202.
6. Meiley, S. Section 4, "Basic Aids and Accessories for Visual Testing": Part 5, "Replication." *Nondestructive Testing Handbook,* second edition: Vol. 8, *Visual and Optical Testing.* Columbus, OH: American Society for Nondestructive Testing (1993): p 108-113.
7. ASTM E 1351, *Standard Practice for Production and Evaluation of Field Metallographic Replicas.* Philadelphia, PA: ASTM International (2001).
8. *ASM Handbook:* Vol. 9, *Metallography and Microstructures.* Materials Park, OH: ASM International (1985, 2004).
9. Marder, A.R. "Replication Microscopy Techniques for NDE." *Metals Handbook,* ninth edition: Vol. 17, *Nondestructive Evaluation and Quality Control.* Materials Park, OH: ASM International (1985): p 52-56.
10. Neubauer, B. and U. Wedel. "Restlife Estimation of Creeping Components by Means of Replicas." *Advances in Life Prediction.* D.A. Woodford and J.R. Whitehead, eds. New York, NY: American Society of Mechanical Engineers (1983): p 307-314.
11. ASME FFS/API 579, *Fitness-for-Service.* New York, NY: ASME International (2008). Dallas, TX: American Petroleum Institute (2000).
12. Parrish, D. Section 4, "Basic Aids and Accessories for Visual Testing": Part 7, "Chemical Aids." *Nondestructive Testing Handbook,* second edition: Vol. 8, *Visual and Optical Testing.* Columbus, OH: American Society for Nondestructive Testing (1993): p 118-124.
13. ASTM E 381, *Standard Method of Macroetch Testing Steel Bars, Billets, Blooms, and Forgings.* West Conshohocken, PA: ASTM International (2006).
14. ASTM E 407, *Standard Practice for Microetching Metals and Alloys.* West Conshohocken, PA: ASTM International (2007).
15. QW 470, "Etching — Processes and Reagents." *ASME Boiler and Pressure Vessel Code*: Section IX*, Welding and Brazing Qualifications.* New York, NY: American Society of Mechanical Engineers (2008).
16. *CRC Handbook of Metal Etchants.* Baton Rouge, LA: CRC Press (1990).
17. *Microengineering of Metals and Ceramics:* Part 2: *Special Replication Techniques, Automation, and Properties.* Baton Rouge, LA: Wiley-VHC (2005).
18. Lewis, R.J. *Sax's Dangerous Properties of Industrial Materials,* tenth edition. Three volumes. New York, NY: Wiley-Interscience (2000).

13

CHAPTER

Visual Testing Glossary

Introduction

Purpose

Standards writing bodies take great pains to ensure that their standards are definitive in wording and technical accuracy. People working to written contracts or procedures should consult definitions referenced in standards when appropriate. For example, persons who work in accordance with standards published by ASTM International are encouraged to refer to definitions in the ASTM standards.[1]

The definitions in this *Nondestructive Testing Handbook* volume should not be referenced for tests performed according to standards or specifications or in fulfillment of contracts. This glossary is provided for instructional purposes. No other use is intended.

On References

Many definitions in this glossary are adapted from other volumes of the *Nondestructive Testing Handbook* series, especially from the second edition's *Visual and Optical Testing* (1993).[2]

Some terms apply generally to nondestructive testing and are not specific to visual testing — terms on subjects such as metallurgy, quality control and personnel qualification. Many of these definitions come from the second edition volume *Nondestructive Testing Overview* (1996)[3]; some are rephrased in the third edition's most recent volume, *Magnetic Testing* (2008).[4]

Entries from other volumes in the *Nondestructive Testing Handbook* series are reprinted but generally not referenced below.

Definitions

A

acceptable quality level (AQL): Maximum percent defective (or the maximum percentage of units with rejectable discontinuities) that, for the purposes of sampling tests, can be considered satisfactory as a process average.

acceptance criterion: Benchmark against which test results are to be compared for purposes of establishing the functional acceptability of a part or system being examined.[4]

acceptance level: Measured value or values above or below which test objects are acceptable, in contrast to *rejection level*.[4]

acceptance standard: (1) Specimen, similar to the product to be tested, containing natural or artificial discontinuities that are well defined and similar in size or extent to the maximum acceptable in the product.[4] (2) Document defining acceptable discontinuity size limits. See also *standard*.

accommodation: Of the eye, adjustment of the lens' focusing power by changing the thickness and curvature of the lens through its movement by tiny muscles.

ACGIH: American Conference of Governmental Industrial Hygienists.

acuity: See *neural acuity, vision acuity*.

adaptive thresholding: Threshold value varying with inconstant background gray level.

adhesive wear: See *wear, adhesive*.

alpha ferrite: Form of pure iron that has a body centered cubic structure stable below 910 °C (1670 °F). Also called *alpha iron*.

alpha iron: See *alpha ferrite*.

ambient light: Light in the environment as opposed to illumination provided by a visual testing system.

ampere (A): SI unit of electric current.[4]

angle: See *field angle*.

angstrom (Å): Disused unit of length. 1 Å = 0.1 nm.

anomaly: (1) In nondestructive testing, a nonrelevant indication. (2) In nondestructive testing, an unintentional or undesired material condition that may qualify as a defect. Compare *defect*; *discontinuity*. Some anomalies, such as inadequate case hardening or rough surface finish, may be defects but, because there is no interruption in the material structure, are not discontinuities.

ASNT Recommended Practice No. SNT-TC-1A: See *Recommended Practice No. SNT-TC-1A*.

ASNT: American Society for Nondestructive Testing.

automated system: Acting mechanism that performs required tasks at a determined time and in a fixed sequence in response to certain conditions or commands.

axial: Of or pertaining to a direction along the length of an oblong object and perpendicular to its radius — for example, down the length of a cylinder. Compare *radial*.

B

background cylinder and difference cylinder: Two devices used to calculate illuminance by using the equivalent sphere illumination technique.[2,5]

binary system: In metallurgy, a two-element alloy system. See also *isomorphous binary system*.

birefringence: Splitting of a light beam into two parts through a translucent material.

black body: Theoretical object that radiates more total power and more power at any wavelength than any other source operating at the same temperature.[2,5]

blackbody: See *black body*.

black light: Term sometimes used for ultraviolet radiation, particularly in the near ultraviolet range of about 320 to 400 nm.

blacklight: See *black light*.

blind spot: Portion of the retina where the optic nerve enters, without rods and cones and hence insensitive to light.[2,6]

blister: Discontinuity in metal, on or near the surface, resulting from the expansion of gas in a subsurface zone. Very small blisters are called *pinheads* or *pepper blisters*.[4]

blotch: (1) An irregularly spaced area of color change on a surface. (2) The nonuniform condition of a surface characterized by such blotches.

blowhole: Hole in a casting or a weld caused by gas entrapped during solidification.[4]

blue light hazard: Danger posed to the eye by long term exposure to high frequency visible light at intensities and durations that may damage the retina.

borescope: Industrial endoscope; a periscope or telescope using mirrors, prisms, lenses, optic fibers or television wiring to transmit images from inaccessible interiors for visual testing. Borescopes are so called because they were originally used in machined apertures and holes such as gun bores. There are both flexible and rigid borescopes.

borescope, angulated: Borescope bent for viewing at forward oblique, right angle or retrospective angles for visual testing of surfaces not accessible with conventional borescopes.

borescope, calibrated: Borescope with gage on external tube to indicate the depth of insertion during a test. Borescopes with calibrated reticles are used to determine angles or sizes of objects in the field when held at a predetermined working distance.

borescope, cave: Multiangulated, periscopic borescope used for remote observation of otherwise inaccessible areas.

borescope, fiber optic: Industrial endoscope, or fiber optic borescope, that uses glass or quartz fibers to transmit light and the optical path to and from the test object.

borescope, indexing: Borescope that can be bent 90 degrees by rotation of a knob after the instrument has been inserted through an aperture. A knob at the eyepiece can rotate the objective head through 360 degrees for scanning a circumferential weld seam.

borescope, micro-: Borescope with an outside diameter generally from 1 to 5 mm (0.04 to 0.2 in.), typically using quartz filaments. Compare *miniature borescope*.

borescope, miniature: Borescope with an outside diameter generally less than 13 mm (0.5 in.). Sometimes called *miniborescope*. See also *microborescope*.

borescope, panoramic: Borescope with a scanning mirror mounted in front of the objective lens system. Rotation of the mirror is adjusted at the ocular end of the instrument to scan in forward oblique, right angle and retrospective directions.

borescope, retrospective: Borescope that looks backward more than 90 degrees from the distal line of interrogation normal to the plane of a conventional objective lens.

borescope, rigid: Borescope that does not bend, typically in order to keep the geometrical optics in alignment through a light train system.

borescope, ultraviolet: Borescope equipped with ultraviolet lamps, filters and special transformers to transmit radiation of ultraviolet wavelengths.

borescope, video: Borescope transmitting image electronically.

borescope, waterproof/vaporproof: Borescope completely sealed and impervious to water or other types of fluid, used for internal tests of liquid, gas or vapor environments.

borescope, wide field: Borescope with rotating objective prism to provide fields of view up to 120 degrees.

borescopy: Viewing or inspection with a borescope.

brinelling: Repeated stripe indentations made by a spherical object. False brinelling refers to a type of surface wear.

burr: Raised or turned over edge occurring on a machined part and resulting from cutting, punching or grinding.[2,5]

burst: In metal, external or internal rupture caused by entrapped gas.

butt weld: Weld that joins the edges of two work pieces in the same plane.

C

calibration: (1) Ratio of the output from a device to a reference input. Knowledge of this ratio helps to infer a device's input from its output. (2) Act of returning an instrument to the parameters and settings of the original equipment manufacturer. (3) Statement of the scale of a device.

candela (cd): Base SI unit of luminous intensity, in a given direction, of a monochromatic radiation source that has a frequency of 5.4×10^{14} Hz and that has a radiant intensity in that direction of $1.464\ mW \cdot sr^{-1}$.[4]

candle: Former name for candela.

cavitation fatigue: Form of pitting, caused by erosion from vibration and movement in liquid environments.

CCD: See *charge coupled device*.

cementite: Iron carbide (Fe_3C), a hard and brittle substance present in steels.

certification: With respect to nondestructive test personnel, the process of providing written testimony that an individual has met the qualification requirements of a specific practice or standard. See also *certified* and *qualified*.

certified: With respect to nondestructive test personnel, having written testimony of qualification. See also *certification* and *qualification*.

CGS system: Obsolete system of measurement units based on the centimeter, gram and second. Compare *SI*.

chafing: See *wear, fretting*.

channels: In biology, mechanisms functioning as band pass filters in the visual cortex of mammals, causing sensitivity to visual stimuli in particular frequencies and ranges.

charge coupled device (CCD): Solid state image sensor. Charge coupled devices are widely used in inspection systems because of their accuracy, high speed scanning and long service life.

check cracking: Surface crack caused by overheating and having cross hatched pattern. See *grinding crack.*

closing: In image processing, dilation followed by erosion. A single pixel by closing connects a broken feature separated by one pixel. See also *opening.*

closure: Process by which a person cognitively completes patterns or shapes that are incompletely perceived.

cocoa: Debris (usually oxides of the contacting metals) of fretting wear, retained at or near the site of its formation — a condition easily identified during visual tests. With ferrous metals, the debris is brown, red or black, depending on the type of iron oxide formed. For this reason, ferrous debris is called *cocoa* or, when mixed with oil or grease, *red mud.*

code: Standard enacted or enforced as a law.

coefficients of the filter: Values in a mask that serves as a filter in image processing.

cold light: Disused word for *fluorescence.*

color: Visual sensation by means of which humans distinguish light of differing hue (predominant wavelengths), saturation (degree to which those radiations predominate over others) and lightness.

color blindness: Deficiency in ability to perceive or distinguish hues.

color discrimination: Perception of differences between two or more hues.

color temperature: Rating of a light source for color vision.

compound microscope: See *microscope, compound.*

cone: Part of the eye; color sensitive photoreceptor at the inner region of the retina. Cones assist with mesopic vision and are responsible for photopic vision. Compare *rod.*

confidence level: Level of assurance for detecting a specified discontinuity size with a specified probability. See also *probability of detection.*

constitution diagram: See *phase diagram.*

contrast: (1) Difference in color or brightness between a test indication and background. (2) Difference between the amount of light reflected or transmitted by an object and by the background in the field of view.

control: See *in control, process control* and *quality control.*

corrosion: Loss or degradation of metal because of chemical reaction.

corrosion, crevice: Corrosion found in tight crevices or pores (pits) and accelerated by galvanic activity from high ion concentrations.

corrosion-erosion: Simultaneous occurrence of erosion and corrosion.

corrosion, fretting: Corrosion facilitated by fretting, particularly where a protective surface has been chafed in a corrosive environment.

corrosion, poultice: Corrosion occurring under a layer of foreign material (for example, under mud in automobile rocker panels).

cosine law: Physical law stating that the illumination of a surface varies as the cosine of the incidence angle. Maximum illumination is obtained where the cosine equals one and when the source is perpendicular to the surface.

crack: (1) Stress induced break, fissure or rupture, sometimes V shaped in cross section and relatively narrow. By convention, a crack is called *linear* if it is at least three times longer than it is wide. (2) Propagating discontinuity caused by fatigue, corrosion or stresses such as heat treating or grinding. May be difficult to detect unaided because of fineness of line and pattern (may have a radial or latticed appearance).

creep: Gradual and permanent change of shape in a metal under constant load, usually at elevated temperature. Occurs in three stages: primary creep, secondary creep and tertiary creep. See also *deformation.*

crevice corrosion: See *corrosion, crevice.*

D

dark adaptation: Process by which the retina becomes adapted to a luminance less than about 0.034 $cd{\cdot}m^{-2}$.[2,6] In dark adaptation, the pupils dilate and the two types of photoreceptors in the retina change chemical balance. After a finite amount of time, possibly 10 min, vision will change from photopic vision to mesopic or scotopic low illumination vision.[4]

dark adapted vision: See *accommodation, visual; scotopic vision.*

defect: Discontinuity whose size, shape, orientation or location (1) makes it detrimental to the useful service of its host object or (2) exceeds an accept/reject criterion of an applicable specification. Some discontinuities do not exceed an accept/reject criterion and are therefore not defects. Compare *crack; discontinuity; indication.*

deformation: Change of shape under load. See also *creep* and *elastic deformation.*

delta ferrite: Solid solution with body centered cubic structure and iron as solvent. Also called *delta iron*.

delta iron: See *delta ferrite*.

depth of field: In photography, the range of distance over which an imaging system gives satisfactory definition when its lens is in the best focus for a specific distance.

dewetting: In soldering, the flow and retraction of solder, caused by contaminated surfaces, dissolved surface coatings or overheating before soldering.

difference cylinder: See *background cylinder*.

dilation: In image processing, the condition of a binary image where the pixel in the output image is a 1 if any of its eight closest neighbors is a 1 in the input image. See also *closing, erosion* and *opening*.

diopter: Unit used to express the resolving power of lenses, equal to the inverse of the length (in meters) of the optical axis.

directional lighting: Lighting provided on the work plane or object predominantly from a preferred direction.[2,6]

direct photometry: Simultaneous comparison of a standard lamp and an unknown light source.[2,6]

direct substitution alloy: Alloy in which the atoms of the alloying element can occupy the crystal lattice spaces normally occupied by the atoms of the parent metal.

direct viewing: (1) Viewing of a test object in the viewer's immediate presence. The term *direct viewing* is used in the fields of robotics and surveillance to distinguish conventional from remote viewing. (2) Viewing of a test object during which the light image is not mediated through a system of two or more lenses (as in a borescope) or transduced through an electronic signal (as with a charge coupled camera). The term *direct viewing* is used in some specifications to mean viewing possibly with a mirror or magnifier but not with a borescope. Compare *indirect viewing; remote viewing*.

direct vision instrument: Device offering a view directly forward. A typical scene is about 20 mm (0.75 in.) wide at 25 mm (1 in.) from the objective lens.

discontinuity: Interruption in the physical structure or configuration of a test object. After nondestructive testing, a discontinuity indication can be interpreted to be a defect.[4,7] Compare *defect; indication*.

discontinuity, artificial: Reference discontinuity such as hole, indentation, crack, groove or notch introduced into a reference standard to provide accurately reproducible indications for determining test sensitivity levels.[4]

discontinuity, inherent: Material anomaly originating from solidification of cast metal. Pipe and nonmetallic inclusions are the most common inherent discontinuity and can lead to other types of discontinuities in fabrication.[2,5]

discontinuity, primary processing: Discontinuity produced from the hot or cold working of an ingot into forgings, rods, bars and other shapes.[2,5]

discontinuity, secondary processing: Discontinuity produced during machining, grinding, heat treating, plating or other finishing operations.[2,5]

discontinuity, service induced: Discontinuity caused by the intended use of the part.

distal: In a manipulative or interrogating system, of or pertaining to the end opposite from the eyepiece and farthest from the person using the system. *Objective; tip*.

E

elastic deformation: Temporary change in shape under a load. The material returns to its original size and shape after the load is removed. Elastic deformation is the state in which most metal components are used in service.

elasticity: Ability of a material to resume its former shape after deformation.

electric arc welding: Joining of metals by heating with electric arc. Also called *arc welding*.

endoscope: Device for viewing the interior of objects. From the Greek words for inside view, the term endoscope is used mainly for medical instruments. Nearly every medical endoscope has an integral light source; many incorporate surgical tweezers or other devices. See *borescope*.

equilibrium diagram: Phase diagram showing the phases present at equilibrium in a material system.

equivalent 20/20 near vision acuity: Vision acuity with remote viewing or other indirect viewing that approximates 20/20 direct viewing closely enough to be considered the same for visual testing purposes.

equivalent sphere illumination: Level of perfectly diffuse (spherical) illuminance that makes the visual task as photometrically visible within a comparison test sphere as it is in the real lighting environment.

erosion: (1) Loss of material or degradation of surface quality through friction or abrasion from moving fluids, made worse by solid particles in those fluids or by cavitation in the moving fluid. See *wear.* (2) In image processing, condition of a binary image where the pixel in the output image becomes a 1 if each of its eight neighbors is a 1 in the input image. See also *closing, dilation* and *opening.*

erosion-corrosion: Simultaneous occurrence of erosion and corrosion.

etch crack: Shallow crack in hardened steel containing high residual surface stresses, produced in an embrittling acid environment.[2,5]

eutectic liquid: Liquid metal having a proportion of metals such that two or more solid phases form at the same temperature during cooling.

eutectic point: Temperature and proportion of metals at which two or more phases of a eutectic liquid form. Compare *eutectoid.*

eutectoid: Similar to eutectic but in a solid system during cooling.

evaluation: Process of deciding the severity of a condition after an indication has been interpreted, to determine whether it meets acceptance criteria.

eye sensitivity curve: Graphic expression of vision sensitivity characteristics of the human eye. In the case of a physical photometer, the curve should be equivalent to the standard observer. The required match is typically achieved by adding filters between the sensitive elements of the meter and the light source. See *photopic vision.*

F

false brinelling: Fretting wear indentations. Compare *brinelling.*

false indication: Test indication that could be interpreted as originating from a discontinuity but that actually originates where no discontinuity exists in the test object. Compare *indication, nonrelevant; defect.*[4]

farsightedness: Vision acuity functionally adequate for viewing objects at a distance, generally farther than arm's length. Also called *hyperopia.* Compare *nearsightedness.*

far vision: Vision of objects at a distance, generally beyond arm's length. Compare *near vision.*

feature extraction: From an enhanced image, derivation of some feature values, usually parameters for distinguishing objects in the image.

ferrite: Solid solution of one or more other elements in alpha iron.

fiber optic borescope: See *borescope, fiber optic.*

fiber optics: Technology of efficient transmission of light through transparent fibers such as glass, quartz and plastic by means of total internal reflection.

fiberscope: Jargon for fiber optic borescope.

field: In video technology, one of two video picture components that together make a frame. Each picture is divided into two parts called *fields* because a frame at the rate of thirty frames per second in a standard video output would otherwise produce a flicker discernible to the eye. Each field contains one half of the total picture elements. Two fields, then, are required to produce one complete picture or frame so the field frequency is sixty fields per second and the frame frequency is thirty frames per second.

field angle: Included angle between those points on opposite sides of a beam axis at which the luminous intensity is 10 percent of the maximum value. This angle may be determined from an illuminance curve or may be approximated by use of an incident light meter.[2,6]

field of view: Range or area where things can be seen through an imaging system, lens or aperture. Compare *depth of field.*

field of vision: Range or area where things can be perceived by eyesight at a point in time, assuming the eye to be immobile.

fillet weld: Weld of approximately triangular cross section joining two surfaces approximately at a right angle to each other.

filter: Processing component or function that excludes a selected kind of signal or part of a signal.

filtering: See *low pass filtering.*

fit up: To secure one or more joint members with special external fixturing in order to prevent movement during welding.[2,5]

flakes: Short discontinuous internal fissures in ferrous metals attributed to stresses produced by localized transformation and/or decreased solubility of hydrogen during cooling usually after hot working. On a fractured surface, flakes appear as bright silvery areas; on an etched surface, flakes appear as short, discontinuous cracks.[2,5]

flaw: Rejectable anomaly.[4] See also *defect.*

fluorescence: Phenomenon of absorption of electromagnetic radiation and its reemission at a lower energy (longer wavelength). In visual testing, fluorescence is typically a response to ultraviolet radiation.

focus: Position of a viewed object and a lens system relative to one another to offer a distinct image of the object as seen through the lens system. See *accommodation* and *depth of field.*

focus, principal plane of: Single plane in focus in a photographic scene.

focusing, automatic: (1) Feature of a camera whereby the lens system adjusts to focus on an object in the field of view. (2) Metaphorical attribute of a borescopic instrument's depth of field (the range of distance in focus). The depth of field is so great in the case of video borescopes that focusing is unnecessary for most applications. Despite the name, no mechanism is actively adjusted. The large depth of field is due both to the small diameter of the lens aperture and to the proximity of the lens to the charge coupled device.

focusing, primary: Focusing by the lens of the image onto a fiber optic bundle at the tip of a probe.

focusing, secondary: Focusing at the eyepiece of a borescope or other optical instrument, specifically the manual refocusing needed when the viewing distance changes.

footcandle (ftc): Disfavored unit of illuminance, where 1 ftc = 1 lm·ft^{-2} = 10.76 lx.

footlambert (ftl): Disfavored unit of luminance, where 1 ftl = 3.426 cd·m^{-2}.

forging crack: Stress induced discontinuity formed during mechanical shaping of metal; see *crack*.

fovea centralis: Region of sharpest vision in the retina, where the layer of blood vessels, nerve fibers and cells above the rods and cones is far thinner than in peripheral regions.

foveal vision: See *photopic vision*.

fracture mechanics: Field of solid mechanics that deals with behavior of cracked bodies subjected to stress and strain.

frame: Complete raster scan projected on a video screen. There are thirty frames per second in a standard video output. A frame may be comprised of two fields, each displaying part of the total frame. See also *field*.

frequency: Number of times per second that a cyclical waveform repeats. The unit of frequency is hertz (Hz).

fretting corrosion: See *corrosion, fretting*.

fretting wear: See *wear, fretting*.

friction oxidation: See *wear, fretting*.

G

galling: Type of surface damage caused by friction.

galvanic series: List of metals, alloys and graphite (a nonmetal) in sequence with the most anodic (easily corroded) in liquids at one end of the list and the most cathodic (least easily corroded) at the other end. For practical reasons, this sequence is compiled using seawater as the electrolyte — 3 to 5 percent sodium chloride and other salts dissolved in water.

gamma iron: Iron with face centered cubic structure formed by slow cooling of delta ferrite. This characteristic lattice structure is stable between 906 °C (1663 °F) and 1391 °C (2535 °F). Also called *austenite*.

gas metal arc welding (GMAW): Inert gas shielded metal joining process that uses a continuous and consumable wire electrode. Also called *MIG* (metal inert gas) *welding*. Compare *gas tungsten arc welding* and *shielded metal arc welding*.

gas tungsten arc welding (GTAW): Inert gas shielded metal joining process that uses a nonconsumable tungsten electrode. Filler material, when needed, is manually fed into the molten weld puddle. Also called *tungsten inert gas (TIG) welding*. Compare *gas metal arc welding; shielded metal arc welding*.

gasket seal: Resilient ring, usually virgin polytetrafluoroethylene (PTFE), in a piping or tubing connection. Compare *interference sealing thread*.

general examination: In personnel qualification, a test or examination of a person's knowledge, typically (in the case of nondestructive testing personnel qualification) a written test on the basic principles of a nondestructive test method and general knowledge of basic equipment used in the method. (According to ASNT's guidelines, the general examination should not address knowledge of specific equipment, codes, standards and procedures pertaining to a particular application.) Compare *practical examination* and *specific examination*.

geometrical optics: Mathematical study of how light rays are reflected and refracted and practical techniques based on such understanding, including the transmission of images by lenses and mirrors. Also called *lens optics*.

glare: Excessive brightness (or brightness varying by more than 10:1 within the field of view) that interferes with observation or interpretation of a test response. Glare may be caused by reflection, whether specular (smooth surface) or diffuse (rough surface), of light or radiation sources.

glare, blinding: Glare so intense that for an appreciable length of time after it has been removed, no object can be seen.[2,6]

glare, direct: Glare resulting from high luminances or insufficiently shielded light sources in the field of view. Direct glare is usually associated with bright areas, such as luminaires, ceilings and windows which are outside the visual task or region being viewed.[2,6]

glare, reflected: Glare resulting from specular reflections of high luminances in polished or glossy surfaces in the field of view. It usually is associated with reflections from within a visual task or nearby areas.[2,6]

gloss meter: Reflectometer used to measure specular reflectance.[2,6]

gnomon: Artifact intended to cast a shadow. The shadow may be used to measure time or distance.

gouge: Surface indentation caused by forceful abrasion or impact or flame cutting. Also called *nick*. Compare *tool mark*.

grain: Solid particle or crystal of metal. As molten metal solidifies grains grow and lattices intersect, forming irregular grain boundaries.

grain boundary: Interface that forms between grains of solidifying metal as the random oriented crystal lattices meet. See *grain*.

gray body: Radiator whose spectral emissivity is uniform for all wavelengths.

gray level: Integer number representing the brightness or darkness of a pixel or, as a composite value, of an image comprised of pixels.

green rot: Form of attack due to simultaneous carburization and oxidation of stainless heating elements common to nickel chromium and nickel chromium iron alloys, especially in furnace environments.

grinding crack: Shallow crack formed in the surface of relatively hard materials because of excessive grinding heat or the brittleness of the material. Grinding cracks typically are oriented 90 degrees to the direction of grinding.[4]

H

Hadfield's steel: Austenitic manganese specialty steel, easily work hardened.

halitation: Rings of light visible around a spot on a video screen where an electron scanning beam is held.

heading: Upsetting wire, rod or bar stock in dies to form parts having some of the cross sectional area larger than the original. Examples are bolts, rivets and screws.[1]

heat affected zone: Portion of base metal not melted during brazing, cutting or welding but with mechanical properties altered by the heat.[4]

heat checking: Surface cracking caused when metal rapidly heated (or cooled and heated repeatedly) is prevented from expanding freely by colder metal below the surface. Friction may produce the heat. Heat checking is sometimes called *thermal fatigue*.

heat wave: Thermally produced variation in flue gas density that distorts images of objects in a firebox.

hot tear: Fracture formed in a cast metal during solidification and due to extensive tensile stress associated with volumetric shrinkage. Hot tears often occur where areas of different thicknesses adjoin.

hue: Characteristic of light at a particular bandwidth; commonly associated with the color's name.

human factors: Factors in the overall test sensitivity based upon mental and physical condition of the inspector, training, experience level and the physical conditions under which the inspector must work.

hyperopia: See *farsightedness*.

hyperthermia: Heating so excessive that it can damage or kill plant or animal cells.

I

illuminance: Intensity of radiant energy (density of luminous flux) on a surface, in the visible light spectrum. Illuminance is measured in lux.

illuminate: Cast light on (something).

illumination: Act of illuminating or state of being illuminated. See also *illuminate*. Compare *illuminance*.[2,6]

image: Visual representation of an object or scene.

image enhancement: Any of a variety of image processing steps, used singly or in combination to improve the detectability of objects in an image.

image guide: Fiber bundle that carries the picture formed by the objective lens at the distal end of a fiber optic borescope back to the eyepiece.

image orthicon: Television tube that uses the photoemission method. Compare *vidicon tube*.

image processing: Actions applied singly or in combination to an image, in particular the measurement and alteration of image features by computer. Also called *picture processing*.

image segmentation: Process in which the image is partitioned into regions, each homogeneous.

in control: Within prescribed limits of process control.

incandescence: Emission of visible radiation as a result of heating.

indication: Nondestructive test response that requires interpretation to determine its relevance.[4] See also *defect; discontinuity; false indication; nonrelevant indication.*

indication, nonrelevant: Indication that has no relation to a discontinuity that might constitute a defect.[1] Test response caused by geometry or by a physical condition that is not a discontinuity.

indication, relevant: Indication from a discontinuity (as opposed to a false indication) requiring evaluation by a qualified inspector, typically with reference to an acceptance standard, by virtue of the discontinuity's size or location.[1]

indirect viewing: Viewing of a test object during which the light image is mediated through a system of two or more lenses (as in a borescope) or transduced through an electronic signal (as with a charge coupled camera). Compare *direct viewing; remote viewing.*

infrared radiation: Electromagnetic radiant energy of wavelengths longer than 770 nm.[2,6]

interference fitted thread: See *interference sealing thread.*

interference objective: In a microscope, a small, metallized glass mounted in contact with the test object and adjustable for tilt to control fringe spacing.

interference sealing thread: Piping seal using a tapered connection made up under great pressure, forcing the mating surfaces together more tightly than is possible with a spiral thread. Compare *gasket seal.*

interlaced scanning: Process whereby the picture appearing on a video screen is divided into two parts. Interlaced scanning reduces flicker by increasing the electron beam's downward rate of travel so that every other line is sent. When the bottom is reached, the beam is returned to the top and the alternate lines are sent. The odd and even line scans are each transmitted at 1/60 s, totaling 1/30 s per frame and retaining the standard rate of 30 frames per second. The eye's persistence of vision allows the odd and even lines to appear as a single image without flicker.

interpretation: Determination of the cause, significance and relevance of test indications.

interstitial alloy: Alloy in which the atoms of the alloying element fit into the spaces between the atoms of the parent metal.

inverse square law: Physical law for a point source of energy. The quantity or strength is inversely proportional to the square of the distance from the origin.

iris: Ring of variable area around the pupil and in front of the lens of the eye. The surface area of the iris adjusts spontaneously to change the amount of light entering the eye.

irradiance: Radiant power falling upon a known surface area at a given angle. Compare *radiance.* See also *radiometer.*

Ishihara™ plates: Trade name for a kind of pseudoisochromatic plates, used for color differentiation vision testing.

isomorphous binary system: Two element alloy system in which both elements are completely soluble in each other in the liquid and the solid states, in all proportions at all temperatures.

J

jaeger eye chart: Eye chart used for near vision acuity examination.

K

kinetic vision acuity: Vision acuity with a moving target. Studies indicate that 10 to 20 percent of visual efficiency can be lost by target movement.

L

laboratory microscope: Conventional compound microscope. See *microscope.*

lambert cosine law: See *cosine law.*

lambertian: Having a surface that reflects light diffusely and uniformly rather than specularly. See *matte.* Most objects have a lambertian surface. Compare *specular.*

lap: Forging discontinuity caused by a folding over of metal. Laps are found in rolled bar stock and at or near diameter changes.[2,5]

laser: Acronym (*l*ight *a*mplification by *s*timulated *e*mission of *r*adiation). A device, the laser that produces a highly monochromatic and coherent (spatial and temporal) beam of radiation.[2]

leaked visible light: Unwanted electromagnetic radiation that has a wavelength between 400 and 800 nm that is generated by a UV-A source but not filtered out of the emission spectrum. Leaked visible light is generally perceived as purple or dark blue light and not accurately measured using a photometric sensor.[4]

lens: Transparent object that refracts light passing through it in order to focus the light.

lens optics: See *geometrical optics.*

light adapted vision: See *photopic vision.*

light guide bundle: Bundle of filaments, usually glass, that carries noncoherent light from a high intensity source through a fiber optic borescope to illuminate an object.

light: Radiant energy that can excite the retina and produce a visual sensation. The visible portion of the electromagnetic spectrum extends from about 400 to 800 nm.[2,6]

lighting, back: Placement of light source and image sensor on opposite sides of the test object, used when the silhouette of a feature is important.

lighting, flash: See *lighting, strobe.*

lighting, front: Placement of light source and image sensor on the same side of the test object.

lighting, strobe: Lighting that flashes intermittently at a rate that may be adjusted and is often perceived as a flicker, used to image moving objects or still objects with potential movement.

lighting, structured: Combining a light source with optical elements to form a line or sheet of light.

light meter: See *photometer.* Compare *radiometer.*

limited certification: Of a person, certified only for specific operations; usually called limited Level (I or II) or designated as having limited certification because they are not qualified to perform the full range of activities expected of personnel at that level of qualification, for a given method.

line pair: Pair of adjacent, parallel lines used to evaluate the resolution of a specific imaging system. See also *minimum line pair.*

lot tolerance percent defective: In quality control, the percent defective at which there is a 10 percent probability of acceptance in a production run.

low pass filtering: In image processing, linear combination of pixel values to smoothen abrupt transitions in a digital image. Also called *smoothing.*

lumen (lm): SI photometric unit of luminous flux, weighted according to the photopic vision response. One lumen equals the light emitted by one candela (cd) point source into one steradian (sr) solid angle ($1 \text{ lm} = 1 \text{ cd} \cdot \text{sr}^{-1}$).

lumen method: Lighting design procedure used for predetermining the relation between the number and types of lamps or luminaires, the room characteristics and the average illuminance on the work plane. It takes into account both direct and reflected flux. Also called *flux method.*[2,6]

luminance: Photometric brightness of a light source defined by the density of its luminous intensity, measured as luminous flux per unit solid angle per unit area in a given direction. Reported in candela per square meter ($\text{cd} \cdot \text{m}^{-2}$).

luminosity: Luminous efficiency of radiant energy.

luminous efficacy: Ratio of the total luminous flux of a light source to the total radiant flux or to the power input. Sometimes called *luminous efficiency.*

luminous efficiency: See *luminous efficacy.*

luminous flux: Radiant energy's time rate of flow. Measured in lumens.

luminous intensity: Measure of a light source's power output per unit solid angle emitted or reflected from a point, when weighted by the photopic spectral luminous efficiency response curve. Luminous intensity is measured in candela. Compare *luminance.*[4]

lux (lx): SI unit of illuminance, equal to one lumen per square meter ($1 \text{ lx} = 1 \text{ lm} \cdot \text{m}^{-2}$).

M

machine vision: Automated system function of acquiring, processing and analyzing images to evaluate a test object or to provide information or interpretation for human interpretation. A typical machine vision system consists of a light source, a video camera, a video digitizer, a computer and an image display.

macula lutea: Irregular, diffuse ring of yellow pigment which partly overlaps the fovea and surrounds it out to around 10 degrees and which absorbs blue light, thus changing the color of the light reaching receptors beneath.

martensite: (1) Acicular (needlelike) microstructure produced by fast cooling or quenching of metals and alloys such as steel. (2) The hard steel with such microstructure produced by fast cooling of austenite. Martensite is a constituent commonly found in quenched steel.

martensite finish temperature: Temperature at which martensite formation is completed as steel cools.

martensite start temperature: Temperature at which martensite starts to form as steel cools.

mask: (1) A spatial filter in the sensing unit of a surface inspection system. (2) An $n \times n$ square matrix with different values that serves as a filter in image processing.

match bend effect: Optical illusion whereby an area of uniform brightness appears to be nonuniform because of contrast with the brightness of an adjacent area.

mathematical morphology: Image processing technique of expanding and shrinking. The basic operators in mathematical morphology are dilation (expanding), erosion (shrinking), opening and closing.

matte: Having a surface that reflects light diffusely rather than at an angle equal to the angle of incidence; not shiny. Also called *lambertian*. The term *matte* is generally applied to smooth surfaces or coatings. Compare *specular*.

mesopic vision: Vision adapted to a level of light between photopic at 3.4×10^{-2} cd·m^{-2} (3.2×10^{-3} cd·ft^{-2}) and scotopic at 3×10^{-5} cd·m^{-2} (2.7×10^{-6} cd·ft^{-2}).

metallograph: Short term for *metallographic microscope*.

metallographic microscope: See *microscope, metallographic*.

metallography: Science and practice of microscopic testing, inspection and analysis of a metal's structure, typically at magnifications from 50× to 2500×.

metallurgical microscope: See *microscope, metallurgical*.

microborescope: See *borescope, micro-*.

microscope: Instrument that provides enlarged images of small objects.

microscope, compound: Conventional microscope, using geometrical optics for magnification. Also called *laboratory microscope*.

microscope, interference: Magnifier using the wavelength of light as a unit of measure for surface contour and other characteristics.

microscope, metallographic: Metallurgical microscope incorporating a camera. Also called a *metallograph*. Most metallographic microscopes share these features: (a) stand with concealed shock absorbers, (b) intense light source, (c) inverted stand so that the test object is face down, (d) viewing screens for prolonged tasks such as dirt count or grain size measurements, (e) bright, dark and polarized illumination options.

microscope, metallurgical: Microscope designed with features suited for metallography.

microscope, phase contrast: Laboratory microscope with two additional optical elements to transmit both diffracted and undiffracted light, revealing refractive index discontinuities in a completely transparent test object.

microscope, polarizing: Microscope with polarizing elements to restrict light vibration to a single plane for studying material with directional optical properties. As fibers, crystals, sheet plastic and materials under strain are rotated between crossed polarizers on the microscope stage, they change color and intensity in a way that is related to their directional properties.

MIG welding: See *gas metal arc welding*.

miniature borescope: See *borescope, miniature*.

miniborescope: Jargon for *miniature borescope*.

minimum line pair: Closest distance that a specific imaging system can resolve between a pair of adjacent, parallel lines (line pair) used to evaluate system resolution.

modulus of elasticity: Ratio between stress and strain in a material deformed within its linear elastic range.

monochromatic: Light from a very small portion of the spectrum.

monochromator: Device that uses prisms or gratings to separate or disperse the wavelengths of the spectrum into one band.

morphology: See *mathematical morphology*.

mottle: Apparently random positioning of metallic flakes that creates an accidental pattern.

multipass weld: Weld made by many passes, one pass at a time.

multiphase alloy: Alloy in which several phases are present.

N

NDE: (1) Nondestructive evaluation. (2) Nondestructive examination. See *nondestructive testing*.

NDI: Nondestructive inspection. See *nondestructive testing*.

NDT: See *nondestructive testing*.

near ultraviolet radiation: Ultraviolet radiation with wavelengths ranging from about 320 to about 400 nm. Formerly called *black light*.

near vision: Vision of objects nearby, generally within arm's length. Compare *far vision*.

nearsightedness: Vision acuity functionally adequate for viewing objects nearby, generally within arm's length. Also called *myopia*. Compare *farsightedness*.

necking down: Localized reduction in area of a specimen or structural member during welding under overload.[2,5]

negative sliding: Rolling and sliding of meshing gears or rollers when the rolling and sliding are in opposite directions.

neural acuity: Ability of the eye and brain together to discriminate patterns from background. Discrimination is influenced by knowledge of the target pattern, by the scanning technique and by the figure-to-ground relationship of a discontinuity. The figure/ground relationship can be referred to as having a level of visual background noise.

nick: Surface indentation caused by forceful abrasion or impact. Also called *gouge*. Compare *tool mark*.

nit: Obsolete unit for measuring luminance, equivalent to one candela per square meter. Abbreviated nt.

noble metals: Cathodic metals (such as gold, platinum and silver), which strongly resist corrosion.

nondestructive evaluation (NDE): Another term for nondestructive testing. In research and academic communities, the word *evaluation* is often preferred because it emphasizes interpretation by knowledgeable personnel.

nondestructive examination (NDE): Another term for nondestructive testing. In the utilities and nuclear industry, *examination* is sometimes preferred because testing can imply performance trials of pressure containment or power generation systems.

nondestructive inspection (NDI): Another term for nondestructive testing. In some industries (utilities, aviation), the word *inspection* often implies maintenance for a component that has been in service.

nondestructive testing (NDT): Determination of the physical condition of an object without affecting that object's ability to fulfill its intended function. Nondestructive test methods typically use an appropriate form of energy to determine material properties or to indicate the presence of material discontinuities (surface, internal or concealed). Sometimes called *nondestructive evaluation, nondestructive examination* or *nondestructive inspection*.

nonferromagnetic material: Material not magnetizable and essentially not affected by magnetic fields.[4]

nonrelevant indication: See *indication, nonrelevant*.

numerical analysis: Technique to generate numbers as the solution to a mathematical model of a physical system; used in place of a closed form analytic expression; usually requires digital computation.[4]

O

objective: In discussion of a lens system (camera, borescope, microscope, telescope), of or pertaining to the end or lens closest to the object of examination — at the end opposite from the eyepiece. *Distal*; *tip*.

OCTG: Oil country tubular goods.

oil country tubular goods: Hollow cylindrical components, such as pipes, used in petroleum wells to case the hole and to convey petroleum and related products.[4]

opening: Image processing operation of erosion followed by dilation. A single opening eliminates isolated single pixels. See also *closing*.

opsin: See *visual purple*.

optic disk: Area in the retina through which the fibers from the various receptors cross the inner (vitreous humor) side of the retina and pass through it together in the optic nerve bundle. This transitional area is completely blind.

optics: Physical science of the transmission of radiation, especially of light. See *geometrical optics*.

organoleptic: Relying on or using sense organs, such as the human eye.

orthicon: See *image orthicon*.

P

parafoveal vision: See *scotopic vision*.

parallax: Apparent difference in position of an imaged point according to two differently positioned sensors.

pass: In welding, a single bead of weld metal along the entire joint or the process of laying down that bead.

pearlite: Platelet mixture of cementite and ferrite in steels or in alpha and beta phases in nonferrous alloys.

peripheral vision: Seeing of objects displaced from the primary line of sight and outside the central visual field.[2,6]

phase: In metallurgy, a physically homogeneous portion of a material system, specifically the portion of an alloy characterized by its microstructure at a particular temperature during melting or solidification.

phase contrast microscope: See *microscope, phase contrast*.

phase diagram: Graph showing the temperature, pressure and composition limits of phase fields in a material system. Also called a *constitution diagram*. Compare *equilibrium diagram*.

photochromic lens: Eyeglass material that automatically darkens to reduce light transmission when exposed to ultraviolet radiation.

photoconduction: Method by which a vidicon television camera tube produces an electrical image, in which the conductivity of the photosensitive surface changes in relation to the intensity of the light reflected from the scene focused onto the surface. Compare *photoemission*.
photoelasticity: Effect of a material's elastic properties on the way that it refracts or reflects light.
photoelectric effect: Emission of electrons from a surface bombarded by sufficiently energetic photons. Such emissions may be used in an illuminance meter and can be calibrated in lux.[2,6]
photoemission: Method by which an image orthicon television camera tube produces an electrical image, in which a photosensitive surface emits electrons when light reflected from a viewed object is focused on the surface. Compare *photoconduction*.
photometer: Device used to measure illuminance. The sensor is filtered such that its response closely matches the spectral responsivity curve of the human eye. In nondestructive testing, photometers measure lux. Compare *radiometer*.
photometric brightness: Luminance of a light source.
photometry: Study and measurement of electromagnetic radiation with approximate wavelengths between 400 and 800 nm, within the human eye's spectral responsivity. See also *photometer*; *photopic vision*; *radiometry*; *relative photometry*.
photon: Particle of light, hypothesized to explain those behaviors of light in which its behavior is corpuscular rather than wavelike.
photopic vision: Average spectral responsivity curve of the human eye when adapted to well lit conditions (greater than 0.034 $cd{\cdot}m^{-2}$). The photopic spectral luminous efficiency response curve is governed by an averaged retinal cone response with sensitivity peaks centered at about 555 nm. Also known as *foveal vision* and *light adapted* vision. Compare *mesopic vision* and *scotopic vision*.[2,6]
photoreceptor: Light sensor.
picture element: See *pixel*.
picture processing: See *image processing*.
pipe: Longitudinal centerline discontinuity inherent in ingots, imparted to some rolled metal and consisting of a concavity or voids.
pitting: Discontinuity consisting of surface cavities. See also *cavitation fatigue*.
pixel: Addressable point in a digital image. The image from a conventional computer is an array of pixels, and each has a numerical value. The higher the number for a pixel, the brighter it is. Formerly called *picture element*.
plane of focus: See *focus, principal plane of*.
platelet: Flat crystallites in certain phases of steel.
polarizing microscope: See *microscope, polarizing*.
porosity: Discontinuity in metal resulting from the creation or coalescence of gas. Very small pores open to the surface are called *pinholes*.[2,5]
positive sliding: Rolling and sliding of meshing gears or rollers when the directions of rolling and sliding are the same.
poultice corrosion: See *corrosion, poultice*.
practical examination: In certification of nondestructive testing personnel, a hands-on examination using test equipment and sample test objects. Compare *general examination*; *specific examination*.
primary creep: First stage of creep, marked by elastic strain plus plastic strain.
principal plane of focus: See *focus, principal plane of*.
probability of detection (PoD): Statistical statement from a specific test procedure indicating how likely a given discontinuity length may be reliably found.
process: Repeatable sequence of actions to bring about a desired result.
process control: Application of quality control principles to the management of a repeated process.
pseudocolor: Image enhancement technique wherein colors are assigned to an image at several gray scale intervals.
pseudoisochromatic plate: Image used for color vision examinations. Each plate bears an image which may be difficult for the examinee to see if his or her color vision is impaired. See also Ishihara™ plates.
psychophysics: Interaction between vision performance and physical or psychological factors. One example is the so-called vigilance decrement, the degradation of reliability based on performing visual activities over a period of time.
pupil: Black aperture in the center of the eye's lens, through which light enters the lens to impinge on the retina.
purple: See *visual purple*.

Q

qualification: Process of demonstrating that an individual has the required amount and the required type of training, experience, knowledge and abilities.[4-6] See also *certification* and *qualified*.

qualified: Having demonstrated the required amount and the required type of training, experience, knowledge and abilities. See also *certified* and *qualification*.

quality: Ability of a process or product to meet specifications or to meet the expectations of its users in terms of efficiency, appearance, reliability and ergonomics.[4-6]

quality assurance: Administrative actions that specify, enforce and verify quality.[4-6]

quality control: Physical and administrative actions required to ensure compliance with a quality assurance program. Quality control may include nondestructive testing in the manufacturing cycle.[4-6]

quality of lighting: Level of luminance in a visual task or environment.

R

radial: Of or pertaining to direction from center of a circle (or a sphere or cross section of a cylindrical object) to its surface, and perpendicular to its axis. Compare *axial*.

radiance: Radiant flux per unit solid angle and per unit projected area of the source. Measured in watts per square meter steradian. Compare *irradiance*.

radiant energy: Energy transmitted through a medium by electromagnetic waves. Also known as *radiation*.

radiant flux: Radiant energy's rate of flow, measured in watts.

radiant intensity: Electromagnetic energy emitted per unit time per unit solid angle. Measured in watts per steradian.

radiant power: Total radiant energy emitted per unit time.

radiometer: Device used to measure irradiance. In nondestructive testing, radiometers are used to measure UV-A output, or leaked visible light, in microwatt per square centimeter ($\mu W \cdot cm^{-2}$). See also *irradiance*. Compare *photometer*.

radiometric photometer: Radiometer for measuring radiant power over a variety of wavelengths.

radiometry: Study and measurement of electromagnetic radiation emitted by a source or falling upon a surface.

raster: Repetitive pattern whereby a directed element (a robotic arm or a flying dot on a video screen) follows the path of a series of adjacent parallel lines, taking them successively in turn, always in the same direction (from top to bottom or from left to right), stopping at the end of one line and beginning again at the start of the next line. Following a raster pattern makes it possible for electron beams to form video pictures or frames and for a sensor-bearing armature to cover a predetermined part of the surface of a test object.

rat's tooth principle: (1) The tendency for hard material on a tooth's front surface to wear more slowly than soft material on the back surface, keeping the edge sharp. (2) Mechanism of wear whereby adjacent hard and soft surfaces wear at different rates, producing a self-sharpening edge.

Recommended Practice No. SNT-TC-1A: Set of guidelines published by the American Society for Nondestructive Testing, for employers to establish and conduct a qualification and certification program for nondestructive testing personnel.[4-6]

recommended practice: Set of guidelines or recommendations.[4-6]

recovery: Reduced stress level and increased ductility of metal after work hardening. See *creep*.

recrystallization: Changes in microstructure and properties upon heating of cold worked metal.

red mud: Debris (usually oxides of the contacting metals) of fretting wear, mixed with oil or grease and retained at or near the site of its formation. See also *cocoa*.

reference standard: Object containing known discontinuities at known distances and representing accept or reject criteria.

reflectance: Ratio of reflected wave energy to incident wave energy. Also known as *reflectivity*.

reflection: General term for the process by which the incident energy leaves a surface or medium from the incident side, without change in frequency. Reflection is usually a combination of specular and diffuse reflection.[2,6]

reflectometer: Photometer used to measure diffuse, specular and total reflectance.

reflector: Device used to redirect the luminous flux from a source by the process of reflection.[2,6]

refraction: Reorientation of radiation's path by the medium through which it passes.

relative photometry: (1) Evaluation of a desired photometric characteristic based on an assumed lumen output of a test lamp. (2) Measurement of an uncalibrated light source relative to another uncalibrated light source.

remote viewing: (1) Indirect viewing of a test object far from the viewer's immediate presence — for example, viewing with telemetry or crawlers. The term *remote viewing* is used in the fields of robotics and surveillance to distinguish conventional from distant viewing tasks. (2) Viewing of a test object during which the light image is mediated through a system of two or more lenses (as in a borescope) or transduced through an electronic signal (as with a charge coupled camera). This use of the term *remote viewing* in some specifications is a misnomer, intended merely to distinguish borescopy from direct viewing. Compare *borescope; direct viewing; indirect viewing.*

replica: Piece of malleable material, such as polyvinyl or polystyrene plastic film, molded to a test surface for the recording or analysis of the surface microstructure.

replica, metallographic: Replica suitable for microscopic examination. See *metallography.*

replication: Method for copying the topography of a surface by making its impression in a plastic or malleable material.

reserve vision acuity: Ability of an individual to maintain vision acuity under poor viewing conditions. A visual system with 20/20 near vision acuity under degraded viewing conditions has considerable reserve vision acuity compared to that of an individual with 20/70 near vision acuity.

resolution: Aspect of image quality pertaining to a system's ability to depict objects, often measured by distinguishing (resolving) a pair of adjacent objects or parallel lines.

resolution test: Procedure wherein a line is detected to verify a system's sensitivity.

resolution threshold: Minimum distance between a pair of points or parallel lines when they can be distinguished as two, not one, expressed in minutes of arc. Vision acuity, in such a case, is the reciprocal of one-half of the period expressed in minutes.[2,6]

resolving power: Ability of vision or other detection system to separate two points. Resolving power depends on the angle of vision and the distance of the sensor from the test surface. Resolving power is often measured using parallel lines. Compare *resolution.*

retina: In the eye, the tissue on back, inside surface of the eyeball, opposite the pupil, where light sensitive rods and cones sense light.

retinene: See *visual purple.*

rhodopsin: See *visual purple.*

robotic system: Automated system programmed to perform purposeful actions in variable sequences.

rod: Retinal receptor that responds at low levels of luminance even down below the threshold for cones. At these levels there is no basis for perceiving differences in hue and saturation. No rods are found in the fovea centralis.[2,6] Concentrated toward the outer region of the retina, rods assist with mesopic vision and are responsible for scotopic night vision. Compare *cone.*

root mean square (rms): Statistical measure of the magnitude of a varying quantity, such as current. Square root of the mean square of a set of measures, usually a time series.

S

sampling, partial: Testing of less than 100 percent of a production lot.

sampling, random partial: Partial sampling that is fully random.

sampling, specified partial: Partial sampling in which a particular frequency or a sequence of sample selection is prescribed. An example of specified partial sampling is the testing of every fifth unit.

saturation: Relative or comparative color characteristic resulting from a hue's dilution with white light.

scalar: Quantity completely specified by a single number and unit.[4]

scaling: (1) Forming a layer of oxidation product on metals, usually at high temperature. (2) Deposition of insoluble constituents on a metal surface, as in cooling tubes and water boilers.[2,5]

scoring: (1) Marring or scratching of any formed part by metal pickup on a punch, die or guide. (2) Reducing the thickness of a part along a line to weaken it purposely at a specific location.[2,5]

scotopic vision: Dark adapted vision, using only the rods in the retina, where differences in brightness can be detected but differences in hue cannot. Vision is wholly scotopic when the luminance of the test surface is below 3×10^{-5} cd·m^{-2} (2.7×10^{-6} cd·ft^{-2}). Also known as *parafoveal vision.* Compare *mesopic vision* and *photopic vision.*

scuffing: Type of adhesive wear.

seam: Linear discontinuity formed by a lack of metal from folds produced by an underfilled pass during metal rolling. Squeezed tight on subsequent passes, the underfill runs parallel to the longitudinal axis of the bar.

second stage replica: Positive replica made from the first cast to produce a duplicate of the original surface.

secondary creep: Second stage of creep, in which deformation proceeds at a constant rate and less rapidly than in primary creep. Essentially an equilibrium condition between the mechanisms of work hardening and recovery. See also *defect* and *discontinuity.*[4]

sensitivity: Ability of a sensor or system to distinguish a signal or indication from background noise. See also *probability of detection.*

sensitization: Precipitation of chromium carbides in the grain boundaries of a corrosion resistant alloy, resulting in intergranular corrosion that would otherwise be resisted.

sensor: Device that detects a material property or mechanical behavior (such as radiation or displacement) and converts it to an electrical signal.

shadow casting: Nondestructive technique of vapor depositing a thin metal film onto a replica at an oblique angle in order to obtain a micrograph of a test surface of an opaque test object.

shear break: Open break in metal at the periphery of a bolt, nut, rod or member at approximately a 45 degree angle to the applied stress. Shear breaks occur most often with flanged products. Also called *shear crack.*[2,5]

shear crack: See *shear break.*

shielded arc welding: Joining of metals by heating them with an electric arc between electrode(s) and the work piece, using an inert gas to shield the electrode(s). See also *gas tungsten arc welding.*

shoulder: Cylindrical metal component (pipe) surface, machined to receive threading indentations but in fact not threaded, where the thread stops on the outside surface of the pipe.

SI (International System of Units): Measurement system in which the following seven units are basic: meter, mole, kilogram, second, ampere, kelvin and candela.

signal electrode: Transparent conducting film on the inner surface of a vidicon's faceplate and a thin photoconductive layer deposited on the film.

signal processing: Acquisition, storage, analysis, alteration and output of digital or analog data.

signal-to-noise ratio: Ratio of signal values (responses that contain relevant information) to baseline noise values (responses that contain nonrelevant information).[4]

signal: Physical quantity, such as voltage, that contains relevant information.[4]

simple magnifier: Microscope having a single converging lens.

smoothing: In image processing, linear combination of pixel values to smooth abrupt transitions in a digital image. Also called *low pass filtering.*

SNT-TC-1A: See *Recommended Practice No. SNT-TC-1A.*

spalling fatigue: See *subcase fatigue.*

specific examination: In certification of nondestructive testing personnel, a written examination that addresses the specifications and products pertinent to the application. Compare *general examination* and *practical examination.*

specific gravity: Unitless ratio of the density of a material divided by the density of water. Water has a density of about 1 g·cm^{-3}, or 1000 kg·m^{-3}.

specification: Set of instructions or standards invoked to govern the properties, results or performance of a specific set of tasks or products.[4-6]

spectral irradiance: Measure of energy emitted by a radiation source as function of wavelength. Units of spectral irradiance are watts per square meter and are often plotted versus wavelength.

spectral power distribution: Radiant power per unit wavelength as a function of wavelength. Also known as *spectral energy distribution, spectral density* and *spectral distribution.*

spectral reflectance: Radiant flux reflected from a material divided by the incident radiant flux.

spectral responsivity: Measure of a photometric or radiometric sensor's sensitivity over a wavelength range of interest, often presented as percent versus wavelength. Photometric sensors should exhibit a bell shaped spectral responsivity curve over the visible light range, whereas radiometric sensors may exhibit a flat or other response curve.

spectral transmittance: Radiant flux passing through a medium divided by the incident radiant flux.
spectrophotometer: Instrument used for spectrophotometry.
spectrophotometry: Measurement of electromagnetic radiant energy as a function of wavelength, particularly in the ultraviolet, visible and infrared wavelengths.
spectroradiometer: Instrument used for spectroradiometry.
spectroradiometry: Measurement of electromagnetic radiant power and spectral emittance, used particularly to examine colors and to measure the spectral emittance of light sources.
spectroscope: Instrument used for spectroscopy.
spectroscopy: Spectrophotometry or spectroradiometry in which the spectrum, rather than being analyzed only by a processing unit, is presented in a visible form to the operator for organoleptic examination.
spectrum: Representation of radiant energy in adjacent bands of hues in sequence according to the energy's wavelengths or frequencies. A rainbow is a well known example of a visible spectrum.
specular: Pertaining to a mirrorlike reflective finish, as of a metal. Compare *lambertian*.
specular reflection: When reflected waves and incident waves form equal angles at the reflecting surface.
speed of light: Speed of all radiant energy, including light, is $2.997\,924\,58 \times 10^8$ m·s^{-1} in vacuum. In all materials the speed is less and varies with the material's index of refraction, which itself varies with wavelength.[2,6]
speed of vision: Reciprocal of the duration of the exposure time required for something to be seen.[2,6]
standard: Object, document or concept established by authority, custom or agreement to serve as a model or rule in the measurement of quantity or the establishment of a practice or procedure.[4,8] See also *reference standard* and *acceptance standard.*
standardization, instrument: Adjustment of instrument readout before use to a specified reference value.[4]
standard observer response curve: See *eye sensitivity curve.*
steel: Iron alloy, usually with less than two percent carbon.
stereo photography: Close range photogrammetric technique involving the capture and viewing of two images of the same object in order to reconstruct a three dimensional image of the object.
stick welding: See *shielded metal arc welding.*
strain: Deflection or alteration of the shape of a material by external forces.
stress: (1) In physics, the action in a material that resists external forces such as tension and compression. (2) Load per unit of area.
stress concentration: Region where force per unit area is elevated, often because of geometric factors or cracks. Also known as a *stress raiser.*
stress raiser: Contour or property change that locally increases stress magnitude.
stress riser: See *stress raiser.*
stringer: In wrought materials, an elongated configuration of microconstituents or foreign material aligned in the direction of working. Commonly, the term is associated with elongated oxide or sulfide inclusions in steel.
subcase fatigue: Fatigue originating below the case depth.
subcase origin fatigue: See *subcase fatigue.*
subsurface fatigue: Fatigue cracking that originates below the surface. Usually associated with hard surfaced or shot peened parts but may occur any time subsurface stresses exceed surface stresses.

T

tarasov etching technique: Way of visually inspecting for the presence of deleterious effects in hardened steels by using specific etching solutions and methods of inspection.
temperature diagram: See *time temperature transformation (TTT) diagram.*
tempering: Process of heating a material, particularly hardened steel to below the austenite transformation temperature, to improve ductility.
tertiary creep: Third stage of creep, marked by steady increase in strain to the point of fracture under constant load.
test object: Physical part or specimen subject to nondestructive testing.
threshold: (1) A value in a phenomenon where a large change of output occurs. (2) Setting of an instrument that causes it to register only those changes in response greater or less than a specified magnitude.[4] See *adaptive thresholding, resolution threshold.*
thresholding: Digital data processing technique that reduces a gray level image into a binary image.
throat, actual: Shortest distance from the root of a fillet weld to its face, as opposed to theoretical throat or weld size.

throat, effective: In fillet welds, the weld throat including the amount of weld penetration but ignoring excess metal between the theoretical face and the actual face.

throat, theoretical: Distance from the beginning of the root of a fillet weld perpendicular to the hypotenuse of the largest right triangle that can be inscribed within the cross section of the fillet weld. Compare *weld size.*

throat, weld: Distance from the root of a fillet weld to its face. Compare *weld size* and *throat, actual.*

TIG welding: Tungsten inert gas welding. See *gas tungsten arc welding.*

time temperature transformation (TTT) diagram: Graph showing time required at any temperature to transform austenite to pearlite, bainite or martensite.

tip: Distal or objective end of a borescope.

tool mark: Shallow indentation or groove made by the movement of manufacturing tools over a surface. Compare *gouge* or *nick.*

trace: Line formed by electron beam scanning from left to right on a video screen to generate a picture.

tracer: In leak testing, a gas that is sensed as it escapes from confinement.

transformation diagram: See *time temperature transformation (TTT) diagram.*

troland: Unit of retinal illuminance equal to that produced by a surface whose luminance is 1 $cd \cdot m^{-2}$ when the pupil measures 1 mm^2.

tubing string: Pipe with which oil or gas has contact as it is brought to the Earth's surface.[4]

tungsten inert gas (TIG) welding: See *gas tungsten arc welding.*

U

ultraviolet borescope: See *borescope, ultraviolet.*

ultraviolet radiation: Electromagnetic radiation with wavelengths between 40 and 400 nm. See also *irradiance* and *UV-A.*

ultraviolet radiometer: Meter, usually calibrated at 365 nm, used in fluorescent liquid penetrant and magnetic particle testing to measure output of ultraviolet lamp.

undercut: Undesirable groove left unfilled by weld metal, created during welding and located in base plate at the toe of a weld.

Unified Numbering System: Alphanumeric system for identifying alloys according to a registry maintained by ASTM International and SAE International.[4]

UV-A: Electromagnetic radiation with wavelengths between 315 and 400 nm. Fluorescent nondestructive testing has historically used ultraviolet energy centered at 365 nm. See also *irradiance* and *radiometer.*

V

video: Pertaining to the transmission and display of images in an electronic format that can be displayed on a monitor or screen.

videoscope: Jargon for *video borescope.* See *borescope, video.*

vidicon tube: Analog television tube that uses the photoconduction method. Compare *image orthicon.*

vigilance decrement: Degradation of reliability during performance of visual activities over a period of time. See also *psychophysics.*

visibility: Quality or state of being perceivable by the eye. In many outdoor applications, visibility is defined in terms of the distance at which an object can be just perceived by the eye. In indoor applications it usually is defined in terms of the contrast or size of a standard test object, observed under standardized view conditions, having the same threshold as the given object.[2,6]

visible light: Radiant energy with a wavelength between 400 and 800 nm as measured in photometric units of lux.

vision: Perception by eyesight. See *far vision, machine vision, mesopic vision, near vision, peripheral vision, photopic vision, scotopic vision,* and *speed of vision.*

vision acuity: Ability to distinguish fine details visually at a given distance. Quantitatively, it is the reciprocal of the minimum angular separation in minutes of two lines of width subtending one minute of arc when the lines are just resolvable as separate.[2,6]

visual acuity: See *vision acuity.*

visual angle: Angle formed by lines drawn from center of eye subtended by an object or detail at the point of observation. It usually is measured in minutes of arc.[2,6]

visual background noise: Formations on or signals from a test object that constitutes the background to a discontinuity. The higher the level of visual background noise, the more difficult it is to distinguish a discontinuity.

visual efficiency: Reliability of a visual system. The term *visual efficiency* uses 20/20 near vision acuity as a baseline in the United States for 100 percent visual efficiency.

visual field: Locus of objects or points in space that can be perceived when head and eyes are fixed. The field may be monocular or binocular.[2,6]

visual perception: Interpretation of impressions transmitted from the retina to the brain in terms of information about a physical world displayed before the eye. Visual perception involves any one or more of the following: recognition of the presence of something (object, aperture or medium); identifying it; locating it in space; noting its relation to other things; identifying its movement, color, brightness or form.[2,6]

visual performance: Quantitative assessment of the performance of a visual task, taking into consideration speed and accuracy.[2,6]

visual purple: Chromoprotein called *rhodopsin*, the photosensitive pigment of rod vision. The mechanism of converting light energy into nerve impulses is a photochemical process in the retina. Chromoprotein is transformed by the action of radiant energy into a succession of products, finally yielding the protein called *opsin* plus the carotenoid known as *retinene*.

visual task: Appearance and immediate background of those details and objects that must be seen for the performance of a given activity. The term *visual task* is a misnomer because it refers to the visual display itself and not the task of extracting information from it. See *visual field.*

visual testing: Method of nondestructive testing using electromagnetic radiation at visible frequencies.

voids: Hollow spots, depressions or cavities. See also *discontinuity*.

volt (V): Measurement unit of electric potential.[4]

VT: Visual testing.

W

wavelength: Distance between repeating values of a wave. For example, the distance from one peak to the next peak on a sine wave.

wear: See *erosion*; *rat's tooth principle*; *wear, adhesive*; and *wear, fretting*.

wear oxidation: See *wear, fretting*.

wear, adhesive: Degradation of a surface by microwelding and consequent fracture due to the sliding of one surface against another. Types include fretting, galling and scuffing.

wear, fretting: Surface degradation by microwelding and microfractures on surfaces rubbing each other. Also called *chafing, friction oxidation* and *wear oxidation*. See also *cocoa* and *false brinelling*.

weld size: Thickness of weld metal — in a fillet weld the distance from the root to the toe of the largest isosceles right triangle that can be inscribed in a cross section of the weld.

weld throat: See *throat, weld.*

welder's flash: Clinical condition, specifically keratoconjunctivitis, commonly caused by overexposure to ultraviolet radiation of welding arc.

white light: Light combining all frequencies in the visible spectrum (wavelengths from 380 to 780 nm) and in equal proportions.

work hardening: Increase in hardness accompanying plastic deformation of a metal. Usually caused in a metal by repeated impacting, bending or flexing. Compare *creep* and *recovery*.

working standard: Work piece or energy source calibrated and used in place of expensive reference standards. In calibrating of photometers, the standard would be a light source.

References

1. ASTM E 1316, *Standard Terminology for Nondestructive Examinations*. West Conshohocken, PA: ASTM International (2007).
2. *Nondestructive Testing Handbook,* second edition: Vol. 8, *Visual and Optical Testing*. Columbus, OH: American Society for Nondestructive Testing (1993).
3. *Nondestructive Testing Handbook,* second edition: Vol. 10, *Nondestructive Testing Overview.* Columbus, OH: American Society for Nondestructive Testing (1996).
4. *Nondestructive Testing Handbook*, third edition: Vol. 8, *Magnetic Testing*. Columbus, OH: American Society for Nondestructive Testing (2008).
5. *EPRI Learning Modules*. Charlotte, NC: Electric Power Research Institute.
6. *IES Lighting Handbook: Reference Volume*. New York, NY: Illuminating Engineering Society of North America (1984).
7. Lovejoy, D. *Magnetic Particle Inspection: A Practical Guide*. New York, NY: Chapman & Hall (1993).
8. TO33B-1-1 (NAVAIR 01-1A-16) TM43-0103, *Nondestructive Testing Methods*. Washington, DC: United States Department of Defense, United States Air Force (June 1984).

Index

D

E

F

G

H

I

S

T

U

Figure Credits

Chapter 1. Introduction to Visual Testing

Figure 13a, 16, 18, 19a — Public domain. Not copyrighted.
Figure 15 — Carnegie Mellon University, Pittsburgh, PA.
Figure 17 — ASME International, New York, NY (ASME library, 1985).
Figure 19b — AMF Tuboscope, Houston, TX (Alfred E. Crouch, 1985).

Chapter 2. Light

Figures 2 to 9, 13b, 14 — Dover Books, New York, NY. Not copyrighted.
Figures 1b, 1c, 10, 11, 16, 18-19 — Illuminating Engineering Society of North America, New York, NY.
Figures 15, 17, 20-21 — National Institute of Standards and Technology, Gaithersburg, MD. Not copyrighted.

Chapter 3. Vision Acuity for Nondestructive Testing

Figure 5 — Not copyrighted (Joel Schneider, 2002).
Figure 6 — Mars Perceptrix, Chappaqua, NY.
Figure 7 — National Vision Research Institute of Australia, Carlton, Victoria, Australia.
Figure 8 — Western Ophthalmics Corporation, Lynnwood, WA.

Chapter 4. Imaging of Visual Tests

Figures 4-5, 7, 9-11, 14, 19, 20 — General Electric, Skaneateles Falls, NY (Trevor Liddell).

Chapter 5. Direct Visual Testing

Figure 4 — KD Marketing, Danaher Tool Group, Sparks, MD.
Figure 5 — After original illustration by Illuminating Engineering Society of North America, New York, NY.
Figures 7, 8 — Edmund Scientific Corporation, Barrington, NJ (1992).
Figure 10 — Bausch and Lomb, Rochester, NY.
Figure 12 — After original illustration by Hommel America. ASNT is unable to contact Hommel America as of 2009.

Chapter 6. Indirect Visual Testing

Figure 8a — Lenox Instrument, Trevose, PA (1987).
Figure 8b — Olympus Corporation, Lake Success, NY (1986).
Figure 9 — Olympus Industrial Systems Europa, Essex, United Kingdom (Christopher I. Collins).
Figures 20-31 — General Electric, Skaneateles Falls, NY (Trevor Liddell).

Chapter 8. Visual Testing of Metals

Figures 4, 5, 10-21, 27 — American Iron and Steel Institute, Warrendale, PA.
Figures 28, 31, 32 — ASM International, Materials Park, OH

Chapter 9. Chemical and Petroleum Applications of Visual Testing

Figures 4-5. American Welding Society, Miami, FL.
Figures 6-9. ASME International, New York, NY.
Figure 10. American Petroleum Institute, Dallas, TX.
Figure 11. Exxon Production Research, Exxon Mobil, Irving, TX.

Chapter 10. Electric Power Applications of Visual Testing

Figures 1-29, 31-34, 37-51 — Electric Power Research Institute, Charlotte, NC.
Figure 35 — Westinghouse Electric Corporation, Pittsburgh, PA.
Figure 36 — General Electric Nuclear, Schenectady, NY.

Chapter 11. Aerospace Applications of Visual Testing

Figures 11, 13 — Federal Aviation Administration. Not copyrighted.